Crystal Growth and Nucleation

University College London, UK
2–4 April 2007

FARADAY DISCUSSIONS
Volume 136, 2007

RSC Publishing

The Faraday Division of the Royal Society of Chemistry, previously the Faraday Society, founded in 1903 to promote the study of sciences lying between Chemistry, Physics and Biology.

EDITORIAL STAFF

Editor
Philip Earis

Deputy editor
Caroline Moore

Publishing assistant
Rachel Dilworth

Team leader, serials production
Gisela Scott

Technical editor
Susan Batten

Publisher
Janet Dean

Faraday Discussions (Print ISSN 1359-6640, Electronic ISSN 1364-5498) is published 3 times a year by the Royal Society of Chemistry, Thomas Graham House, Science Park, Milton Road, Cambridge, UK CB4 0WF.
Volume 136 ISBN: 0 85404 952 5
ISBN-13: 978 0 85404 952 3

2007 annual subscription price: print+electronic £494, US $934; electronic only £445, US $840. Customers in Canada will be subject to a surcharge to cover GST. Customers in the EU subscribing to the electronic version only will be charged VAT. All orders, with cheques made payable to the Royal Society of Chemistry, should be sent to RSC Distribution Services, c/o Portland Customer Services, Commerce Way, Colchester, Essex, UK CO2 8HP.
Tel +44 (0) 1206 226050;
E-mail sales@rscdistribution.org

If you take an institutional subscription to any RSC journal you are entitled to free, site-wide web access to that journal. You can arrange access via Internet Protocol (IP) address at www.rsc.org/ip. Customers should make payments by cheque in sterling payable on a UK clearing bank or in US dollars payable on a US clearing bank. Periodicals postage is paid at Rahway, NJ and at additional mailing offices. Airfreight and mailing in the USA by Mercury Airfreight International Ltd., 365 Blair Road, Avenel, NJ 07001, USA.

US Postmaster: send address changes to *Faraday Discussions*, c/o Mercury Airfreight International Ltd., 365 Blair Road, Avenel, NJ 07001. All despatches outside the UK by Consolidated Airfreight.

PRINTED IN THE UK

Faraday Discussions documents a long-established series of *Faraday Discussion* meetings which provide a unique international forum for the exchange of views and newly acquired results in developing areas of physical chemistry, biophysical chemistry and chemical physics.

ORGANISING COMMITTEE, Volume 136

Chairs
C R A Catlow (Royal Institution, UK)
N H de Leeuw (UCL, UK)

J Anwar (Bradford, UK)
R J Davey (Manchester, UK)
K J Roberts (Leeds, UK)
P R Unwin (Warwick, UK)

FARADAY STANDING COMMITTEE ON CONFERENCES

Chair
C D Bain (Durham, UK)

K J Edler (Bath, UK)
A J Orr-Ewing (Bristol, UK)
G Jackson (Imperial, UK)
A Rodger (Warwick, UK)

© The Royal Society of Chemistry 2007. Apart from fair dealing for the purposes of research or private study, or criticism or review, as permitted under the Copyright, Designs and Patents Act 1988 and Related Rights Regulations 2003, this publication may only be reproduced, stored or transmitted, in any form or by any means, with the prior permission in writing of the Publishers or in the case of reprographic reproduction in accordance with the terms of licences issued by the Copyright Licensing Agency in the UK. US copyright law applicable to users in the USA. The Royal Society of Chemistry takes reasonable care in the preparation of this publication but does not accept liability for the consequences of any errors or omissions.
Reprinted 2009
Royal Society of Chemistry: Registered Charity No. 207890.

⊚The paper used in this publication meets the requirements of ANSI/NISO Z39.48-1992 (Permanence of Paper).

Crystal Growth and Nucleation

Faraday Discussions
www.rsc.org/faraday_d

A General Discussion on Crystal Growth and Nucleation was held at University College London, UK on 2nd, 3rd and 4th April 2007.

RSC Publishing is a not-for-profit publisher and a division of the Royal Society of Chemistry. Any surplus made is used to support charitable activities aimed at advancing the chemical sciences. Full details are available from www.rsc.org

CONTENTS

ISSN 1359-6640; ISBN 0-85404-952-5
ISBN-978-085404-952-3

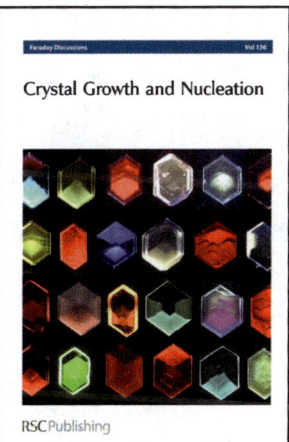

Cover
See Theresa Bullard, John Freudenthal, Serine Avagyan and Bart Kahr, *Faraday Discuss.*, 2007, **136**, 231–245.
A 'light quilt' assembled from potassium hydrogen phthalate crystals doped with luminophores for monitoring growth. Most crystals prepared by Dr S.-H. Jang.

Image reproduced by permission of Professor Bart Kahr, from *Faraday Discuss.*, 2007, **136**, 231.

INTRODUCTORY LECTURE

9   **Spiers Memorial Lecture**
    Lessons from biomineralization: comparing the growth strategies of mollusc shell prismatic and nacreous layers in *Atrina rigida*
    Fabio Nudelman, Hong H. Chen, Harvey A. Goldberg, Steve Weiner and Lia Addadi

PAPERS AND DISCUSSIONS

27 **Sintering, crystallisation and biodegradation behaviour of Bioglass®-derived glass–ceramics**
    Aldo R. Boccaccini, Qizhi Chen, Leila Lefebvre, Laurent Gremillard and Jérôme Chevalier

45 **The formation of nanoscale structures in soluble phosphosilicate glasses for biomedical applications: MD simulations**
    Antonio Tilocca, Alastair N. Cormack and Nora H. de Leeuw

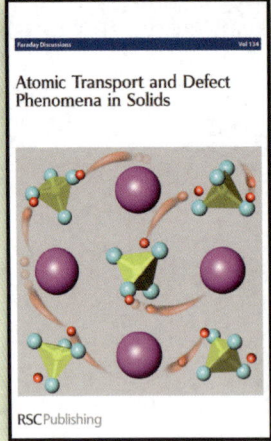

57	**Towards an atomic-scale understanding of crystal growth in solution** Elias Vlieg, Menno Deij, Daniel Kaminski, Hugo Meekes and Willem van Enckevort
71	**A multi-technique approach for probing the evolution of structural properties during crystallization of organic materials from solution** Colan E. Hughes, Said Hamad, Kenneth D. M. Harris, C. Richard A. Catlow and Peter C. Griffiths
91	**An examination of polymorphic stability and molecular conformational flexibility as a function of crystal size associated with the nucleation and growth of benzophenone** Robert B. Hammond, Klimentina Pencheva and Kevin J. Roberts
107	**General Discussion**
125	**Insights into the crystal growth mechanisms of zeolites from combined experimental imaging and theoretical studies** Ben Slater, Tetsu Ohsuna, Zheng Liu and Osamu Terasaki
143	**Crystal growth in nanoporous framework materials** Michael W. Anderson, Jonathan R. Agger, L. Itzel Meza, Chin B. Chong and Colin S. Cundy
157	**New insights into the formation of microporous materials by *in situ* scattering techniques** Gopinathan Sankar, Tatsuya Okubo, Wei Fan and Florian Meneau
167	**Cocrystal architecture and properties: design and building of chiral and racemic structures by solid–solid reactions** Tomislav Friščić and William Jones
179	**The nucleation of inosine: the impact of solution chemistry on the appearance of polymorphic and hydrated crystal forms** Renato A. Chiarella, Amy L. Gillon, Rebecca C. Burton, Roger J. Davey, Ghazala Sadiq, Anthony Auffret, Marina Cioffi and Christopher A. Hunter
195	**Membrane protein crystallization in lipidic mesophases. A mechanism study using X-ray microdiffraction** Vadim Cherezov and Martin Caffrey
213	**General Discussion**
231	**Test of Cairns-Smith's 'crystals-as-genes' hypothesis** Theresa Bullard, John Freudenthal, Serine Avagyan and Bart Kahr
247	**Precipitation of α L-glutamic acid: determination of growth kinetics** Jochen Schöll, Christian Lindenberg, Lars Vicum, Jörg Brozio and Marco Mazzotti
265	**Precursor structures in the crystallization/precipitation processes of $CaCO_3$ and control of particle formation by polyelectrolytes** J. Rieger, T. Frechen, G. Cox, W. Heckmann, C. Schmidt and J. Thieme
279	**Does supercooled liquid Si have a density maximum?** Masahito Watanabe, Masayoshi Adachi, Tetsuya Morishita, Kensuke Higuchi, Hidekazu Kobatake and Hiroyuki Fukuyama
287	**Simulating ice nucleation, one molecule at a time, with the 'DFT microscope'** Angelos Michaelides
299	**Nucleation in alkali metal chloride solution observed at the cluster level** Akihiro Wakisaka
309	**General Discussion**
329	**The effect of oxygen-containing reagents on the crystal morphology and orientation in tungsten oxide thin films deposited *via* atmospheric pressure chemical vapour deposition (APCVD) on glass substrates** Geoffrey Hyett, Christopher S. Blackman and Ivan P. Parkin

345 **Stabilization of metastable phases in spatially restricted fields: the case of the Fe_2O_3 polymorphs**
Martí Gich, Anna Roig, Elena Taboada, Elies Molins, Caroline Bonafos and Etienne Snoeck

355 **Using *in situ* synchrotron radiation wide angle X-ray scattering (WAXS) to study $CaCO_3$ scale formation at ambient and elevated temperature**
Tao Chen, Anne Neville, Ken Sorbie and Zhong Zhong

367 **Nucleation and control of clathrate hydrates: insights from simulation**
C. Moon, R. W. Hawtin and P. Mark Rodger

383 **Crystallization of carbon tetrachloride in confined geometries**
Adil Meziane, Jean-Pierre E. Grolier, Mohamed Baba and Jean-Marie Nedelec

395 **General Discussion**

CONCLUDING REMARKS

409 **Crystal growth and nucleation: tracking precursors to polymorphs**
Patrick R. Unwin

ADDITIONAL INFORMATION

417 **Poster titles**
421 **List of Participants**
423 **Index of Contributors**

PAPER

Spiers Memorial Lecture
Lessons from biomineralization: comparing the growth strategies of mollusc shell prismatic and nacreous layers in *Atrina rigida*

Fabio Nudelman,[a] Hong H. Chen,[b] Harvey A. Goldberg,[b] Steve Weiner[a] and Lia Addadi*[a]

Received 23rd March 2007, Accepted 30th March 2007
First published as an Advance Article on the web 13th April 2007
DOI: 10.1039/b704418f

The mollusc shell prismatic layer of *Atrina rigida* is composed of an assemblage of large and relatively perfect single calcite crystals, embedded in an organic matrix. A key to elucidating basic mechanisms of mineralization is understanding the structures of the matrix, the mineral and the relations between them. The matrix that envelopes each prism (the inter-prismatic matrix) is composed mainly of glycine-rich proteins, while the matrix inside each prism (intra-crystalline matrix) is composed of a network of chitin fibers. Prisms grow by deposition of mineral particles on the chitin fibers. The mineral particles are associated with highly acidic proteins from the Asprich family, which presumably stabilize an amorphous mineral precursor. We infer that once in contact with the already formed crystalline prism, the particles crystallize by epitaxial nucleation. In nacre, sheets of β-chitin are interspaced by silk-like proteins in a hydrated gel-like state. β-Chitin forms a scaffold onto which the acidic proteins are adsorbed. Some of these are organized into a crystal nucleation site, where nucleation of aragonite, supposedly from colloidal amorphous calcium carbonate particles, is induced. Comparing the mechanisms of growth of the nacreous and prismatic layers can help to understand the underlying strategies of formation of mineralized structures.

1. Introduction

Mollusc shells are composed of calcium carbonate, in the form of calcite or aragonite. Some shells have layers composed of calcite and layers composed of aragonite. The mineral constitutes 95–99% of the shell weight, and the organic material 0.1–5%.[1] The organic material is composed mainly of proteins and polysaccharides that are assembled into a 3-dimensional organic matrix framework[2] and are thought to provide the microenvironment where mineral deposition occurs.[3] Some of these macromolecules are also occluded inside the mineral phase,[4,5] where they presumably exert direct control over crystal nucleation, crystal growth, polymorph type and morphology.[6–8]

---

[a] *Department of Structural Biology, Weizmann Institute of Science, Rehovot 76100, Israel. E-mail: lia.addadi@weizmann.ac.il*
[b] *CIHR Group in Skeletal Development and Remodeling, Schulich School of Medicine & Dentistry, University of Western Ontario, London, Ontario, Canada, N6A 5C1*

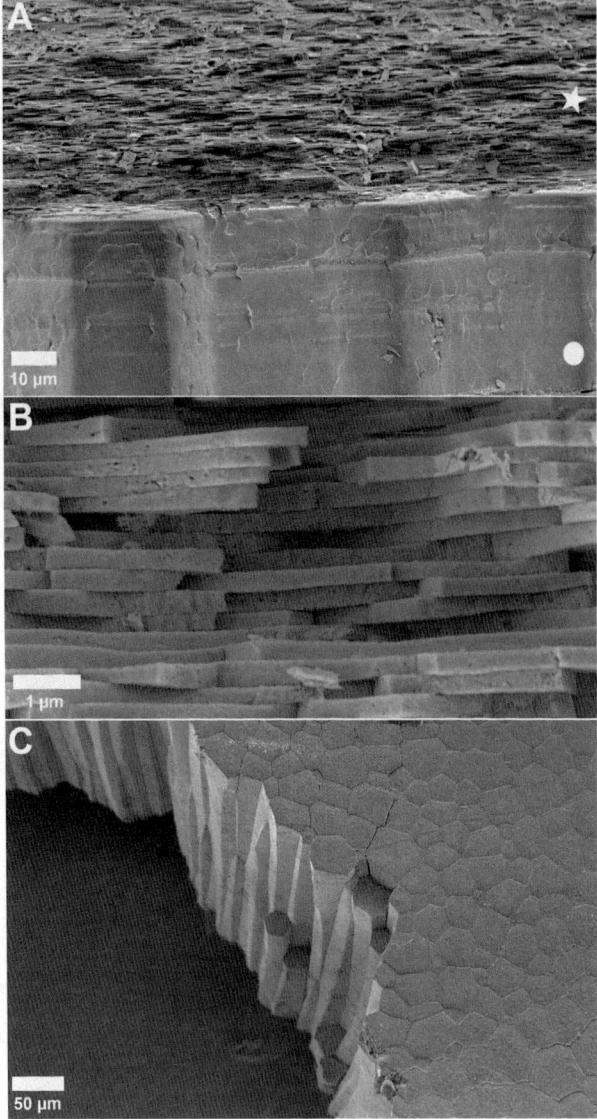

Fig. 1 A: Scanning electron micrograph of a cross-section of the shell of *Atrina rigida*, showing the nacreous (white star) and prismatic (white circle) layers. B: Scanning electron micrograph of a fracture surface of the cross-section of the nacreous layer. C: Scanning electron micrograph of the surface of the prismatic layer.

The shell of the bivalve *Atrina rigida* possesses an inner aragonitic nacreous layer and an outer calcitic prismatic layer (Fig. 1A). In general the nacreous layer is composed of polygonal tablets about 5–15 μm in diameter, arranged in continuous parallel lamellae, typically 0.5 μm thick, separated by sheets of interlamellar organic matrices (Fig. 1B).[9–12] The nacreous layer organic matrix has 3 major components: (1) β-chitin, which is the main constituent of the inter-lamellar matrix, forms a scaffold where crystals nucleate and grow;[13,14] (2) silk-like proteins,[13,15] present in a hydrated gel-like state, presumably fulfil the task of space-fillers prior to mineral deposition, also providing a hydrophobic microenvironment that contributes to the control of crystal formation;[16,17] (3) an assembly of acidic glycoproteins,[18,19] some of which specifically nucleate aragonite when adsorbed on chitin.[8,19] These major components do not function in isolation, however, and their assembly and interaction are essential for the correct regulation of crystal nucleation, growth, morphology, and polymorphism. In vitro experiments have demonstrated that the assembly of all the components is important to induce preferred nucleation of aragonite over calcite.[8]

Much of what is known to date about mollusc shell formation is from studies of the nacreous layer. Here we focus on the structure and formation processes of the prismatic layer. Compared to nacre, the structural properties of the prismatic layer and the components controlling its formation are not well understood. The calcite prisms of mollusc shells are usually several hundreds of micrometers long, surrounded by a thick layer of organic matrix.[20–23] Atrina has a so-called "simple prismatic structure",[24] where the prisms are unusually large, with a diameter of 20–50 µm in diameter and hundreds of microns in length (Fig. 1C). Individual prisms diffract X-rays as single crystals, with their c-axes perpendicular to the shell surface.[4] Based on their coherence lengths and degrees of mosaic spread they are almost as well ordered as geological calcite.[4] Dauphin et al. suggested that in the prismatic layer from Pinna nobilis,[25,26] which is closely related to Atrina rigida, the growth of prisms has a periodic rhythmic deposition, which results in growth lines perpendicular to the long axis of the prism.[27]

A number of individual proteins have been sequenced from the matrix of the prismatic layers of various molluscs, although their functions have not been identified. The organic layer enveloping each prism (inter-prismatic matrix) is the main component of the organic matrix and is composed of glycine-rich proteins.[22] One glycine-rich protein has been identified to date, MSI31, from the prismatic layer of the shell of Pinctada fucata, which may be a constituent of the organic matrix framework.[28] Other proteins have also been extracted and characterized from the organic matrix (the inter and/or intra-prismatic matrix). A few unusually acidic proteins have been identified to date: caspartin, from the shell of Pinna nobilis;[29] aspein, from the shell of Pinctada fucata;[30] and Asprich, a family of highly acidic proteins from the water-soluble matrix from the shell of Atrina rigida, which possess 39–71% Asp and 10–14% Glu.[31] Their roles in mineral formation in the prismatic layer are not yet known. Other identified proteins are prismalin-14[32] and calprismin.[29] To date chitin has not been identified as a component of the organic matrix of the prismatic layer.

In the present study we investigated the structure of the organic matrix of the prismatic layer and characterized the inter- and intra-prismatic matrices. In particular, we identified the presence of chitin and mapped the location of Asprich within the prisms. We attempt to relate the structures and locations of these matrix components to their respective functions, formulating a tentative mechanism of prism growth.

2. Experimental

2.1 Shell preparation

Fresh shells of the bivalve Atrina rigida from the east coast of Florida were purchased from Gulf Specimen Marine Laboratories Inc. (Panacea, FL, USA) and were stored dry at 4 °C.

Fragments of the outer calcitic prismatic layer were either broken from the shell edge or broken and then mechanically separated from the inner aragonitic nacreous layer, when samples from inside the pallial line were required. The fragments were washed overnight with 10% NH_4OH solution followed by extensive washes with Milli-Q water before processing or observation.

Single prisms were isolated from ground fragments (particle size between 400 µm and 1 mm), by suspending in 4% sodium hypochlorite solution for 3 days at room temperature on a rocking table. The dispersed prisms were extensively washed with Milli-Q water, lyophilized and stored dry at −20 °C.

For scanning electron microscopy samples were dehydrated in an ethanol series (50, 70, 96 and 100%), critical point dried and coated with 4 nm chromium (see below).

2.2 Organic matrix extraction

Total organic matrix (inter- and intra-crystalline) was extracted from ground fragments of the prismatic layer (particle size <250 µm). Intra-crystalline organic matrix was extracted from isolated single prisms (see above). Demineralization by cationic ion exchange resin was adapted from Albeck et al.[33] Ground fragments of the prismatic layer (total) or isolated prisms in 5 g batches were suspended in 0.025% sodium azide solution and sealed inside a dialysis membrane (3500 MW cut-off). The membrane was placed inside a 50 mL tube which was filled to about one fifth of its volume with pre-washed cationic ion exchange resin (Dowex 50 × 8, mesh 50–100, Sigma-Aldrich) and the tube was filled with 0.025% sodium azide solution. The tube was rotated slowly and continuously in a propeller-like motion at room temperature until decalcification was complete. The solution outside the dialysis membrane was changed twice a day. After complete decalcification the fractions were dialyzed against Milli-Q water for 48 h at 4 °C with four changes of water. The fractions were then centrifuged at 4000 rpm and the precipitate (the water-insoluble fraction) separated from the supernatant (the water-soluble fraction). The precipitate was washed 3 times with Milli-Q water and both fractions were lyophilized and stored dry at −20 °C.

For amino acid analysis, aliquots of soluble and insoluble fractions from the total and intra-crystalline extracts were hydrolyzed with 6 M HCl vapor in vacuum for 24 h. Following evaporation of the HCl, hydrolysates were analysed with an automatic amino acid analyser (HP Aminoquant system).

2.3 Expression of GFP-chitin binding protein

Expression plasmids pQE31 coding for recombinant green fluorescent protein-tagged chitin binding protein[34] (CBGFP) were kindly provided by Dr Ingrid M. Weiss (Regensburg University, Regensburg, Germany), and were expressed as described in ref. 34. Briefly, the plasmids were transformed into competent E. coli XL-1 Blue (Strategene) and positive transformants were selected by resistance to ampicilline. Expression and purification of the recombinant GFP-chitin binding protein was performed using Ni-NTA resin according to the instructions for soluble 6xHis-tagged proteins (Qiagen).

2.4 Fluorescence labeling with CBGFP

Single prisms were etched with 10% acetic acid for 5 min and extensively washed with Milli-Q water. Labeling with CBGFP was performed as described in ref. 34, with modifications. Etched prisms were incubated with 0.5% bovine serum albumin (Sigma) in saline solution (150 mM NaCl, 20 mM KCl, 10 mM $CaCl_2$, 10 mM $MgCl_2$) at room temperature for 1 h with shaking. CBGFP or GFP without the chitin binding domain were added to the samples to reach final concentrations of 20 µg mL^{-1}, and incubated for 1 h at room temperature. Unbound CBGFP or GFP was removed by washing 3 times with saline solution containing 0.1% Tween 20 (Sigma) for 15 min, followed by 2 washes with saline solution for 10 min. Samples were rinsed with Milli-Q water and mounted on glass slides with elvanol for fluorescence microscopy.

CBGFP labeling was also performed on etched prisms previously treated with chitinase from *Streptomyces griseus* (Sigma). The prisms were incubated with 0.2 U of the enzyme in 0.05 M phosphate buffer pH 6.3 at room temperature for 16 h with shaking.[35] For electron microscopy, prisms were dehydrated in an ethanol series (50, 70, 96 and 100%), critical point dried and coated with 4 nm chromium (see below).

2.5 Cloning and expression of Asprich

Asprich (His-22) cDNA obtained from the *Atrina* cDNA library[31] was cloned into the pET28 expression vector (Novagen). The vector added an N-terminal 6xHis tag and a TEV cleavage site to the Asprich cDNA sequence. The amino acid sequence is the same as that reported previously (protein sequence "c").[31] The recombinant DNA procedures were carried out using methods described by Sambrook *et al.*[36] The coding sequence was confirmed by DNA sequencing.

Plasmids were first transfected into competent *E. coli* XL-10 cells, colonies were picked and sequence confirmed. *E. coli* BL21 (DE3) were transformed with the expression plasmids, and grown in phosphate-buffered Super Broth (SB) supplemented with 15 µg mL^{-1} kanamycin and 0.4% glucose to an absorbance of 0.6–0.9. After induction with 2 mM isopropyl-β-D-thiogalactopyranoside, cultures were grown for a further 4 h, the cells were recovered by centrifugation, resuspended and then sonicated in denaturing binding buffer (5 mM imidazole, 0.5 M NaCl, 0.02 M Tris–HCl, 6 M urea, pH 7.9) as described.[37] The extract was applied to a His-bind column (Novagen), and Asprich eluted with 0.5 M imidazole-containing elution buffer. Fractions containing Asprich were pooled and subjected to purification by fast protein liquid chromatography (FPLC) using established protocols.[38] Proteins were purified by chromatography on a Q-Sepharose Fast Flow column followed by gel filtration on a Superdex 200PG column (GE Biosciences). Purification of Asprich was monitored by electrophoresis on 12.5% SDS polyacrylamide gels, stained with Stains-all and silver nitrate as described previously.[39] Fractions containing Asprich were dialyzed against 0.01 M ammonium bicarbonate buffer, lyophilized in aliquots, and protein content determined by amino acid analysis. Yield of intact Asprich (∼95% purity) was approximately 400 µg L^{-1} of initial culture media.

2.6 Polyclonal antibodies against Asprich

New Zealand rabbits were injected subcutaneously with 100 µg of Asprich. Three weeks later they were boosted subcutaneously with the protein (100 µg) and 3 weeks after that were boosted intramuscularly with protein (100 µg). The rabbits were bled 10 days later and the presence of antibodies was tested. These experiments conformed to the US National Institute of Health's ethical guidelines. The specificity of polyclonal antibodies was tested with ELISA in which Asprich was reacted with diluted antibodies (1 : 100, 1 : 200, 1 : 500 and 1 : 1000).

2.7 Immunolabeling

Immunolabeling was performed on etched single prisms and on fragments of the growing shell edge. Single prisms were etched with 10% acetic acid for 1–5 min, followed by extensive washing with Milli-Q water. The prisms were then incubated with 0.5% bovine serum albumin in phosphate buffered saline (PBS, Sigma) for 1 h at room temperature with shaking. Either serum containing polyclonal antibodies against Asprich diluted 1 : 100 or pre-immune serum diluted 1 : 100 in PBS was added to the samples for 1 h. Unbound antibodies were removed by washing 3 times with PBS + Tween 20 (0.1%) for 15 min. The secondary antibody was applied to the samples for 1 h. We used either rhodamine-conjugated goat anti-rabbit (Jackson, diluted 1 : 100 in PBS + 0.1% Tween 20) for fluorescence microscopy, or gold-labeled goat anti-rabbit (Jackson, 18 nm gold particles, diluted 1 : 10 in PBS + 0.1%

Tween 20) for scanning electron microscopy. Unbound secondary antibodies were removed by washing with **PBS** + Tween 20 (0.1%) 3 times for 15 min, followed by 2 washes with **PBS** for 10 min, and then the samples were rinsed in water. For fluorescence microscopy the samples were mounted on glass slides in Milli-Q water. No drying of the slides was observed under the microscope. For scanning electron microscopy samples were dehydrated in an ethanol series (50, 70, 96 and 100%) and critical point dried. Samples were coated either with 2 nm of chromium or with coarse carbon. Chromium coating was used to map the location of Asprich where high resolution was needed. However, as such coating results in poor contrast between the organic matrix, the mineral and the gold particles, for statistical analysis of the amount of gold particles found on the samples, we used coarse carbon coating.

2.8 Optical microscopy

Fluorescently labeled samples were viewed with epifluorescence microscopy using a Zeiss optical microscope equipped with a video camera attached to an LIS-700 integration amplifier (Applitec-Holon, Israel) that allows amplification of the image intensity up to 3000 times. Rhodamine-labeled samples were excited at 550 nm and CBGFP-stained samples were excited at 494 nm.

2.9 Electron microscopy

Critical point drying was performed using a CPD-030 critical point dryer (Bal-Tec). Chromium sputtering was done in a K757X Chromium Sputter (Emitech). Carbon coating was done using an SPI-Module carbon coater. The samples were visualized either in the LEO-Supra 55 VP FEG SEM (Zeiss) or in the high-resolution SEM Ultra 55 (Zeiss, Oberkochen, Germany).

3. Results

In the prismatic layer, each calcitic prism is surrounded by an envelope of water-insoluble proteinaceous matrix. This framework for the mineral prisms can be readily observed once the mineral is completely dissolved (Fig. 2A). When this inter-prismatic matrix is removed with sodium hypochlorite, the whole structure disintegrates into individual prisms (Fig. 2B). Etching of the mineral with acetic acid is not uniform, but reveals zones of more dense and less dense fibers perpendicular to the long axis of the prism (Fig. 2C), similar to those observed in prisms from the shell of *Pinna nobilis*.[27,40] At higher magnifications, this intra-crystalline network of fibers does not appear to have any local order (Fig. 2D). The mineral after slight etching, appears to be composed of distinct spheres of size 40–50 nm (Fig. 2D), in agreement with previous observations by Dauphin.[41]

3.1 Study of the growing prismatic layer

Fragments of the growing prismatic layer were removed from the edge of shells of live animals that were flash-frozen, and were observed in the SEM after critical point drying. The forming surfaces can be divided into four stages of growth based on the density of mineral particles. Fig. 3A shows a small area of a growing prism, where numerous rounded particles of 50–100 nm size appear to be deposited on top of a smoother surface. The latter is not continuous, but rather interrupted by cracks that could have been caused by drying of the sample. In the backscattered electron image (Fig. 3B), the contrast between the particles and the underlying mineral surface is very small, indicating that they are also composed of mineral, possibly together with organic matrix. In Fig. 3A, many mineral particles envelope the fibers (white arrows). The area shown in Fig. 3C and D, has a much higher density of particles. Most of the particles are not spherical, and resemble small calcite crystals. In Fig. 3E and F, where the inter-prismatic envelope can be observed (white star),

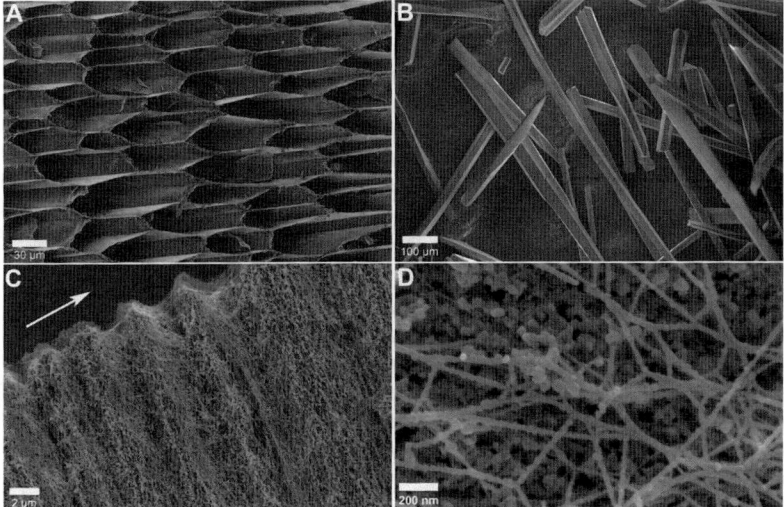

Fig. 2 A: Scanning electron micrograph of an oblique section of the inter-prismatic matrix (the prism envelopes) after complete decalcification. B: Scanning electron micrograph of individual prisms dispersed after treatment with sodium hypochlorite. C: Scanning electron micrograph of an etched single prism, showing the cyclic growth pattern. The white arrow points in the direction of the long axis of the prism. D: Higher magnification of C, where a network of fibers that permeates the prisms was exposed by partially dissolving the mineral.

the adjacent prisms appear to be formed of densely packed particles. Note the difference in contrast between the organic and mineral phases in the back scattered mode (Fig. 3F). Fig. 3G and H show the edges of two fully formed prisms separated by an inter-prismatic matrix layer. The mineral phase is homogeneous and continuous. The difference in the thicknesses of the inter-prismatic envelopes in Fig. 3E and G may be a result of the crystallization process, although they do appear to have variable thicknesses throughout the prismatic layer. Fig. 3 shows that mineral accretion in the prismatic layer occurs *via* the formation and growth of mineral particles.

3.2 Protein composition

Proteins were extracted from the shell by complete decalcification with cationic ion exchange resin. The average yield of organic material is 5.28% of shell weight (Table 1), out of which 5% comprises the total water-insoluble fraction (consisting of both inter- and intra-crystalline matrices) and 0.28% the total water-soluble fraction. The intra-crystalline organic matrix is composed of 0.28% of insoluble and 0.085% of soluble material. Most of the organic material present in the prismatic layer is thus insoluble in water, with the inter-prismatic matrix comprising ∼4.72% and the intra-prismatic ∼0.28% of the shell weight.

Amino acid composition analysis shows that the proteins present in the total insoluble fraction are rich in Gly (48.17%). These are the main constituents of the prism envelopes. They may be related to the framework proteins characterized from the prismatic layer of *Pinctada fucata*.[28] The total soluble fraction and the intra-crystalline soluble fraction have similar amino acid compositions. In both fractions Asx and Glx together comprise more than 70 mol% of the protein. In mollusc shells most of these residues are in the aspartate and glutamate form, rather than asparagines and glutamines.[19] Interestingly, the intra-crystalline insoluble fraction is also very rich in Asx, but not in Glx. These acidic proteins are extremely soluble,

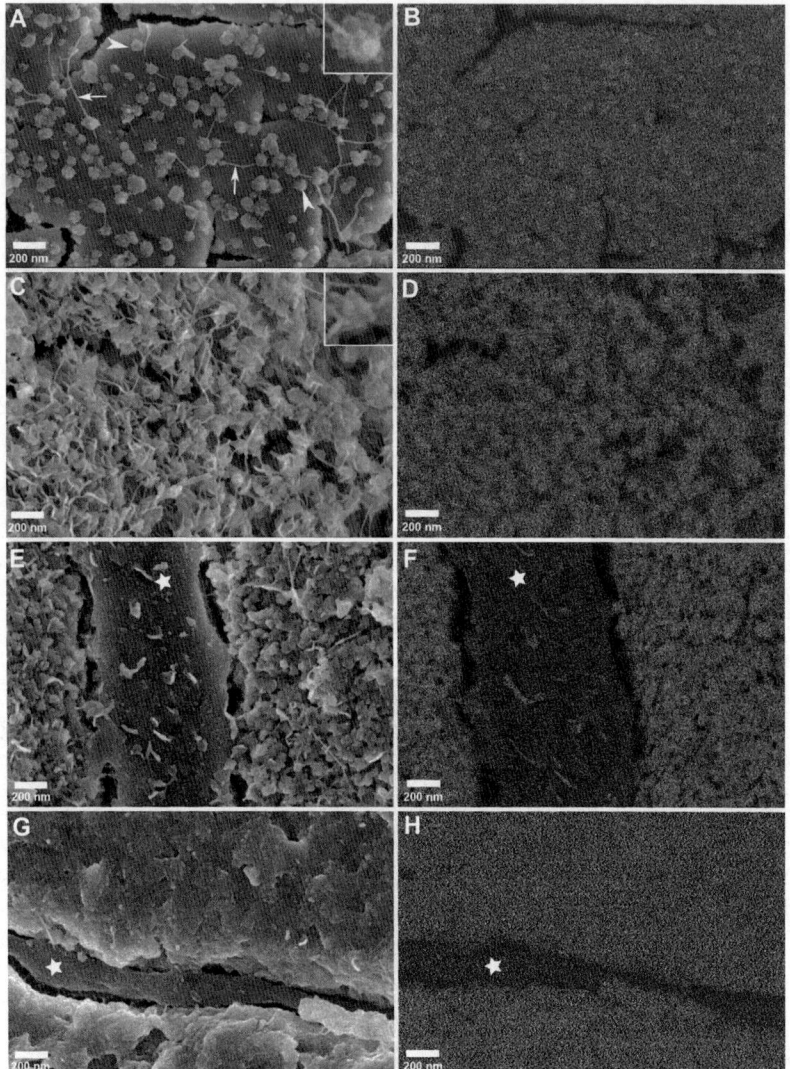

Fig. 3 SEM images of different areas of the growing prismatic layer representing different stages in the formation of a prism. A, C, E, G, secondary electrons images, and B, D, F, H, backscattered electrons images. A, B: Initial deposition of globular mineral particles ranging from 50–100 nm (white arrowheads) on an already formed layer of mineral. Fibers are visible (white arrows), going in and out of the mineral particles and of the mineral substrate. Inset in A: Higher magnification of a mineral particle. C, D: Density of particles and of organic matrix increases. Inset in C: Higher magnification of a mineral particle. E, F: Image of two adjacent prisms separated by the inter-prismatic matrix (white star). Mineral phase is still composed of densely packed particles. G, H: Image of two fully mineralized adjacent prisms, separated by the inter-prismatic envelope (white star). The mineral is continuous and there are no indications of globular particles.

and the fact that they are present in the insoluble fraction suggests that they are strongly bound to the insoluble matrix.

N-acetylglucosamine was detected in the intra-crystalline insoluble fraction, but not in the total matrix fraction. Based on the known N-acetylglucosamine content in chitin, and the relative sensitivity of N-acetylglucosamine measurement by amino acid analysis, chitin is either absent or present at negligible levels in the inter-

Table 1 Amino acid compositions of water-soluble and water-insoluble fractions from the total and intra-crystalline matrices extracted from the prismatic layer

Amino acid	Total fraction(average)/ mol%		Intra-crystalline fraction(average)/ mol%	
	Soluble	Insoluble	Soluble	Insoluble
Asx	53.90	3.34	50.74	52.38
Ser	3.80	5.94	3.12	4.95
Glx	20.68	1.72	29.61	9.04
Gly	4.50	48.17	2.86	6.24
His	0.17	0.93	0	0.80
Arg	0.41	2.1	0	1.60
Thr	1.61	1.23	1.29	2.50
Ala	9.07	3.71	8.68	8.25
Pro	0.89	2.02	0.62	2.39
Tyr	0.48	8.69	0	1.57
Val	2	5.43	1.78	2.74
Met	0	0.08	0	0.48
Lys	0.63	0.92	0.29	2.17
Ile	0.44	2.46	0.21	1.31
Leu	0.99	9.58	0.6	2.25
Phe	0.43	3.67	0.21	1.35
N-acetylglucosamine	—	Absent	—	Present
Weight fraction (%)	0.28	5	0.085	0.28

crystalline matrix envelopes, implying that they are composed mostly of Gly-rich proteins. The presence of *N*-acetylglucosamine in the intra-crystalline matrix raises the possibility that chitin is present, possibly as a constituent of the dense network of fibers that forms part of this matrix.

3.3 Identification of chitin

To further investigate the possible presence of chitin inside single prisms, we used a recombinant GFP-tagged chitin-binding protein (CBGFP). Single prisms were isolated, etched, fluorescently labeled with the CBFGP and monitored with fluorescence microscopy.[34]

Etched individual prisms are labeled with the CBGFP (Fig. 4A), indicating that chitin is part of the intra-crystalline insoluble matrix and is distributed throughout the prism. Labeling was not completely homogeneous, but displayed a striated pattern that could be due to preferential etching of certain areas of the prism, resulting in regions with more exposed organic matrix. Labeling by GFP alone was not observed (Fig. 4A, inset), showing that binding is indeed through the chitin binding domain and not due to the GFP itself. When the prisms were treated with chitinase, no fluorescent labeling was observed (Fig. 4B), concomitant with the removal of all the fibers that were exposed on the surface of the mineral (Fig. 4C). These results clearly demonstrate that chitin is indeed present within the prisms, in the form of an intra-crystalline network of fibers.

A close examination of the chitin fibers shows that they are not completely smooth, but are decorated with globular structures distributed along the length of the fiber (Fig. 5A, white arrowheads). Upon treatment of the etched prisms with sodium hypochlorite, these globular structures disappear and the fibers become completely smooth, while retaining their original thickness (Fig. 5B). Note that chitin is resistant to treatment by NaOCl. These observations indicate that the globular structures are not composed of chitin, but are probably composed of a proteinaceous material associated with the chitin. They may have been either

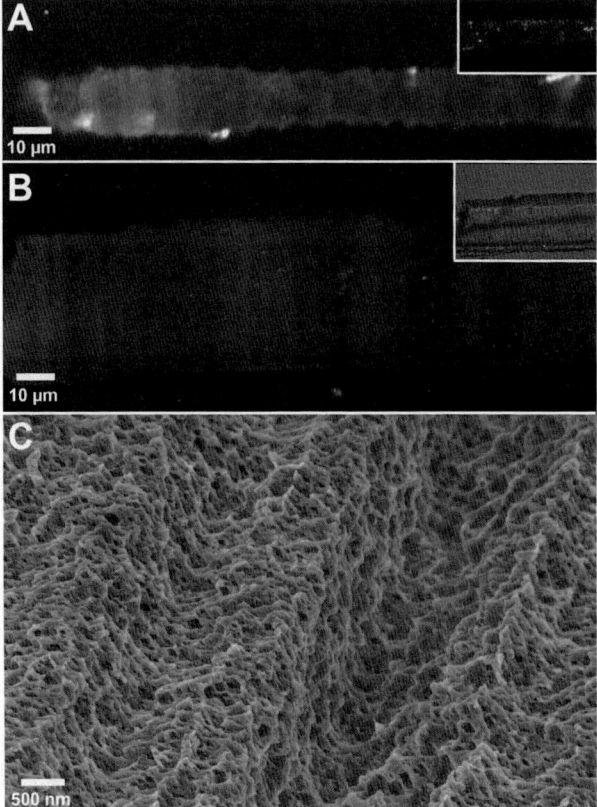

Fig. 4 A: Epifluorescence micrographs of single prisms etched with acetic acid and stained with GFP-tagged chitin-binding protein (CBGFP). Inset shows part of a sample stained with GFP as control. B: Fluorescence micrographs of single prisms etched with acetic acid and stained with CBGFP after treatment with chitinase. Inset shows part of a sample observed under bright field. C: Scanning electron micrograph of an etched single prism after treatment with chitinase; no fibers are visible on the prism surface (compare to 2D).

specifically deposited on the fibers during the formation of the organic framework, or they may have been associated with the mineral phase and adhered to the fibers after dissolution of the former.

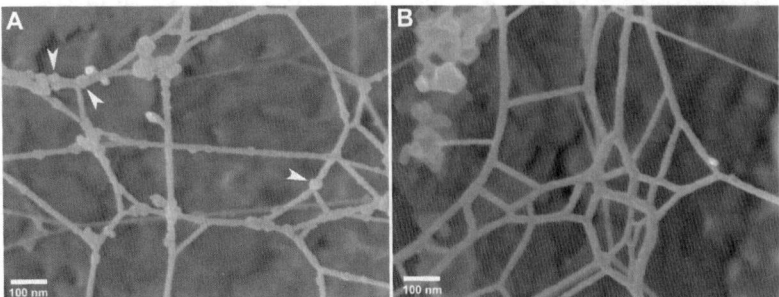

Fig. 5 A: Scanning electron micrograph of a single prism etched with acetic acid, showing the intra-prismatic fibers at high magnification. Globular structures that are distributed along the fibers are indicated (white arrowheads). B: The prism was treated with sodium hypochlorite after etching and the fibers are completely smooth.

3.4 Localization of Asprich

Asprich proteins from the prismatic layer of *Atrina rigida* were identified and sequenced.[31] We used polyclonal antibodies raised against one of the Asprich family of proteins and mapped its location within the prisms using both fluorescent and gold-labeled secondary antibodies. Fig. 6A shows an etched single prism fluorescently labeled with antibodies against Asprich, demonstrating that these proteins are present throughout the prism and are an integral component of the intra-crystalline organic matrix. The labeling pattern is similar to that of the chitin binding protein. Prisms treated with the pre-immune serum (Fig. 6B) did not show any labeling. It has to be noted that etching was done without fixatives, and since these proteins are very soluble in water, the fact that they are still present indicates that they are strongly associated either to the mineral or to the organic phase.

Further investigation of the location of Asprich within the prisms was performed using SEM. Prisms were observed after incubation with the anti-Asprich antibody and decoration with gold-labeled secondary antibodies. These experiments were performed both on etched single prisms and on fragments of the prismatic layer taken from the shell edge where the layer is still growing and the process of mineralization is active. Fig. 6C shows an SEM image of a prism etched for 5 min, where fibers and mineral are evident. The white dots were inserted to clearly mark the location of the gold particles, identified using the backscattered electrons detector (not shown, compare with Fig. 6D and E). Asprich is found mostly

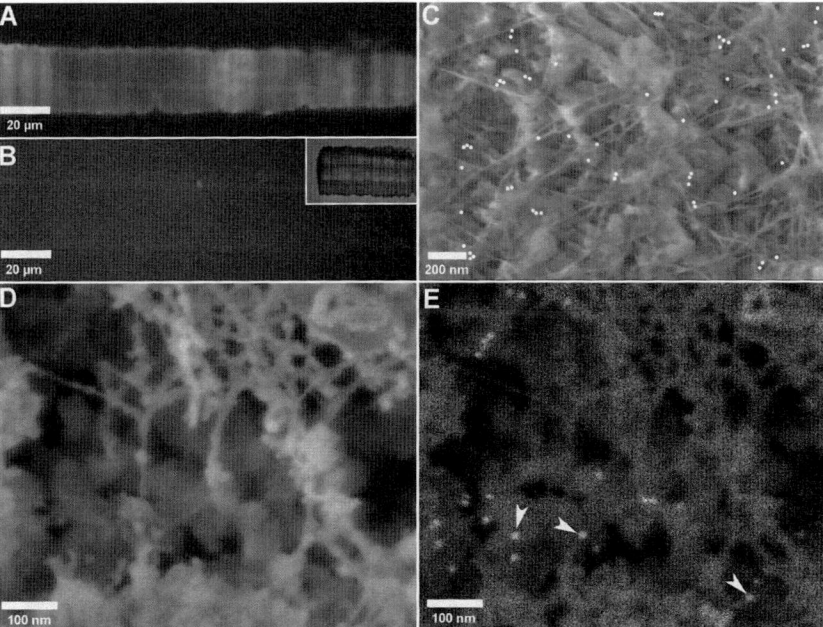

Fig. 6 A: Epifluorescence micrograph of a single prism etched with acetic acid and labeled with polyclonal antibodies raised against Asprich. B: Epifluorescence micrograph of a single prism etched with acetic acid and labeled with pre-immune serum. Inset shows part of the sample observed under bright field. C: SEM image (secondary electrons detector) of an etched prism gold-labeled with polyclonal antibodies raised against Asprich. White dots were inserted to clearly mark the location of the gold particles, and hence of Asprich, identified through the backscattered electron detector (not shown). D: SEM image (secondary electrons detector) of an area of a growing prism gold-labeled with polyclonal antibodies raised against Asprich. E: Backscattered electrons detector image of D, showing the gold particles (white arrowheads) that mark the distribution of Asprich over the mineral. Contrast between the mineral and the organic matrix is very low due to the coating with 2 nm of chromium.

Table 2 Distribution of gold particles per μm² on single prisms of the prismatic layer etched for different times, and on the growing edge, after immuno-labeling with antibodies against Asprich. Measurements were performed on carbon-coated samples

	Immune serum	Pre-immune serum
Single prisms non etched	3.0 ± 2.2	1.0 ± 0.6
Single prisms etched for 1 min	5.1 ± 2.7	0.5 ± 0.5
Single prisms etched for 5 min	14 ± 7	0.3 ± 0.1
Growing edge	102 ± 35	14 ± 6

associated with the mineral, but in some cases also localized directly on the fibers. Table 2 summarizes the distribution of gold particles per μm² on single prisms that underwent etching for different times (no etching, 1 and 5 min). The amount of gold particles found in the single prisms is directly proportional to the etching time, indicating that more protein is exposed to the antibodies as the mineral is dissolved. The colloidal gold particles are mainly associated with the mineral (70%), and less so with the fibers (30%).

Immuno-labeling on growing fragments of the prismatic layer showed that Asprich is mainly associated with the mineral particles and only sporadically on the fibers (Fig. 6D, secondary electrons detector; Fig. 6E, backscattered electrons detector). Moreover, the amount of Asprich on fragments of the growing layer was significantly higher than on etched prisms (Table 2).

The difference in labeling between the post-immune and pre-immune sera is significant both in the etched prisms and in the growing edge, highlighting the specificity of the antibodies and hence the specific location of Asprich.

4. Discussion

We have characterized aspects of the biochemical composition, the location and the relation to the mineral phase of the inter- and intra-prismatic matrices from the prismatic layer of the shell of the bivalve *Atrina rigida*. The inter-prismatic insoluble matrix—the prism envelope—is composed mainly of Gly-rich proteins. In contrast, the intra-crystalline insoluble matrix is composed mainly of chitin fibers. The amino acid compositions of the proteins of the total and intra-crystalline soluble matrices are very similar, indicating that they may, in part, be composed of the same proteins. Among the acidic proteins, we mapped the location of Asprich, and observed that it is associated mainly with the mineral particles, and to a lesser extent with the chitin fibers. From this information we derive some mechanistic conclusions on the formation of the prismatic layer and compare them to suggested mechanisms for formation of nacre.

Even though the biochemical composition of the prismatic layer matrices is similar to that of nacre, their structures and locations within the mineral phase are quite different. We will discuss each component individually, comparing and contrasting the organic matrices from the two layers.

4.1 Chitin

In the nacreous layer of molluscs, β-chitin forms into sheets that are located between aragonite crystal layers. These sheets are organized into an ordered scaffold onto which the aragonite-nucleating proteins are adsorbed and where the aragonite crystals nucleate and grow.[17] The β-chitin fibrils and the protein polypeptide chains of the nucleating proteins are aligned respectively with the *a* and *b* crystallographic axes of the aragonite crystals,[42] suggesting that chitin is in part responsible for the orientation of the crystals in the *ab* plane. Moreover, *in vitro* experiments have demonstrated that β-chitin is essential for inducing the formation of aragonite rather than calcite.[8] β-Chitin in nacre therefore functions as a scaffold and regulates crystal orientation, either directly or indirectly.

In the prismatic layer of *Atrina rigida* we observed that chitin is not located between prisms, but is mainly intra-crystalline. It forms a dense network of fibers with no apparent orientation or organization. In agreement with previous studies,[27,40] we also observed that although the fibers are distributed throughout the length of a prism, there are layers with more dense fibers and layers with less dense fibers. This pattern probably reflects the layered mode of growth of a prism (see below).

4.2 Framework proteins

In the nacreous layer, the framework proteins are rich in Ala and Gly.[43,44] Pereira-Mouries *et al.*[45] observed that such silk-like proteins leach out from fractured nacre upon incubation in water. We imaged this process in the environmental scanning electron microscope (ESEM) under controlled humidity, and concluded that these proteins form hydrated gels. We subsequently suggested that the silk is initially present in a hydrated gel-like state filling the space between two layers of chitin before mineralization.[16,17,45] As the mineral fills this space, the silk-like proteins are pushed aside by the crystal and in the mature shell overlie the chitin sheets.[16] The silk is a mild inhibitor of crystal nucleation.[46] Its function during mineral formation is thus possibly to provide a hydrophobic environment, such that nucleation of the aragonite crystals will occur only at the nucleation site, when induced by the acidic proteins adsorbed on the chitin sheets (see below).

In the prismatic layer the framework proteins are rich in Gly.[22] The protein MSI31 that was sequenced from the prismatic layer of the shell of *Pinctada fucata*,[28] is also Gly-rich, but not homologous to silk.[47] These proteins are the main components of the inter-prismatic matrix. They are observed to form prior to mineral formation[21,48] and hence their function appears to be exclusively related to providing a framework for mineralization. ESEM experiments performed exactly as in nacre, show that the Gly-rich proteins from the prismatic layer of *Atrina rigida* do not have a tendency to form hydrogels, but remain solid and homogeneous even after prolonged exposure to water, without any visible leaching of material (results not shown). Moreover, their relatively high tyrosine content is suggestive of a protein cross-linking process by quinone-tanning, similar to that of the periostracum.[2,22]

4.3 Acidic proteins

The acidic proteins from the nacreous layer of *Atrina rigida* are composed of approximately 43 mol% acidic amino acids[49] and are found, at least in part, adsorbed on the chitin sheets where some are involved in nucleation.[50] Indeed, some of these proteins are capable of specifically nucleating aragonite *in vitro*, when adsorbed onto chitin in the presence of silk fibroin gel.[8,49,51]

The acidic proteins from the prismatic layer of *Atrina rigida* are considerably more acidic than those in nacre (~75 mol% acidic amino acids). These proteins are present in the soluble fractions of both the inter- and intra-prismatic matrices. Surprisingly, the intra-crystalline insoluble matrix is also acidic (Table 1), indicating that some of the acidic proteins are stably associated to another component, possibly chitin. This makes the complex water-insoluble. Indeed, the chitin fibers exposed after partial dissolution of the mineral were decorated with globular structures of proteinaceous composition (Fig. 5).

Asprich is more closely associated with the mineral (Fig. 6) than with the chitin fibers: it is associated with the 50–100 nm sized mineral particles that are deposited on the growth front of the prism (Fig. 3). It is conceivable therefore that the mineral binds to chitin through Asprich, and once the mineral is dissolved the protein remains on the fibers. In such a scenario it is difficult to envision that Asprich is involved in induction of crystal nucleation in the same manner as some of the nacre proteins are.

In conclusion, even though both the nacreous and prismatic layers contain very acidic proteins, these differ in the proportion of acidic amino acids, in their location, and in the manner in which they affect crystal nucleation, which is probably indicative of different functions in the regulation of mineralization in the two layers.

4.4 Mineral

The calcite of a mature prism of *Atrina rigida* is very ordered at the atomic scale. Each prism diffracts X-rays as a single crystal with coherence lengths (around 450 nm) and degrees of mosaic spread comparable to that of large crystals of geological calcite.[4] This high degree of order is consistent with each prism crystal forming from a single nucleation event. The growth of a prism occurs in the direction of its c axis by deposition of layers.[27,52] Etching of the mineral phase of mature prisms exposes a pattern of small particles associated with organic material.[26] We observed the initial deposition of 50–100 nm sized mineral particles on the prism ab plane and their subsequent growth and fusion into dense layers (Fig. 3). This mode of growth would very likely result in polycrystalline prisms, if the particles were already crystalline prior to deposition. This paradox can be conceivably resolved, if we assume that the mineral crystallizes only after deposition.

To date there is no direct evidence of the presence of amorphous calcium carbonate (ACC) in the nacreous and prismatic layers. However, there are indirect indications in support of this hypothesis: (i) larval mollusc shells were shown to form from an ACC precursor;[53] (ii) the prism mineral after etching exhibits a texture of rounded particles (50–100 nm, Fig. 2D) that is consistent with calcium carbonate crystals grown from amorphous mineral precursor phases. Similar etching patterns were observed in the nacreous layer,[16] and most significantly in sea urchin spines,[54] larval spicules[55,56] and in calcitic sponge spicules,[57] where the mineral phases do form from amorphous precursors;[54] (iii) Asprich inhibits calcite nucleation *in vitro*, and appears to induce and stabilize the deposition of transient amorphous calcium carbonate (Politi *et al.*, in preparation). It is also conceivable that Asprich mediates the process further by anchoring the particles to the chitin fibers, thus effectively providing the 'logistics' needed for contact to occur between the particles and the prism surface. This scenario is schematically depicted in Fig. 7. It should be regarded as a working hypothesis that clearly needs to be further investigated.

4.5 Broader perspective

The composition, structure and 3-dimensional assembly of the organic matrix are clearly parameters that control the whole process of mineral formation. The organic matrices from the prismatic and nacreous layer of *Atrina rigida* are each structured so that the former results in the formation of large calcitic prisms, while the latter generates tablet-like aragonitic crystals.

The two calcium carbonate polymorphs have intrinsically different growth behaviors. Aragonite crystals preferably grow along the c axis, forming small thin needles with no cleavage planes. Calcite, on the other hand, tends to form larger rhomb-shaped crystals that cleave easily along the {10.4} planes.[5] The organic matrix framework of the nacre is structured in such a way that the aragonite crystals grow as single crystals that are not thicker than 0.5 μm, and usually 10–20 μm wide.[9,11,58] The thickness of the crystal is controlled by the thickness of the space between two sheets of chitin. The crystal grows first vertically along its c axis until it reaches the upper layer of chitin and then expands laterally until it meets another crystal.[59] Calcitic prisms on the other hand are rather large, being several hundreds of micrometers long (Fig. 1C and 2B). Other organisms, such as echinoderms and sponges, which build large single crystals of calcium carbonate and use them as

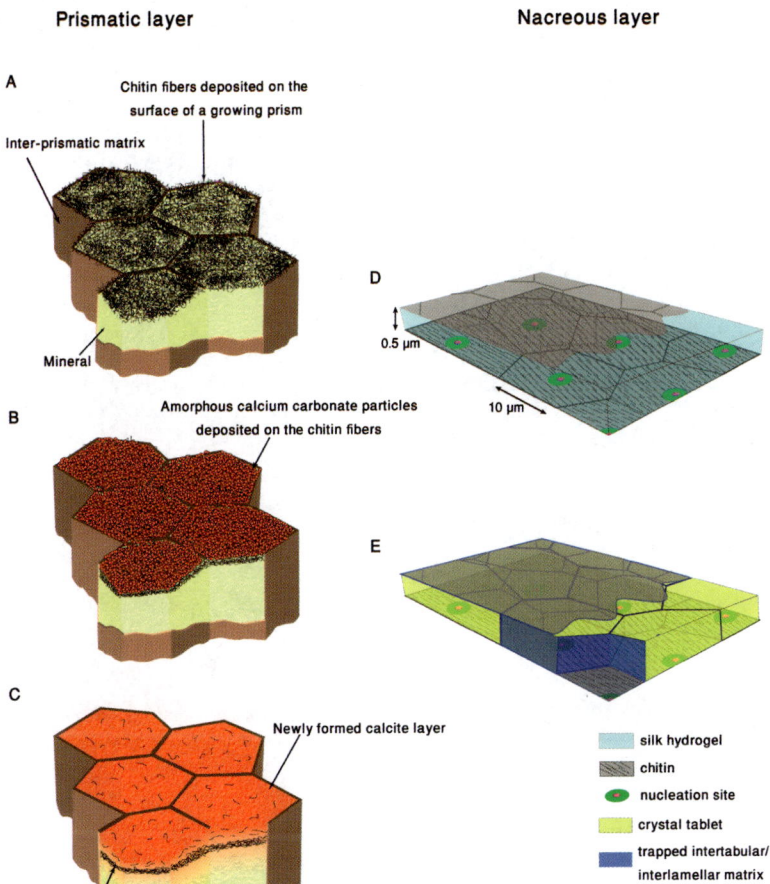

Fig. 7 Proposed mechanisms of growth for the prismatic (A–C) and nacreous (D and E) layers. A. Prism growth starts by deposition of a dense meshwork of chitin fibers on top of an already formed mineral layer. B. Particles of ACC stabilized by Asprich are deposited on the chitin fibers. C. ACC crystallizes by epitaxial nucleation once in contact with the growing mineral, thus forming a new layer of calcite and occluding the chitin fibers within the crystal. The formation of the inter-prismatic matrix precedes the growth of the mineral. Successive iterations of this cycle result in growth lines, with zones of more dense and less dense chitin fibers. D. Nacreous layer scheme before mineralization. Formation of the nacre starts by assembly of the organic matrix. Two sheets of β-chitin are interspaced by the silk-like proteins in a hydrated gel-like state. β-Chitin forms a scaffold onto which the acidic proteins are adsorbed and form the crystal nucleation site, while the silk-like proteins inhibit non-specific crystallization. E. Nacre scheme after mineralization. Nucleation of aragonite, supposedly from colloidal mineral particles, is induced by the acidic proteins on the crystal nucleation site. As the mineral tablets grow the silk-like proteins are pushed aside and become compressed between two mineral tablets. D and E are reproduced from Addadi et al.[16] with permission from Wiley-VCH Verlag GmbH & Co.

skeletal parts, also selected calcite,[5] and evolved different mechanisms to control its brittleness. In sponges and echinoderms, the crystals use intra-crystalline water-soluble proteins for crack stopping.[5,60] These proteins are adsorbed on the growing crystals and end up intercalated in the mosaic block structure at distances of

~150 nm along specific crystallographic planes.[4] In addition to the intra-crystalline proteins, in the prismatic layer, chitin fibers are interspersed inside the crystal forming a stable meshwork with a mesh size of several hundreds of nanometers. It is almost incredible that a major framework component such as chitin can be located inside a crystal at very high density, while the mineral phase is still continuous and the prisms are single crystals with high crystallinity. It is probable that the network of chitin fibers within the prisms affects their mechanical properties, such that their brittleness will be further reduced.

In contrast to nacre, which is always composed of aragonitic single crystal tablets, not all mollusc shell prismatic layers have the same structure. Some prismatic layers are composed of polycrystalline aragonite.[24] Most are composed of calcite. The individual prisms may be single crystals or polycrystalline.[24,61] It will be interesting to study the composition and structure of their intra-crystalline organic matrices and determine their functions in the different scenarios. Ultimately, it will lead to understanding how organisms build and shape the mineralized structures in a manner most suitable for their needs.

Acknowledgements

We thank Dr Ingrid M. Weiss for kindly providing the plasmid encoding the chitin binding protein; Dr Naama Kessler for her help in the expression of the chitin-binding protein; Dr Tamar Unger for cloning of Asprich; and Dr Eugenia Klein for her help in the electron microscopy work. We thank the Minerva Foundation and the Kimmelman Center for Biomolecular Structure and Assembly, Weizmann Institute, for financial support. L.A. is the incumbent of the Dorothy and Patrick Gorman professorial chair of Biological Ultrastructure, and S.W. is the incumbent of the Dr Trude Burchardt professorial chair of structural biology.

References

1. P. E. Hare and P. H. Abelson, *Carnegie Inst. Washington Year Book*, 1965, **64**, 223.
2. H. A. Lowenstam and S. Weiner, *On Biomineralization*, Oxford University Press, New York, 1989.
3. S. Weiner, W. Traub and H. A. Lowenstam, in *Biomineralization and Biological Metal Accumulation*, ed. P. Westbroek and E. W. de Jong, Reidel Publishing Company, 1983, p. 205.
4. A. Berman, J. Hanson, L. Leiserowitz, T. F. Koetzle, S. Weiner and L. Addadi, *Science*, 1993, **259**, 776.
5. S. Weiner and L. Addadi, *J. Mater. Chem.*, 1997, **7**, 689.
6. L. Addadi and S. Weiner, *Proc. Natl. Acad. Sci. U. S. A.*, 1985, **82**, 4110.
7. L. Addadi, J. Moradian, E. Shay, N. G. Maroudas and S. Weiner, *Proc. Natl. Acad. Sci. U. S. A.*, 1987, **84**, 2732.
8. G. Falini, S. Albeck, S. Weiner and L. Addadi, *Science*, 1996, **271**, 67.
9. N. Watabe, *J. Ultrastruct. Res.*, 1965, **12**, 351.
10. C. Gregoire, *J. Biophys. Biochem. Cytol.*, 1957, **3**, 797.
11. K. Wada, *Bull. Natl. Pearl Res. Lab.*, 1968, **13**, 1561.
12. S. Wise, *Eclogae Geol. Helv.*, 1970, **63**, 775.
13. S. Weiner and W. Traub, *FEBS Lett.*, 1980, **111**, 311.
14. C. Jeunieux, *Chitine et Chitinolyse*, Masson, Paris, 1963.
15. S. Weiner, Y. Talmon and W. Traub, *Int. J. Biol. Macromol.*, 1983, **5**, 325.
16. L. Addadi, D. Joester, F. Nudelman and S. Weiner, *Chem. Eur. J.*, 2006, **12**, 981.
17. Y. Levi-Kalisman, L. Addadi and S. Weiner, *J. Struct. Biol.*, 2001, **135**, 8.
18. M. A. Crenshaw, *Biominer. Res. Rep.*, 1972, **6**, 6.
19. S. Weiner, *Calcif. Tissue Int.*, 1979, **29**, 163.
20. C. Gregoire, *J. Biophys. Biochem. Cytol.*, 1961, **9**, 395.
21. H. Nakahara and G. Bevelander, *Calcif. Tissue Res.*, 1971, **7**, 31.
22. H. Nakahara, M. Kakei and G. Bevelander, *Jpn. J. Malacol.*, 1980, **39**, 167.
23. J. P. Cuif, D. Flamand, B. Frerotte, A. Chabin and A. Raguideau, *C. R. Acad. Sci., Ser. II*, 1987, **304**, 475.
24. J. G. Carter, *Skeletal Biomineralization: Patterns, Processes and Evolutionary Trends*, Van Nostrand Reinhold, New York, 1990.

25 J. P. Cuif, P. Gautret and F. Marin, in *Phylogeny of Mineralization in Biological Systems*, ed. S. Suga and H. Nakahara, Springer-Verlag, Tokyo, 1991, p. 391.
26 Y. Dauphin, *J. Biol. Chem.*, 2003, **278**, 15168.
27 Y. Dauphin, J. P. Cuif, J. Doucet, M. Salome, J. Susini and C. T. Williams, *Mar. Biol.*, 2003, **142**, 299.
28 S. Sudo, T. Fujikawa, T. Nagakura, T. Ohkubo, M. Sakaguchi, K. Tanaka, K. Nakashima and T. Takahashi, *Nature*, 1997, **387**, 563.
29 F. Marin, R. Amons, N. Guichard, M. Stigter, A. Hecker, G. Luquet, P. Layrolle, G. Alcaraz, C. Riondet and P. Westbroek, *J. Biol. Chem.*, 2005, **280**, 33895.
30 D. Tsukamoto, I. Sarashina and K. Endo, *Biochem. Biophys. Res. Commun.*, 2004, **320**, 1175.
31 B. A. Gotliv, N. Kessler, J. L. Sumerel, D. E. Morse, N. Tuross, L. Addadi and S. Weiner, *ChemBioChem*, 2005, **6**, 304.
32 M. Suzuki, E. Murayama, H. Inoue, N. Ozaki, H. Tohse, T. Kogure and H. Nagasawa, *Biochem. J.*, 2004, **382**, 205.
33 S. Albeck, S. Weiner and L. Addadi, *Chem. Eur. J.*, 1996, **2**, 278.
34 I. M. Weiss and V. Schonitzer, *J. Struct. Biol.*, 2006, **153**, 264.
35 L. R. Berger and D. M. Reynolds, *Biochim. Biophys. Acta*, 1958, **29**, 522.
36 J. Sambrok, E. F. Fritsch and T. Maniatis, *Molecular Cloning: A Laboratory Manual*, Cold Spring Harbor Laboratory Press, Cold Spring Harbor, NY, 1989.
37 C. E. Tye, K. R. Rattray, K. J. Warner, J. A. Gordon, J. Sodek, G. K. Hunter and H. A. Goldberg, *J. Biol. Chem.*, 2003, **278**, 7949.
38 H. A. Goldberg and J. Sodek, *J. Tissue Cult. Methods*, 1994, **16**, 211.
39 H. A. Goldberg and K. J. Warner, *Anal. Biochem.*, 1997, **251**, 227.
40 J. P. Cuif, Y. Dauphin, A. Denis, D. Gaspard and J. P. Keller, *C. R. Hebd. Seances Acad. Sci., Ser. D*, 1980, **290**, 759.
41 Y. Dauphin, *Comp. Biochem. Physiol., A*, 2002, **132**, 577.
42 S. Weiner and W. Traub, *Philos. Trans. R. Soc. London, Ser. B*, 1984, **304**, 421.
43 C. Gregoire, in *Chemical Zoology*, ed. M. Florkin and B. T. Scheer, Academic Press, New York, 1972, p. 45.
44 V. R. Meenakshi, P. E. Hare and K. M. Wilbur, *Comp. Biochem. Physiol., B*, 1971, **40B**, 1037.
45 L. Pereira-Mouries, M. J. Almeida, C. Ribeiro, J. Peduzzi, M. Barthelemy, C. Milet and E. Lopez, *Eur. J. Biochem.*, 2002, **269**, 4994.
46 O. Cohen, M. Sc. Thesis, Weizmann Institute of Science, 2003.
47 P. A. Guerette, D. G. Ginzinger, B. H. Weber and J. M. Gosline, *Science*, 1996, **272**, 112.
48 A. G. Checa, A. B. Rodriguez-Navarro and F. J. Esteban-Delgado, *Biomaterials*, 2005, **26**, 6404.
49 B. A. Gotliv, L. Addadi and S. Weiner, *ChemBioChem*, 2003, **4**, 522.
50 F. Nudelman, B. A. Gotliv, L. Addadi and S. Weiner, *J. Struct. Biol.*, 2006, **153**, 176.
51 A. M. Belcher, X. H. Wu, R. J. Christensen, P. K. Hansma, G. D. Stucky and D. E. Morse, *Nature*, 1996, **381**, 56.
52 J. P. Cuif, Y. Dauphin, A. Denis, D. Gaspard and J. P. Keller, *C. R. Seances Acad. Sci., Ser. D*, 1980, **290**, 759.
53 I. M. Weiss, N. Tuross, L. Addadi and S. Weiner, *J. Exp. Zool.*, 2002, **293**, 478.
54 Y. Politi, T. Arad, E. Klein, S. Weiner and L. Addadi, *Science*, 2004, **306**, 1161.
55 J. Aizenberg, G. Lambert, L. Addadi and S. Weiner, *Adv. Mater.*, 1996, **8**, 222.
56 E. Beniash, L. Addadi and S. Weiner, *J. Struct. Biol.*, 1999, **125**, 50.
57 I. Sethmann, R. Hinrichs, G. Worheide and A. Putnis, *J. Inorg. Biochem.*, 2006, **100**, 88.
58 K. Wada, *Bull. Natl. Pearl Res. Lab.*, 1961, **7**, 703.
59 H. Nakahara, *Jpn. J. Malacol.*, 1979, **38**, 205.
60 L. Addadi and S. Weiner, *Angew. Chem., Int. Ed. Engl.*, 1992, **31**, 153.
61 K. Wada, *Bull. Natl. Pearl Res. Lab.*, 1956, **1**, 1.

PAPER

Sintering, crystallisation and biodegradation behaviour of Bioglass®-derived glass–ceramics†

Aldo R. Boccaccini,‡[a] Qizhi Chen,‡[a] Leila Lefebvre,[b] Laurent Gremillard[b] and Jérôme Chevalier[b]

Received 13th November 2006, Accepted 5th February 2007
First published as an Advance Article on the web 8th May 2007
DOI: 10.1039/b616539g

Sintering and crystallisation phenomena in powders of a typical bioactive glass composition (45S5 Bioglass®) have been investigated in order to gain further understanding of the processes involved in the fabrication of Bioglass® based glass–ceramic scaffolds for tissue engineering applications. *In situ* experiments in an environmental scanning electron microscope with a heating stage were carried out to follow the morphology of Bioglass® particles during sintering and crystallisation. Optimal processing parameters for the manufacture of Bioglass® based glass–ceramic scaffolds by the foam-replica technique were determined. To assess the *in vitro* performance and bioactivity of Bioglass®-derived glass–ceramic scaffolds, the biodegradation of samples in simulated body fluid (SBF) was investigated using various techniques, including SEM, TEM, XRD and EDX. The mechanism of interaction of the glass–ceramic surface with SBF was determined, which involves (i) preferential dissolution at glass/crystal interfaces, (ii) break-down of crystalline particles into very fine grains through preferential dissolution at crystal structural defects, and (iii) amorphisation of the crystalline structure by introduction of point defects produced during ion exchange. The present report thus offers for the first time a complete assessment of the processing parameters, microstructure, and *in vitro* performance of Bioglass® derived glass–ceramic scaffolds intended for bone tissue engineering.

1. Introduction

Tissue engineering uses a scaffold made of a biocompatible natural or synthetic material to induce the regeneration of damaged or injured tissue in the host body.[1] When using a synthetic scaffold, its use should be as short lived as possible, at the same time the scaffold must maintain its viability and integrity long enough for the cells to produce their own extracellular matrix.[2] Ideally, the scaffolds should be highly porous, support/foster cells, temporarily provide mechanical support and degrade at rates appropriate to tissue regeneration, *i.e.* matching the growth rate of new tissue.[3,4]

[a] *Department of Materials, Imperial College London, Prince Consort Road, London, UK, SW7 2BP*
[b] *National Institute of Applied Science, Materials Department, UMR CNRS 551020, Avenue Albert Einstein, 69621, Villeurbanne Cedex, France*

† The HTML version of this article has been enhanced with colour images.
‡ These two authors contributed equally to this paper.

Bioactive glasses are a class of bioactive materials, *i.e.* materials which elicit a special response on their surface when in contact with biological fluids, leading to strong bonding to living tissue.[5] In the field of bone engineering, bioactivity is defined as the ability of the material to bond to bone tissue *via* the formation of a bone-like hydroxyapatite layer on its surface.[6] Due to a number of attractive properties for use in tissue engineering and regeneration, *e.g.* enhanced angiogenesis[7,8] and up-regulation of specific genes that control the osteoblast cell cycle,[9,10] there is increasing effort in the use of bioactive glasses in tissue engineering applications.[3,11–13] Bioglass® (type 45S5) of composition (mol%) 46.1 SiO_2, 24.4 Na_2O, 26.9 CaO and 2.6 P_2O_5, has been frequently considered the material of choice for tissue engineering applications based on its long history of clinical applications[14] and superior bioactivity.[15] There have been, however, problems with the manufacture of highly porous scaffolds (*i.e.* foams) by sintering of 45S5 Bioglass® powders. It is recognised that obtaining porous bioactive glasses with controlled mechanical and biological properties necessitates the optimisation of the sintering conditions. It is well known that sintering of glass powders may be strongly affected by phase transformations (nucleation and growth of crystalline phases) occurring upon heating.[16–18]

Indeed the major hurdle in the production of highly porous Bioglass®-based foam-like scaffolds has been caused by the following apparently irreconcilable issues of this glass:

(a) It has been reported that crystallisation of 45S5 Bioglass® turns this glass into an inert material.[16]

(b) Full crystallisation of the glass occurs prior to significant densification by sintering;[17] and

(c) Extensive densification is required to strengthen the solid phase, *i.e.* the struts in the foam-like structure, which would otherwise be made of loosely bonded particles and thus be too fragile to handle.

According to these three factors, to obtain a mechanically competent 45S5 Bioglass®-based scaffold in a foam-like structure, one must sinter it at a high temperature where extensive densification of the struts can occur. The bioactivity of the 45S5 Bioglass®-based foam, however, is expected to be severely retarded or even suppressed by crystal nucleation and growth occurring before densification. Consequently, it is assumed that to maintain the bioactivity of 45S5 Bioglass®, one should sinter the foam at a relatively low temperature at which crystallisation (nucleation and growth) does not take place or does not occur to a great extent. Under these conditions, sufficient densification by sintering will not happen, and a very fragile scaffold made of loosely packed 45S5 Bioglass® particles would be produced.

Contrary to that common assumption, research by Clupper, Zanotto and Hench and coworkers[17–21] has revealed that the nucleation and growth of a crystalline phase (*e.g.* $Na_2Ca_2Si_3O_9$) slightly decreased the kinetics but it did not suppress the formation of a hydroxyapatite layer, which is the marker for bioactivity. Moreover it is recognised that the bioreaction kinetics of a highly porous network can be very different from that of a dense product of the same chemical composition due to the high surface area associated with high porosity (*e.g.* in foams). Based on this knowledge, we have in our previous work[22–24] independently developed new types of 45S5 Bioglass®-derived glass–ceramic scaffolds. A unique feature of these novel scaffolds is that the mechanical strong crystalline phase can transform into a degradable amorphous phase under biological conditions, which does not usually occur in dense materials at body temperature. It was hypothesized that this transformation effectively combines mechanical strength (associated with the crystalline phase) with tailorable biodegradability (associated with the amorphous phase) in the scaffold.[25] However, there is a lack of careful investigation of the interaction between densification and crystallisation (nucleation and growth) kinetics during the manufacture of Bioglass® based foams as well as on the biodegradation mechanisms associated with the dissolution of the crystalline phase.

A fundamental understanding of these phenomena would provide a solid base for the design of the scaffold glass–ceramic microstructure for tailored degradation kinetics and mechanical competence of the scaffolds. The investigation of these phenomena is the primary objective of this work, focusing on the experimental investigation of: (i) sintering and crystallisation of Bioglass® powders (processing issues) and (ii) biodegradation aspects of Bioglass® based glass–ceramic scaffolds (*in vitro* performance).

2. Materials and methods

2.1. Sintering and crystallisation experiments

2.1.1 Materials. High purity SiO_2, Na_2CO_3, $CaCO_3$ and P_2O_5 powders were weighed and mixed to obtain 45S5 Bioglass® (45 SiO_2, 24.5 CaO, 24.5 Na_2O, 6 P_2O_5 in wt%). 45S5 Bioglass® is the original bioactive glass composition developed by Hench *et al.* in 1971.[26] The powders were melted in a Pt crucible for 4 h at 1400 °C with a decarbonatation step (5 h at 950 °C). The melt was then quenched in water and ground in ethanol to a fine powder (mean particle size: 1 µm). Quenching was fast enough to retain a completely amorphous material, as verified by X-ray diffraction (XRD). The composition was also checked by chemical analysis, to ascertain that no impurities were present after the different preparation steps.

2.1.2 *In situ* investigation of transformations during heating. The physical and morphological transformations of the glass powder during heating were followed by simultaneous thermal gravimetric analysis (TGA-DTA) (SETARAM TG-DTA 92, Caluire, France) and environmental scanning electron microscopy (ESEM, FEI, XL30, Eindhoven, Netherlands). For TGA-DTA, 20 mg of powder were heated in a 20% N_2–80% O_2 atmosphere, at 5 °C per minute up to 1300 °C. An innovative ESEM with an *in situ* heating stage was used. In order to get an easier observation, larger Bioglass® particles (about 10 µm in diameter) were introduced into a Pt crucible. The set-up was heated up to 950 °C (rate 5 °C min^{-1}) in the microscope. Measurements were conducted with a 1.9 Torr (250 Pa) water vapour pressure.

2.1.3 Characterization of thermally treated powders. Several thermal treatments, for 5 min (heating rate 5 °C min^{-1}) in the range of 550–950 °C, were carried out on the glass powder. Two types of investigations were then performed: X-ray diffraction (XRD) (RIGAKU vertical diffractometer, Kent, England) and Fourier transform infrared analysis (FTIR, Nicolet Magna-IR 550 spectrometer, Madison, Wisconsin). XRD was used to identify the crystalline phases and their evolution with increasing temperature. Samples were obtained by pressing a mixture of 45S5 Bioglass® powder with an amorphous binder (PEG 1500) into 1 cm diameter pieces. These pieces were then analysed with Cu Kα radiation from 10 to 90° (2θ) at 2° min^{-1} for crystalline phase determination and 0.2° min^{-1} for Rietveld analysis. Rietveld analysis was performed with the software WINPLOTR® to evaluate the lattice parameters of the main phase. The diffractograms measured at different temperatures were compared to theoretical diffractograms of $Na_2CaSi_2O_6$ established from the structure determined by Oshato and Maki.[27] FTIR analyses were performed to identify the nature of the chemical bonds between atoms. The samples were small pellets of 0.5 cm diameter obtained by pressing the glass powder with potassium bromide.

2.1.4 Sintering behaviour. The sintering behaviour of the glass powders was followed using two methods: dilatometry (SETARAM TMA 92, Caluire, France) and ESEM with *in situ* heating stage. Dilatometry measurements were carried out from room temperature to 950 °C on pellets obtained by uniaxially pressing Bioglass® powders (200 MPa). The heating rates (1 and 5 °C min^{-1}) were chosen to be compatible with the processing of porous scaffolds, as reported elsewhere.[22–24] For the ESEM investigation, a small amount of powder was introduced into the Pt crucible and heated up to 950 °C (5 °C min^{-1}) in the microscope. The neck formation by viscous flow between glass particles of different sizes (3 or 20 μm diameter) was followed as a function of temperature and time.

2.2 Scaffold fabrication and biodegradation investigation

2.2.1 Scaffold production. For this part of the study, a commercially available melt-derived 45S5 Bioglass® powder (NovaMin Technology Inc., Florida, USA) was used. The mean particle size of the glass powder was in the range 10–20 μm. Bioglass® scaffolds were processed using a classical foam-replication technique, which has been described in detail elsewhere.[22–24] Different types of fully reticulated polyurethane foams (60 ppi, pores per inch, from Recticel, Corby, UK, and 40 ppi, Plastiform's, France) were used as sacrificial templates. The replication method involves preparation of green bodies of ceramic (or glass) foams by coating a polymer (*e.g.* polyurethane) foam with a ceramic (or glass) slurry.[22] The polymer, having the desired pore structure, simply serves as a sacrificial template for the ceramic coating. The polymer template is immersed in the slurry, which subsequently infiltrates the structure and ceramic (glass) particles adhere to the surfaces of the polymer. Excess slurry is squeezed out leaving a more or less homogeneous coating on the foam struts. After drying, the polymer is slowly burned out at temperatures between 300–500 °C in order to minimise damage to the ceramic (glass) coating. Once the polymer has been removed, the ceramic (or glass) network is sintered to a desired density. The process replicates the macrostructure of the starting sacrificial polymer foam and results in a rather distinctive and well-defined microstructure within the struts.

Fig. 1 shows the heat treatment program used to sinter the green bodies. The burning condition of the polyurethane foam was set at 400 °C for 1 h and at 550 °C for 5 h for the 60 and 40 ppi foams, respectively. The heating and cooling rates were determined to be 2 and 5 °C min^{-1}, respectively.

2.2.2. Assessment of biodegradation and bioactivity in simulated body fluid. The bonding capability of a biomaterial to bone tissue is referred to as bioactivity and it has been associated with the formation of a carbonated hydroxyapatite (HA) layer on the surface of the material when implanted or in contact with biological fluids.[6] Hence, the bioactivity of a potential biomedical device can be assessed *in vitro* in simulated body fluid (SBF) *via* monitoring the formation of HA on its surface. While in contact with SBF the bioactive glass–ceramic structure undergoes degradation and this aspect of the *in vitro* scaffold behaviour is also investigated here.

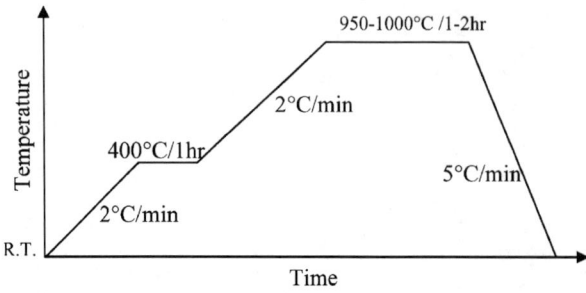

Fig. 1 Heat treatment program designed for burning-out polyurethane templates (60 ppi) and sintering 45S5 Bioglass® foams.[22] The optimal sintering conditions were determined to be 950–1000 °C and sintering time 1–2 h.

In this work, the standard *in vitro* procedure described by Kokubo *et al.*[28] was used. The foams were immersed in 75 ml of acellular SBF in clean conical flasks, which had previously been washed using HCl and deionised water. The conical flasks were placed inside an incubator at controlled temperature of 37 °C. The pH of the solution was maintained constant at 7.25. The size of all samples for these tests was 10 mm × 10 mm × 10 mm. Two samples were extracted from the SBF solution after given times of 0.5, 1, 2, and 4 weeks. The SBF was replaced twice a week because the cation concentration decreased during the course of the experiments, as a result of the changes in the chemistry of the samples. Once removed from the incubation, the samples were rinsed gently, firstly in pure ethanol and then using deionised water, and left to dry at ambient temperature in a desiccator for further characterisation.

2.2.3. **Characterisation.** The microstructure of the foams was characterised using a JEOL 5610LV scanning electron microscope (SEM), before and after immersion in SBF. Samples were gold-coated and observed at an accelerating voltage of 15–20 kV. Energy dispersive X-ray (EDX) spectra (Kα line) were collected at 20 kV in a field emission gun (FEG) SEM (Leo15). They were processed using an INCA (Oxford instruments) program, using standard reference spectra. At least 5 measurements were taken from each condition investigated. For TEM investigation, selected foams before or after immersion in SBF were embedded in epoxy resin at room temperature, followed by ultrathin sectioning. The foils were examined in a JEOL 2000FX transmission electron microscope (TEM) at 200 kV. EDX spectra (Kα line) were collected at 200 kV in the TEM. The data were processed using an INCA (Oxford instruments) program, using standard reference spectra. Five measurements were taken from each condition investigated. Selected foams were also characterised using XRD analysis with the aim to assess the crystallinity after sintering and formation of HA crystals on strut surfaces after different times of immersion in SBF. The foams were first ground to a powder. Then 0.1 g of the powder was collected for XRD analysis. A Philips PW 1700 Series automated powder diffractometer was used, employing Cu Kα radiation (at 40 kV and 40 mA) with a secondary crystal monochromator. Data were collected over the range of $2\theta = 5-100°$ using a step size of 0.04° and a counting time of 25 s per step.

3. Results and discussion

3.1 Transformation processes during heating Bioglass® powder

TGA results presented in Fig. 2 show a total of 2.8% weight reduction, which can be divided into two main weight losses: evaporation of free water at 100 °C and loss of –OH groups at 400 °C. The small apparent weight loss at 610 °C may be attributed to the onset of crystallisation. On the same figure, DTA shows an endothermic effect at $T_{g1} = 550$ °C caused by the glass transition, followed by an exothermic crystallisation peak beginning at $T_{c1} = 610$ °C. A second small endothermic effect, related hereafter to another glass transition, is observed at $T_{g2} = 850$ °C. Finally, melting takes place in the 1070–1278 °C range. Two endothermic peaks (maximum signal, respectively at $T_{m1} = 1192$ °C and $T_{m2} = 1235$ °C) may be attributed to the melting of two different crystalline phases. The results are in agreement with those presented by Chatzistavrou *et al.*,[29] obtained at a heating rate of 10 °C min^{-1} using powder of size 20–63 µm.

Fig. 3 shows ESEM micrographs of the surface of a Bioglass® particle taken during *in situ* heating. Fig. 3a, taken below the glass transition (T_{g1}), exhibits the features of a faceted ground powder. A rounding of the particle occurs at $T_{g1} = 550$ °C (Fig. 3b) confirming the glass transition temperature measured *via* DTA. This stage is followed by the creation of domains on the surface at $T_s = 570$ °C (Fig. 3c). This may be attributed to the glass-in-glass phase separation expected when two high valence ions such as Si^{4+} and P^{5+} are present simultaneously in a glass[30,31] (each ion

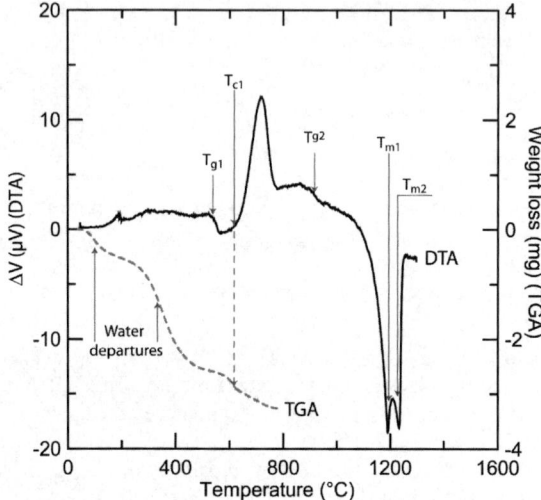

Fig. 2 Results of the simultaneous thermal gravimetric analysis-differential thermal analysis (TGA-DTA) of Bioglass® powder, showing the different transformation temperatures.

type tending to concentrate in a separate phase). Phase separation is then followed by partial crystallisation, characterized by the formation of 'craters' on the surface of the particles due to the contraction associated with crystallisation (Fig. 3d, taken at 680 °C). At 800 °C, Fig. 3e shows the migration of a second phase with a different morphology and structure. The present micrograph was obtained *via* charge contrast imaging (the contrast is due to local variations of the electrical conductivity at the surface of the sample). The last micrograph (Fig. 3f) shows the resulting surface after thermal treatment at 950 °C (it was taken at room temperature, after cooling, to ensure thermal stability and better contrast and resolution). On this micrograph, the separation between two types of domains is seen to be more pronounced, with a spinodal-like morphology.

Fig. 4a shows the XRD spectra of the powder before and after thermal treatments between 550 and 950 °C. The result obtained for the powder before any treatment confirms the amorphous nature of Bioglass® after quenching and grinding. In agreement with the results obtained by TGA-TDA analysis, powders treated above 600 °C are partially crystallised. The stoichiometry of 45S5 Bioglass® is closer to

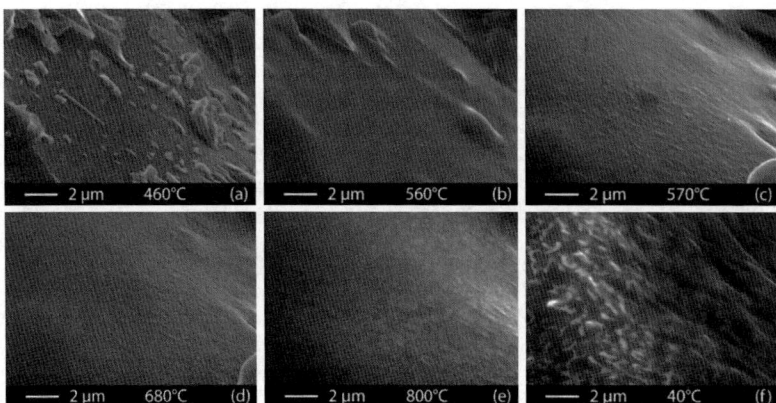

Fig. 3 ESEM micrographs taken at different temperatures during *in situ* heating at the surface of one Bioglass® particle.

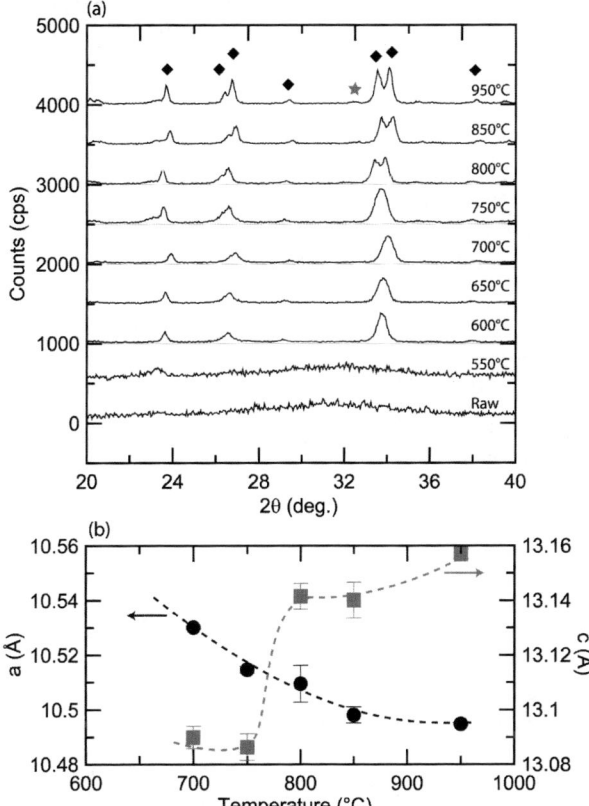

Fig. 4 (a) Diffractograms of raw and thermally treated 45S5 Bioglass®. At temperatures below 600 °C Bioglass® remains amorphous. At higher temperature, the crystallization takes place by the formation of $Na_2CaSi_2O_6$ (♦: PDF 77-2189). At 800 °C, the two major peaks of this phase begin to be dissociated while the crystallization of silicorhenanite (★: PDF 32-1053, isostructural to apatite) begins. (b) Evolution of the lattice parameters (*a* and *c*) of $Na_2CaSi_2O_6$ with temperature.

$Na_2CaSi_2O_6$ than to $Na_2Ca_2Si_3O_9$, consequently the Rietveld refinement method led to better agreement between the experimental diffractograms and the $Na_2CaSi_2O_6$ calculated diffractogram. The crystalline phase is thus identified as $Na_2CaSi_2O_6$, as shown previously by Lin et al.[32] An apparent broadening of the main peak (at 33.75°) occurs at 750 °C, which is in fact the beginning of the separation into two peaks, still both attributed to $Na_2CaSi_2O_6$. This separation is clearly observed at 800 °C and intensifies at higher temperatures. Concurrently, in the 800–950 °C range, a new small peak is observed, attributed to a second crystalline phosphate phase (silicorhenanite: $Na_2Ca_4(PO_4)2SiO_4$).

As found in previous investigations,[22,25] changing the heating rate to 2 °C min^{-1} and the sintering temperature to 1000 °C, led to a different crystalline structure, characterised by the presence of, predominantly, the phase $Na_2Ca_2Si_3O_9$. This crystalline phase has also been found by Jun et al.[33] when sintering Bioglass® powders at 1000 °C for 5 h and in similar studies by Arstilla et al.[34] working with powders of large particle size (>300 μm). Further evidence of the crystallisation behaviour of 45S5 Bioglass® powders and its dependence on glass particle size at higher heating rates (10 °C min^{-1}) has been presented by Chatzistavrou et al.[35] Previous studies by Peitl et al.[19] have also confirmed the formation of $Na_2Ca_2Si_3O_9$ in bulk bioactive glass samples while the complex crystallisation of bioactive glasses

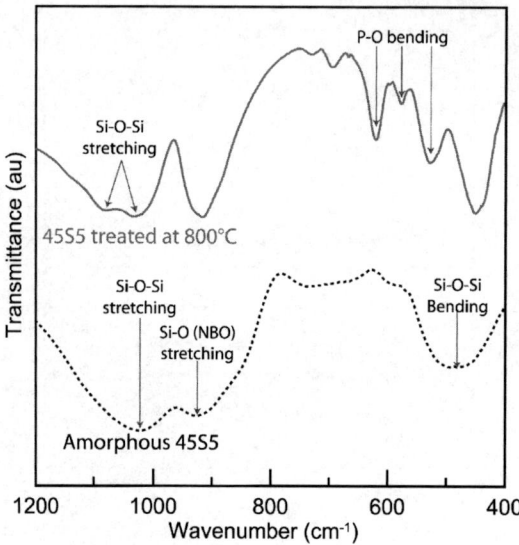

Fig. 5 FTIR spectra taken on the Bioglass® powder before and after thermal treatments at 800 °C.

synthesised by wet chemistry (rather than melt-derived glasses) has been presented by Rizkalla et al.[36]

In the present investigation, more precise XRD analyses (at lower scan speed and better angular resolution) were performed on powders treated between 750 and 950 °C in order to clarify the separation between the two main peaks of the main crystalline phase. A linear dependency of the angular separation of these two peaks ($\Delta 2\theta$) versus temperature was measured. Thus it can be inferred that the peak separation is due to a progressive variation of lattice parameters rather than to a phase change. A Rietveld refinement was applied to the different diffractograms to measure the lattice parameters (a and c). Fig. 4b shows a gradual decrease of a with temperature, while a steep increase of c occurs around 800 °C, when the secondary crystalline phase appears.

FTIR spectra of the powder before and after thermal treatments at 800 °C are shown in Fig. 5. The main absorption bands for the amorphous Bioglass® are observed at 1024, 926 and 480 cm^{-1}, and attributed to Si–O–Si and Si–O stretching modes and to the Si–O–Si bending mode, respectively.[29] These bands are generally observed in amorphous silica glasses.[37,38] The band at 600 cm^{-1} is related to amorphous phosphate. The FTIR spectra of the powder treated at 800 °C reveals the combination of isolated tetrahedral Si (splitting of the broad band at 1024 cm^{-1} into two bands). At the same time, the number of non-bonding oxygen (NBO) linked to Si increases (band at 926 cm^{-1}). This increase is due to the crystallization of the major phase Na$_2$CaSi$_2$O$_6$,[39] in agreement with the XRD data. The presence of a crystalline calcium phosphate (apatite-like) phase is suggested by the new bands at 620, 580 and 530 cm^{-1} since these peaks are attributed to the P–O bending vibration.[29] This is also in agreement with XRD results and the identification of the silicorhenanite phase (Fig. 4a).

3.2 Sintering behaviour

Fig. 6 shows results related to the sintering process. The dimensional changes (thickness of the pellet) occurring during sintering were measured by dilatometry. From the knowledge of the initial density of the powder compact before thermal

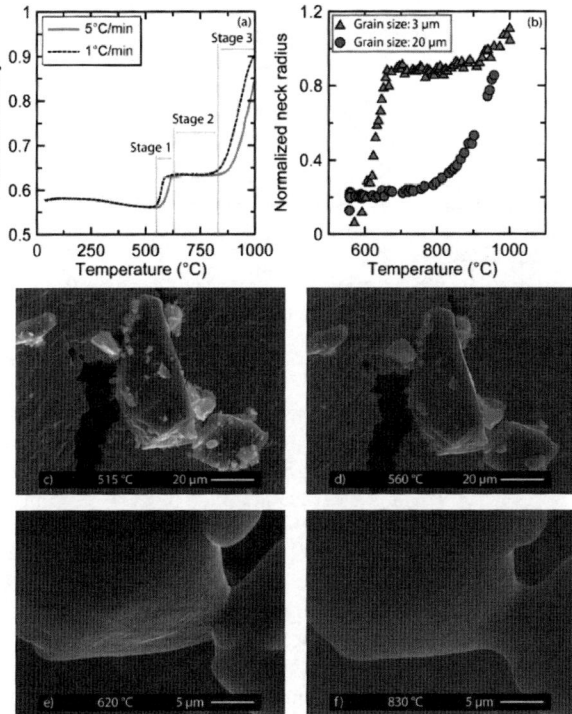

Fig. 6 Sintering behaviour of Bioglass® powder: (a) densification of a powder compact *versus* temperature, for two heating rates, (b) neck formation between two particles *versus* temperature (heating rate: 5 °C min^{-1}) for two different grain sizes, (c–f) ESEM micrographs taken during *in situ* heating before neck formation (c) and during the three different sintering stages (d–f).

treatment, and assuming isotropic shrinkage, the density was plotted *versus* temperature in Fig. 6a. Three stages can be distinguished: fast densification above 550 °C (T_{g1}), then a plateau from 610 °C (T_{c1}) to 850 °C (T_{g2}), followed by a second important densification above 850 °C. Densification by sintering therefore operates between T_{g1} and T_{c1} and above T_{g2}. The saturation of densification due to crystallization has been described for other glasses in the literature.[40–42] This observation is confirmed by the evolution of the diameter of the neck between two particles (Fig. 6b) with temperature: the three stages are also apparent, and their temperature spans are the same as those observed by dilatometry. Each stage of the densification process can be correlated with the features observed during *in situ* sintering of two glass particles (Fig. 6). No sintering occurs below the glass transition temperature (Fig. 6c), while densification starts after the first glass transition (T_{g1}), when the glass is softened enough to allow sintering by viscous flow and particles start to be connected through sintering necks (Fig. 6d). Just after the glass transition (T_{g1}), the glass starts to decompose (glass phase separation discussed above) and then to crystallize, both phenomena leading to viscosity increase. The viscosity becomes too high and sintering is actually stopped at ∼610 °C. This explanation of the second stage (plateau) is supported by Fig. 6e, that shows the crystallisation at the surface of the particle at 620 °C. The transition between the two first stages is usually abrupt: the saturation of densification starts immediately once the crystalline phase forms a continuous, percolating network in the glass. The temperature of the transition depends on the heating rate and on the particle size, since these factors affect the crystallisation behaviour as discussed in the previous section. In terms of microstructure control, it should therefore be possible to obtain various degrees of densification (*i.e.* of porosity) for the same crystallinity ratio by changing the particle

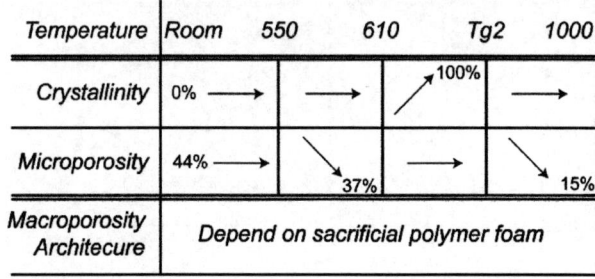

Fig. 7 Evolution of microstructure during sintering of Bioglass® powders for fabrication of scaffolds, indicating the major parameters controlling scaffold performance.

size or the sintering rate. At temperatures above T_{g2}, the amorphous part of the partially crystallized material is soft enough for viscous flow sintering to occur (Fig. 6f). It is also possible that liquid phase sintering through diffusion by dissolution/precipitation takes place. It is not clear at this stage of the investigation, however, if viscous flow or liquid phase sintering is the controlling mechanism. This issue will be the subject of further research. However, it is likely that the low quantity of remaining amorphous phase will not allow such a significant shrinkage to occur, thus it is suggested that temperatures higher than T_{g2} are high enough for diffusion to occur, which is supported by the smooth transition between the second and third stages (the transition extends over ~50 °C). This last stage of sintering allows complete densification of the powder compact.

3.3 Bioglass® derived glass–ceramic scaffolds

3.3.1 Microstructural characterisation. Few studies have reported the preparation of highly porous Bioglass® scaffolds from melt-derived powders.[3] The technique developed recently for fabrication of highly porous Bioglass® foam-like glass–ceramic scaffolds is based on the replication method.[22–24] The present study on the evolution of the microstructure during sintering and crystallisation of Bioglass® powders shows the versatility of porous Bioglass® processing, since both crystallinity and porosity can be controlled, almost independently. Fig. 7 summarizes the major transformations occurring in Bioglass® powder compacts during heating and their effects on the scaffold's final properties. Regarding the applications of the scaffolds in bone tissue engineering, three parameters will have a major influence on mechanical and biological behaviour:

(i) The amount and size of micropores in the foam struts, leading to different surface topographies, which will influence protein and cell adhesion.

(ii) The architecture and the amount of macropores, which control the penetration of the fluids and cells.

(iii) The crystallinity of the Bioglass® derived glass–ceramic microstructure, which controls scaffold bioactivity.

These three parameters also determine the mechanical strength and structural integrity of the scaffolds.[25]

An example of a Bioglass® based glass–ceramic scaffold, processed using the foam-replication technique, is given in Fig. 8. The high pore interconnectivity of the foam is observed both by X-ray tomography and SEM images, Fig. 8a and b, respectively. The thermal cycle was adjusted to remove completely the sacrificial polyurethane scaffold without losing the sample integrity, and to obtain dense struts in the sintered porous structure. The first condition imposes a slow heating (1–2 °C min^{-1}) to an intermediate temperature (400–550 °C) followed by a plateau. The second condition imposes a thermal treatment above T_{g2} to decrease or

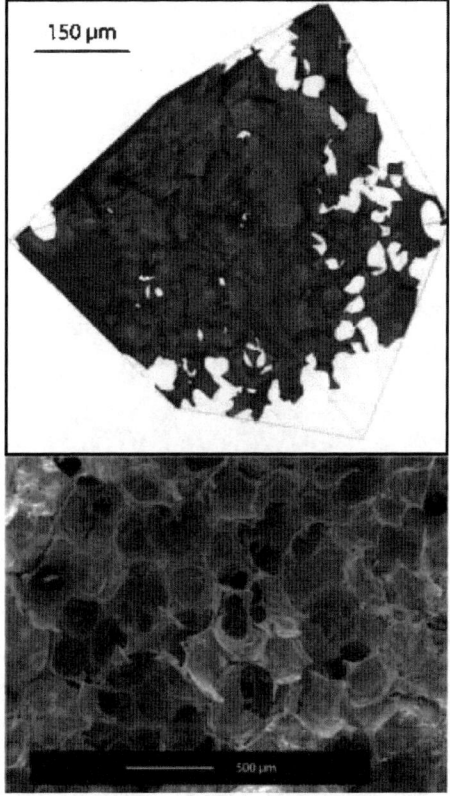

Fig. 8 Typical microstructure of Bioglass® derived scaffolds processed *via* foam replication technique and sintering at 950–1000 °C: (a) X-ray tomography 3D reconstruction and (b) SEM micrograph.

eliminate micro-porosity. Thus a temperature above the stage 2 to stage 3 transition (Fig. 6a) should be used (950–1000 °C).

3.3.2 Scaffold degradation behaviour in SBF

XRD analysis. Fig. 9 shows the XRD spectra for the investigated scaffolds. The "as-received" spectrum represents the results for the commercial amorphous 45S5 Bioglass® powder used to fabricate scaffolds in as-received condition. The spectrum is equivalent to that presented in Fig. 4a for the Bioglass® powder synthesised in the present study. The diffusive peak was caused by the short range ordering of the silicate structure in the glass. Before immersion in SBF, the samples exhibited sharp diffraction peaks, which were identified as diffractions of the $Na_2Ca_2Si_3O_9$ phase (marked with triangles) using the standard PDF #22.1455. As mentioned above, this is the crystalline phase commonly reported in studies on Bioglass® derived glass–ceramics sintered at temperatures > 950 °C.[18,19,22]

After immersion in SBF for 3 days, the sample showed a reduction of crystallinity. After two weeks in SBF, the sharp diffraction peaks of the crystalline phase $Na_2Ca_2Si_3O_9$ entirely disappeared from the XRD spectra, leaving HA peaks (marked with solid squares) overlapped on a new broad halo. This result is in agreement with previous observations that the crystalline phase $Na_2Ca_2Si_3O_9$ in Bioglass®-based glass–ceramic scaffolds decomposes into an amorphous calcium phosphate upon immersion in SBF.[22] XRD analysis also revealed that the

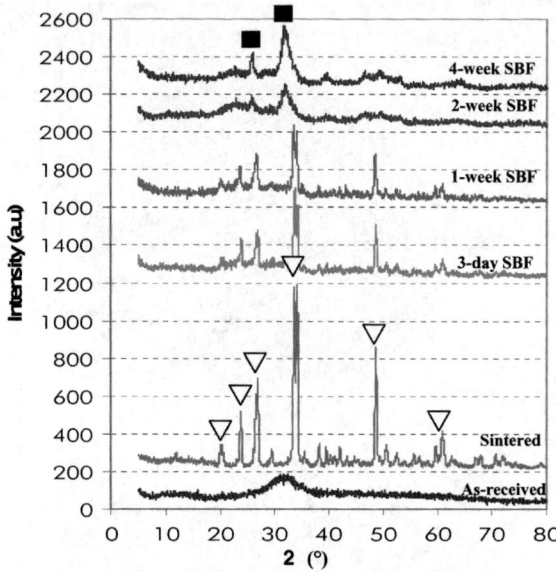

Fig. 9 X-Ray diffraction spectra of scaffolds sintered at 1000 °C for 1 h followed by immersion in simulated body fluid for indicated periods of time (∇: $Na_2Ca_2Si_3O_9$ crystalline phase; ■: hydroxyapatite). Reprinted from *Acta Biomaterialia*, Q.-Z. Chen, K. Rezwan, V. Françon, D. Armitage, S. N. Nazhat, F. H. Jones and A. R. Boccaccini, Surface functionalization of Bioglass®-derived porous scaffolds, Copyright 2007, DOI: 10.1016/j.actbio.2007.01.008, with permission from Elsevier.

transformation of the crystalline phase $Na_2Ca_2Si_3O_9$ to an amorphous phase in the present Bioglass®-derived glass–ceramic foams was completed after 2 weeks of immersion in SBF.

Microscopy investigation. Fig. 10 illustrates the microstructural evolution upon immersion in SBF of the strut surfaces. Fine crystalline particles (~ 0.5 μm in diameter) embedded in the glass matrix are the feature of the as-sintered microstructure (Fig. 10a). After soaking in SBF for up to 1 week, HA-like bulbs were precipitated on the surface of foam struts in which the polycrystalline phase was clearly observed (Fig. 10b). However, a significant change occurred in the surface morphology after immersion in SBF for 2 weeks, *i.e.* the polycrystalline microstructure was replaced by an amorphous matrix containing HA crystallites (Fig. 10c). The above observation was in agreement with the results of XRD analysis (Fig. 9). The boundaries of crystallites were unclear in the as-sintered samples (Fig. 10a) because they were embedded in the glass matrix. However, the boundaries between the crystalline phase and the glass matrix became sharp after 3 days in SBF (Fig. 10b). This effect is due to the faster dissolution of the glass phase compared to the crystalline phase and the fact that the interfaces between the glass matrix and crystalline particles were favourable sites for dissolution.

The results of the TEM investigation are presented as images (Fig. 11) and selected area diffraction patterns (SADP) (Fig. 12). TEM observations confirmed that the crystalline particles (~ 0.5 μm in diameter, Fig. 11a) that formed during the sintering process broke down into smaller particles (Fig. 11b) upon immersion in SBF. In Fig. 11b, which is a TEM image of a sample immersed in SBF for one week, most fine particles have already transformed to an amorphous phase, leaving a few fine crystalline particles, as confirmed by SADP (Fig. 12b). EDX analysis presented in Fig. 13 showed diverse results, depending on the crystallinity of the individual particles. The composition of the crystalline particles is seen to be closer to the

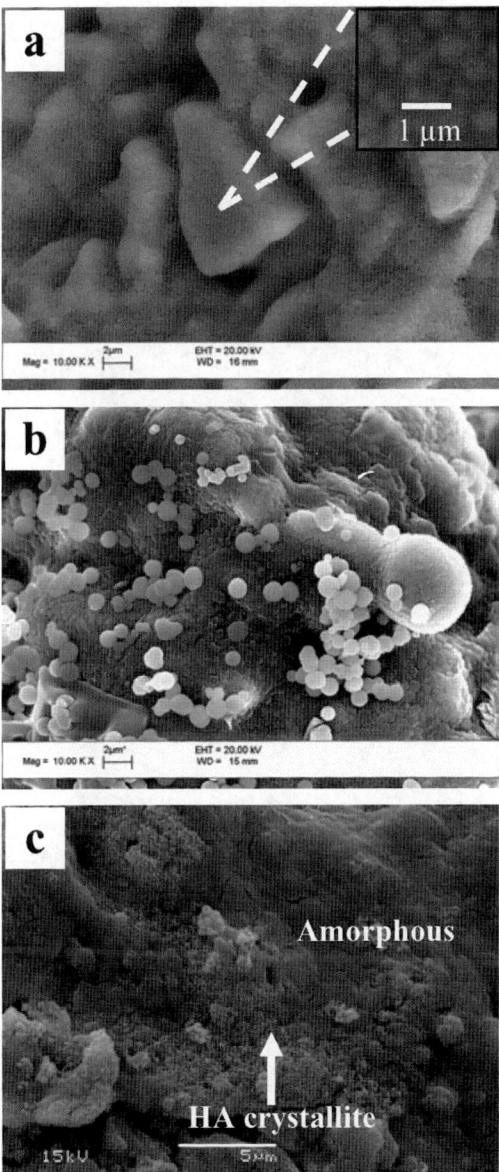

Fig. 10 Surface microstructure of struts of foams sintered at 1000 °C for 1 h followed by immersion in SBF for (a) 0, (b) 1 and (c) 2 weeks. Reprinted from *Acta Biomaterialia*, Q.-Z. Chen, K. Rezwan, V. Françon, D. Armitage, S. N. Nazhat, F. H. Jones and A. R. Boccaccini, Surface functionalization of Bioglass®-derived porous scaffolds, Copyright 2007, DOI: 10.1016/j.actbio.2007.01.008, with permission from Elsevier.

original composition of Bioglass®, whereas the composition of the amorphous particles appears more similar to hydroxyapatite or tricalcium phosphate compositions. A subsequent degradation phenomenon observed was the transformation of the fine crystalline particles into an amorphous calcium phosphate (Fig. 11c and 12c) which occurred after two weeks in SBF. This amorphous calcium phosphate transformed further into crystalline hydroxyapatite (Fig. 11d and 12d) with increasing incubation time in SBF.

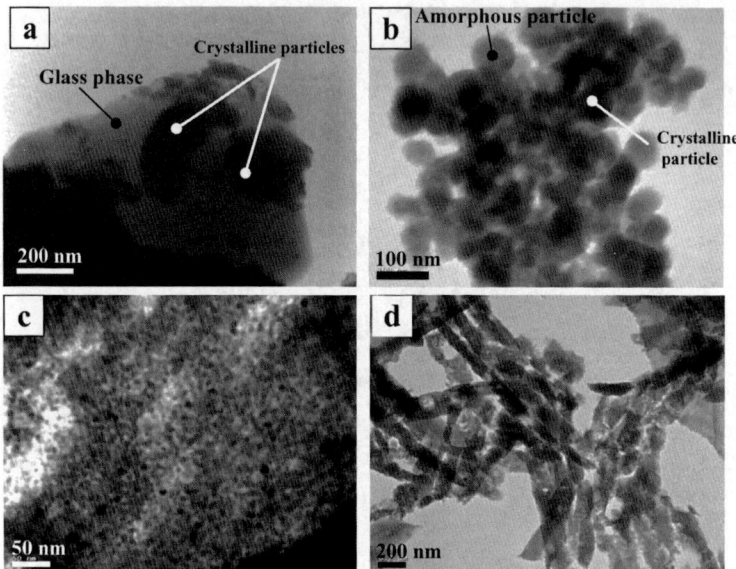

Fig. 11 Microstructure of foam struts by transmission electron microscopy. (a) Crystalline particles embedded in the amorphous matrix in as-sintered condition. (b) Large crystalline particles broke down into fine particles of 50 nm in diameter, which were the main feature in the foam after incubation in simulated body fluid (SBF) for one week. In the sample immersed in SBF for two weeks, the major microstructure was (c) an amorphous calcium phosphate embedded with very fine particles of ∼5 nm in diameter, incorporating (d) hydroxyapatite crystalline nanofibrils.

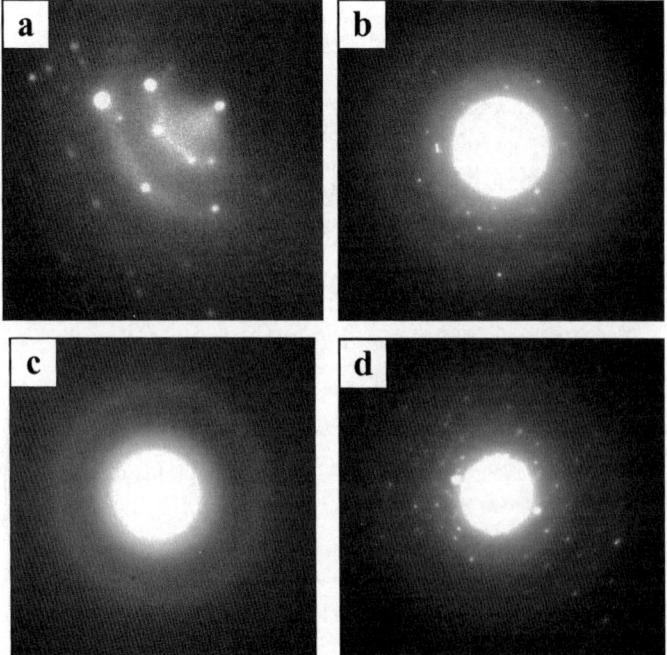

Fig. 12 Selected area diffraction patterns of the regions imaged in Fig. 11, corresponding to: (a) Fig. 11a, (b) Fig. 12b showing amorphous halo rings and crystalline diffraction spots, (c) Fig. 11c, showing only amorphous halo rings, and (d) Fig. 11d, showing crystalline diffraction spots.

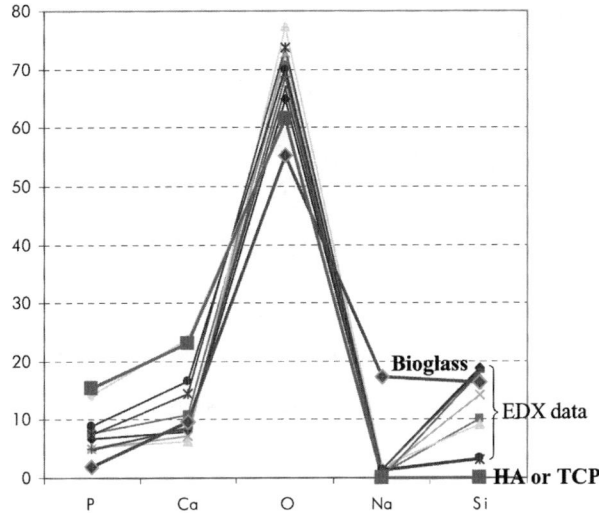

Fig. 13 Composition of fine particles shown in Fig. 11b analyzed by EDX in TEM. The nominal compositions of Bioglass® and of two calcium phosphates (hydroxyapatite or tricalcium phosphate) are shown for comparison.

Degradation mechanisms in SBF. Based on the experimental evidence presented, the mechanisms of degradation of Bioglass®-based glass–ceramic scaffolds can be elucidated, as illustrated in Fig. 14. We suggest that the degradation starts with the

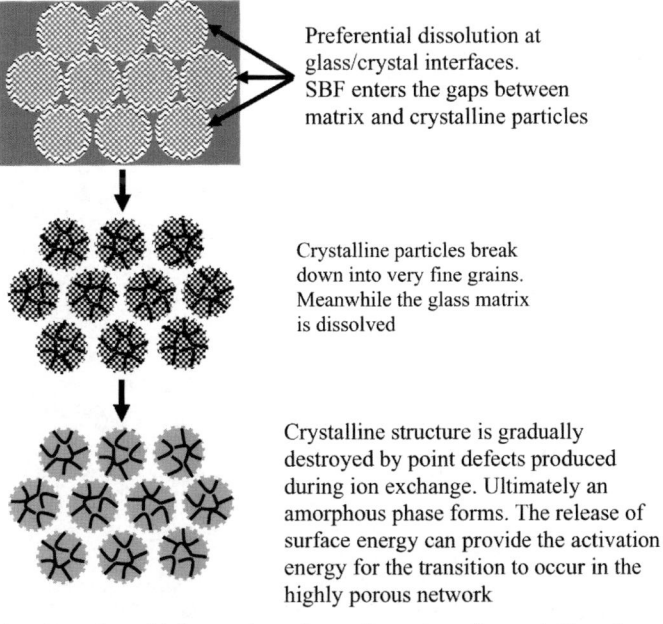

Fig. 14 A proposed model for accelerated transformation of a crystalline phase (Na_2Ca_2-Si_3O_9) to an amorphous phase in the present Bioglass®-based glass–ceramic scaffolds after immersion in simulated body fluid. The model explains the TEM observation of surface morphologies, as shown in Fig. 11b and c. Modified from *Acta Biomaterialia*, Q.-Z. Chen, K. Rezwan, V. Françon, D. Armitage, S. N. Nazhat, F. H. Jones and A. R. Boccaccini, Surface functionalization of Bioglass®-derived porous scaffolds, Copyright 2007, DOI: 10.1016/j.actbio.2007.01.008, with permission from Elsevier.

preferential dissolution at the interfaces between the glass matrix and the crystallites ($Na_2Ca_2Si_3O_9$). When the glass–ceramic samples are soaked in SBF, the fluid can quickly penetrate into the gaps, and ion (especially Na^+) leaching from the crystalline phase takes place immediately throughout the material. As for the dissolution of an individual crystallite, it is likely that the ion exchange preferentially occurs at certain favourable micro locations within the crystalline phase (*e.g.* dislocations and subgrain boundaries), and that this process leads to the breaking down of the as-sintered crystalline particles into very fine grains (Fig. 11b). Moreover, ion exchange will produce a large amount of point defects, which in turn distort the periodic structure of the crystalline phase. This distortion of the lattice can develop to such a level that an amorphous structure forms eventually (Fig. 11c and 12c). Certainly the above steps do not necessarily occur separately from one another; simultaneous occurrence of two, three or even four steps is likely to be the actual case.

The above mechanisms were proposed based on widely accepted facts (*e.g.* material interfaces are corroded preferentially, amorphous phases dissolve faster than their crystalline counterparts) and considering the results of the present XRD, EDX, SEM and TEM analyses. The mechanisms can explain the microstructure evolution in the Bioglass®-based glass–ceramics after incubation in SBF (Fig. 11) and the structural changes with increasing immersion time in SBF (Fig. 9).

4. Conclusions

The present study has considered the processing of Bioglass® based glass–ceramic scaffolds by sintering melt-derived Bioglass® powder. The interaction of viscous flow densification and crystallisation phenomena, which occur during heat treatment of the 'green bodies', was elucidated by *in situ* ESEM and standard thermal characterisation methods. The optimal temperatures and heating rates required to design the heat treatment cycle for the production of Bioglass® based glass–ceramic scaffolds by the replica technique were determined. The degradation behaviour of 45S5 Bioglass®-based glass–ceramic scaffolds was investigated by immersion studies in simulated body fluid. Experimental evidence by SEM, TEM and XRD support a mechanism of degradation involving: (1) preferential dissolution at glass/crystal interfaces, (2) break-down of crystalline particles into very fine grains through preferential dissolution at crystal structural defects, and (3) amorphorisation of the crystalline structure by point defects produced during ion exchange. The results of this study contribute to the overall task of optimising the production of Bioglass® based scaffolds with required biodegradation effect and mechanical competence. The measurement of mechanical properties of scaffolds upon degradation in SBF is the focus of current research.

Acknowledgements

The ESEM observations were made in a consortium of laboratories (Clyme) with the technical support of Mr Gilbert Thollet. The X-ray microtomography investigation was carried out under supervision of Dr Eric Maire.

References

1 J. P. Vacanti and C. A. Vacanti, The history and scope of tissue engineering, in *Principles of Tissue Engineering*, ed. R. P. Lanza, R. Langer and J. P. Vacanti, Academic Press, California, 2nd edn, 2000, pp. 3–7.
2 R. M. Nerem, Tissue engineering: The hope, the hype, and the future, *Tissue Eng.*, 2006, **12**(5), 1143–1150.
3 J. R. Jones and A. R. Boccaccini, Cellular ceramics in biomedical applications: tissue engineering, in *Cellular Ceramics: Structure, Manufacturing, Processing and Applications*, ed. M. Scheffler and P. Colombo, Wiley-VCH Verlag GmbH & Co. KgaA, Weinheim, 2005, vol. 1, pp. 550–573.

4   D. W. Hutmacher, Scaffolds in tissue engineering bone and cartilage, *Biomaterials*, 2000, **21**(24), 2529–2543.
5   L. L. Hench and H. A. Paschall, Direct chemical bond of bioactive glass-ceramic materials to bone and muscle, *J. Biomed. Mater. Res. Symp.*, 1973, **4**, 25–42.
6   L. L. Hench and J. Wilson, Surface-active biomaterials, *Science*, 1984, **226**(4675), 630–636.
7   R. M. Day, A. R. Boccaccini, S. Shurey, J. A. Roether, A. Forbes, L. L. Hench and S. M. Gabe, Assessment of polyglycolic acid mesh and bioactive glass for soft-tissue engineering scaffolds, *Biomaterials*, 2004, **25**, 5857–5866.
8   R. M. Day, Bioactive glass stimulates the secretion of angiogenic growth factors and angiogenesis *in vitro*, *Tissue Eng.*, 2005, **11**(5–6), 768–777.
9   I. D. Xynos, A. J. Edgar, L. D. K. Buttery, L. L. Hench and J. M. Polak, Gene-expression profiling of human osteoblasts following treatment with the ionic products of Bioglass (R) 45S5 dissolution, *J. Biomed. Mater. Res.*, 2001, **55**(2), 151–157.
10 I. D. Xynos, M. V. J. Hukkanen, J. J. Batten, L. D. Buttery, L. L. Hench and J. M. Polak, Bioglass (R) 45S5 stimulates osteoblast turnover and enhances bone formation *in vitro*: Implications and applications for bone tissue engineering, *Calcif. Tissue Int.*, 2000, **67**(4), 321–329.
11 K. Rezwan, Q. Z. Chen, J. J. Blaker and A. R. Boccaccini, Biodegradable and bioactive porous polymer/inorganic composite scaffolds for bone tissue engineering, *Biomaterials*, 2006, **27**(18), 3413–3431.
12 J. A. Roether, A. R. Boccaccini, L. L. Hench, V. Maquet, S. Gautier and R. Jerome, Development and *in vitro* characterisation of novel bioresorbable and bioactive composite materials based on polylactide foams and Bioglass (R) for tissue engineering applications, *Biomaterials*, 2002, **23**(18), 3871–3878.
13 C. T. Laurencin and H. H. Lu, Polymer-ceramic composites for bone-tissue engineering, , in *Bone Engineering*, ed. J. E. Davies, Em Squared Incorporated, Toronto, Canada, 2000, pp. 462–472.
14 L. L. Hench, Bioceramics, a clinical success, *Am. Ceram. Soc. Bull.*, 1998, **77**(7), 67–74.
15 L. L. Hench and J. M. Polak, Third-generation biomedical materials, *Science*, 2002, **295**(5557), 1014–1017.
16 P. Li, Q. Yang, F. Zhang and T. Kokubo, The effect of residual glassy phase in a bioactive glass-ceramic on the formation of its surface apatite layer *in vitro*, *J. Mater. Sci. Mater. Med.*, 1992, **3**(6), 452–456.
17 D. C. Clupper and L. L. Hench, Crystallization kinetics of tape cast bioactive glass 45S5, *J. Non-Cryst. Solids*, 2003, **318**(1–2), 43–48.
18 O. Peitl, G. P. LaTorre and L. L. Hench, Effect of crystallization on apatite-layer formation of bioactive glass 45S5, *J. Biomed. Mater. Res.*, 1996, **30**(4), 509–514.
19 O. Peitl, E. D. Zanotto and L. L. Hench, Highly bioactive P_2O_5–Na_2O–CaO–SiO_2 glass-ceramics, *J. Non-Cryst. Solids*, 2001, **292**(1–3), 115–126.
20 D. C. Clupper, J. J. Mecholsky, G. P. LaTorre and D. C. Greenspan, Sintering temperature effects on the *in vitro* bioactive response of tape cast and sintered bioactive glass-ceramic in tris buffer, *J. Biomed. Mater. Res.*, 2001, **57**(4), 532–540.
21 D. C. Clupper, J. J. Mecholsky, G. P. LaTorre and D. C. Greenspan, Bioactivity of tape cast and sintered bioactive glass-ceramic in simulated body fluid, *Biomaterials*, 2002, **23**(12), 2599–2606.
22 Q. Z. Chen, I. D. Thompson and A. R. Boccaccini, 45S5 Bioglass (R)-derived glass-ceramic scaffolds for bone tissue engineering, *Biomaterials*, 2006, **27**(11), 2414–2425.
23 A. R. Boccaccini and Q. Z. Chen, *Process for Preparing Bioactive Glass Scaffolds*, UK filing number GB 0516157.5, 2005.
24 R. Zenati, G. Fantozzi, J. Chevallier and A. Mourad, *Porous Bioglass and Preparation Thereof*, French Patent no FR2005/001921, 2005.
25 Q. Z. Chen and A. R. Boccaccini, Coupling mechanical competence and bioresorbability in Bioglass®-derived tissue engineering scaffolds, *Adv. Eng. Mater.*, 2006, **8**, 285–289.
26 L. L. Hench, R. J. Splinter and W. C. Allen, Bonding mechanisms at the interface of ceramic prosthetic materials, *J. Biomed. Mater. Res. Symp.*, 1971, **2**(Part 1), 117–141.
27 H. Oshato and I. Maki, Structure of $Na_2CaSi_2O_6$, *Acta Crystallogr., Sect. C*, 1985, **C41**, 1575–1577.
28 T. Kokubo, K. Hata, T. Nakamura and T. Yamamura, Apatite formation on ceramics metals and polymers induced by a CaO-SiO_2-Based glass in simulated body fluid, , in *Bioceramics 4*, ed. W. Bonfield, G. W. Hastings and K. E. Tanner, Butterworth-Heinemainn, London, 1991, pp. 113–120.
29 X. Chatzistavrou, T. Zorba, E. Kontonasaki, K. Chrissafis, P. Koidis and K. M. Paraskevopoulos, Following bioactive glass behaviour beyond melting temperature by thermal and optical methods, *Phys. Status Solidi A*, 2004, **201**(5), 944–951.
30 R. H. Doremus, *Glass Science*, Wiley, New York, 1994, p. 48.

31 H. Rawson, *Inorganic Glass Forming Systems*, Academic Press, New York, 1967, p. 11.
32 C. C. Lin, L. C. Huang and P. Shen, $Na_2CaSi_2O_6$-P_2O_5 based bioactive glasses. Part 1: Elasticity and structure, *J. Non-Cryst. Solids*, 2005, 1–9.
33 I. K. Jun, Y. H. Koh and H. E. Kim, Fabrication of a highly porous bioactive glass-ceramic scaffold with a high surface area and strength, *J. Am. Ceram. Soc.*, 2006, **89**, 391–394.
34 H. Arstilla, L. Froeberg, L. Hupa, E. Vedel, H. Ylanen and M. Hupa, The sintering range of porous bioactive glasses, *Glass Technol.*, 2005, **46**, 138–141.
35 X. Chatzistavrou, E. Kontonasaki, K. Chrissafis, T. Zorba, P. Koidis and K. M. Paraskevopoulos, Surface and bulk contributions in the crystallisation process of a bioactive glass, *Key Eng. Mater.*, 2006, **309–311**, 313–316.
36 A. S. Rizkalla, D. W. Jones, D. B. Clarke and G. C. Hall, Crystallisation of experimental bioactive glass compositions, *J. Biomed. Mater. Res.*, 1996, **32**, 119–124.
37 M. R. Filgueiras, G. La Torre and L. L. Hench, Solution effects on the surface reactions of three bioactive glass compositions, *J. Biomed. Mater. Res.*, 1993, **27**, 445–453.
38 E. Kontonasaki, T. Zorba, L. Papadopoulou, X. Chatzistavrou, K. Paraskevopoulos and P. Koidis, Hydroxy carbonate apatite formation on particulate bioglass *in vitro* as a function of time, *Cryst. Res. Technol.*, 2002, **37**, 1165–1171.
39 J. M. Gomez-Vega, E. Saiz, A. P. Tomsia, G. W. Marshall and S. J. Marshall, Bioactive glass coatings with hydroxyapatite and Bioglass® particles on Ti-based implants. 1. Processing, *Biomaterials*, 2000, **21**(2), 105–111.
40 T. J. Clark and J. S. Reed, Kinetic process involved in the sintering and crystallisation of glass powders, *J. Am. Ceram. Soc.*, 1986, **69**, 837–846.
41 M. O. Prado, C. Fredericci and E. D. Zanotto, Non-isothermal sintering with concurrent crystallization of polydispersed soda-lime-silica glass beads, *J. Non-Cryst. Solids*, 2003, **331**, 157–167.
42 A. R. Boccaccini, W. Stumpfe, D. M. R. Taplin and C. B. Ponton, Densification and crystallization of glass powder compacts during constant heating rate sintering, *Mater. Sci. Eng., A*, 1996, **A219**, 26–31.

PAPER

The formation of nanoscale structures in soluble phosphosilicate glasses for biomedical applications: MD simulations†

Antonio Tilocca,*[a] Alastair N. Cormack[b] and Nora H. de Leeuw[a]

Received 5th December 2006, Accepted 24th January 2007
First published as an Advance Article on the web 10th April 2007
DOI: 10.1039/b617540f

The occurrence of chain-like fragments and rings in phosphosilicate glasses of known bioactivity was examined using classical molecular dynamics simulations, in order to reveal the possible effect of such nanostructures on the bioactive behaviour. Highly bioactive compositions display a large fraction of non-crosslinked, nonlinear chains of tetrahedra, which are not present in bio-inactive compositions. The low(er) energetic cost associated with the direct release of these silicate fragments into solution can assist the fast partial dissolution observed for bioactive glass compositions. Loss of bioactivity when the silicate content increases is accompanied by the transformation of these chains into small closed rings (3- to 8-membered), which appear to protect the silicate network from fast dissolution.

1. Introduction

Silicate glass compositions containing variable amounts of P_2O_5, CaO and Na_2O are used in restorative and regenerative medical applications, which exploit their biocompatibility and their ability to form a surface layer of bone-like hydroxyapatite (HAp) upon implant in the body or upon contact with simulated body fluid (SBF).[1–3] The interaction of the apatite layer growing on the implant with biomolecules and cells provides a strong, stable interface with the existing living tissues (bioactive fixation); depending on the composition, silicate glasses are able to bond to both hard (bone) and soft (muscles) tissues (class A bioglasses), to hard tissues only (class B bioglasses) or to neither soft nor hard tissues (bio-inactive).[1] Bone bonding is enhanced by the equivalence of the formed HAp to the inorganic portion of the bone, whereas bonding to soft tissues also involves the attachment of collagen fibrils to the glass surface. The tissue regeneration ability of bioactive glasses has also been proven, which leads to possible applications as third-generation biomaterials for tissue engineering.[4–7]

The initial stages of the bioactive fixation mechanism are known and involve the partial dissolution of the glass, with loss of Na cations and pH increase, hydrolysis of Si–O–Si bridges and release of silica to the solution, followed by formation of a silica-rich surface layer (depleted of other cations) through which Ca and phosphate ions are released and combine with additional Ca and PO_4 ions from the solution to form the HAp layer.[1]

[a] *Department of Chemistry, University College London, London, UK WC1H 0AJ*
[b] *New York State College of Ceramics, Alfred University, Alfred NY 14802, USA*

† The HTML version of this article has been enhanced with colour images.

The rate of HAp formation on the surface is the main indicator of the glass bioactivity; this rate is determined by the first dissolution stages mentioned above, which many experimental studies have focused on, revealing several key aspects of the bioactive behaviour.[8–14] Fast release of Na cations is an important step by which the solution pH is increased and the hydrolysis of the silicate network started, and release of Ca and P from the glass is known to enhance the rate of HAp formation. However, the bioactivity of Na-free, as well as that of Ca- and P-free silicate glasses highlights the crucial role played by the hydrated silica-rich layer in the nucleation and crystallisation of HAp.[15,16] Indeed, the bio-inactivity of melt-derived glasses with more than 60% SiO_2 may reflect their reduced ability to develop this reactive silica-rich layer, as a result of lower solubility and slower dissolution rates.[10] The central role of the silica-rich layer is probably connected to the high density of surface silanol (Si–OH) groups, which are essential for HAp formation in a physiological environment;[11,16] Si–OH are created through the hydrolysis of Si–O–Si bonds and release of $Si(OH)_4$ or larger silicate fragments.[12] In other words, the loss of silica indirectly enhances the bioactive behavior through the surface Si–OH groups generated in the Si–O–Si hydrolysis process. The soluble silica species are also thought to play a direct role as nucleation centers for the precipitation of calcium phosphate;[17,18] new applications of bioactive glasses as scaffolds for *in vitro* tissue engineering require the direct action of released silica and calcium in activating genes which induce osteoblast proliferation.[7] The concentration of Si released in solution displays a rapid initial increase for bioactive compositions, before approaching a constant value;[8,14] the higher bioactivity of sol–gel glasses is generally linked to the more effective release of soluble silica,[19] and, analogously, the lower bioactivity of class B than class A melt-derived bioglasses is related to the low or zero rate of silica dissolution.[8]

The close correlation between activation energy for Si release and bioactivity[12] further confirms that structural and textural factors enhancing silica release also enhance rate of HAp formation and bioactivity. This activation energy depends on the mobility of small silica chains and rings in the bulk;[12] it has been proposed that the high bioactivity of structures containing silicate chains decreases when these chains condense upon heating to form small (three- or four-membered) or even larger rings, whose presence enhances the glass stability and slows down the dissolution.[20]

Based on these observations, an important step towards unraveling the structure–bioactivity relationship of these materials would be to discover which kind of chains and rings are formed in the silicate network of bioglasses, and correlate them to the degree of bioactivity. An efficient and accurate way to pursue this goal is provided by computer simulations. In particular, classical molecular dynamics (MD) simulations using empirical potential models[21] are widely used to investigate the structural and dynamical properties of modified binary glasses;[22–24] the simulation of multicomponent glasses is less advanced, due to the intrinsic difficulty in obtaining reliable potentials for these systems. Although parameter-free *ab initio* MD simulations[25] can be efficiently used to model glasses, in most cases the initial glass structure has to be obtained by classical MD,[26–29] and the medium-range structure (such as ring and chain distributions) of the "classical" glass is largely maintained in the *ab initio* run. In other words, the benefits of AIMD in terms of accuracy are basically lost if the focus is on the medium-range structure, unless one is able to perform a full *ab initio* melt-and-quench glass generation. On the other hand, we have recently developed a reliable interatomic potential for multicomponent silicate glasses, which approximately include polarization effects through a shell-model approach.[30] This potential is used here to model two typical compositions of bioactive silicate glasses, and a bio-inactive one. In the following sections we analyze the structure of these three different compositions in terms of their tendency to form chain-like and small ring structures, and discuss the relevancy for the bioactive behaviour.

2. Computational methods

Classical MD simulations were carried out with the DL_POLY code[31] using an ionic interatomic potential model recently developed in our group to model silicate glasses incorporating Na, Ca and PO_4 ions.[30,32] A shell-model approach[30,33,34] is used in the potential to effectively include polarization effects: the total charge Z of the ion is split between a core (of charge $Z + Y$) and a shell (of charge $-Y$), which are coupled by a harmonic spring. The core-shell dynamics are controlled through the adiabatic shell method:[34] a small fraction of the core mass (0.2 a.u. for oxygen ions in this work) is shifted to the shells, which move according to conventional equations of motion and follow the ionic motion adiabatically. As described in ref. 30, a damping term was added to the core-shell harmonic forces in order to improve the energy conservation and the stability of the trajectories over relatively long time scales.

The potential form and parameters are reported in Table 1. Besides the damped harmonic interaction with the corresponding core, the oxygen shells interact with each other and with Si, Na, Ca and P cations through a short-range Buckingham term, whereas Coulombic forces act between all species, which bear full formal charges. In addition, three-body screened harmonic potentials are used to control the intra-tetrahedral O–Si–O and O–P–O angles. The parameters describing Na–O and Ca–O short-range interactions were previously fitted to the structure of crystalline silicate phases as identified in typical bioglass and glass–ceramics compositions;[30] keeping these parameters fixed, the P–O pair interaction parameters have been fitted to the structure of sodium and calcium phosphates. Unlike the silicate case, where the effective O–Si–O force constant had to be increased with respect to the original value obtained for quartz in order to improve the silicon coordination number for glasses,[30,35] for the phosphate group this adjustment was

Table 1 The interatomic potential

	Buckingham potential		
	$Ae^{-r/\rho} - Cr^{-6}$		
	A/eV	ρ/Å	C/eV Å^6
O–O	22 764.30	0.14900	27.88
Si–O	1283.91	0.320520	10.661580
Na–O	56 465.3453	0.193931	0.0
Ca–O	2152.3566	0.309227	0.099440
P–O	1120.09133	0.334772	0.0

	Three-body potential		
	$1/2k(\theta - \theta_0)^2 e^{-(r(T-O)/\rho\ +\ r'(T-O')/\rho)}$		
	k_{3B}/eV rad^{-2}	θ_0/°	ρ/Å
O–Si–O'	100.0	109.47	1.0
O–P–O'	50.0	109.47	1.0

	Core-shell potential	
	$1/2kr^2$	
	k/eV Å^{-2}	Y/e
O_c–O_s	74.92	2.8482

not needed and the three-body O–P–O interaction parameters, taking into account a screening exponential factor, are essentially the same as those used in previous MD simulations of silicon aluminophosphates.[36] A small time step is required to control the high frequency of the core-shell spring: 0.2 fs was used in our case, which led to fluctuations of less than 0.005% and no overall drift in the total energy. Cubic periodic boundary conditions are used with a cut-off of 8 Å for the short-range interactions, as well as the Ewald summation of the long-range Coulomb interactions, with a real-space cut-off of 12 Å.

The three glass compositions examined here (BG45, BG55 and BG65 in the following) contain 45, 55 and 65 SiO_2 in weight%, a constant low amount (6% in weight) of P_2O_5, and equal weight percentages (24.5, 19.5, 14.5 wt%, respectively) of Na_2O and CaO. BG45 and BG55 are experimentally found to be bioactive, whereas BG65 is not. The number of atoms (~ 1500) and cell sides (~ 28 Å) were adjusted to reproduce the glass density at room temperature. The melt-derived glass structures were obtained using a standard MD procedure:[30] starting from a random arrangement of ions in the supercell, the system was heated and held at 3500 K for 60 ps in the NPT ensemble, ensuring a suitable melting of the sample. The liquid was then continuously cooled to 1000 K at a nominal cooling rate of 10 K ps^{-1}, in another constant-pressure run of 250 ps, followed by a constant-volume run of 70 ps where the system was cooled to RT with its volume gradually adjusted to the final value, corresponding to the experimental density. The resulting glass structure was then used in a final NVT trajectory of 200 ps, the last 150 ps of which were included in the structural analysis. In each case, two different glass samples, obtained by starting the melt-quench procedure from different random initial configurations, were generated and their structural properties averaged in order to improve the statistical weight of each composition.

The distribution of ring sizes was obtained following the algorithm for primitive ring search proposed by Yuan and Cormack.[37] The search for primitive rings (*i.e.*, rings which cannot be decomposed into smaller rings) was carried out on each final configuration of the MD trajectories, where we have checked that no significant changes in the ring size distributions occur during the dynamics. The length distribution of silicate chains was also evaluated; a chain of length N is defined as a sequence of N interconnected Q^2 silicate tetrahedra, where Q^n is a Si atom bonded to n bridging oxygens (BOs). The chain could be either incorporated in the silicate network on both ends, or terminated by a Q^1 silicate; no isolated silicate chains with Q^1 on both ends were found. In both the ring and chain analyses, we performed the search by considering both Si and P atoms (denoted T in the following) as possible chain/ring formers. Indeed, especially for the glasses with higher silica content, a significant fraction of Q^1 and Q^2 phosphorus atoms are found, which can be incorporated into chains (as Q^1 chain-terminator and Q^2) and rings (Q^2 only). In order to locate chains based on the definition above, the following efficient algorithm, derived from the general ring search algorithm mentioned above[37] was devised: (i) build the list of T nodes linked (through a bridging oxygen) to each T node; (ii) identify all possible chain ends in the original cell, that is either Q^1 species, or Q^2 linked to Q^3/Q^4; (iii) for each chain end (source node), build a shortest distance map in which each other node is ranked according to the lowest number of T–O–T links which separate it from the source node; (iv) identify a Q^2 node linked to the source node and, using the shortest distance map above, determine the shortest path connecting them, that is the shortest sequence of interconnected T nodes leading from the source node to the other end; (v) if all the nodes in this shortest path are Q^2, then a new chain has been found. No crosslink between different chains is possible. As for the ring search algorithm, no significant changes in the distribution of chain lengths are observed along the dynamics, and the search for chains was performed on the final configuration of the dynamics.

3. Results and discussion

A first insight into the structural differences between the three compositions is evident from the Q^n distributions reported in Table 2. BG45 has a more open structure, due to the lower silica content, with prevalence of Q^2 silicates and almost no fully interconnected Q^4. The distribution shifts towards higher n for BG55 and BG65, which are dominated by Q^3 silicates, with a significant fraction of Q^4 silicates for BG65. Most phosphate groups are isolated orthophosphates for BG45, whereas the dominant phosphate groups in BG55 and BG65 are Q^1 pyrophosphates, with three NBO and one BO linking P to a neigbouring Si. This distribution shows that P becomes partially incorporated in the silicate network as the SiO_2 concentration increases, and a significant fraction (35%) of Q^2 metaphosphates with two P–O–Si links are formed for BG65. The silicate network connectivity (NC) in the last column of Table 2 represents the average number of BOs per Si atom; if P atoms were also included in the average, then the overall NC for BG45, BG55 and BG65 would be 1.9, 2.6 and 3.09, respectively.

The emerging picture is that bioactivity requires a fragmented, open network, with very little or no fully interconnected Q^4 silicate. The dissolution of the latter species is energetically unfavourable, as it requires breaking 3–4 Si–BO bonds, hence their negative impact on the glass bioactivity. On the other hand a Q^2 silicate can be released after breaking either two Si–BO bonds, or only one when a fragment with a Q^1 termination on the other end is released. Therefore the significant fraction (20%) of Q^1 silicate present in BG45 should play a key role in the network dissolution processes.

The release of phosphate groups, even though not crucial since PO_4 ions are already present in the physiological environment, is known to enhance the rate of HAp formation.[8] Moreover, glass compositions containing phosphorus are more soluble and show a faster release of silica in solution.[8] Isolated orthophosphate groups can be directly released without breaking any chemical bond; thus, the superior bioactivity of BG45 is likely to be related to the high availability of these species, whereas for BG65 the release of phosphate requires the hydrolysis of one or two P–O–Si links.

3.1 Chain structures

The presence in the glasses of small, low-connectivity silicate fragments, which can be easily released by breaking a small number of Si–O bonds, should greatly enhance the solubility of these materials. For instance, a sequence of several interconnected Q^2 species, with no crosslinks to the three-dimensional glass network, could be detached and released by breaking only two end Si–O bonds.

Fig. 1 shows that only a few "dimers" (2-membered chains) are present in our BG65 samples, whereas a higher number of these short chains composed of 2–4 species is found in BG55. The BG45 structure features a much larger fraction of short chains, and also displays longer chains with up to nine interconnected species. Considering an average Si–Si distance of 3 Å, the length of a linear 9-membered chain would approach the side of our simulation cell, thus introducing problems related to the effect of periodic boundary conditions on the actual length of the

Table 2 Q^n distribution (%) and silicate network connectivity (NC)

	Q^n (Si)						Q^n (P)			NC
	Q^0	Q^1	Q^2	Q^3	Q^4	Q^5	Q^0	Q^1	Q^2	
BG45	1	20	53	24	2	0	65	33	2	2.07
BG55	0	3	31	53	13	0	34	59	7	2.77
BG65	0	0	9	58	32	1	9	56	35	3.24

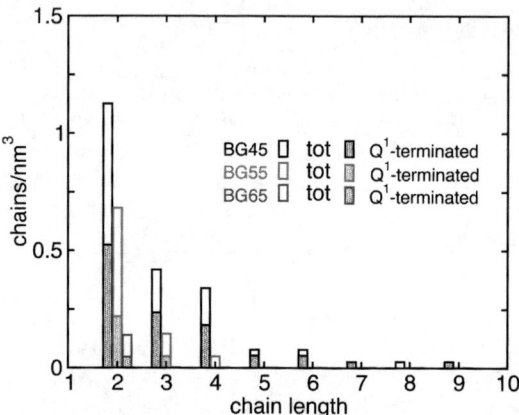

Fig. 1 Calculated number of chains (per unit volume) of length 2–10. For each chain length, the left, central and right bars correspond to BG45, BG55 abd BG65, respectively. The shaded areas highlight the Q^1-terminated fraction of the overall number of chains.

chain. However, direct inspection of the shape of such longer chains (one of which is shown in Fig. 2) shows that they assume a folded configuration, and do not extend along the full cell length.

The fraction of chains with a Q^1 termination is also highlighted in Fig. 1; approximately half of the short (n = 2–4) chains are Q^1-terminated, and these fragments are much more numerous for BG45, as a consequence of both the larger number of chains and the higher total fraction of Q^1 in the glass. The presence of these more soluble fragments, which can be detached and released by breaking only a single Si–O bond, is thus another possible reason for the higher bioactivity of BG45.

The overall picture is summarized in Fig. 3, where only the Si and P tetrahedra which are incorporated in chain fragments of any length are plotted. The figure clearly marks the transition from a structure dominated by these fragments, as BG45, to an almost total lack of them in BG65 (whose highly cross-linked network occupies most of the space left in the cell). The figure also allows us to reveal some participation of phosphorus in the formation of linear chains; despite the low overall number of phosphate group in our systems, we find that about 20% of all P atoms

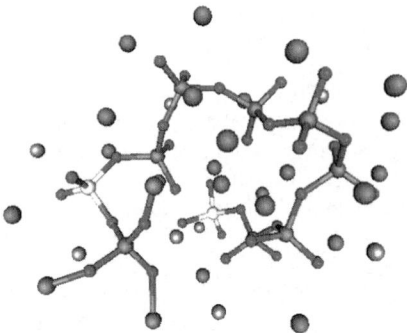

Fig. 2 Structure of a silicate chain found in BG45, containing 9 tetrahedral units. Only the atoms close to the chains are shown. The silicate chain is shown as ball-and-stick, while dark and light isolated spheres represent Na and Ca ions, respectively. The two Si chain ends are highlighted: the chain is Q^1-terminated on one end, whereas the other end is a Q^2 linked to a Q^4 Si.

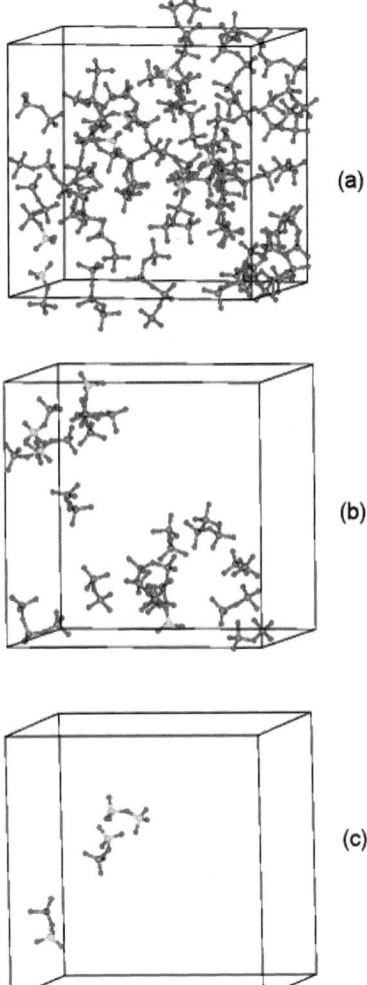

Fig. 3 Chain structure of BG45 (a), BG55 (b) and BG65 (c): only the SiO$_4$ and PO$_4$ tetrahedra incorporated in chain fragments of any length are shown.

are incorporated in chains for BG45, mostly as terminal Q^1 groups in both short and long chains, whereas 15 and 9% are the corresponding fractions of P incorporated in the short chains formed in BG55 and BG65 glasses. These fractions should be compared with the corresponding fractions of 52, 27 and 3% of total Si incorporated in chains, reflecting the general trend revealed in Fig. 3.

The main message is that loss of bioactivity, as is found in BG65, is accompanied by an almost complete lack of chain-like fragments, thus involving a considerably higher activation barrier to the detachment and release of soluble silica.

3.2 Ring structures

The condensation of surface silanol groups leads to the formation of 2-, 3-, and 4-rings which are often thought to be involved in the bioactive process. In particular, the particular geometry of the three-membered (3M) trisiloxane rings, which yields the characteristic D2 peak in the vibrational spectrum,[41,42] has been proposed to enhance their ability to form nucleation sites for the deposition of calcium and phosphate ions.[38,40] Highly strained and reactive disiloxane rings as well as 4-rings can also be involved in the bioactive mechanism.[39] Although the distribution and overall concentration of rings in the bulk does not necessarily reflect the situation at the surface, examining the ring structures formed in the bulk glasses could provide insight into their properties, and enhance our general picture of the structure of these materials and its relation to the bioactivity. The ring size distributions (RSD), represented as number of rings per T atom (T = Si/P), are shown in Fig. 4. The increasing silica content going from BG45 to BG65 allows each T atom to be involved in a much larger number of small and medium-size rings (up to 10-membered): on average, each T atom in BG65 is involved in 3.6 rings of size between 3 and 10, whereas the number drops to 1.4 and 0.2 for BG55 and BG45. The distributions peak at $n = 5$ for BG55 and BG65, whereas 4-rings are the most common ring size in BG45. A non negligible amount of 3-rings is also observed in these glasses, whereas no disiloxane rings are formed in the bulk. It is interesting to observe that, compared to BG45 and BG55, BG65 shows a much higher number of rings of all sizes, except for the 3-rings, which could point to a lower tendency of this bio-inactive composition to form these active sites on the glass surface. Only a negligible amount of phosphorus is present in the rings.

Typical geometrical arrangements of rings comprising between three and six Si atoms are shown in Fig. 5: the planar arrangement of the 3-rings is evident, whereas for $n = 4$ the ring conformation already tends towards a puckered structure. A more detailed insight into the structure of rings can be obtained by determining the distribution of O–Si–O, Si–O–Si and Si–Si–Si angles separately for each different size: the distributions in Fig. 6 were calculated from the full MD trajectories, including in each calculation only the Si/P and O atoms belonging to rings of a specified size. The figures denote a substantial distortion in the tetrahedra forming 3-rings, with O–Si–O and Si–O–Si angles of 107.5 and 125°, respectively, both lower than the corresponding values of 109 and 133–135° measured for the larger 4- and 5-rings. The Si–Si–Si angle distributions also highlight the very different conformation of 3-rings, which show a single sharp peak at 60°, completely absent from the Si–Si–Si distribution of larger rings: 4- and 5-rings are characterized by broader Si–Si–Si angle distributions shifted to greater angles, with two maxima at 80 and 90° for 4-rings and additional peaks/shoulders at 109 and 120° for 5-rings, denoting puckered, non-planar conformations.

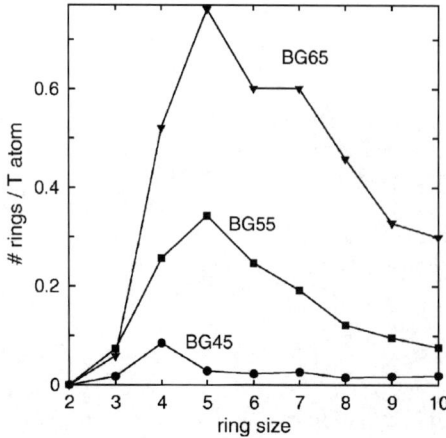

Fig. 4 Ring size distributions of BG45 (circles), BG55 (squares) and BG65 (triangles).

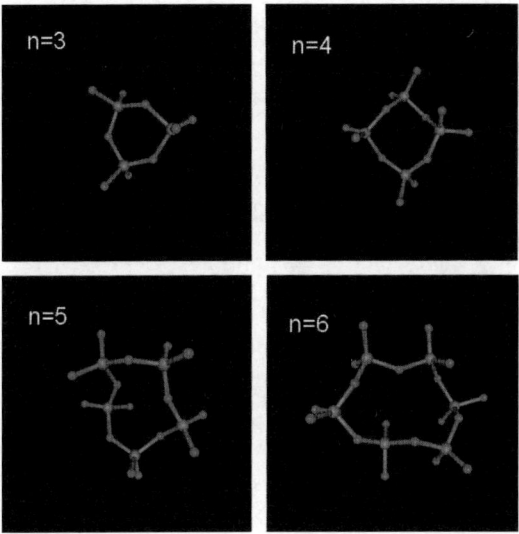

Fig. 5 Ball-and-stick structure of 3- to 6-rings, extracted from the final configuration of the BG65 trajectory. Si and O are represented as large and small spheres, respectively.

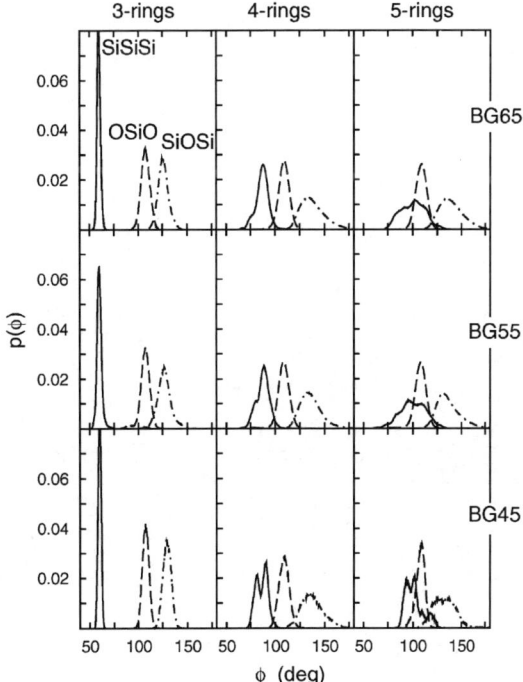

Fig. 6 Angle distributions of O–Si–O (dashed line), Si–O–Si (dot-and-dashed line) and Si–Si–Si (full line) calculated for 3-, 4- and 5-rings (left to right panels) and BG45, BG55, BG65 compositions (bottom to top panels).

4. Conclusions

The transition from highly bioactive to bio-inactive compositions of silicate glasses is related to the combination of several effects, which involve modifications to the surface structure and composition, changes to the energy barriers for ionic migration, and other complex modifications. We have investigated the changes which occur at the molecular level in the bulk structure, focusing on the features of chain fragments and rings. The structural analysis shows that the structure of highly bioactive compositions can be decomposed into many non-crosslinked chains and fragments of length between 2 and 9 tetrahedra; these structural units are not present in bio-inactive compositions such as BG65. On the other hand, a much larger number of small rings (size between 3 and 8 members) can form in the bulk of BG65, than in typical bioactive compositions. Therefore we can identify several factors which greatly enhance the resorption rate of bioactive compositions: the lower energetic cost of releasing small and large silicate fragments into solution leads to a fast disruption of the glass network and allows the solvent to effectively penetrate the interior regions of the glass, in turn leading to a faster hydrolysis of additional Si–O–Si bonds. When these chain fragments are partially or completely transformed into rings, as occurs in higher silica compositions, the result is a protective effect against dissolution and a lower bioactivity.

Acknowledgements

Prof. R. Hill (Imperial College, London) is gratefully acknowledged for fruitful discussions. EPSRC (Grant No. GR/S77714/01) and Royal Society (University Research Fellowship to A. T.) are gratefully acknowledged for financial support. Computer resources were provided by the MOTT2 facility (EPSRC Grant GR/S84415/01) run by the CCLRC e-Science Centre, and by the HPCx service *via* the UK's HPC Materials Chemistry Consortium funded by EPSRC grant EP/D504872.

References

1. *An Introduction to Bioceramics*, ed. L. L. Hench and J. Wilson, World Scientific, Singapore, 1993.
2. L. L. Hench, *Science*, 1980, **208**, 826.
3. L. L. Hench and J. Wilson, *Science*, 1984, **226**, 630.
4. L. L. Hench, *Biomaterials*, 1998, **19**, 1419.
5. L. L. Hench and J. M. Polak, *Science*, 2002, **295**, 1014.
6. I. D. Xynos, M. V. J. Hukkanen, J. J. Batten, L. D. Buttery, L. L. Hench and J. M. Polak, *Calcif. Tissue Int.*, 2000, **67**, 321.
7. I. D. Xynos, A. J. Edgar, L. D. K. Buttery, L. L. Hench and J. M. Polak, *J. Biomed. Mater. Res.*, 2001, **55**, 151.
8. O. Peitl, E. D. Zanotto and L. L. Hench, *J. Non-Cryst. Solids*, 2001, **292**, 115.
9. M. M. Pereira, A. E. Clark and L. L. Hench, *J. Biomed. Mater. Res.*, 1994, **28**, 693.
10. M. Ogino, F. Ohuchi and L. L. Hench, *J. Biomed. Mater. Res.*, 1980, **14**, 55.
11. S. B. Cho, F. Miyaji, T. Kokubo, K. Nakanishi, N. Soga and T. Nakamura, *J. Mater. Sci. Mater. Med.*, 1998, **9**, 279.
12. D. Arcos, D. Greenspan and M. Vallet-Regi, *J. Biomed. Mater. Res.*, 2003, **65A**, 344.
13. M. Cerruti, D. Greenspan and K. Powers, *Biomaterials*, 2005, **26**, 1665.
14. M. Cerruti, D. Greenspan and K. Powers, *Biomaterials*, 2005, **26**, 4903.
15. K. Ohura, T. Nakamura, T. Yamamuro, T. Kokubo, Y. Ebisawa, Y. Kotoura and M. Oka, *J. Biomed. Mater. Res.*, 1991, **25**, 357.
16. T. Kokubo, H.-M. Kim and M. Kawashita, *Biomaterials*, 2003, **24**, 2161.
17. M. Cerruti, A. Perardi, G. Cerrato and C. Morterra, *Langmuir*, 2005, **21**, 9327.
18. M. M. Pereira, A. E. Clark and L. L. Hench, *J. Am. Ceram. Soc.*, 1995, **78**, 2463.
19. L. L. Hench, J. R. Jones and P. Sepulveda, in *Future Strategies for Tissue and Organ Replacement*, ed. J. M. Polak, L. L. Hench and M. Kemp, World Scientific, Singapore, 2002.
20. D. Arcos, D. C. Greenspan and M. Vallet-Regi, *Chem. Mater.*, 2002, **14**, 1515.
21. M. P. Allen and D. J. Tildesley, *Computer Simulations of Liquids*, Clarendon Press, Oxford, 1987.
22. C. Huang and A. N. Cormack, *J. Chem. Phys.*, 1991, **95**, 3634.

23 W. Smith, G. N. Greaves and M. J. Gillan, *J. Chem. Phys.*, 1995, **103**, 3091.
24 J. Horbach, J. W. Kob and K. Binder, *J. Phys. Chem. B*, 1998, **103**, 4104.
25 R. Car and M. Parrinello, *Phys. Rev. Lett.*, 1985, **55**, 2471.
26 A. Tilocca and N. H. de Leeuw, *J. Mater. Chem.*, 2006, **16**, 1950.
27 D. Donadio, D. M. Bernasconi and F. Tassone, *Phys. Rev. B*, 2004, **70**, 214205.
28 T. Charpentier, S. Ispas, M. Profeta, F. Mauri and C. J. Pickard, *J. Phys. Chem. B*, 2004, **108**, 4147.
29 A. Tilocca and N. H. de Leeuw, *J. Phys. Chem. B*, 2006, **110**, 25810.
30 A. Tilocca, N. H. de Leeuw and A. N. Cormack, *Phys. Rev. B*, 2006, **73**, 104209.
31 W. Smith and T. R. Forester, *J. Mol. Graphics*, 1996, **14**, 136.
32 A. Tilocca, A. N. Cormack and N. H. de Leeuw, *Chem. Mater.*, 2007, **19**, 95.
33 B. G. Dick and A. W. Overhauser, *Phys. Rev.*, 1958, **112**, 90.
34 P. J. Mitchell and D. J. Fincham, *J. Phys.: Condens. Matter*, 1993, **5**, 1031.
35 M. J. Sanders, M. Leslie and C. R. A. Catlow, *J. Chem. Soc., Chem. Commun.*, 1984, 1271.
36 G. Sastre, D. W. Lewis and C. R. A. Catlow, *J. Phys. Chem.*, 1996, **100**, 6722.
37 X. Yuan and A. N. Cormack, *Comput. Mater. Sci.*, 2002, **24**, 343.
38 N. Sahai and J. A. Tossell, *J. Phys. Chem. B*, 2000, **104**, 4322.
39 L. L. Hench and J. K. West, *Annu. Rev. Mater. Sci.*, 1995, **25**, 37.
40 N. Sahai and M. Anseau, *Biomaterials*, 2005, **26**, 5763.
41 A. E. Geissberger and F. L. Galeener, *Phys. Rev. B*, 1983, **28**, 3266.
42 S. B. Cho, K. Nakanishi, T. Kokubo, N. Soga, C. Ohtsuki and T. Nakamura, *J. Biomed. Mater. Res.*, 1996, **33**, 145.

Towards an atomic-scale understanding of crystal growth in solution

Elias Vlieg, Menno Deij, Daniel Kaminski, Hugo Meekes and Willem van Enckevort

Received 19th December 2006, Accepted 5th February 2007
First published as an Advance Article on the web 8th May 2007
DOI: 10.1039/b618566p

Our understanding of crystal growth continues to increase thanks to progress in theoretical models, computer simulations and experimental techniques. A discussion of the state-of-the-art in morphology prediction and of the determination of the solid–liquid interface structure using X-ray diffraction shows, however, that there is still a large gap between experiment and theory. We expect that computer modelling, in the form of both Molecular Dynamics simulations and first-principle calculations, will play a crucial role in filling this gap.

Introduction

Whether one can claim to understand crystal growth depends strongly on one's perspective. At a simple level, crystal growth in solution can be understood to be the result of the supersaturation in a solution. Also in industrial crystallization a relatively simple power law may be adequate to describe the average size of a growing crystal population in time. However, if one aims at a fundamental understanding at an atomic level, many questions remain concerning the various mechanisms in the growth process. We will here discuss recent progress in this understanding as a result of theoretical and computational methods and modern experimental techniques. This will show that the gap between the quite detailed experimental information that is becoming available from realistic growth systems and the more simplified picture typically employed in computational and theoretical models, though getting smaller, is still large. Closing this gap requires a major effort in computer modelling and is an important challenge for the future.

Fig. 1 schematically depicts a crystal in a solution with a chemical potential that is higher than that of the crystal. We will not discuss nucleation here, since this is even less understood and is a research field of its own. The growth can be divided into two parts: (1) the transport of the growth units towards the crystal–solution interface and (2) the incorporation into the crystal at the surface. The description of mass transport does usually not require an atomic-scale description and a continuum model is therefore appropriate. Mass transport can be quite complex, involving the interplay of convection, diffusion and, possibly, forced flow, but overall this is quite well understood. Software is available to model this under realistic conditions.

IMM Solid State Chemistry, Radboud University Nijmegen, Toernooiveld 1, 6525 ED, Nijmegen, The Netherlands

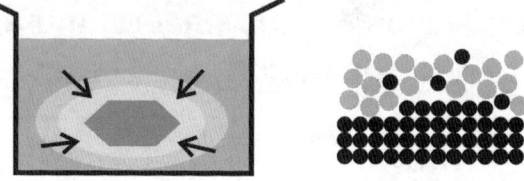

Fig. 1 A schematic of crystal growth in solution. Mass transport in the solution can be described on a macroscopic scale. The actual incorporation of growth units at the solid–liquid interface requires an atomic-scale description.

The second step, involving the interface is much less understood and does require a description at the atomic level. A growth unit (which may contain a shell of water/solution molecules) has to diffuse through the solution layers near the crystal surface. These quasi-liquid layers provide kinetic barriers of a different kind than the bulk liquid, because of their interaction with the crystal surface. The liquid will be more ordered than the bulk liquid and may even be chemisorbed to the crystal. If the growth unit contains a water/solution shell, this needs to be partly removed before contact with the crystal surface can be made. The crystal surface itself may also deviate from its bulk structure and thus present a different bonding geometry than expected. Kinks and steps are the most important growth sites. Growth units may either be incorporated at these sites directly from the solution or first diffuse over the surface. The step and kink density depend on the step and kink free energy, which will be influenced by the solvent used. This description is still simplified, because parameters like temperature, electrochemical potential, defect concentration and impurities/additives can all play a role. Even when ignoring these parameters, a full description of solution growth is a formidable task.

Let us now try to evaluate how well we understand these atomic-scale processes. In this context it is useful to make a distinction in system complexity and system size. Simple model systems are of course better understood than growth under realistic conditions. Fig. 2 provides a scheme with complexity and size as parameters. The essential ingredients for understanding crystal growth are contained in the seminal paper by Burton, Cabrera and Frank from 1952.[1] This has formed the basis of a host of subsequent theoretical papers, including computer simulations. The focus of these

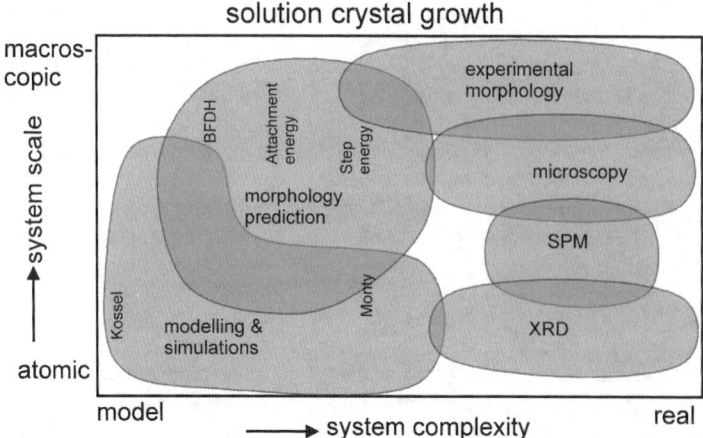

Fig. 2 A diagram showing the realms of experimental, theoretical and computational methods for investigating crystal growth in solution and using system complexity and system scale as parameters. Experimental methods always deal with real systems, while modelling has to simplify reality. A number of specific models that are mentioned in the text are indicated by vertical text.

papers, however, was on simple model systems, in particular on the Kossel crystal.[2] The Kossel crystal has been essential in elucidating many aspects of crystal growth and continues to play this role.[3] At this simple model level, no specific features occur at larger length scales and thus the left side of our complexity-scale diagram is fully covered by theory and computer simulations. More realistic systems pose a greater challenge. Several model systems with more complexity or different bonding than the Kossel case have been investigated. 'Non-Kossel' models containing more than one growth unit per unit cell are found to be quite challenging.[4,5]

More realism is also aimed for in Molecular Dynamics computer simulations in which model potentials describe the interactions between molecules. Such simulations can provide not only an approximate description of the solid–liquid interface structure, but also estimate kinetic barriers as was demonstrated in the pioneering work of Liu *et al.* on the urea–water interface.[6] More simulations of this type have since been done, but computer power remains a bottle neck and thus the system size needs to be small. Even more demanding are full quantum-chemical calculations of the structure of a solid–liquid interface using *e.g.* density-functional theory and such results are starting to appear only now.[7] All in all it can be stated that computer modelling is slowly gaining territory in the complexity direction, but only on small systems and with several parameters from real systems still left out.

While having overlap with computer modelling and simulations, the realm of morphology prediction can be treated as somewhat separate and has a different goal: predicting the shape of a growing crystal. This field lies in the middle section of our diagram. Progress in this area will be discussed in the next section.

Experimental observations of crystal growth are always on 'real' systems, but of course one can study model systems that are chosen for their simplicity. Even the simplest experimental model system, however, will be far more complex than the Kossel crystal. Thus all experimental observations are located at the right side of our complexity-scale diagram. Crystal morphology on a macroscopic scale has been studied for centuries and a lot of phenomenological knowledge is available about the influence of parameters like temperature, solution and impurities on the morphology.[8,9] Visual inspection, or, more quantitatively, an optical goniometer, is sufficient at this level of detail. Optical microscopy brings the length scale down by several orders of magnitude and electron microscopy allows even smaller details to be visualized. At the μm length scale, the crystal morphology is often 'surprisingly' similar to that at the macroscopic scale. For a fundamental understanding, it is therefore more relevant that microscopy can reveal, under favourable conditions, the growth mechanisms (spirals, 2D nucleation) and the shape of surface steps. Optical methods based on phase-shift interferometry have proven to be very powerful and continue to be refined for crystal growth applications.[10]

When scanning-probe microscopy (SPM) became available, it revolutionized several branches in science and crystal growth was no exception. The increased resolution enables the visualization of local structure in great detail, including step fluctuations and impurities. SPM experiments can be performed *in situ*, although real growth is often too fast for the technique and thus fairly low driving forces are needed. At the lower-right corner of our diagram we finally find X-ray diffraction, because this technique can provide information on the interface structure with the highest structural resolution. By its very nature, diffraction always provides an average picture and thus microscopy (local, moderate resolution) and diffraction (global, high resolution) provide information that is complementary in an ideal way. Recent progress in X-ray diffraction will be discussed in a separate section. Note that Fig. 2 also illustrates the well-known fact that a full experimental understanding requires the information from several techniques.

The next two sections describe in more detail recent progress in our group on morphology prediction and X-ray diffraction. This will show that despite this progress, there is still a long way to go before theories and crystal growth include all parameters that are experimentally known to be important.

Morphology prediction

Here we discuss our recent progress in morphology prediction, *i.e.*, in theories that aim to predict the shape of a growing crystal. This is an area lying between the detailed simulations of model systems and the world of real crystals. Predicting the morphology requires predicting the growth velocity of all relevant faces (*hkl*), from which the morphology is derived using a kinetic Wulff plot.[11] Theories for morphology prediction are based on the bulk crystallographic structure of a material and are thus more complex than *e.g.* the Kossel system. Using pragmatic simplifications, the theories are frequently applied in industrially-relevant situations.

The oldest theory was developed in several steps by Bravais, Friedel, Donnay and Harker some 100 years ago.[12–14] This BFDH-theory ignores all details and simply states that the growth velocity of a facet is inversely proportional to the lattice spacing d_{hkl}. This 'explains' that facets on real crystals have typically low values for the Miller indices (*hkl*). The BFDH theory works quite well for isotropic crystal structures, *e.g.*, metals, but fails when the crystal structure becomes more complex. The bulk crystallographic structure was fully included in the Hartman–Perdok theory from the 1950s.[15] This theory uses all the important bonds between the different growth units in the crystal. The essential structure of a crystal is represented in a crystal graph, showing the bonds between the various growth units, see Fig. 3.[16] In specific directions, one can construct so-called periodic-bond chains (PBC's) of growth units connected by bonds. A connected net is a 2D network consisting of two (or more) non-parallel, interconnected periodic bond chains. A stable crystal face, called a flat or F-face, will be parallel to such a connected net. The morphology of a crystal will mainly consist of these F-faces. The morphological importance of an

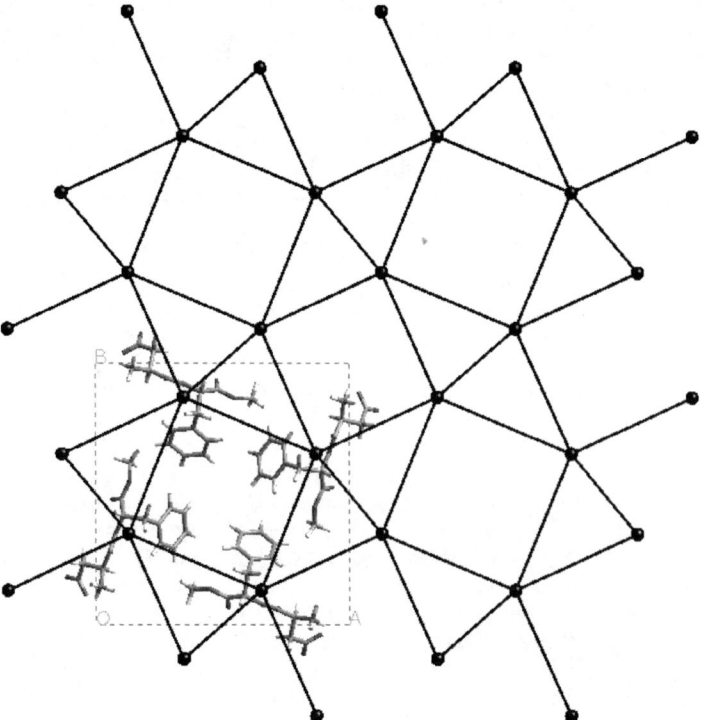

Fig. 3 2D projection of a crystal graph. The lower-left corner shows a full crystal structure (of aspartame). In a crystal graph, each growth unit is replaced by a node and the bonds between growth units are represented by an effective bond. The nodes plus bonds form the crystal graph.

F-face in the Hartman–Perdok theory is derived from its attachment energy, *i.e.*, the energy released when a complete layer with this orientation is crystallised.[17] The growth rate is assumed to be proportional to the attachment energy. A software module called Facelift was developed that finds all possible connected nets in a given crystal graph and that can subsequently calculate the corresponding attachment energies.[18] Using this, it is straightforward to calculate the attachment-energy morphology. For complex crystal structures, the number of F faces can be very large, but typically only a small set of the most stable ones determine the morphology.

While highly successful, the Hartman–Perdok theory has been found to fail in several instances, typically involving structures with a low step energy.[18] Since steps are well-known to be the dominant factor in crystal growth, it is not surprising that the classic Hartman–Perdok approach should fail, because only the attachment energy is considered. It is thus logical to develop a theory based on step energies and that is precisely what we have achieved recently.

Such a theory requires the determination of the energy of a single step and this is non-trivial except for simple model systems. The energy of a step can be defined as the difference in energy of a terrace with the step and that without the step. For the simple crystal structure shown in Fig. 4a, one can immediately see that this is well defined and that the step energy is that of a single bond. For more complex crystal structures, however, this is not so simple. Fig. 4b shows a crystal model with two different atoms and two different bond strengths. When making a single step on such a surface, the top and bottom terrace are no longer vertically aligned, even though they are crystallographically equivalent. The result of subtracting the step-free terrace energy from that with the step, now depends on where one stops counting

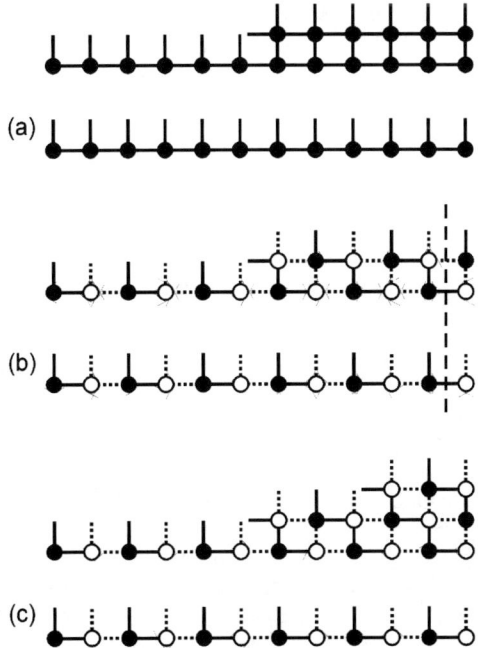

Fig. 4 Models for a stepped surface, projected along the step, together with a flat reference surface. The step energy is defined as the energy of the stepped surface minus that of the flat surface. (a) Kossel model. (b) Model with two different atoms and two different bonds. The calculated energy difference with respect to the flat surface depends on the boundary where one stops counting. (c) Same model as in (b), but now with two steps. This removes the ambiguity in subtracting the reference surface.

behind the step. Using the entire Fig. 4b, the energy difference is found to be that of a single dashed bond. However, when counting until the vertical dashed line, the result is a single solid bond. The ambiguity in this particular example can be resolved using the scheme in Fig. 4c. When taking two steps, the top-most terrace is again vertically aligned with the bottom one and subtracting the flat terrace is unambiguous. It is always possible to make the two steps identical as has already been done in Fig. 4c. The total energy of the two steps is found to be the sum of a solid and a dashed bond and thus the energy of a single step is half this value. The same principle can be applied to any crystal structure, but is more easily done by using a geometrical construction using a single step only.[19] Our method thus disagrees with the notion that only the sum energy of two opposite steps is a well-defined quantity.[20] Of course the absolute energy of a step does not exist, since it has to be determined with respect to a reference. For crystal growth, the most convenient reference is the flat surface. A further refinement would be to use the step *free* energy, because that is the real quantity of interest. We will not do that and use the step energy as a measure of the step free energy.

The possibility to determine the energy of a single step has also consequences for the estimate of surface roughening. If the step (free) energy of a crystal surface is zero, the surface will be rough. On a macroscopic scale, it will grow so fast that in most cases it disappears from the growth morphology. The condition for a surface to be flat, *i.e.* not rough, is usually given as:

$$E_{\text{step,up}} + E_{\text{step,down}} > 0, \qquad (1)$$

because a step up requires also a step down.[21] According to this relation, even if the energy of the step up is negative, the sum of the energies can still be positive and the surface can still be flat. This is, however, not generally true. Using the single step energy as defined above, it is easy to give an example of a system that is thermally rough while still satisfying condition (1). Assume that the sum of two opposite steps is positive and that one of the step energies is negative. If the surface has three-fold symmetry, like on the {111} surface of cubic crystals, then one can make a triangular island of which all sides have negative step energy. The opposite step with positive energy is not involved in constructing such an island. Generating such islands thus costs no energy and this surface will be rough. A more general relation for a surface to be flat is therefore:

$$\oint E_{\text{step}}(\boldsymbol{u})\mathrm{d}\boldsymbol{u} > 0, \qquad (2)$$

where $\boldsymbol{u}$ denotes the step direction of a 2D nucleus with equilibrium shape. Using a Wulff plot to find the island with the lowest energy, one can use eqn (2) to determine whether a surface will be rough or flat. In order to use this as a criterion for thermal roughening, one should of course use the surface free energy.

We have developed a computer code called STEPLIFT that can determine all possible steps on a surface and their energy.[22] The procedure starts where the connected net determination of FACELIFT stops. Steps are constructed by the combination of the connected nets of two non-parallel F-faces, one representing the step terraces, the other the step edge. The structure of a specific step can be viewed in STEPLIFT, but the large number of possible steps makes an automated evaluation highly desirable. This is done by determining the broken bond step energy which, as stated before, largely determines the growth properties of a face. With many possible steps, each in principle with a different energy, we need to specify what is meant by 'THE' step energy. In order to arrive at a step energy value that is characteristic for a specific orientation, we use the average step energy of a 2D island with the lowest edge energy, *i.e.*, of a 2D island with equilibrium shape. Such an island can be constructed from all calculated step energies using the 2D Wulff construction. We will denote this average step energy by $E_{\text{step},hkl}$.

With the step energy well-defined, we next need to relate this to the growth velocity of each face. One very simple choice is to mimic the use of the attachment energy in the Hartman–Perdok theory and state that the growth rate is inversely proportional to the step energy. We can do much better, however, since we can use the more precise expressions from crystal growth theory. The growth velocity depends on the growth mechanism, *i.e.*, whether growth proceeds through 2D nucleation or spiral growth. (In the case of rough growth the specific facet usually disappears from the growth morphology.) We will limit ourselves here to 2D nucleation, for which the following expression holds:[23]

$$R_{hkl} \propto \beta_{\text{step},hkl}\left(\frac{\Delta\mu}{kT}\right)^{5/6} \exp\left(\frac{-VE^2_{\text{step},hkl}}{3kTd_{hkl}\Delta\mu}\right), \quad (3)$$

with $\beta_{\text{step},hkl}$ the step kinetic coefficient, $\Delta\mu$ the driving force, V the volume of a growth unit and d_{hkl} the step height. If we assume that $\beta_{\text{step},hkl}$ is approximately constant, *i.e.*, its variations are small compared to the factors in the exponent, we have an expression for the relative growth rate of which all parameters are known. Compared with the earlier theories for morphology prediction, we even have a model that includes the driving force and the temperature. Fig. 5 summarizes the various steps to arrive at a morphology using the step energy model.

We give here one example of the application of the step energy model, for the morphology of a polymorph of venlafaxine®, a pharmaceutical compound that works as an anti-depressant. The experimental morphology is shown in Fig. 6a. From the crystal structure[24] the bond energies can be calculated using the Dreiding force field. For this case, the energies are scaled to yield the correct dissolution enthalpy in heptane, the actual growth solution. The crystal graph contains seven different bonds. After finding the various connected net orientations using FACE-LIFT, the morphology was calculated using both the attachment energy and the step energy model. Table 1 lists the corresponding energies for the orientations that are most relevant for the present discussion.

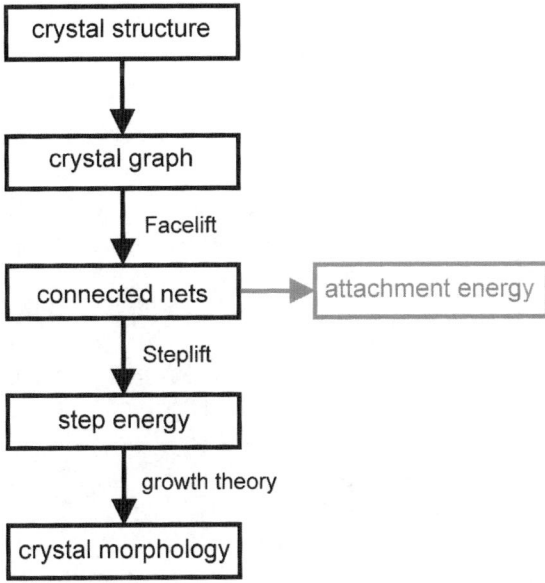

Fig. 5 The various steps in the determination of the step-energy morphology. The Hartman–Perdok theory has the same starting point, but uses the connected nets to calculate the attachment energy as a measure for the morphological importance.

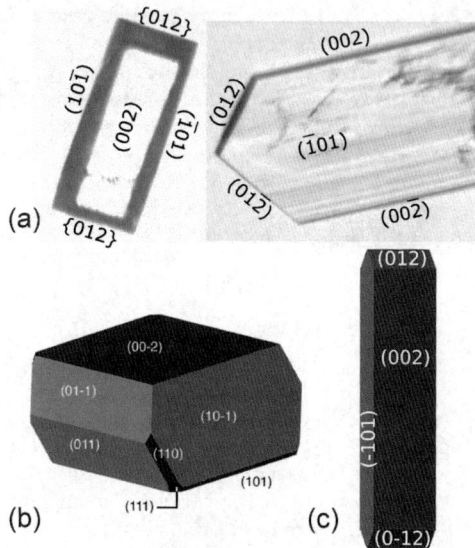

Fig. 6 The morphology of a specific polymorph of venlafaxine®. (a) The experimental, (b) attachment energy and (c) step energy morphology. The step energy morphology agrees much better with the experiment than the attachment energy.

Fig. 6b shows that the attachment energy morphology is quite different from the experimental one. The {01$\bar{2}$} orientation is observed but not predicted, while the {111}, {110}, {101} and {01$\bar{1}$} orientations are predicted but not observed. Furthermore, the {$\bar{1}$01} facet is less prominent than predicted. Fig. 6c shows the step energy morphology for $\Delta\mu/kT = 4$. This agrees much better with the experiment: the three experimental facets are found and the shape agrees also quite well. Table 1 shows the characteristic difference between the two methods for the {01$\bar{1}$} facet. The attachment energy in this direction is small and thus this orientation is predicted to be important in the morphology. This facet, however, is found to be a case with very low step energy and thus the step energy method correctly predicts this facet to be absent. This is a clear case where the usual reciprocal relationship between attachment and step energies fails.[17] The same is true for needle-shaped morphologies, where the step energy in the needle direction is found to be close to zero.[25] For all these cases, the step energy method predicts a vastly improved morphology.

The choice for the value used for the driving force in the step energy prediction is not obvious. For venlafaxine® a value of $\Delta\mu/kT = 4$ gives good agreement with experiment and lowering this value gives a more elongated shape, again in agreement with experiment. We find, however, that the theoretical values of the driving force required for the calculations are typically much higher than the experimental ones.

Table 1 The connected net orientations of venlafaxine® with the lowest attachment energies and the highest average step energies

(hkl)	E_{att}/kcal mol^{-1}	E_{step}/cal mol^{-1} Å^{-1}
{002}	−9.52	798.9
{$\bar{1}$01}	−14.82	452.9
{01$\bar{2}$}	−19.17	302.2
{111}	−25.03	150.2
{110}	−24.22	91.0
{101}	−20.01	73.3
{01$\bar{1}$}	−19.04	25.2

The origin of this discrepancy is not clear, but could be due to a spiral growth mechanism for the real crystal. This needs further investigation.

Atomic-scale structure at the solid–liquid interface

Next we discuss experiments on the structure of the growth interface. Most models for crystal growth assume that the crystal surface has the same structure as the bulk. Several experiments performed in the last few years have shown that this is often not true. Two main techniques are available to determine the *in situ* structure of solid–liquid interfaces: scanning-probe microscopy (SPM) and X-ray diffraction. SPM, often in the form of atomic-force microscopy (AFM), has (literally) provided a much better picture of the structure and role of steps. This has been highly relevant for a better understanding of crystal growth.[26,27] For the determination of the atomic-scale structure, however, X-ray diffraction (XRD) is the most suitable technique.[28] XRD is widely used for the structure determination of crystal surfaces.[29,30] Initially this was mainly carried out in a vacuum, but the large penetrating power of X-rays also enables their use for non-vacuum environments such as solid–liquid interfaces. The strong X-ray beams from a synchrotron radiation source are required for these experiments. Experiments are difficult, because the surface signal is about a million times less than that from the bulk and the signal needs to be detected against a background coming from the liquid and other sources. For this reason, only highly-ordered crystals with atomically flat surfaces can be investigated. The system to be studied has to be selected with this in mind.

Over the last years we have studied the solid–liquid interface structure of potassium-dihydrogen-phosphate (KH_2PO_4, KDP) with increasing levels of detail. The growth morphology of this crystal, that is grown from aqueous solution, is determined by two facets: {101} and {100}. After determining the surface termination and relaxations,[31] we also observed the ordering in the interfacial liquid.[32] As an illustration of the current state-of-the-art for a crystal growth system, we present here our recent results on the pH dependence of the interface structure. From macroscopic observations, the growth velocity of the {101} face is known to exhibit quite a strong dependence on the pH value, with a maximum for a stoichiometric solution (pH = 4.4), while there is little pH dependence for the {100} face.[33,34]

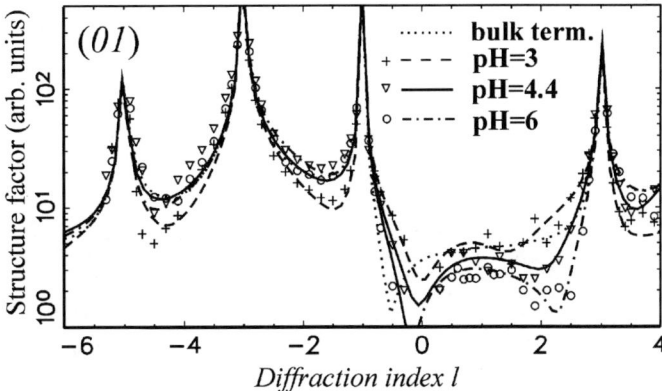

Fig. 7 The (01) crystal truncation rod of the KDP{101} surface as measured using *in situ* X-ray diffraction. The structure factor amplitude is plotted *versus* the diffraction index *l*. The high values for $l = -5, -3, -1$ and 3 correspond to bulk reflections. In between these, the data have the highest surface sensitivity. Data (symbols) are shown for three pH values. The curves correspond to the best fits using a model including crystal surface relaxations and ordering in the interfacial liquid. The dotted curve is a calculation for a bulk terminated surface. The differences between the various curves are small, but significant.

Using the DUBBLE beam line at the European Synchrotron Radiation Facility (ESRF), we obtained extensive data sets in the form of so-called crystal truncation rods.[35] These consist of diffuse intensity connecting the bulk Bragg peaks in the direction perpendicular to the surface. Fig. 7 shows one such rod for the {101} face measured for pH values of 3, 4.4 and 6. The surface/interface sensitivity is highest (and the intensity weakest) in between the bulk Bragg peaks and the figure shows that indeed in these regions there is a significant difference between the three conditions. We are thus able to detect changes in the interfacial structure as a function of pH. A fit of the full data set requires a model describing both relaxations in the top most crystal layers and ordering in the first few liquid layers. The fit results are shown in Fig. 7 as well. Such a fit is very reliable for the highly ordered crystalline side of the interface; for the {101} face we find that the surface terminates in a K^+

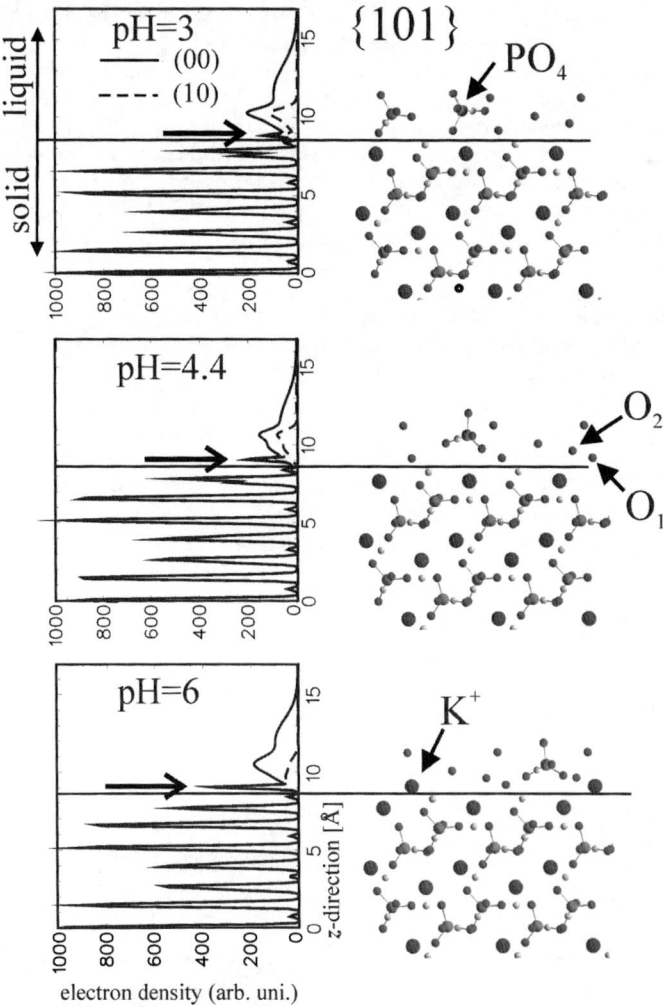

Fig. 8 The interface structure of KDP{101} for three pH values. On the left the projected electron density is shown, which depends on the specific Fourier component considered. The specular or (00) rod shows the full density, the (10) rod only the density with significant in-plane crystalline order. The largest variation in the experiment is indicated by the arrows. On the right a scheme of the actual interface structure is shown. For decreasing pH value (going from bottom to top), K^+ ions are more-and-more replaced by H_3O^+ ions.

layer with a small relaxation of 0.10, 0.06 and −0.04 Å for the pH equal to 3, 4.4 and 6, respectively. The relaxation in the PO_4^{3-} layer directly underneath is even smaller. The structure in the liquid is more difficult to determine, because it contains K^+, $PO_4H_2^-$ and H_2O groups and is less ordered. This makes a unique assignment of atomic positions difficult. Nevertheless, when a large data set of sufficient quality is obtained, structural trends can be derived. A convenient way to show this is in the form of the calculated electron density projected on the surface normal, see Fig. 8. The amount of electron density that is visible for a particular diffraction rod depends on the parallel momentum transfer.[32] For the specular rod there is no parallel momentum transfer and all perpendicular order (layering) in the liquid is visible. Other rods only probe layers that exhibit lateral order. We find approximately three ordered layers in the liquid. The most obvious change as a function of pH in the projected density is that the density in the first liquid layer is strongly decreasing for decreasing pH (indicated by arrows in Fig. 8). From a full analysis, we conclude that at pH = 6 a partial K^+ layer, with the ion essentially at a bulk-extrapolated position, is responsible for this high density. In the bulk of the crystal the binding is such that a positive ion is favoured at this location and this apparently remains true at the solid–liquid interface. We attribute the decrease in density to an increasing replacement of K^+ by H_3O^+, because the concentration of the latter increases (by definition) by a factor 1000. Its concentration remains low with respect to that of K^+, but the possibility to form additional H-bonds near the interface makes H_3O^+ more favourable.[36]

A similar analysis for the {100} face shows no significant changes in the XRD data as a function of pH and thus the interface structure remains constant. The main feature we derive from the analysis is an H-bond between an oxygen from the topmost PO_4^{3-} layer and a water molecule, see Fig. 9. This compensates for a broken H-bond due to the crystal termination. This new bond will not be affected by the change in pH, which explains the insensitivity of the {100} face to the pH value. The chemisorbed water molecule is found to be well-ordered.[37]

The structural changes at the {101} face and the absence of such changes at the {100} face as a function of pH correlate with the fact that the corresponding changes in the macroscopic growth velocity are large for the {101} and small for the {100} face. Unfortunately, we cannot directly translate the structural changes at the {101} face into changes in the expected growth velocity. There appears to be a monotonous change in K^+ versus H_3O^+ bonding, but the maximum growth velocity is found for

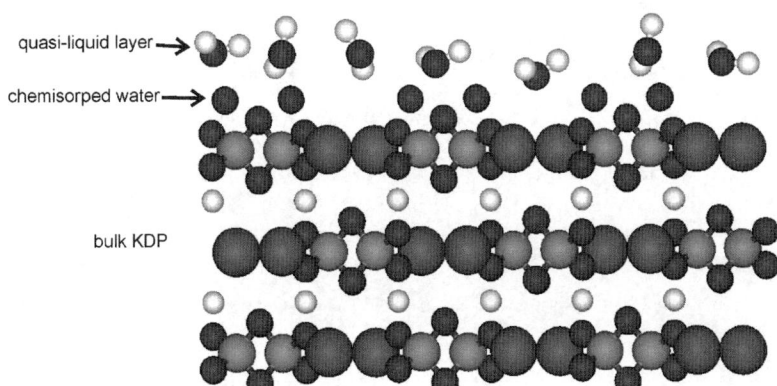

Fig. 9 Side view of the KDP(100) surface, showing the chemisorbed water molecules at the interface that compensates the broken hydrogen bonds due to the surface termination. The water molecule is represented by a single oxygen atom, because X-rays are very insensitive to hydrogen atoms. The data analysis also shows order in the form of layering in the second water layer.

the middle pH value of 4.4. The actual growth for both faces involves several kinetic barriers and these will depend on the species present at the interface.

Closing the gap?

When comparing the two recent results presented above in morphology prediction and on the interface structure of KDP, it is clear that there is a large gap in the level of detail included in the two methods. The morphology prediction assumes a bulk-terminated crystal, without surface relaxations and it ignores the solution and possible effects of chemisorption. The main variable in the experimental results on KDP, the pH value of the solution, is not a parameter in the theory. It is by no means obvious how to include such effects. Interaction with the solution will generally change the stability of steps and the change in energy will depend on the step structure. If a method to estimate such interactions was available, this could be included in the step energy method and a modified morphology could be calculated.

It is our belief that the development of more sophisticated growth theories needs the inspiration obtained from computer simulations. In fact, the step energy method was developed after performing Monte Carlo simulations using the Monty program[38] that uses the same crystal graph, but also includes the effects of kinks and step entropy. Such simulations showed that the step energy is the dominant parameter and thus support our theory that ignores the less important effects.

Understanding the effects of surface relaxations or solvent interactions requires simulations beyond a lattice model. This is slowly happening along two directions. On the one hand first-principle calculations can help in understanding the equilibrium solid–liquid structure. Such calculations have recently been performed for KDP[7] and were found to be in promising agreement with the experimental results. The other approach is through Molecular Dynamics simulations using effective potentials to describe the interactions between the various components in the growth system. The growth interface of KDP is still beyond the reach of current methods, but for a limited number of systems involving less components, encouraging results have been obtained.[6,39–41] These realistic calculations are very costly and the system size has to be quite small. Nevertheless, in this area a genuine contact between experiment and modelling is emerging.

We hope and expect that progress will continue to be made in both first-principle calculations and in MD simulations. This will help in understanding growth kinetics and will guide the development of more detailed growth theories. In this way the overlap between theory/modelling and experiment will increase together with our understanding of crystal growth at the atomic scale.

Acknowledgements

We thank the DUBBLE staff for assistance during the XRD experiments. This work was made possible by financial support from Synthon B. V. and the Council for Chemical Sciences of the Netherlands Organisation for Scientific Research (CW-NWO).

References

1. W. K. Burton, N. Cabrera and F. C. Frank, *Philos. Trans. R. Soc. London, Ser. A*, 1951, **243**, 299–358.
2. G. H. Gilmer and P. Bennema, *J. Appl. Phys.*, 1972, **43**, 1347–1360.
3. H. M. Cuppen, H. Meekes, W. J. P. van Enckevort, E. Vlieg and H. J. F. Knops, *Phys. Rev. B*, 2004, **69**, 245–404.
4. A. A. Chernov, *J. Mater. Sci.*, 2001, **12**, 437–449.
5. H. M. Cuppen, H. Meekes, W. J. P. van Enckevort and E. Vlieg, *Surf. Sci.*, 2004, **571**, 41–62.
6. X. Y. Liu, E. S. Boek, W. J. Briels and P. Bennema, *Nature*, 1995, **374**, 342–345.
7. D. J. Carter, A. L. Rohl and J. D. Gale, *J. Chem. Theor. Comput.*, 2006, **2**, 797–800.

8   H. E. Buckley, *Crystal growth*, Wiley, New York, 1951.
9   J. Nyvlt and J. Ulrich, *Admixtures in crystallization*, VCH, Weinheim, 1995.
10 G. Sazaki, K. Tsukamoto, S. Yai, M. Okada and K. Nakajima, *Cryst. Growth Des.*, 2005, **5**, 1729–1735.
11 G. Wulff, *Z. Krist. Mineral.*, 1901, **34**, 449.
12 A. Bravais, *Etudes Cristallographiques*, Paris, 1866.
13 G. Friedel, *Bull. Soc. Fr. Mineral.*, 1907, **30**, 326.
14 J. D. H. Donney and G. Harker, *Am. Mineral.*, 1937, **22**, 446–467.
15 P. Hartman and W. G. Perdok, *Acta Crystallogr.*, 1955, **8**, 49–52.
16 P. Bennema, in *Handbook of Crystal Growth*, ed. D. T. J. Hurle, North-Holland, Amsterdam, 1993, vol. 1B, pp. 477–582.
17 P. Hartman and P. Bennema, *J. Cryst. Growth*, 1980, **49**, 145–156.
18 R. F. P. Grimbergen, H. Meekes, P. Bennema, C. S. Strom and L. J. P. Vogels, *Acta Crystallogr.*, 1998, **A54**, 491–500.
19 M. A. Deij, J. Los, H. Meekes and E. Vlieg, *J. Appl. Crystallogr.*, 2006, **39**, 563–570.
20 N. Akutsu and Y. Akutsu, *J. Phys. Soc. Jpn.*, 1995, **64**, 736–756.
21 H. van Beijeren and I. Nolden, in *Structure and Dynamics of Surfaces II*, ed. W. Schommers and P. van Blanckenhagen, Springer, Berlin, 1987, pp. 259–300.
22 M. A. Deij, H. Meekes and E. Vlieg, *Cryst. Growth Des.*, submitted.
23 J. P. van der Eerden, in *Handbook of Crystal Growth*, ed. D. T. J. Hurle, Elsevier Science Publisher, Amsterdam, 1993, vol. 1, pp. 307–475.
24 M. A. Deij, J. van Eupen, H. Meekes, P. Bennema and E. Vlieg, to be published.
25 H. M. Cuppen, G. Beurskens, S. Kozuka, K. Tsukamoto, J. M. M. Smits, R. de Gelder, R. F. P. Grimbergen and H. Meekes, *Cryst. Growth Des.*, 2005, **5**, 917–923.
26 T. A. Land, T. L. Martin, S. Potapenko, G. T. Palmore and J. J. De Yoreo, *Nature*, 1999, **399**, 442–445.
27 M. Plomp, W. J. P. van Enckevort and E. Vlieg, *J. Cryst. Growth*, 2000, **216**, 413–427.
28 E. Vlieg, *Surf. Sci.*, 2002, **500**, 458–474.
29 R. Feidenhans'l, *Surf. Sci. Rep.*, 1989, **10**, 105–188.
30 I. K. Robinson, in *Handbook on Synchrotron Radiation*, ed. G. S. Brown and D. E. Moncton, North-Holland, Amsterdam, 1991, vol. 3, pp. 221–266.
31 S. A. de Vries, P. Goedtkindt, S. L. Bennett, W. J. Huisman, M. J. Zwanenburg, D.-M. Smilgies, J. J. De Yoreo, W. J. P. van Enckevort, P. Bennema and E. Vlieg, *Phys. Rev. Lett.*, 1998, **80**, 2229–2232.
32 M. F. Reedijk, J. Arsic, F. F. A. Hollander, S. A. de Vries and E. Vlieg, *Phys. Rev. Lett.*, 2003, **90**, 066103.
33 L. N. Rashkovich and G. T. Moldazhanova, *J. Cryst. Growth*, 1995, **151**, 145–152.
34 S. K. Sharma, S. Verma, B. B. Shrivastava and V. K. Wadhawan, *J. Cryst. Growth*, 2002, **244**, 342–348.
35 I. K. Robinson, *Phys. Rev. B*, 1986, **33**, 3830–3836.
36 D. Kaminksi, N. Radenovic, M. A. Deij, W. J. P. van Enckevort and E. Vlieg, *Phys. Rev. B*, 2005, **72**, 245404.
37 D. Kaminksi, N. Radenovic, M. A. Deij, W. J. P. van Enckevort and E. Vlieg, *Cryst. Growth Des.*, 2006, **6**, 588–591.
38 S. X. M. Boerrigter, G. P. H. Josten, J. van de Streek, F. F. A. Hollander, J. Los, H. M. Cuppen, P. Bennema and H. Meekes, *J. Phys. Chem. A*, 2004, **108**, 5894–5902.
39 M. Brunsteiner and S. L. Price, *J. Phys. Chem. B*, 2004, **108**, 12537–12546.
40 C. Stoica, P. Verwer, H. Meekes, P. J. C. M. van Hoof, F. M. Kaspersen and E. Vlieg, *Cryst. Growth Des.*, 2004, **4**, 765–768.
41 S. Piana, M. Reyhani and J. D. Gale, *Nature*, 2005, **438**, 70–73.

A multi-technique approach for probing the evolution of structural properties during crystallization of organic materials from solution

Colan E. Hughes,[a] Said Hamad,[b] Kenneth D. M. Harris,*[a] C. Richard A. Catlow*[b] and Peter C. Griffiths[a]

Received 13th November 2006, Accepted 29th January 2007
First published as an Advance Article on the web 16th May 2007
DOI: 10.1039/b616611c

We are engaged in a multidisciplinary study of fundamental aspects of the crystallization of organic molecular materials from solution, focusing on polymorphic systems under the recognition that such systems represent an ideal opportunity for obtaining a systematic understanding of competing pathways in crystallization processes. The range of techniques employed in this work are sensitive to structural properties on different length scales and are thus appropriate for mapping the changes that occur at different stages of the crystallization process, starting from the early aggregation events in solution (probed by solution-state NMR and molecular dynamics simulations, including studies of diffusion properties), leading to the growth of molecular aggregates (probed by small-angle neutron scattering), then the emergence of solid microcrystals dispersed in the crystallization solution (probed by small-angle neutron scattering and solid-state NMR) and finally the formation of the bulk solid crystalline phase (probed by powder X-ray diffraction). This paper reports preliminary results on the application of this multi-technique approach to study the crystallization of glycine (which has three known polymorphic forms under ambient conditions) from aqueous solution.

1. Introduction

Although the phenomenon of polymorphism in crystalline solids was first raised in the scientific literature over 180 years ago,[1] there has been a huge upsurge of activity in this field in the last ten years or so, driven both by fundamental scientific curiosity and by industrial necessity. In the case of molecular solids, polymorphism arises when a given type of molecule is able to form different crystal structures.[2–4] Although different polymorphs have the same chemical composition, their solid-state properties are generally different as a consequence of their different crystal structures. A wide variety of strategies exist for producing different polymorphic forms of a given molecule, including conventional crystallization from solution using different solvents, different crystallization conditions (*e.g.*, temperature) or

---

[a] *School of Chemistry, Cardiff University, Park Place, Cardiff, Wales, UK CF10 3AT. E-mail: HarrisKDM@cardiff.ac.uk*
[b] *Davy Faraday Research Laboratory, The Royal Institution of Great Britain, 21 Albemarle Street, London, UK W1S 4BS. E-mail: richard@ri.ac.uk*

crystallization in the presence of additives that promote the nucleation of a specific polymorphic form.

The occurrence of polymorphism is of importance from both fundamental and applied scientific viewpoints. Fundamentally, studies of polymorphic systems represent an ideal opportunity for obtaining a systematic understanding of structure–property relationships for crystalline materials and (of more direct relevance to the present paper) a systematic understanding of competing pathways in crystallization processes. In recent years, there has been considerable interest in being able to find and characterize as many polymorphs as possible of a given molecule of interest. In the industrial context, for example, the motivation is to ensure that the active molecule of interest (*e.g.*, drug or pigment) is produced and marketed in the polymorphic form that has the most desirable properties for the targeted application. It is then essential that the desired polymorph can be produced reliably and reproducibly on scale-up, and that it remains stable during subsequent processing and marketing. However, controlling the production process such that a specific desired polymorph is obtained reproducibly and reliably over a sustained period of time can be associated with significant challenges. Given these issues, the quest to produce and fully characterize all accessible polymorphic forms of a given molecule of interest, and to control their production, has become an area of intense activity within pharmaceuticals and other industries in recent years.

To date, knowledge of the experimental procedure required to produce a given polymorphic form has arisen largely through empirical experience, rather than premeditated design. From knowledge (either from experiment or computer prediction) of the crystal structure of a targeted polymorphic form, there are no universal strategies for deducing the specific crystallization procedure that will lead uniquely and reproducibly to this polymorphic form. In part, the crucial issue is to understand fundamentals of crystallization processes, rather than the properties of the bulk solid phase of the targeted polymorphic form, and to understand the evolution of events that take place throughout the crystallization process, from the initial aggregation of molecules in solution to the subsequent stages of crystal nucleation and growth. Moreover, in the case of polymorphic systems, a key issue is to understand the competition between the formation of the different polymorphic forms under a given set of experimental conditions and, therefore, not only to understand how to promote the formation of the desired polymorphic form but also how to suppress the competitive formation of all other polymorphic forms.

It is well known that crystallization processes for polymorphic systems can be difficult to control, and the question of *which* polymorph is produced in a given crystallization experiment is often governed by factors (*e.g.*, inadvertent seeding) that are beyond the control of the experimenter. Thus, it is common for metastable polymorphs to be produced at the expense of the thermodynamically stable form, or for mixtures of polymorphs to be obtained.[4] The irreproducibility inherent in the crystallization of polymorphs is most emphatically demonstrated by the many documented cases of "disappearing" polymorphs.[3]

Elementary theories[5] of crystal nucleation and growth from solution invoke models in which molecular aggregates of different sizes co-exist in solution. For small aggregates (representing early stages of the process towards crystallization) the Gibbs energy of the aggregate increases as the size of the aggregate increases until a maximum in Gibbs energy is reached, representing the formation of the "critical nucleation entity". Further increase in the size of the aggregate beyond this critical size leads to a decrease in Gibbs energy and crystal growth ensues. For a crystallization process to form the critical nucleation entity and then to pass over this energy maximum on the route to crystal growth is analogous in many respects to the concept of a transition state in chemical reactions. As crystallization processes are generally under kinetic control, the outcome of a given crystallization process should depend more on the relative rates of formation of the critical nucleation entities for the different polymorphs, rather than on the relative thermodynamic stabilities of the bulk solid phases of the polymorphs that are ultimately produced following crystal growth. Clearly, changes in the experimental conditions for the crystallization process may change the nature and relative energies of the critical nucleation entities for the different polymorphs and may therefore change the outcome of the crystallization process.

Recognizing the importance of understanding crystallization processes in relation to the phenomenon of polymorphism, we are engaged in a multidisciplinary study of fundamental aspects of the crystallization of organic molecular materials from solution. The different techniques employed in this work are sensitive to structural properties on different length scales and are thus appropriate for mapping the changes that occur at different stages of the crystallization process. The set of techniques encompasses solution-state NMR spectroscopy, molecular dynamics (MD) simulation, small-angle neutron scattering (SANS) and solid-state NMR spectroscopy (carried out *in situ* on the heterogeneous liquid/solid system that is formed during crystallization). Importantly, the MD simulations are carried out for solute/solvent systems that faithfully reproduce the conditions of concentration and temperature that are employed in the parallel series of experimental crystallization studies. This suite of techniques has been selected to provide complementary information on different stages of the crystallization process, starting from (a) the early aggregation events in solution (probed by solution-state NMR and MD simulations), leading to (b) the growth of molecular aggregates (probed by SANS), then (c) the formation of solid microcrystals dispersed in the crystallization solution (probed by SANS and solid-state NMR), and finally (d) the formation of the bulk solid crystalline phase (probed by powder X-ray diffraction). Additional insights into the evolution of molecular aggregation in solution have been established from diffusion measurements using NMR techniques, in conjunction with quantitative information on diffusion properties established from the MD simulations.

While emphasizing the merits of exploiting a multi-technique approach of the type employed in the present research programme, it is relevant to note that some of these techniques have previously been used individually to probe aspects of crystallization processes, including solution-state ^{1}H NMR studies of 2,6-difluoro-N-(2,6-dimethyl-4-methoxyphenyl)benzamide and sulfamerazine,[6] combined small-angle X-ray scattering and wide-angle X-ray diffraction studies of 2,6-dibromo-4-nitroaniline[7] (in which the crystallization process was monitored in a novel flow system), and small-angle X-ray scattering studies on glycine.[8] Extensive studies employing grazing incidence X-ray diffraction have also been carried out[9] to understand crystal nucleation processes at the air–water interface.

In polymorphism research, a prototypical system that has received wide-ranging attention is glycine, for which three polymorphic forms (see section 2) are known under ambient conditions. Initially, our application of the multi-technique approach outlined above has been focused on the crystallization of glycine from aqueous solution, for which there is considerable interest, *inter alia*, in understanding factors that control the formation of different polymorphic forms under different experimental conditions. Here we present preliminary studies from this work, with our studies of the different stages (a) to (d) of the crystallization process in this system discussed in sections 4.1 to 4.4, respectively. While focusing on the results obtained for this specific system, we also emphasize that the strategy of combining appropriate experimental and computational techniques for probing crystallization processes is more general, and the application of such an approach is strongly advocated for other systems as well.

2. Background to polymorphism of glycine

Under ambient conditions, three polymorphic forms of glycine are known and are denoted the α, β and γ polymorphs.[10] The γ polymorph is the thermodynamically stable form under ambient conditions and the order of stability under these conditions is $\gamma > \alpha > \beta$.[10-13] The α polymorph crystallizes spontaneously from supersaturated aqueous solutions at neutral pH, whereas the γ polymorph crystallizes in both basic (pH > 8.9) and acidic (pH < 3.8) aqueous solutions, but is not obtained spontaneously from neutral aqueous solution under normal conditions.[11,14] This observation suggests that nucleation of the γ polymorph does not occur in neutral aqueous solution under ambient conditions, or at least that, under these conditions, nucleation of the γ polymorph does not compete successfully with nucleation of the α polymorph (*i.e.*, nucleation of the γ polymorph is too slow to be observed in competition with nucleation of the α polymorph). Reported strategies to induce nucleation of the γ polymorph include the addition of electrolytes such as sodium chloride, sodium fluoride and sodium nitrate, or the application of polarized laser radiation.[15] Addition of methanol or ethanol to an aqueous solution of glycine is reported[16] to promote the formation of the β polymorph. Although many papers describe crystallization methods for the three polymorphs, the results are often not reproducible (particularly with regard to formation of the β polymorph[11]), suggesting that the nucleation and crystal growth in this system is influenced by factors that may be difficult to control.

In neutral aqueous solution, glycine exists mainly in the zwitterionic form $H_3N^+CH_2COO^-$, and it is estimated that non-zwitterionic glycine molecules NH_2CH_2COOH have a population of only *ca.* 4.4 ppm.[17] In the crystal structures of all three polymorphs, glycine also exists in the zwitterionic form.

In the present work, we focus exclusively on the α and γ polymorphs of glycine, in part because our experimental studies have been unsuccessful in yielding the β polymorph, in spite of repeating published procedures for the formation of the β polymorph. In the crystal structure of the α polymorph[18] (Fig. 1a; space group $P2_1/n$), glycine molecules are arranged in sheets parallel to the *ac*-plane. Within the sheet, pairs of glycine molecules are connected to each other by two hydrogen bonds in a centrosymmetric arrangement, subsequently called the "double hydrogen-bonded dimer" (Fig. 1b). There are no hydrogen bonding interactions between glycine molecules in adjacent sheets. On the other hand, the crystal structure of the γ polymorph[19] (Fig. 1c; space group $P3_2$) is non-centrosymmetric, with all molecules aligned with the molecular dipole oriented in the same direction with respect to the *c*-axis. In both crystal structures, each N–H bond of the NH_3^+ group is engaged in an N–H$\cdots$O hydrogen bond to the carboxylate group of a neighbouring molecule; for each carboxylate group, one oxygen atom is involved in two N–H$\cdots$O hydrogen bonds, and the other oxygen atom is involved in one N–H$\cdots$O hydrogen bond.

Several recent papers have sought to shed light on the formation of the α and γ polymorphs of glycine under different crystallization conditions. Towler *et al.*[14] proposed that the effect of acid, base or sodium salts in inducing formation of the γ polymorph is due to "self poisoning", while Aber *et al.*[20] demonstrated that a dc electric field induces nucleation of the γ polymorph as a result of the polar nature of the crystals. However, an important observation on γ glycine, which was reported[21] in 1961 but has since received very little attention, is that "*from heavy water solutions the γ-form crystallizes more easily without adding any substance except the seeding crystals*". Although some subsequent papers[11,22] have mentioned this observation, there have been no reported attempts to carry out systematic investigations of this issue. This issue is discussed in more detail in section 4.4 of the present paper.

A number of papers have suggested that glycine tends to form hydrogen-bonded dimers in aqueous solution. In the early 1930s, several studies[23,24] were carried out on the depression of freezing point of water due to glycine, which led to the conclusion[24] that there is a significant extent of dimerization. Diffusion measurements by Gouy interferometry[25] also suggested that clustering of glycine molecules occurs in aged, super-saturated solutions. More recently, Myerson and co-workers[8] have claimed, on the basis of small angle X-ray scattering data, that there is a large population of dimers in saturated aqueous solutions of glycine. However, several aspects of this paper remain open to question, which will be discussed in detail in due

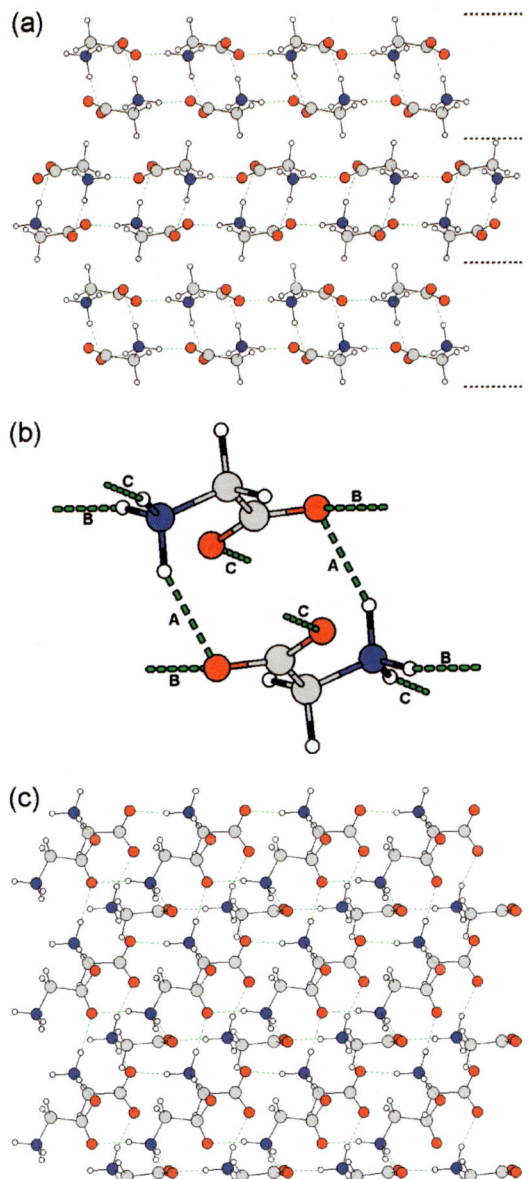

Fig. 1 (a) The crystal structure of α-glycine viewed along the *a*-axis, with hydrogen bonds shown in green and the interface between hydrogen bonded sheets shown as black, dashed lines. (b) The double hydrogen-bonded dimer motif in the crystal structure of α-glycine, with hydrogen bonds labelled A (2.121 Å, 153.93°), B (1.832 Å, 168.45°) and C (1.728 Å, 169.32°). (c) The crystal structure of γ-glycine viewed along the *a*-axis, with hydrogen bonds shown in green.

course.[26] The suggested prevalence of dimers in solution has been used as the basis of a model for crystallization of glycine from neutral aqueous solution, in which it is proposed that the preferential formation of the α polymorph arises because the double hydrogen-bonded dimer motif is present in the crystal structure of this polymorph, whereas the more thermodynamically stable γ polymorph does not contain this motif. The enhanced formation of the β polymorph in solutions containing either methanol or ethanol has also been explained on the basis that these molecules disrupt the formation of dimers in solution.[16]

As discussed below, an important requirement for some of the techniques employed in the present work (particularly SANS and solution-state NMR) is to control the ^{1}H : ^{2}H isotopic ratio in the system. In this regard, it is important to recognize that, in aqueous solution, the three hydrogen sites of the NH_3^+ group will exchange rapidly with the solvent and thus the ^{1}H : ^{2}H ratio of the NH_3^+ group of glycine and the ^{1}H : ^{2}H ratio of the water will rapidly equilibrate to approximately the same value (under the assumption of equal partitioning of ^{1}H and ^{2}H between the glycine and water molecules). Thus, for example, dissolution of a sample of glycine with natural isotopic abundances (*ca.* 99% ^{1}H; 1% ^{2}H) in deuterated water D_2O will lead to a situation in which the ^{1}H : ^{2}H ratios for the ammonium group of glycine and the water molecules are approximately equal. Clearly, the actual value of the ^{1}H : ^{2}H ratio will depend on the amounts of glycine and D_2O used to prepare the solution (*i.e.*, the concentration of the solution). The hydrogen sites of the CH_2 group of glycine, on the other hand, do not undergo appreciable exchange with the solvent in neutral aqueous solution.

3. Experimental details

3.1 General aspects of sample preparation

The sample of glycine with natural isotopic abundances (denoted h_3-glycine) used in our work was obtained from Sigma-Aldrich. This sample was shown by powder X-ray diffraction to be a mixture of the α and γ polymorphs. A monophasic sample of the α polymorph of h_3-glycine (denoted α-h_3-glycine) was prepared by crystallization from a 5.0 M solution of h_3-glycine in H_2O. A monophasic sample of the γ polymorph of h_3-glycine (denoted γ-h_3-glycine) was prepared by crystallization from a solution of h_3-glycine (3.33 M) and NaCl (2.1 M) in H_2O.

A sample of glycine ^{13}C-labelled (99%) on both carbon sites of the molecule (denoted $^{13}C_2$-glycine) was obtained from Cambridge Isotopes Ltd, and was also shown by powder X-ray diffraction to be a mixture of the α and γ polymorphs.

A sample of glycine deuterated in the exchangeable hydrogen sites (*i.e.*, $D_3N^+CH_2COO^-$; denoted d_3-glycine) was prepared by five successive crystallizations of glycine (originally h_3-glycine) from D_2O (with dissolution carried out at 80 °C, corresponding to a 5.33 M solution, followed by solvent evaporation). Samples of d_3-glycine prepared in this way have been shown by powder X-ray diffraction to be a mixture of the α and γ polymorphs, with the γ polymorph predominant. To obtain monophasic samples of α-d_3-glycine and γ-d_3-glycine, the crystals in the mixture were separated manually, based on recognizing the distinct physical appearances of the two polymorphs. Powder X-ray diffraction confirmed that the manual separation carried out in this way led to a monophasic sample of each polymorph.

3.2 Solution-state NMR

Solution-state ^{13}C NMR experiments of ^{13}C-glycine were carried out on a Bruker Avance 500 spectrometer (11.7 T magnet). The ^{13}C NMR spectra were recorded as a function of concentration and temperature. For these experiments, solutions were prepared by dissolving $^{13}C_2$-glycine in D_2O. The ^{13}C-glycine sample had natural ^{1}H/^{2}H abundances, and ^{1}H/^{2}H exchange therefore takes place with the D_2O as described in section 2. The resultant level of deuteration for the exchangeable hydrogens in the system (*i.e.*, the ammonium group of glycine and the water molecules) ranged from 91.2% for the 3.6 M solution to 99.9% for the 0.05 M solution. The level of deuteration was found to have no significant effect on the NMR spectrum.

Solution-state NMR diffusion measurements were performed on a Bruker AMX360 spectrometer (8.46 T magnet) using a 5 mm diffusion probe (Cryomagnet Systems, Indianapolis) and a Bruker gradient spectroscopy accessory. Data were recorded using a pulsed-gradient stimulated echo sequence.[27] These experiments were carried out on a solution prepared by dissolution of d_3-glycine in D_2O, to ensure complete deuteration of all exchangeable hydrogens in the system.

3.3 Powder X-ray diffraction

Powder X-ray diffraction measurements were carried out at 298 K on Station 9.1 at the SRS facility, Daresbury Laboratory, using an X-ray wavelength of 0.999227 Å and with samples packed in capillary tubes (0.5 mm diameter). Data were recorded for six samples: monophasic samples of α-h_3-glycine, α-d_3-glycine, γ-h_3-glycine and γ-d_3-glycine, and physical mixtures of α-h_3-glycine/α-d_3-glycine and γ-h_3-glycine/γ-d_3-glycine. For these experiments, monophasic samples of α-h_3-glycine, α-d_3-glycine, γ-h_3-glycine and γ-d_3-glycine were prepared using the procedures described in section 3.1. The physical mixtures were prepared by mixing approximately equal amounts of the h_3-glycine and d_3-glycine isotopomers of a given polymorph.

As discussed in section 4.4, this set of powder X-ray diffraction experiments was designed to allow a comprehensive comparison between the h_3-glycine and d_3-glycine isotopomers of each polymorph. For each of the samples α-h_3-glycine, α-d_3-glycine, γ-h_3-glycine, γ-d_3-glycine, unit cell refinement was carried out from the powder X-ray diffraction data using the Le Bail profile fitting technique,[28] within the GSAS program package.[29] In each case, the unit cell parameters reported from single-crystal neutron diffraction of the appropriate polymorph were used as the starting point for the refinement, which involved optimization of the unit cell parameters, the line shape and line width parameters, and the zero-point shift. High-quality fits were obtained in all cases, with final R_{wp} values in the range 0.09–0.1.

3.4 Small angle neutron scattering

SANS measurements were performed on the LOQ diffractometer at the ISIS Facility (Rutherford Appleton Laboratory). The samples were contained in 2 mm path length, UV-spectrophotometer grade, quartz cuvettes (Hellma) and mounted in aluminium holders on top of an enclosed, computer-controlled sample changer. Sample volumes were approximately 0.4 cm^3. Temperature control was achieved using a thermostatted circulating bath pumping fluid through the base of the sample changer, with temperature stability ±0.5 °C. Measurement times were *ca.* 40–60 min.

Measurements were made on two detectors covering the ranges q = 0.09–0.29 Å^{-1} and q = 0.12–1.50 Å^{-1}. All scattering data were (a) normalized for sample transmission, (b) background corrected using data for a quartz cell filled with D_2O (which also removes contributions to instrumental background arising from vacuum windows, *etc.*) and (c) corrected for the linearity and efficiency of the detector response using the instrument-specific software package. The data were put onto an absolute scale by reference to the scattering from a partially deuterated polystyrene blend.

Our SANS experiments focused on two samples: (a) a 5.0 M solution prepared directly within the sample cell by dissolving h_3-glycine (mixture of the α and γ polymorphs) in D_2O (resulting in a deuteration level for the exchangeable hydrogens of 88.2%) and (b) a 6.0 M solution prepared by dissolving d_3-glycine (monophasic sample of the α polymorph, obtained as described in section 3.1) in D_2O, and thus with full deuteration of all exchangeable hydrogens in the system.

3.5 Computer simulation

MD simulations were performed on aqueous solutions of glycine corresponding to concentrations of 0.5, 1.0, 2.0 and 3.6 M. The solubility of glycine in water at room temperature is 2.72 M, so the set of concentrations studied represent both supersaturated and undersaturated solutions. The MD simulations employed a cubic cell with sides of length $ca.$ 31 Å and containing 1000 water molecules and 9, 18, 36 and 64 glycine molecules (in the zwitterionic form) for the 0.5, 1.0, 2.0 and 3.6 M solutions, respectively. A time step of 0.6 ps was used and the simulations were carried out in the NPT ensemble using the Nosé–Hoover barostat and thermostat, maintaining the pressure at 1 atm and fixing the temperature. For each concentration, MD simulations were run at four different (fixed) values of temperature: 20, 30, 40 and 50 °C. Each system was given an equilibration time of 100 ps, followed by 3 ns of production. The simulations were carried out using the DL_POLY2 program.[30]

The AMBER potential[31] was used to model the glycine molecules, as this potential has been shown previously[13] to successfully model the crystal structures of the three polymorphs of glycine and their relative energy differences. The water molecules were modelled using the TIP3P potential,[32] which has been used successfully in several studies.[33] Since the TIP3P potential is included in the AMBER force field, it can also be used to model interactions between glycine and water molecules (without resorting to the use of Lorentz–Berthelot rules). Preliminary calculations of the self-diffusion coefficients of glycine in water for a range of temperatures and concentrations provide results within an order of magnitude of the experimental values, giving additional support to the validity of the potentials for modelling glycine–water systems. Detailed analysis of these results will be discussed in a future paper.[26]

4. Results and discussion

4.1 Stage (a): early aggregation in solution

4.1.1 Computer simulation studies. To test the validity of the potential used to model glycine–water interactions, we have compared the radial distribution functions (RDFs) for selected atoms in glycine with the oxygen atoms of water molecules (denoted O_w). Neutron diffraction measurements by Kameda et al.[34] have shown that the ammonium group of glycine in 5.0 M aqueous solution is coordinated to 3.0 ± 0.6 water molecules and that the distance between the nitrogen atom in glycine and O_w is 2.85 ± 0.05 Å. All our simulations at different concentrations and temperatures give very similar results, and the 1.0 M solution at 20 °C is discussed here as a typical example of the RDFs obtained. Fig. 2a shows the RDF and coordination numbers for the $N \cdots O_w$ interactions. Taking into account the uncertainty in the definition of the first hydration shell, the average number of O_w atoms surrounding the N atom is $ca.$ 3.2, in good agreement with the value determined experimentally. The $N \cdots O_w$ distance at the peak of the RDF is 2.92 Å, which is also very close to the experimental value.

Neutron diffraction data are also available on the hydration around the CH_2 group of glycine,[35] and show that in a 4 M solution, the distance between the hydrogen atom in the methylene group (denoted H_C) and O_w is 2.64 ± 0.01 Å, and the number of water molecules in the first hydration shell is $ca.$ 2. Our simulations predict $H_C \cdots O_w$ distances (see Fig. 2b) of 2.89 Å, with 2.3 water molecules in the first hydration shell of each H_C atom. Since there are two H_C atoms in the glycine molecule, our simulations suggest that the total hydration number of the methylene group is between 2.3 and 4.6, although it is likely that in the latter case one water molecule will be present in the hydration shells of both H_C atoms. The larger difference between the experimental and computational results might arise from the very flat nature of the first peak in the RDF. Previous simulation studies[36,37] have predicted that the first hydration shell of the methylene group ranges from 1.34 to 6.1.

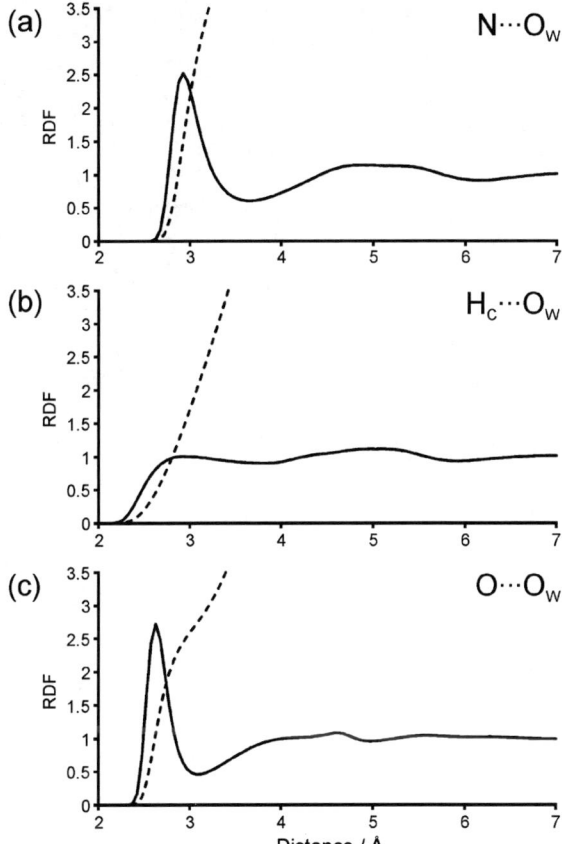

Fig. 2 Radial distribution functions (solid line) and coordination numbers (dashed line) determined for a 1.0 M solution at 20 °C, showing (a) $N \cdots O_w$, (b) $H_C \cdots O_w$ and (c) $O \cdots O_w$ interactions.

For the interaction of the carboxylate group of glycine with water molecules, a previous *ab initio* study[38] has found that the number of water molecules in the first hydration shell is *ca.* 4.7. Our simulations suggest that the total number of water molecules is *ca.* 4.5, which is close to the above value and also to the value 5.1 predicted from a Monte Carlo simulation study.[36] From our simulations, the $O \cdots O_w$ distance is *ca.* 2.63 Å (see Fig. 2c). From the good agreement between these computational and experimental results, we are confident that the interaction between glycine and water molecules is described well by our interatomic force fields.

The results from our MD simulations have been analyzed to find the extent of aggregation of the glycine molecules in aqueous solution as a function of both concentration and temperature, with aggregation of glycine molecules defined in terms of the formation of recognizable hydrogen bonding interactions. To assess the most appropriate geometric criteria for the definition of hydrogen bonding in this system, we have carried out the analysis of molecular aggregates in the MD simulations using nine different definitions for hydrogen bonds, corresponding to three distance criteria ($H \cdots O$ distance less than 1.8, 2.0 and 2.2 Å) combined with three angular criteria ($N-H \cdots O$ angle greater than 120, 140 and 160°).

The numbers of "monomers" (*i.e.*, glycine molecules not hydrogen bonded to any other glycine molecule), "dimers" (*i.e.*, pairs of glycine molecules hydrogen bonded to each other, but not to any other glycine molecule), "trimers", *etc*, have been examined using the different geometric criteria discussed above. With regard to the distance criterion, we find that while many pairs of glycine molecules have H···O distances between 1.8 and 2.0 Å, relatively few have H···O distances between 2.0 and 2.2 Å, perhaps due to the presence of water molecules making such arrangements rare. The dependence upon the angular cut-off value is less marked, with fewer pairs of molecules having N–H···O angles between 120 and 140° than between 140 and 160°. On this basis, we might be inclined to choose 2.0 Å/140° as sensible cut-off values. However, the crystal structure of the α polymorph (Fig. 1) contains the double hydrogen-bonded dimer motif in which the hydrogen bonds have an O···H distance of 2.121 Å and N–H···O angle of 153.93°. Thus, in order to ensure that dimers with a similar structure to that found in the α-glycine crystal structure are included within our geometric criterion, we focus on the results obtained with 2.2 Å/140° as the cut-off values.

Fig. 3 shows the percentage of glycine molecules present in solution as monomers, dimers and trimers, using the 2.2 Å/140° cut-off values. Clearly, as the concentration is increased, the fraction of monomers decreases, whereas the fractions of dimers and trimers increase. At each concentration, no significant variation with temperature is observed in the range 20–50 °C investigated. At 3.6 M, approximately 45% of glycine molecules are hydrogen bonded to other glycine molecules, with approximately half of these molecules (20–25% of the total) in dimers and one quarter (*ca.* 10% of the total) in trimers. The remainder of the glycine molecules are present in aggregates larger than trimers. Of the experimental methods employed in our work

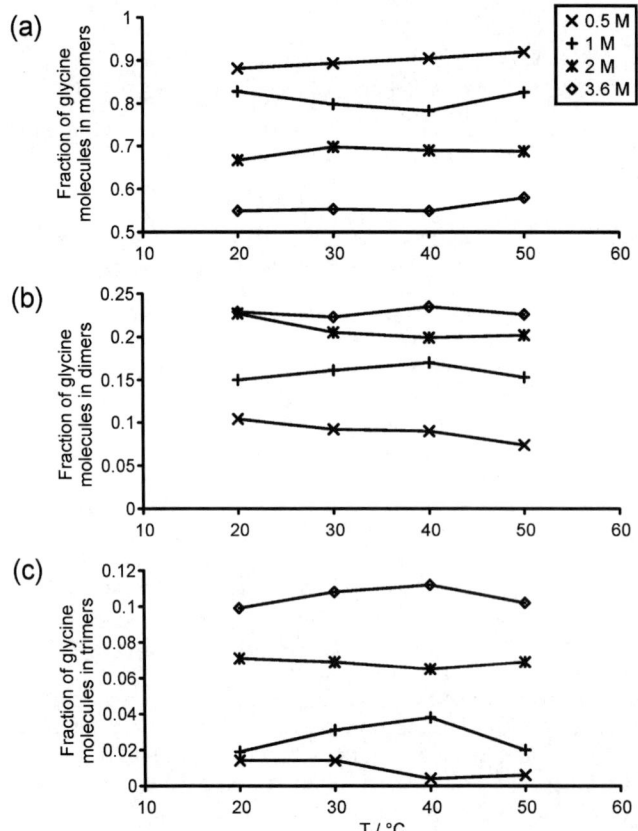

Fig. 3 Fraction of glycine molecules present as (a) monomers, (b) dimers and (c) trimers for concentrations of 0.5, 1.0, 2.0 and 3.6 M as a function of temperature.

to study aggregation, pulsed-gradient stimulated echo NMR is sensitive to the presence of monomers or small aggregates of the type inferred from the MD simulations, whereas SANS is more sensitive to larger aggregates.

No appreciable concentration of double hydrogen-bonded dimers (*i.e.*, resembling the centrosymmetric dimers found in the crystal structure of the α polymorph) are observed in the MD simulations at the lowest concentration (0.5 M) studied, and only a few parts per thousand of such dimers are observed in the simulations at 1.0 M . For each concentration, these conclusions hold for all temperatures studied. As may be expected, a slight increase in the percentage of glycine molecules present in such dimers is observed at higher concentrations (ranging from 1.6 to 1.8% for the different temperatures studied at 2.0 M and ranging from 3.0 to 3.6% for the different temperatures studied at 3.6 M). However, at all concentrations the percentage of such double hydrogen-bonded dimers is significantly lower than both the percentage of glycine dimers linked by a single hydrogen bond and the percentage of glycine trimers.

At 0.5 M, only 10% of glycine molecules are hydrogen bonded to other glycine molecules, with the vast majority present as dimers and less than 2% of the total present as trimers. The double hydrogen-bonded dimer arrangement is not observed at all at this concentration. This observation is in agreement with a recent study[39] using dielectric relaxation spectroscopy to probe glycine solutions in concentrations ranging from 0 to 2.6 M, which has shown that glycine–glycine interactions are very weak under these conditions. Recent MD simulations of glycine–water solutions,[37] carried out using a different force field, suggest very similar behaviour with regard to glycine–glycine interactions, with most of the dimers again observed to form only a single hydrogen bond in a linear arrangement.

4.1.2 Experimental studies. First, we focus on our solution-state ^{13}C NMR studies, the motivation for which has been to use chemical shift measurements as a means of probing changes in the average environment experienced by glycine molecules in aqueous solution as a function of concentration and temperature. In order to remove the necessity of including a reference compound such as TMS in the solution, we have measured not the absolute chemical shifts but the *difference* ($\Delta\delta$) between the chemical shifts of the two ^{13}C nuclei in the ^{13}C$_2$-glycine molecule. Fig. 4 shows the variation of $\Delta\delta$ as a function of temperature (Fig. 4a) and concentration (Fig. 4b). The error in measuring the positions of the peaks for calculation of chemical shifts is estimated to be ±0.004 ppm. For comparison, from high-resolution solid-state ^{13}C NMR spectra recorded at 20 °C for the polymorphs of glycine, the value of $\Delta\delta$ is 132.9 ppm for the α polymorph and 131.9 ppm for the γ polymorph.

The plots of $\Delta\delta$ *versus* temperature (at fixed concentration) reveal a linear dependence over the temperature range studied, with very similar values of gradient (ranging between −0.0101 and −0.0107 ppm K^{-1}) at each concentration. The plots of $\Delta\delta$ *versus* concentration (at fixed temperature) show that, for concentrations greater than 0.5 M, the value of $\Delta\delta$ increases linearly with concentration. This linear dependence is observed at all four temperatures studied and the slopes of the lines are very similar in each case. However, for concentrations less than 0.5 M, the value of $\Delta\delta$ falls below this line for all four temperatures studied. These trends may imply that the observed dependence of $\Delta\delta$ on concentration and temperature is linked to interactions between glycine molecules, as the extent of aggregation is expected to decrease with increasing temperature and decreasing concentration. Thus, $\Delta\delta$ may be interpreted as a measure of the extent of aggregation in this system, with a higher value of $\Delta\delta$ corresponding to a greater extent of aggregation. The fact that the values of $\Delta\delta$ determined from solid-state NMR for both the α and γ polymorphs are higher than any of the solution-state measurements is consistent with this assertion.

An alternative solution-state NMR strategy to investigate the early stages of aggregation of glycine molecules in aqueous solution has focused on diffusion

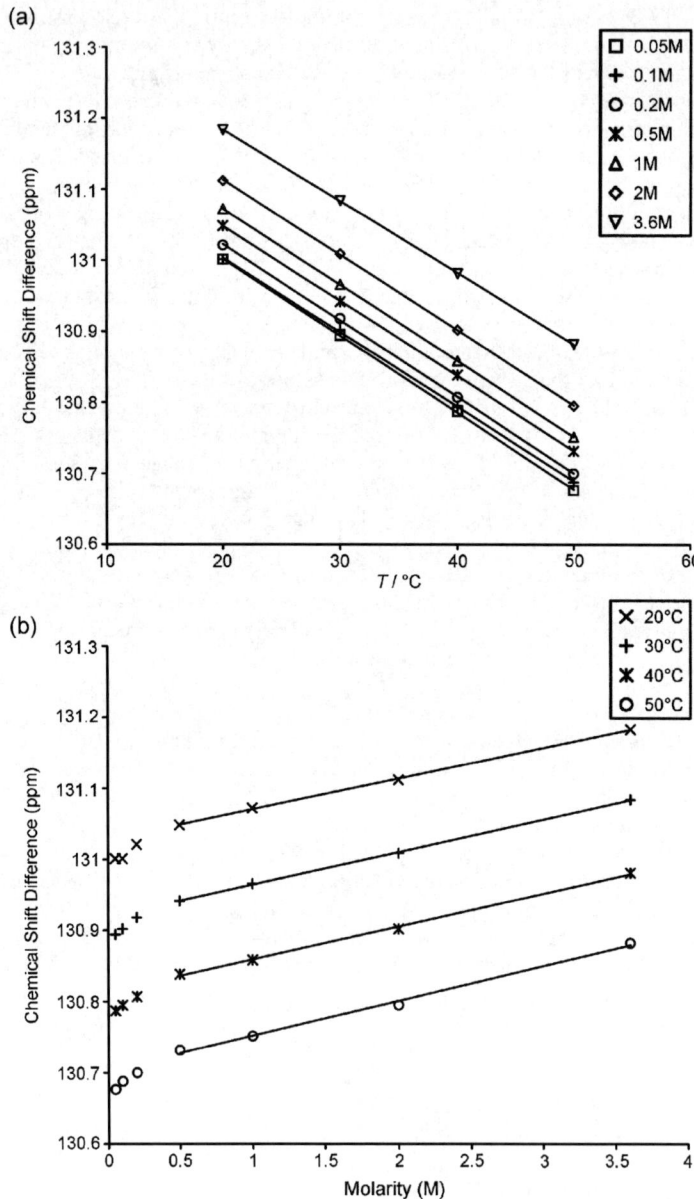

Fig. 4 The difference ($\Delta\delta$) between the chemical shifts of the two ^{13}C resonances for glycine in aqueous solution as a function of (a) temperature (top line, 3.6 M; bottom line, 0.05 M) and (b) concentration (top line, 20 °C; bottom line, 50 °C). The lines are least squares fits of the data. In (b), only the data points for 0.5 M, 1.0 M, 2.0 M and 3.6 M were included in the fit.

measurements, based on the fact that the measured self-diffusion coefficient should exhibit a well-defined monotonic decrease as the size of aggregates increases.[40] Our results (Fig. 5) show the expected trends, with the measured self-diffusion coefficient decreasing as concentration is increased (at fixed temperature), or as temperature is decreased (at fixed concentration). Our values of self-diffusion coefficient are of the same order of magnitude as (but *ca.* 15% lower than) those reported by Ma *et al.*[41] for glycine in H_2O; the differences may be ascribed largely to the fact that D_2O (as

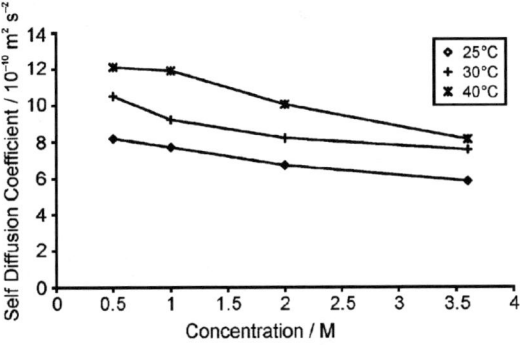

Fig. 5 Self-diffusion coefficients measured by pulsed-gradient stimulated echo NMR as a function of concentration.

used in our study) has a higher viscosity. As self-diffusion coefficients can also be extracted from MD simulations, the next stage of our work in this area will focus on a detailed comparison of self-diffusion coefficients extracted from the simulations and from our experimental studies.

4.2 Stage (b): growth of aggregates in solution

In principle, studies of SANS data in the low q region provide access to information on the size and shape of aggregates in a size range ($N \geq 10 \pm 5$) that is towards the upper end of the range of aggregate sizes that can be probed by MD simulations. In the present MD simulations (section 4.1), however, the primary modes of aggregation were observed as dimers and trimers, and the SANS data are not sensitive to the detection of such aggregates. Instead, SANS is more appropriate for probing larger aggregates representing a later stage of the aggregation process.

In the present work, two separate sets of SANS measurements have been carried out. In the first set, a 5.0 M solution of h_3-glycine in D_2O was studied over the q range 0.09–1.52 Å^{-1} as a function of temperature, increasing from 25 to 45 °C. At the lower temperatures, there is evidence that the scattering arises from aggregates of glycine molecules (Fig. 6a). The intensity of this low-angle scattering decreases as temperature is increased (as shown by the decrease in the integral with increasing temperature in Fig. 6(b)), suggesting that dissolution of these aggregates occurs on increasing temperature.

It is relevant to consider the types of information on the size and shape of aggregates that may be elucidated from such data, particularly when, as in the case of aggregates formed during crystallization, the shape of the scattering entity is not known and may indeed be irregular and time-dependent. Under such circumstances, the data in the low q region may be analysed in a model-free fashion to yield an estimate of the size of the scattering entity.[42] Provided the system is dilute, the form factor, $P(q)$, may be approximated at small q as

$$P(q) = q^{-D} \exp\left(-\frac{q^2 R^2}{K}\right). \quad (1)$$

For spherical particles, $D = 0$, R is the radius and $K = 5$. For long thin rigid cylindrical particles, $D = 1$, R is the cross-sectional radius and $K = 4$. For large rigid disc-like particles, $D = 2$, R is the thickness and $K = 12$. For non-interacting particles, the intensity of scattering, $I(q)$, is given by

$$I(q) = sP(q) + B_{\text{inc}}, \quad (2)$$

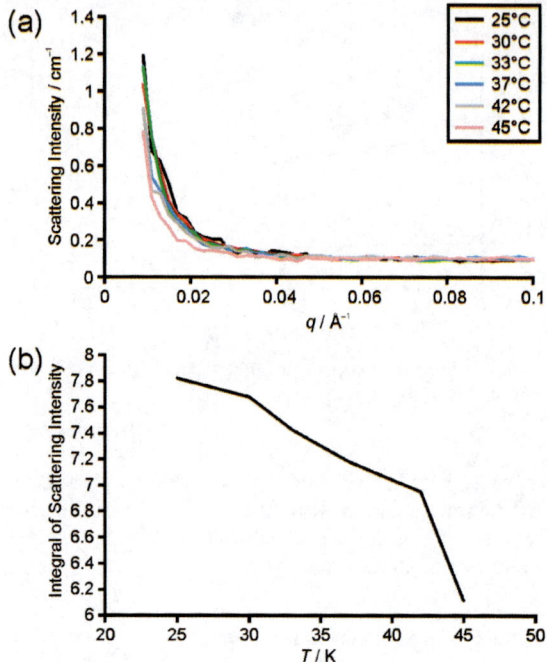

Fig. 6 SANS data for a 5.0 M solution of d_3-glycine in D_2O at six different temperatures: (a) low q region, (b) relative integrals between q = 0.009 and 0.100 Å^{-1}.

where s is a constant and B_{inc} is the incoherent background. Thus, if the scattering entities in the system are described by a well-defined value of D, and therefore conform to a particular gross morphology, eqn (1) and (2) may be used as the basis for fitting the scattering data. In the present work, we find that the model for discs, with D = 2, gives moderately good fits to the SANS data. From the fits to the data recorded under each different set of experimental conditions investigated, the values of thickness range from 75 to 181 Å. However, no well-defined trends are observed as a function of temperature.

A more complex model for fitting SANS data[43] assumes cylindrical particle shape (with radius R and length L) and gives rise to the form factor

$$P(q) = \int_0^{\pi/2} \left(\frac{\sin(\frac{1}{2}qL\cos\gamma) 2J_1(qR\sin\gamma)}{\frac{1}{2}q^2 \sin\gamma \cos\gamma} \right)^2 \sin\gamma \, d\gamma. \quad (3)$$

Fitting our data using this model gives values of L in the range 75–150 Å (similar to the values of thickness obtained for the disc model discussed above) and values of R in the range 275–450 Å. Considering the fits to the data recorded under each different set of experimental conditions investigated, no correlations were found between the values of L and R, nor between either of these values and temperature.

It is expected that, under the conditions of the experiments reported here, the molecular aggregates present during the early stages of the crystallization process should be described by relatively broad distributions of both size and shape, and moreover that the nature of these distributions may also be time-dependent. Under such circumstances, the fact that no well-defined trends are observed in the fitted parameters is perhaps not surprising.

4.3 Stage (c): emergence of crystalline particles in solution

The second aspect of our analysis of the SANS data has concentrated on the high q region, seeking to identify the polymorphic form of the first microcrystalline particles formed during the crystallization process, focusing on a 6.0 M solution of d_3-glycine in D_2O. In our experiments, the d_3-glycine–D_2O solution was stored for several hours at *ca.* 80 °C to ensure complete dissolution and was then placed in the heating rack of the SANS instrument, which was initially at 75 °C. During the transfer of the sample to the heating rack, the temperature may have dropped momentarily, evidence for which comes from the fact that the first SANS measurement at 75 °C has a Bragg peak in the region around $q = 1$ Å^{-1}. On the assumption that the sample was not fully equilibrated before starting the measurement at 75 °C, the data at this temperature were therefore not considered further. Subsequent measurements were made at 55, 35 and 15 °C, with each measurement acquired over approximately one hour. Between measurements, the sample was left to equilibrate at the next temperature for about four hours.

Fig. 7 shows the data recorded in the region around $q = 1$ Å^{-1} at each temperature. At 55 and 35 °C, the peak at $q \approx 1$ Å^{-1} is virtually absent, suggesting that no microcrystals of sufficient size were present in the solution at these temperatures. At 15 °C, however, a single Bragg peak has emerged at $q = 1.025$ Å^{-1}, indicating that the crystallization process has advanced towards the production of microcrystals. The exact value of q, and the absence of other peaks in this region, indicates that these crystals are the γ polymorph. Thus, within the scattering range studied, there are three predicted reflections for the α polymorph [(1, 0, 0) at $q = 1.050$ Å^{-1}, (1, 0, −1) at 1.344 Å^{-1}; (2, −1, 0) at 1.420 Å^{-1}] and one predicted reflection for the γ polymorph [(0, 2, 0) at 1.029 Å^{-1}], with the quoted values of q determined from the published crystal structures[18,19] of the α and γ polymorphs at 298 K. The peak at $q = 1.025$ Å^{-1} is sufficiently narrow to exclude the co-existence of any significant amount of any peak at 1.050 Å^{-1} due to the α polymorph.

The size of the particles that give rise to the observed diffraction peak can be estimated from the line width at half maximum using the Scherrer equation, assuming that the observed line width is dominated by particle size effects.[44] However, the measured line width is approximately equal to the instrumental resolution (0.056 Å^{-1}), and therefore, in the present case, particle size effects do not significantly influence the observed line width. Thus, we may only conclude that the particles are sufficiently large that they do not contribute any significant line broadening beyond the instrumental line width (given the known instrumental resolution in the present case, we estimate that particles below *ca.* 400 Å in size would have a discernible effect on the line width). Under these circumstances, we are

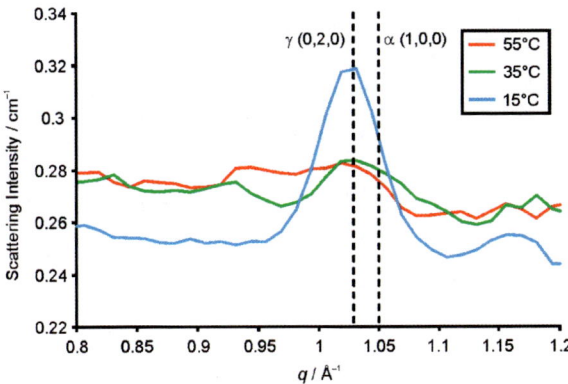

Fig. 7 SANS data for a 6.0 M solution of d_3-glycine in D_2O at three different temperatures.

clearly unable to assign an actual estimate of the size of particles contributing to the observed diffraction peak.

We note that the formation of the γ polymorph from a crystallization solution in which all exchangeable hydrogens are deuterated, as observed from our SANS measurements, is completely consistent with the results now discussed in section 4.4.

4.4 Stage (d): characterization of bulk crystalline phases

Here we focus on isotope effects in the crystallization of the polymorphs of glycine from solution. In the course of our sample preparations, we have observed that crystallization of glycine from D_2O reproducibly promotes the formation of the γ polymorph, without requiring acidity, alkalinity or the presence of any additives (*e.g.*, salts), which are reported (see section 2) to be necessary for crystallization of the γ polymorph from H_2O. As noted in section 2, Iitaka[21] noted previously the preferential formation of the γ polymorph from D_2O, although he specifically observed that D_2O promoted γ-glycine formation under conditions of seeding of the crystallization solution. In contrast, the work carried out here has involved crystallization without seeding.

It is important to note that our experiments described above represent two extreme situations, in which the exchangeable hydrogens in the crystallization system correspond either to natural $^1H/^2H$ isotopic abundances (favouring formation of the α polymorph) and full deuteration (favouring formation of the γ polymorph), respectively. We are currently carrying out investigations of the effect of the percentage of deuteration on the polymorphic outcome for crystallization of glycine from aqueous solution, representing systematic variation of the $^1H : {}^2H$ isotope ratio between these extremes. The results from these experiments will be reported in due course.

To assess further effect of the $^1H : {}^2H$ isotope ratio on the polymorphs of glycine, we have investigated the small differences that are expected to exist between the structural properties of the h_3-glycine and d_3-glycine isotopomers of each polymorph. Fig. 8a shows selected regions of the powder X-ray diffraction patterns for the three samples (*i.e.*, pure h_3-glycine, pure d_3-glycine and a physical mixture of crystals of h_3-glycine and d_3-glycine) of the α polymorph and for the corresponding three samples of the γ polymorph. For each polymorph, shifts in peak positions are clearly observed between the h_3-glycine and d_3-glycine samples, indicating differences in unit cell parameters. The powder X-ray diffraction patterns for the h_3-glycine/d_3-glycine physical mixtures for each polymorph are shown in Fig. 8b, and demonstrate clearly that differences in unit cell parameters for the h_3-glycine and d_3-glycine isotopomers are detectable. For the h_3-glycine/d_3-glycine physical mixtures, the peaks in the powder X-ray diffraction patterns are significantly broader than those in the powder X-ray diffraction patterns for each individual isotopomer and in some cases peak splittings are observed. The importance of recording the powder X-ray diffraction patterns for the h_3-glycine/d_3-glycine physical mixtures is that they provide a comparison between the different isotopomers of a given polymorph under experimental conditions (including temperature) that are guaranteed to be identical (by virtue of *simultaneous* measurement) for each isotopomer.

Le Bail analysis was carried out on the powder X-ray diffraction data for α-h_3-glycine, α-d_3-glycine, γ-h_3-glycine and γ-d_3-glycine, leading to the refined unit cell parameters given in Table 1. The largest difference in a single unit cell parameter between the h_3-glycine and d_3-glycine samples occurs for the α polymorph, with the *b*-axis *ca.* 0.01 Å shorter for α-d_3-glycine than for α-h_3-glycine (corresponding to a shortening of the periodic repeat distance perpendicular to the sheets of glycine molecules in the crystal structure). Overall, however, the volume per molecule in the crystal structure increases upon deuteration for the α polymorph, whereas it decreases upon deuteration for the γ polymorph.

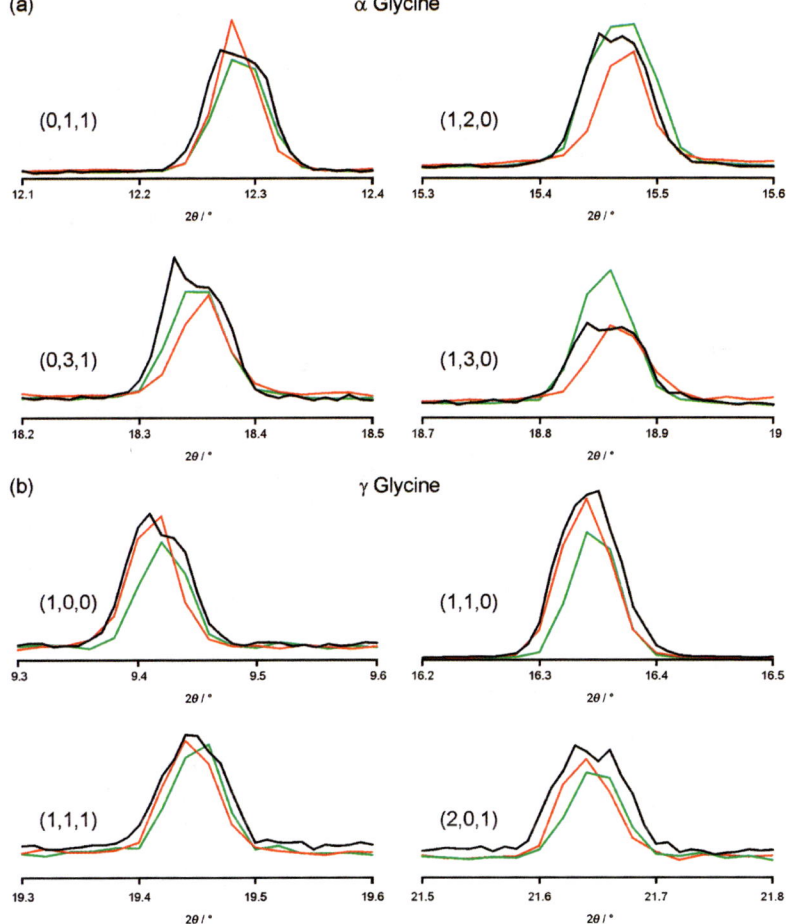

Fig. 8 Powder X-ray diffraction patterns of (a) α-glycine and (b) γ-glycine, showing selected peaks that have significantly different positions for the h_3-glycine (green) and d_3-glycine (red) isotopomers of each polymorph. The powder X-ray diffraction pattern for the physical mixture of the h_3-glycine and d_3-glycine isotopomers of each polymorph is shown in black.

Thus, the effect of deuteration on the unit cell volume and hence crystal density is opposite for the two polymorphs and this may be an important factor underlying the apparent shift in preference for crystallization of the γ polymorph for the deuterated system.

Table 1 Unit cell parameters determined from Le Bail fitting of the powder X-ray diffraction data

Sample	a/Å	b/Å	c/Å	β/°	Volume per molecule/Å^3
α-h_3-Glycine	5.10037(29)	11.9568(6)	5.45976(26)	111.7578(31)	77.310(10)
α-d_3-Glycine	5.10157(34)	11.9467(6)	5.46323(27)	111.7182(30)	77.333(13)
γ-h_3-Glycine	7.03198(19)	—	5.47839(14)	—	78.202(5)
γ-d_3-Glycine	7.02987(18)	—	5.48064(14)	—	78.187(5)

5. Concluding remarks

Although the results presented in this paper represent the preliminary stages of our investigations on the crystallization of glycine from aqueous solution, it is nevertheless clear that some significant insights on this system are emerging from this research. Most importantly, these preliminary studies demonstrate the advantages (and indeed the necessity) of adopting a multi-technique strategy for investigating crystallization processes, particularly by exploiting the complementarity of the different techniques in order to probe different stages of the same process. It is clear from previous literature that the crystallization of glycine has been much studied but as yet remains relatively poorly understood. However, building upon the results obtained to date, we are confident that our continuing studies of the crystallization of glycine (following, *inter alia*, specific future directions highlighted in sections 4.1–4.4) will lead to a more consolidated understanding of the nature of this system. In addition, applications of the same experimental and computational strategy to investigate a range of other crystallization systems, particularly those of relevance to the phenomenon of polymorphism, are also planned.

Acknowledgements

This research is funded by the *Control and Prediction of the Organic Solid State* project supported by the Basic Technology Program of the U.K. Research Councils. We are grateful to Drs Mingcan Xu and Alison Paul for discussions, and we thank Richard Heenan (ISIS facility, Rutherford Appleton Laboratory) and Mark Roberts (SRS facility, Daresbury Laboratory) for experimental assistance with SANS and powder X-ray diffraction measurements, respectively.

References

1 E. Mitscherlich, *Ann. Chim. Phys.*, 1821, **19**, 350–419.
2 (*a*) J. D. Dunitz, *Pure Appl. Chem.*, 1991, **63**, 177–185; (*b*) J. Bernstein, *J. Phys. D: Appl. Phys.*, 1993, **26**, B66–B76; (*c*) J. D. Dunitz, *Acta Crystallogr., Sect. B*, 1995, **51**, 619–631; (*d*) M. R. Caira, *Top. Curr. Chem.*, 1998, **198**, 163–208; (*e*) J. Bernstein, *Polymorphism in Molecular Crystals*, Oxford University Press, Oxford, 2002; (*f*) R. J. Davey, *Chem. Commun.*, 2003, 1463–1467; (*g*) J. Bernstein, *Chem. Commun.*, 2005, 5007–5012.
3 J. D. Dunitz and J. Bernstein, *Acc. Chem. Res.*, 1995, **28**, 193–200.
4 J. Bernstein, R. J. Davey and J.-O. Henck, *Angew. Chem., Int. Ed.*, 1999, **38**, 3441–3461.
5 (*a*) M. Volmer and A. Weber, *Z. Phys. Chem.*, 1926, **119**, 227–301; (*b*) L. Farkas, *Z. Phys. Chem.*, 1927, **125**, 236–242; (*c*) R. Becker and W. Döring, *Ann. Phys. (Berlin)*, 1935, **24**, 719; (*d*) F. Kuhrt, *Z. Phys.*, 1952, **131**, 185–204; (*e*) J. Lothe and G. M. Pound, *J. Chem. Phys.*, 1962, **36**, 2080–2085; (*f*) H. Reiss, *Adv. Colloid Interface Sci.*, 1977, **7**, 1–66; (*g*) D. W. Oxtoby, *Acc. Chem. Res.*, 1998, **31**, 91–97; (*h*) E. Ruckenstein and Y. S. Djikaev, *Adv. Colloid Interface Sci.*, 2005, **118**, 51–72.
6 A. Spitaleri, C. A. Hunter, J. F. McCabe, M. J. Packer and S. L. Cockroft, *CrystEngComm*, 2004, **6**, 489–493.
7 H. G. Alison, R. J. Davey, J. Garside, M. J. Quayle, G. J. T. Tiddy, D. T. Clarke and G. R. Jones, *Phys. Chem. Chem. Phys.*, 2003, **5**, 4998–5000.
8 S. Chattopadhyay, D. Erdemir, J. M. B. Evans, J. Ilavsky, H. Amenitsch, C. U. Segre and A. S. Myerson, *Cryst. Growth Des.*, 2005, **5**, 523–527.
9 I. Weissbuch, M. Lahav and L. Leiserowitz, *Cryst. Growth Des.*, 2003, **3**, 125–150.
10 G. L. Perlovich, L. K. Hansen and A. Bauer-Brandl, *J. Therm. Anal. Calorim.*, 2001, **66**, 699–715.
11 E. V. Boldyreva, V. A. Drebushchak, T. N. Drebushchak, I. E. Paukov, Y. A. Kovalevskaya and E. S. Shutova, *J. Therm. Anal. Calorim.*, 2003, **73**, 409–418.
12 E. V. Boldyreva, V. A. Drebushchak, T. N. Drebushchak, I. E. Paukov, Y. A. Kovalevskaya and E. S. Shutova, *J. Therm. Anal. Calorim.*, 2003, **73**, 419–428.
13 S. L. Price, S. Hamad, A. Torrisi, P. G. Karamertzanis, M. Leslie and C. R. A. Catlow, *Mol. Simul.*, 2006, **32**, 985–997.
14 C. S. Towler, R. J. Davey, R. W. Lancaster and C. J. Price, *J. Am. Chem. Soc.*, 2004, **126**, 13347–13353.
15 B. A. Garetz, J. Matic and A. S. Myerson, *Phys. Rev. Lett.*, 2002, **89**, 175501.

16　I. Weissbuch, V. Y. Torbeev, L. Leiserowitz and M. Lahav, *Angew. Chem., Int. Ed.*, 2005, **44**, 3226–3229.
17　I. G. Darvey, *Biochem. Educ.*, 1995, **23**, 141–143.
18　P.-G. Jönsson and Å. Kvick, *Acta Crystallogr., Sect. B*, 1972, **28**, 1827–1833.
19　Å. Kvick, W. M. Canning, T. F. Koetzle and G. J. B. Williams, *Acta Crystallogr., Sect. B*, 1980, **36**, 115–120.
20　J. E. Aber, S. Arnold, B. A. Garetz and A. S. Myerson, *Phys. Rev. Lett.*, 2005, **94**, 145503.
21　Y. Iitaka, *Acta Crystallogr.*, 1961, **14**, 1–10.
22　(*a*) K. S. Kunihisa, *J. Cryst. Growth*, 1974, **23**, 351–352; (*b*) A. Y. Lee, I. S. Lee, S. S. Dette, J. Boerner and A. S. Myerson, *J. Am. Chem. Soc.*, 2005, **127**, 14982–14983.
23　(*a*) M. Frankel, *Biochem. Z.*, 1930, **217**, 378–388; (*b*) G. A. Anslow, M. L. Foster and C. Klinger, *J. Biol. Chem.*, 1933, **103**, 81–92.
24　W. C. M. Lewis, *Chem. Rev.*, 1931, **8**, 81–165.
25　A. S. Myerson and P. Y. Lo, *J. Cryst. Growth*, 1990, **99**, 1048–1052.
26　C. E. Hughes, S. Hamad, P. C. Griffiths, K. D. M. Harris and C. R. A. Catlow, manuscript in preparation.
27　P. T. Callaghan, *Principles of Nuclear Magnetic Resonance Microscopy*, Oxford University Press, Oxford, 1991.
28　A. Le Bail, H. Duroy and J. L. Fourquet, *Mater. Res. Bull.*, 1988, **23**, 447–452.
29　A. C. Larson and R. B. Von Dreele, *Los Alamos National Laboratory Report*, 2004, LAUR 86-748.
30　W. Smith and T. R. Forester, *J. Mol. Graphics*, 1996, **14**, 136–141.
31　W. D. Cornell, P. Cieplak, C. I. Bayly, I. R. Gould, K. M. Merz, D. M. Ferguson, D. C. Spellmeyer, T. Fox, J. W. Caldwell and P. A. Kollman, *J. Am. Chem. Soc.*, 1995, **117**, 5179–5197.
32　W. L. Jorgensen, J. Chandrasekhar, J. D. Madura, R. W. Impey and M. L. Klein, *J. Chem. Phys.*, 1983, **79**, 926–935.
33　(*a*) E. S. Ferrari, R. C. Burton, R. J. Davey and A. Gavezzotti, *J. Comput. Chem.*, 2006, **27**, 1211–1219; (*b*) B. Hess and N. F. A. van der Vegt, *J. Phys. Chem. B*, 2006, **110**, 17616–17626.
34　Y. Kameda, H. Ebata, T. Usuki, O. Uemura and M. Misawa, *Bull. Chem. Soc. Jpn.*, 1994, **67**, 3159–3164.
35　M. Sasaki, Y. Kameda, M. Yaegashi and T. Usuki, *Bull. Chem. Soc. Jpn.*, 2003, **76**, 2293–2299.
36　M. Mezei, P. K. Mehrotra and D. L. Beveridge, *J. Biomol. Struct. Dyn.*, 1984, **2**, 1–27.
37　M. G. Campo, *J. Chem. Phys.*, 2006, **125**, 114511.
38　K. Leung and S. B. Rempe, *J. Chem. Phys.*, 2005, **122**, 184506.
39　T. Sato, R. Buchner, S. Fernandez, A. Chiba and W. Kunz, *J. Mol. Liq.*, 2005, **117**, 93–98.
40　W. S. Price, Applications of pulsed gradient spin-echo NMR diffusion measurements to solution dynamics and organization in *Diffusion Fundamentals: Basic Principles of Theory, Experiment and Applications*, ed. J. Kärger, F. Grinberg and P. Heitjans, Leipzig University Press, Leipzig, 2005, pp. 490–508.
41　Y. Ma, C. Zhu, P. Ma and K. T. Yu, *J. Chem. Eng. Data*, 2005, **50**, 1192–1196.
42　(*a*) A. Guinier, *Ann. Phys.*, 1939, **12**, 161–237; (*b*) A. Guinier and G. Fournet, *Small Angle Scattering of X-rays*, Wiley, New York, 1955.
43　I. Livsey, *J. Chem. Soc., Faraday Trans. 2*, 1987, **83**, 1445–1452.
44　H. P. Klug and L. E. Alexander, *X-ray Diffraction Procedures*, John Wiley & Sons, New York, 1974.

An examination of polymorphic stability and molecular conformational flexibility as a function of crystal size associated with the nucleation and growth of benzophenone†

Robert B. Hammond, Klimentina Pencheva and Kevin J. Roberts*

Received 20th November 2006, Accepted 29th January 2007
First published as an Advance Article on the web 8th May 2007
DOI: 10.1039/b616757h

The polymorphic behaviour of the aromatic ketone, benzophenone, which is a conformationally flexible molecule and forms crystal structures dominated by van der Waals intermolecular interactions, is examined. Crystallization of this material from the undercooled molten state yields the two known polymorphic forms, *i.e.* the stable α-form and the metastable β-form. The relative, energetic stabilities are examined using both crystal lattice and molecular conformational modelling techniques. Examination of nano-sized faceted molecular clusters of these forms, with cluster sizes ranging from 3 to 100 molecules, reveals that at very small cluster size (<5 molecules) the relative energetic stability of clusters representative for the two forms become very similar, indicating that for high melting undercooling (*i.e.* small critical cluster size for nucleation) crystallization of the metastable β-phase becomes more likely. Detailed analysis of the variation in molecular conformations within the simulated molecular clusters reveals more disordered three-dimensional structures at small compared to larger cluster sizes. The conformational disorder was found to be higher for the metastable β-form. This observation, together with the lower stability of clusters for this form is indicative of the difficulty in achieving crystallization of the metastable β-form from the melt, which requires a considerable undercooling.

1. Introduction

The formation, during crystallization, of polymorphic forms and their relative stabilities represents an interesting and topical area in solid state chemistry. An important reason for this is that molecular-scale effects are expected to have a significant impact in directing the final solid-form following crystallization. It is well known that the fundamental molecular and energetic steps associated with the nucleation, and subsequent growth of organic crystals from the liquid-phase mother liquor involve surface, edge and kink-site adsorption coupled with varying degrees of molecular structural changes. These reflect a reduction in intermolecular coordination that accompanies the various stages of incorporation of a molecule into the crystal surface during the growth process. Crystallization of conformationally

Institute of Particle Science and Engineering, School of Process, Environmental and Materials Engineering, University of Leeds, Leeds, UK LS2 9JT

† The HTML version of this article has been enhanced with colour images.

flexible molecules can also involve significant changes in intramolecular energies, particularly in cases where the crystallizing molecular species undergoes a degree of conformational change when transferring from the liquid to the solid state. In such cases, for example, crystal field effects can cause changes in the molecular conformation with respect to the more unconstrained state for the molecular species experienced in the mother liquor.

An example of the above phenomena lies in the case of the crystallization and polymorphic behaviour of benzophenone (Fig. 1(b)); an aromatic ketone $(C_6H_5C=O)_2$ which is a particularly interesting molecule for fundamental study of the impact of crystallization conditions on molecular and solid state chemistry. Historically, this compound is one of the first recorded examples of polymorphism in the organic solid state chemistry literature through the work of Grossner[1] and Burger and Bloom[2] (see review by Kutzke et al.[3]). Benzophenone is also interesting since it offers little or no opportunities for hydrogen bonding and thus can be viewed as purely a van der Waals solid, easily amenable to examination via molecular and crystallographic modelling techniques.

Benzophenone is known to crystallize in two polymorphic forms. The stable α-form[4-6] (Fig. 1(a)) crystallizes in an orthorhombic lattice with space group $P2_12_12_1$, $Z = 4$, with a melting point of 321 K and cell parameters $a = 10.28$ Å, $b = 12.12$ Å, $c = 7.99$ Å whilst the metastable β-form[3] (Fig. 1(c)) is monoclinic with space group $C2/c$, $Z = 8$, a melting point of 297 K and cell parameters $a = 16.22$ Å, $b = 8.15$ Å, $c = 16.66$ Å, $\beta = 112.91°$. The transformation of the metastable β- into the stable α-phase is monotropic and destructive and can be very easily initiated by the application of mechanical load and/or through contact with the stable α-phase.[3] The transformation frequently occurs spontaneously at room temperature. There are no known conditions under which the β-form is more stable than the α-form.

An interesting feature of this compound is its molecular flexibility, afforded via the potential for the rotation of the two phenyl rings about the central carbonyl group. Indeed, the molecular conformation in the solid state form becomes distorted with respect to that expected for the relaxed molecule in the gas phase. This means that the intermolecular van der Waals interactions are able to overcome, to a small extent, those acting intramolecularly. This, in turn, implies that a change in the molecular conformation is likely to be associated with the crystallization process.[5,7] The torsion angles (ϕ_1 and ϕ_2) of a molecule packed in the crystal structure (A), and gas phase optimized molecule (B) were previously calculated[5] as shown in Fig. 2. It has been reported elsewhere[8] that the configuration associated with a minimal conformational energy has torsion-angles values of 30 and 30°, respectively. Our previous work[5] shows that at $\phi_1 = \phi_2 = 0$ (or 180°) degrees there is a intramolecular clash due to hydrogen–hydrogen non-bonded contacts (see Fig. 1(b)) and that this clash is relieved when one or both of the torsion angles are increased. It was also shown that when the molecule is in the gas phase (B) it has a conformation located in the region manifesting no clash (increased values of the torsions). The molecule when packed in the crystal lattice (A) has a conformation located in a region of slightly higher (less favorable) intramolecular energy. In this work,[5] it was also demonstrated that the extent of such conformational changes is likely to differ from one crystal habit surface (hkl) to another, depending on the surface termination of the crystal structure in the habit plane concerned, i.e. through a face-specific surface reconstruction process, associated with molecular conformational change, that forms an additional energy barrier to the crystal growth process. Similar effects have also been found in the related compound benzil.[9]

Additionally, effects due to conformational flexibility can be expected to take place during bulk 3-D nucleation but have not, as of yet, attracted significant study. However, progress has recently been made in the development of a protocol for building nano-sized and faceted molecular clusters via the development of a polyhedral molecular-packer program POLYPACK. This program has been successfully applied to model the surface[10,11] and solubility[12] properties of particulate

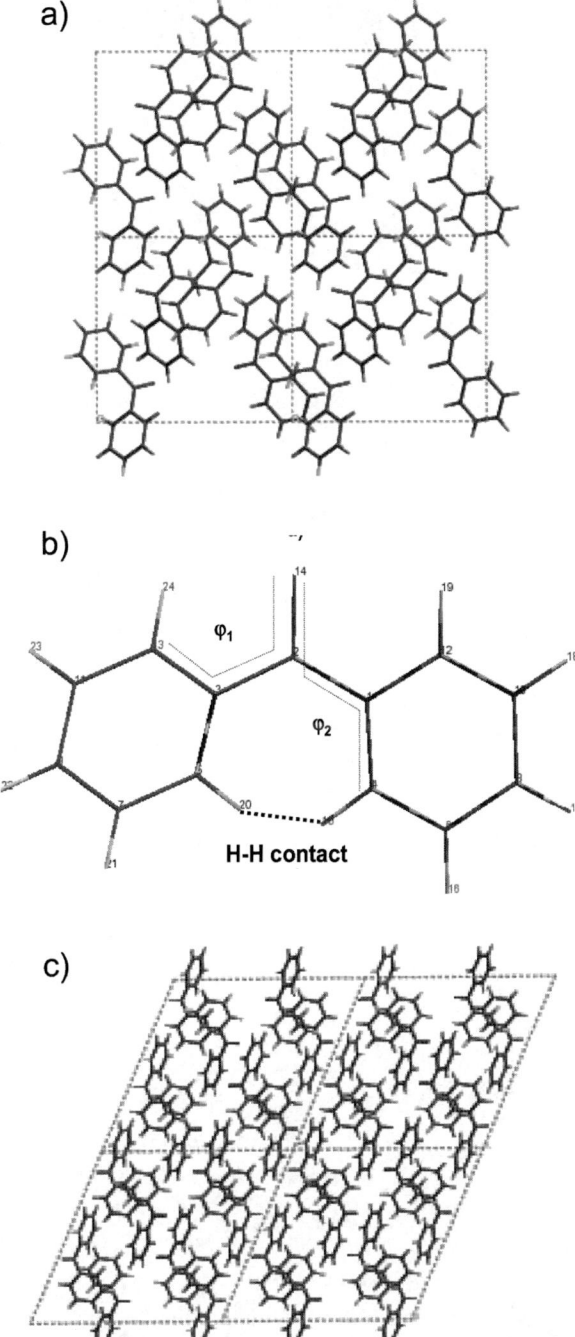

Fig. 1 Benzophenone (b, center) is an aromatic ketone with a degree of conformational flexibility around the central carbonyl group. Its crystal structures (stable form a, top and metastable c, bottom) are dominated purely by van der Waals intermolecular forces. The two torsion angles under consideration in this study are highlighted (b).

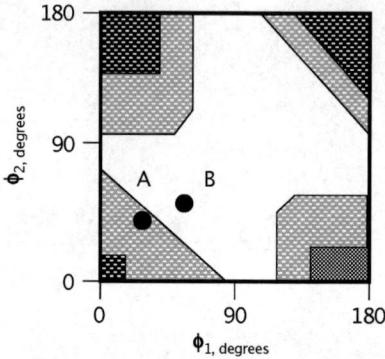

Fig. 2 Torsion map built from measured torsion angles of benzophenone molecules, after ref. 5, shows the greater molecular clash due to a close H–H contact is observed (A) within a molecule packed into the crystal structure whereas the clash is relieved from the free molecule (B). Figure reproduced by kind permission of Institute of Physics Publishing.[5]

clusters together with their polymorphic stability as a function of particle size.[13] In the latter case, cluster stabilities for two polymorphs of the hydrogen-bonded system L-glutamic acid[13] have been studied. In this, the early stages of nucleation were modelled by constructing suitable clusters of defined size based upon the compound's root crystalline structure. The shape of the emerging nuclei was taken to be that manifested in the growth morphology, either that predicted *via* morphological simulations or as observed experimentally. The attraction of this approach is that the clusters that are built can be optimized *via* molecular energy minimization and relaxation techniques and, through this, models of the expected structure for the early stages of the crystal growth process can be obtained. The work revealed that for L-glutamic acid, at small cluster sizes, the metastable β-form becomes surface stabilized, hence becoming more thermodynamically stable than the known stable α-form. However, whether such size-dependent polymorphism effects are also important for non-hydrogen bonded system requires further study, hence providing the context for this paper. In this, a detailed investigation of the cluster stability of the two contrasting polymorphs of the van der Waals compound benzophenone, together with a more detailed analysis of molecular conformation as a function of cluster size, is provided.

2. Computational details

The overall methodology adopted in this study is summarised in Fig. 3.

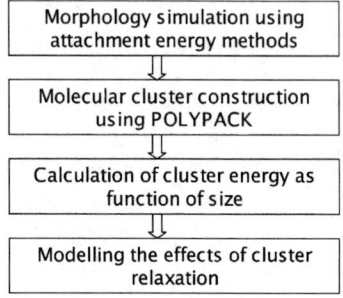

Fig. 3 Schematic representation of the overall methodology adopted for the cluster modelling of benzophenone.

2.1 Potential function selection and morphological simulation

Predicting the dominant intermolecular interactions and through this, the expected crystal morphology requires careful selection of a suitable potential function (force field).[14] In this work the atom–atom approximation[15] together with the Scheraga potential function[16] was used to model the van der Waals interactions with the total potential energy (V_{ij}) being calculated using combined Lennard-Jones 12-6 and coulomic function, thus:

$$V_{ij} = -\frac{A}{r_{ij}^6} + \frac{B}{r_{ij}^{12}} + \frac{q_i q_j}{D r_{ij}} \qquad (1)$$

where A and B are parameters for describing a particular atom–atom interaction, q_i and q_j are the fractional charges on atoms i and j separated by distance r and D is the dielectric constant.

The accuracy of the potential function used in this study was tested through comparison of the calculated lattice energy with the experimentally determined lattice energy, where the latter was derived from experimental sublimation enthalpy data.[17] The calculated lattice energies also need to be consistent with the order of the relative polymorphic stabilities for this material.

The known crystal structures were optimized using the potential function whilst allowing the hydrogen atoms to relax, reflecting the well known issue that the determination of the exact hydrogen atom positions from X-ray crystallography is at best approximate. The partial atomic charges were calculated using the MOPAC software[18] with the MNDO method. The crystal structures for the stable α- and metastable β-polymorph were taken from the work of Girdwood[7] and Kutzke and co-workers,[3] respectively.

2.2 Morphology prediction

Given the fact that the external crystal morphology relates to the intermolecular forces between crystallizing entities, the morphological prediction involved calculation of the dominant intermolecular interaction energies by a summation process[15] as implemented in the HABIT98 program.[19] Through this approach, slice (E_{sl}) and attachment energies (E_{att}) were determined. The latter is defined as the energy released on the addition of a growth layer of thickness d_{hkl} and energy E_{sl} to the (hkl) surface, forming a part of the total lattice energy (E_{lat}) thus:

$$E_{att} = E_{lat} - E_{sl} \qquad (2)$$

The crystal habit faces expected to be dominant were identified from the crystal lattice geometry, using the Bravais Freidel Donnay Harker (BFDH) rules (see *e.g.* ref. 20). The attachment energy can be taken as a measure of the relative crystal growth rates[21] for the crystal habit faces hence, through this, permitting the prediction of the crystal morphology.

2.3 Construction of molecular clusters as a function of cluster size

Molecular clusters of different sizes were built using the POLYPACK program.[10,13] POLYPACK is a system-specific toolbox for constructing faceted molecular clusters with shapes corresponding to the particle morphology. The program can be used for determination of a range of particulate properties, such as the number of bulk and surface molecules within the particle, surface area and volume, surface molecular roughness, R_{wp} parameter, radial distribution function (RDF), torsion angle distribution *etc*. The program uses, as an input, a list of the Miller indices of the faces likely to appear in the crystal habit as determined from the BFDH rules, together with centre-to-face distances for each face. The methodology is flexible in terms of the selection method used for the morphological shape and hence, in principle, the

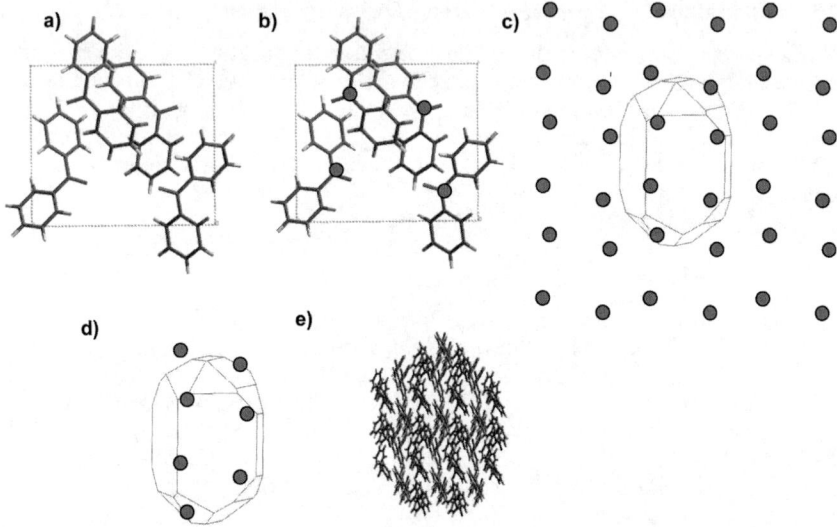

Fig. 4 Schematic diagram showing the method for building shaped molecular clusters used in the POLYPACK programme,[10] presented for the example of the stable α-form of benzophenone where the molecules in the unit cell form a starting point (a), then the centres of gravity of these molecules are calculated thus removing the molecules (b). The centres of gravity are shifted in x, y and z directions of the Cartesian space to build an 3-dimentional array of molecular centre of mass points (c), the centre of which a polyhedra of defined size is positioned with the centres of gravity of molecules outside the polyhedra being deleted (d) and the atomic coordinates again being recalculated (e).

molecular clusters can be built using different approaches for crystal shape prediction. The particle shape was defined using an irregular polyhedral approach, adopting an algorithm previously described[22] involving the calculation of the Cartesian co-ordinates of the polyhedra's corners.

Polyhedra of different sizes were modelled using an expansion coefficient parameter, which was used to scale the inputted central-to-face distances of the initial polyhedra to new values to expand the polyhedron to a different size. The algorithm adopted to create shaped molecular clusters is shown schematically in Fig. 4 where the centres of gravity (mass weighted co-ordinates), shown as dots, of all the molecules in the unit cell were first calculated from the atomic coordinates (a, b) and shifted in the three space directions. A three-dimensional array of molecular centres of gravity was thus created and the predicted polyhedral shape, with a defined size, was placed in the centre of this lattice, in a manner that maximized the number of the molecules inside the polyhedron (c). The centres of gravity of the molecules outside the polyhedron were deleted (d) and the atomic coordinates for all the cluster molecules were then recalculated (e).

The energetic stabilities of the shaped molecular clusters built this way were calculated using the Scheraga[16] force field parameters together with the Open Force Field module in the Cerius^2 program. The cluster structures were subjected to energy minimization in which the molecular conformations were maintained constant and only the cluster packing was allowed to vary (intermolecular optimization) after which the whole cluster structure was relaxed, allowing the molecular conformation to change (intramolecular optimization). Hereinafter, the cluster structures resulting from these calculations will be referred to as minimized and relaxed structures, respectively.

These simulations are conducted at 0 K and therefore, by interference, the entropic factors are not considered explicitly and the Gibbs free energy can be equated to the internal energy (enthalpy). In order to consider the temperature effects it would be necessary to conduct molecular dynamics simulations, an aspect which will be addressed in future work.

It is also noteworthy that the modelling approach adopted in this study also does not include the effects of solution binding at the cluster surfaces which may also play an important role in terms of any surface reconstruction effects. However, the latter are known to be relatively small for molecular organic crystals[23] compared to inorganic materials.[24]

2.3 Conformational analysis

The conformational analysis of the relaxed molecular clusters was carried out *via* measurements of the intramolecular torsion angles, ϕ_1, ϕ_2, with respect to the orientation of the plane of the two phenyl rings and the carbonyl group for an isolated benzophenone molecule using the Cerius^2 software.

It should be noted, however, that the selection of a molecule on the cluster surface or in the bulk of the cluster could be rather subjective as the variation in the torsion angles between the molecules within one molecular cluster is expected to be significant. For this reason, the torsion angle distribution for relaxed molecular clusters for both polymorphic forms and for different cluster sizes was examined.

The POLYPACK program allows the torsion angles of interest in each molecule within the cluster to be determined from which the variation in molecular conformation within the clusters can be assessed *via* calculation of a variance parameter (VAR):

$$\text{VAR} = \frac{\sum_{i=1}^{N}(T_i - T_{\text{crystal}})}{N-1} \quad (3)$$

where (T_{crystal}) is the torsion angle of a molecule packed in the crystal structure and (T_i) is a torsion angle of the ith molecule in the cluster, with N being the number of cluster molecules. The variance parameter was determined by summing up all the differences between the torsion angle of each cluster molecule and the torsion angle of a molecule in the bulk of the respective crystal structure. Hence, a larger variance parameter value reflects a greater variance in the torsion angle distribution, hence a higher degree of molecular structural disorder and, concomitantly a reduced crystallinity within the cluster.

The change of the cluster structure, with respect to the bulk crystal structure, was also assessed *via* calculation of an R_{wp} factor, *i.e.* as typically used in X-ray structure optimization, thus:

$$R_{\text{wp}} = \sum_{i=1}^{N} \frac{1}{N}\sqrt{(x_i - x_i^{\text{opt}})^2 + (y_i - y_i^{\text{opt}})^2 + (z_i - z_i^{\text{opt}})^2} \quad (4)$$

where (x_i, y_i, z_i) are the coordinate of an atom and N is the number of atoms in the system. The coordinates of the atoms in the optimized structure are marked with an 'opt' superscript. It is well understood, at least intuitively, that the change after optimization will not be homogeneous throughout the entire cluster structure, especially for larger cluster sizes where the molecular conformations for the surface and bulk molecules will be expected to be more sharply differentiated. The bulk of the structure will most probably behave differently from the surface layer and the latter, of course, will differ depending on the cluster size.

It was expected that the relaxation of the cluster structures would result in larger changes being manifested in small clusters and, in particular, in the surface layer of the clusters. The effect of cluster relaxation on the atomic positions for the benzophenone system was studied using a 'polyhedral wave method'. In this, the clusters were divided at equal intervals in terms of a radial distance from the geometric centre of the cluster, and for each layer the R_{wp} factor was calculated as

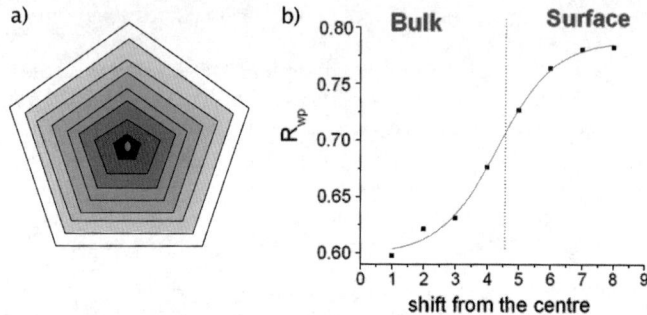

Fig. 5 Schematic representation of the polyhedral wave cluster scanning method adopted for assessing the structural change due to optimization. Smaller shifts from the center of the cluster corresponds to the darker area (bulk of the cluster) in the schematic 2D representation of the cluster whilst a larger shift locates a position further away from the center or at the surface cluster layer (represented in lighter shades (a). For each layer the R_{wp} value was calculated and plotted against the number of steps from the center of the cluster as shown schematically (b).

shown schematically in Fig. 5. The R_{wp} parameter was expected to be close to zero for structures which did not undergo a significant change in terms of the atomic positions when optimized. Hence, this parameter can be used as a measure of the crystallinity for the structure of the relaxed cluster and can be related to the ability of the system to crystallize, *i.e.* its ability to form critical crystalline nuclei with structures corresponding to the bulk crystal structure of the material. The results from such calculations could thus be related, for example, to the experimentally determined assessment of the nucleation barrier, *e.g. via* measurement of the metastable zone width (MSZW) preceding spontaneous crystallisation and thus providing information about the relative ease with which the system modelled is able to crystallise.

3. Results and discussion

3.1 Force field selection and intermolecular bond strength

The calculated lattice energies of the two forms of benzophenone, based on crystal structures, following optimization, are given in Table 1. These reveal, as expected, more stable lattice energy for the stable α-structure, hence the correct polymorph stability ordering for benzophenone. As expected for a van der Waals solid, the electrostatic contribution to the lattice energy was found to be very small *ca.* 3% of the lattice energy.

The values for the lattice energies obtained were consistent with experimentally determined lattice energy[5] of −23.93 kcal mol^{-1} demonstrating the broad applicability of the Scheraga force field to model the benzophenone crystal structures.

Examination of the eight strongest, individual intermolecular interactions, revealed that these account for 72.0 and 71.7% of the lattice energies for the stable α- (Table 2) and metastable β-forms (Table 3), respectively. The distribution of interactions differ somewhat between the two polymorphic structures being more

Table 1 Lattice energy and contributions to the lattice energy calculated using the Cerius^2 Open Force Field module and direct summation for both polymorphic forms of benzophenone, units kcal mol^{-1}

Polymorphic form	$E_{crystal}$	VdW	Elst
Orthorhombic α-form	−20.84	−20.20	−0.64
Monoclinic β-form	−20.50	−19.92	−0.58
Experimental	−23.93	—	—

Table 2 The top eight intermolecular bonds together with a breakdown into the respective energetic contributions for the stable α-polymorphic form. N is the multiplicity for the interaction, the distance (Å) is between the molecular centres of mass and the energy units are in kcal mol^{-1}

Bond type	N	Distance	Attractive	Repulsive	Coulombic	Total
a	2	5.32	−4.31	1.53	−0.05	−2.83
b	2	6.94	−2.27	0.69	−0.19	−1.77
c	2	7.01	−2.28	0.75	−0.05	−1.58
d	2	7.37	−1.90	0.60	−0.02	−1.33

Table 3 The top eight intermolecular bonds together with a breakdown into the respective energetic contributions for the metastable β-polymorphic form. N is the multiplicity for the interaction, the distance (Å) is between the molecular centres of mass and the energy units are in kcal mol^{-1}

Bond type	N	Distance	Attractive	Repulsive	Coulombic	Total
A	1	5.32	−4.87	1.53	−0.10	−3.45
B	2	5.86	−3.84	1.29	−0.09	−2.64
C	1	6.37	−3.12	1.11	−0.04	−1.97
D	1	7.35	−1.68	0.60	−0.15	−1.23
E	1	7.47	−1.38	0.45	−0.13	−1.06
F	2	9.06	−1.27	0.45	−0.03	−0.85

asymmetric for the metastable β-form which has a single strong A intermolecular bond, two strong B bonds and a weaker C bond compared to the two pairs (a and b) of strong bonds in the solid state structure of the stable α-form in the top four bonds. This pattern reflects the more asymmetric inter-molecular packing in the monoclinic unit cell of the metastable β-form compared to the orthorhombic unit cell of the stable α-form, and is reproduced in the next four intermolecular bonds (c and d; and D, E and F; for the two polymorphic forms respectively).

3.2 Morphological prediction

The predicted morphology of the stable α-form of benzophenone was found to be dominated by large {110} {011} and {101} faces and smaller {020} and {111} forms as shown in Fig. 6(a) with the observed crystal shape shown in Fig. 6(b). The predicted morphology of the metastable β-form of benzophenone is dominated by large {002} and {−111} forms and smaller {200}, {−202} and {111} crystal forms. The crystal morphology observed experimentally for the metastable α-form is shown in Fig. 6(e). The values for the associated calculated attachment energies are given in Table 4.

For both polymorphic forms, however, the match between the predicted and observed morphologies was not perfect. In the stable α-form the {101} and {011} forms were over-emphasized leading to suppression of {002}, and {021} forms. Similarly in the metastable β-form the {−111}, {002} and {200} forms were completely overestimated. The reason for these observations for the stable α-form is well-documented[5] reflecting differences in molecular-plane packings of the different habit surfaces. For the stable α-form the {110} and {020} faces are known to have very open packing structures, thus facilitating molecular conformational changes during crystal growth. In contrast the {011} surface is found to have much tighter binding sites. The latter provide much more energetically favourable inter-molecular interactions, which in turn, increases the surface attachment energy, hence decreasing the morphological importance. Detailed analysis for the metastable

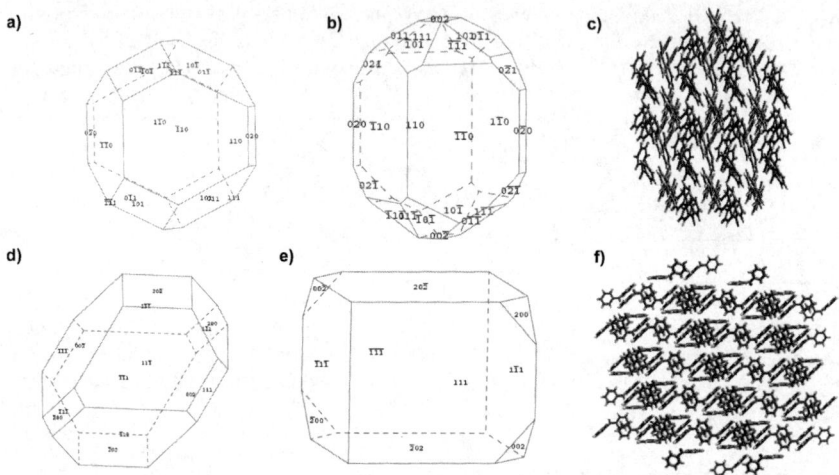

Fig. 6 Morphological predication of the stable α- (a) and metastable β- (d) forms of benzophenone carried out using the attachment energy model. The crystal shapes were modified to match the experimentally observed crystal morphology for the stable (b) and metastable (e) forms. Shaped molecular clusters were built in POLYPACK adopting experimentally observed crystal shapes for the stable (c) and metastable (f) forms.

β-form has not, as yet, been carried out, albeit the same broad features observed for the stable α-form seem to be replicated.

For example, examination of the surface chemistry of (−111) crystal face (Fig. 7) of the metastable β-form revealed that the phenyl moieties are aligned perpendicular to the growth surface in a close-packed surface structural arrangement. In the bulk structure, this orientation is expected to be stabilized *via* close van der Waals interactions with the phenyl rings of the neighbouring molecules. Hence, the surface molecules, in the absence of full intermolecular coordination, would be likely to relax presenting a greater opportunity for interatomic interaction and therefore, in a manner similar to the stable α-form, increasing the surface attachment energy and decreasing the morphological importance of the {−111} face of the metastable β-form.

This situation is closely analogous to the case of the related compound benzil $(C_6H_5C{=}O)_2$ where the morphology prediction revealed a predominance of an (0003) habit surface despite the fact that this face was not observed in the

Table 4 Normalised surface attachment energies, calculated using HABIT, for the habit faces of benzophenone with scaling factor for the (110) and (−202) crystal surfaces of the stable α- and metastable β-forms being 8.49 kcal mol^{-1} and 11.49 kcal mol^{-1}, respectively, together with adjusted centre-to-face distances to match the observed crystal shape. D-spacing (d) and face multiplicity (M) are also provided

α stable form					β metastable form				
Crystal face	M	d	Attachment energy	Observed distance	Crystal face	M	d	Attachment energy	Observed distance
110	4	7.84	1.00	1.00	−202	2	6.78	1.00	1.00
101	4	6.31	1.24	1.55	002	2	7.52	0.70	1.70
001	4	6.67	1.14	1.50	200	2	7.47	1.11	1.70
020	2	6.06	1.43	1.38	−111	4	6.98	0.89	1.90
111	8	5.95	1.36	1.46	111	4	6.04	1.16	1.17
021	4	4.82	1.46	1.48					
002	2	3.99	1.65	1.66					

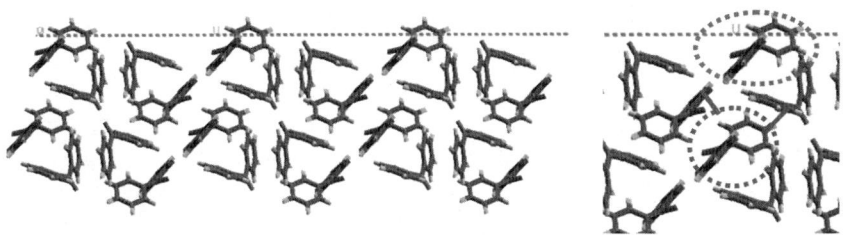

Fig. 7 Molecular packing of the (−111) crystal surface from the predicted morphology for the metastable form of benzophenone, which is not presented in the experimentally observed morphology. This crystal face reveals a close-packed molecular arrangement, which is stabilized in the bulk structure through close van der Waals interactions (right), however at the crystal surface this structure is expected to increase the number of atom–atom interactions through surface reconstruction, thus increasing the effective attachment energy and reducing the morphological importance of this face.

experimental morphology. Walker[9] carried out modelling studies of molecular relaxations on the (0003) surface of benzil to understand the effect of molecular conformation and these simulations, supported by experimental surface characterisation using NEXAFS spectroscopy,[25] were consistent with this being an effect of surface relaxation.

Hence, mindful of the above considerations, in this work we have chosen the observed crystal morphologies as a starting position for the building of the shaped molecular clusters as a function of their size.

3.3 Polymorphic cluster stabilities

Faceted molecular clusters, with shapes corresponding to the observed morphology for both polymorphic forms, were built in POLYPACK and are shown in Fig. 6(c) and (f) for the stable α- and metastable β-forms, respectively. The calculated cluster energies are plotted as a function of the cluster size (Fig. 8). These results reveal that at small cluster size, *ca.* 5 molecules, the metastable β-form has very similar cluster energies to those of the stable α-form (a), an effect also found for the minimized (b) and conformationally relaxed (c) cluster structures. These results clearly show that the energetic stabilities of the clusters for the two polymorphic forms of benzophenone, built as a function of cluster size, become very close as the size decreases (d). However, a well-resolved crossing point, where below a given cluster size the metastable β-form would perhaps become more stable than the stable α-form, was not observed, in contrast to our recent studies of L-glutamic acid.[13] This result can be rationalized, readily, with experimental observations as variation in solution cooling rates, and hence supersaturation, are well-known to effect changes when crystallizing polymorphs[26] and the metastable form of L-glutamic acid is known to have long-term stability under ambient conditions. In contrast, the metastable β-form of benzophenone is very difficult to prepare, requiring very low temperatures and a high degree of melt undercooling[3] and also readily transforms with little provocation into the stable α-form. Hence, the results from the modelling studies, which do not predict much enhanced stability for small cluster size for the metastable β-form with respect to the stable α-form of benzophenone, seem to bear out these experimental observations.

3.4 Molecular conformation analysis

The two torsion angles of interest are shown in Fig. 1(b) and the results of the conformational analysis are shown in Fig. 9. When the molecular and crystal structures were optimized one of the torsion angles increased, hence relieving the intramolecular clash due to hydrogen–hydrogen close contacts. The torsion angles of

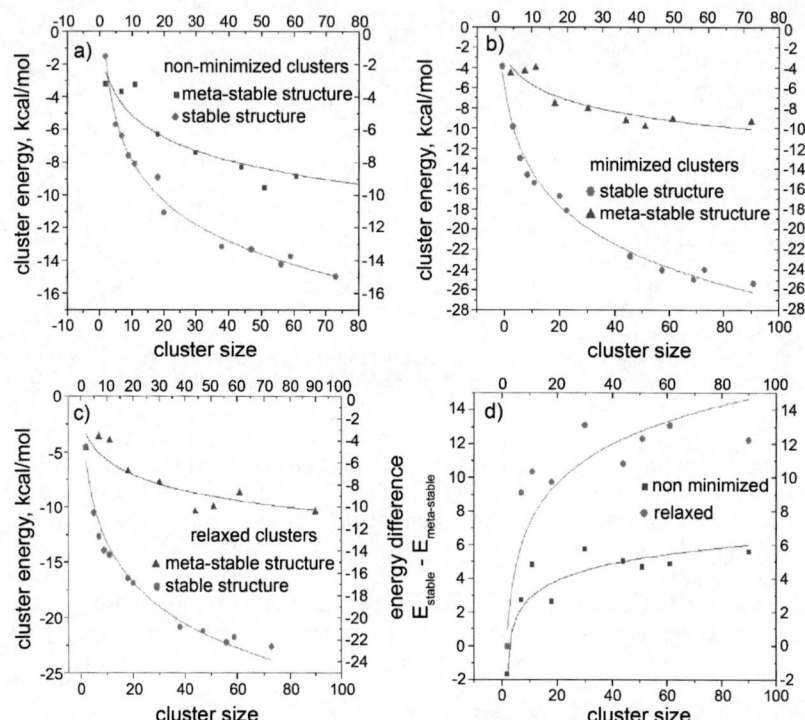

Fig. 8 Energetic polymorphic stability of different sized benzophenone molecular clusters: (a) non-minimized, (b) minimized and (c) relaxed, revealing that the cluster stability of the two forms is very close at small cluster size and increase when the cluster size becomes larger. Energy difference ($E_{stable} - E_{metastable}$) increases as a function of cluster size (d) and is larger for the relaxed clusters.

a molecule in the bulk of a cluster were found to be close to those of the optimized structure but still greater than those values. The torsion angles of the cluster's surface molecules showed a greater reduction in the molecular clash due to the lack

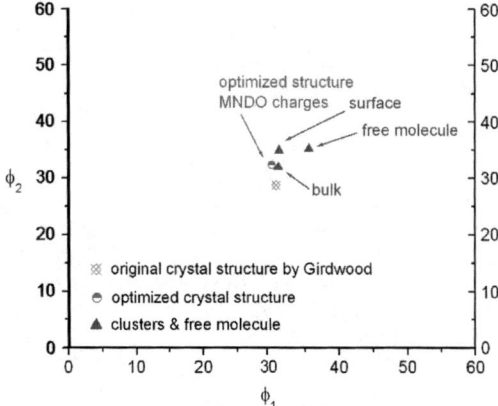

Fig. 9 Torsion map for a benzophenone molecule using the crystal structure by Girdwood[7] was built by measuring the torsion angles in Cerius[2]. The plot shows that the molecular clash due to close H–H contact is less for the freely optimized molecule and increases for the molecules at the cluster surface, in the bulk of the cluster and for those in the bulk crystal structure (b).

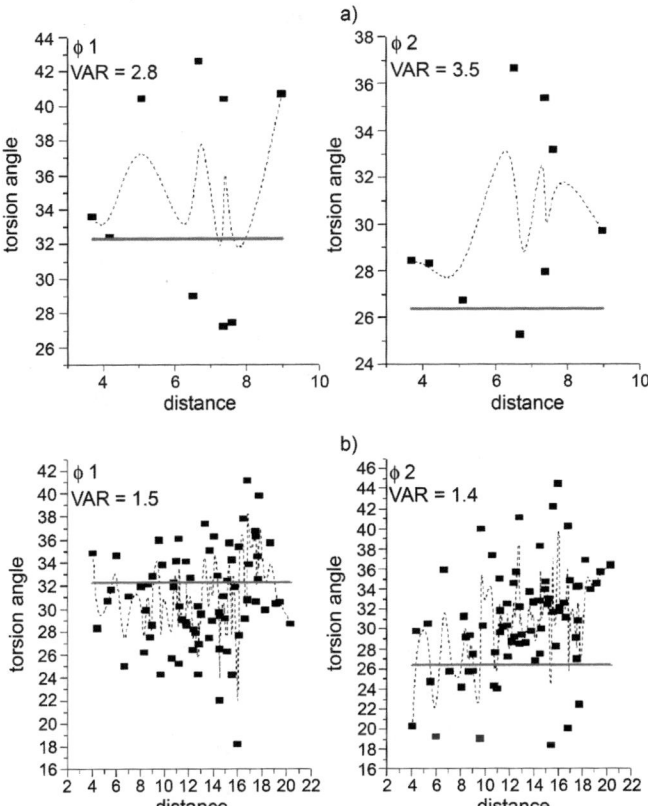

Fig. 10 Torsion angle distribution for the stable α-form of benzophenone revealing much more disordered distribution at smaller cluster size. When the cluster size becomes larger the torsions of the constituent molecules begin to cluster around the line representing the torsion angle of a molecule packed in the bulk crystal structure: (a) 9 molecules; (b) 73 molecules.

of neighbouring molecules and hence the absence of non-bonded contacts at the cluster surface. The torsion angles associated with a free molecule were found to be greater still, thus reflecting the total absence of an intermolecular clash compared to molecules belonging to the surface and the bulk regions of the molecular cluster.

The torsion angular distributions within two different cluster sizes, *ca.* 8 and 70 molecules, for the α- and β- polymorphic forms were compared using the variance parameter calculated using eqn (3). When the cluster size increased the variance parameter was found to decrease, as expected. Examination of the torsion angle distribution of the two forms of benzophenone, plotted in Fig. 10 and 11, revealed the cluster molecules derived from the stable α-form to be characterized by less disorder with respect to their torsion angles compared to those of the same size cluster derived for the metastable β-form. This difference was found to be much more significant at small cluster sizes, for example at similar cluster sizes (8 molecules for the metastable β-form and 9 molecules for the stable α-form) the variance parameters were found to be 12.5 and 2.8, respectively. From this, it can be concluded that the molecules in the metastable α-form clusters are more conformationally labile than those in the stable α-form. This higher range of conformational flexibility of the molecule in the metastable β-phase reflects the fact that most of the molecules within the cluster have torsion angles very far from those in the bulk crystal structure.

The results of torsion angle analysis revealed that the molecular conformational disorder for the two polymorphic forms showed the same trend with respect to the

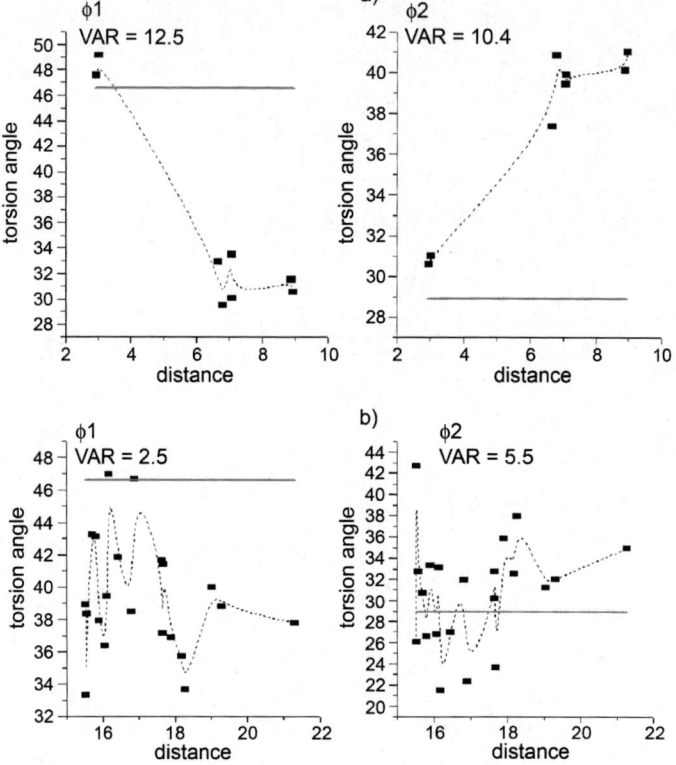

Fig. 11 Torsion angle distribution for the metastable β-form of benzophenone revealing much more disordered distribution at smaller cluster size. Both torsions of the metastable form molecules show much larger disorder compared to the stable one, especially when the cluster size is small: (a) 8 molecules; (b) 61 molecules.

two torsion angles. The greater conformational variation of the molecules in the metastable β-form clusters could possibly relate to the experimental observation[3] that this particular form is more difficult to prepare than the stable α-form and that the inherent molecular disorder present in the small cluster sizes representative of the on-set of the nucleation process perhaps forms a structural barrier restricting the molecular self-assembly needed for easy crystallization. Clearly further, and more extensive, work is needed to quantify these effects.

An R_{wp} parameter was calculated (using eqn (4)) for clusters of both polymorphic forms of benzophenone to illustrate the effect of cluster relaxation on the change in the atomic coordinates with respect to their original crystallographic positions. As an example, R_{wp} was calculated for two molecular clusters with 61 and 69 molecules, for the stable α- and metastable β-polymorphic forms, respectively. The resultant R_{wp} values were plotted as a function of a series of radial steps outward from the cluster center *via* the polyhedral wave method, described in section 2.3, to explore the variation in the molecular conformation throughout the cluster (Fig. 12). The results demonstrated that there was only slight variation in the amount of change in the atomic coordinates after relaxation throughout the molecular clusters for both polymorphic forms and, hence, both plots show very limited correlation with the molecular position within a given cluster. These results are not really that unexpected and are, perhaps, consistent with the lack of a hydrogen bonded network for this system, which in the absence of any surface hydrogen bond breaking, allows for a somewhat more homogeneous change in the atomic positions of the cluster

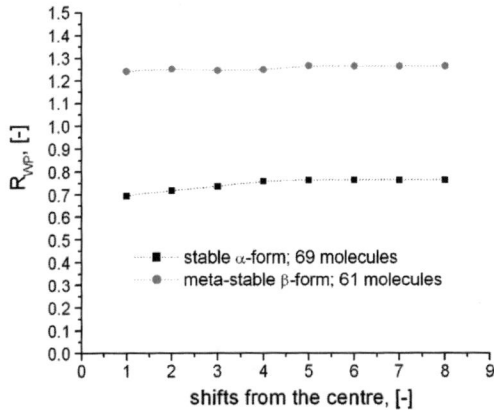

Fig. 12 R_{wp} parameters calculated for clusters of the stable α- and metastable β-polymorphic forms with 69 and 61 molecules, respectively. The change of the atomic coordinates throughout the molecular cluster upon relaxation is not significant for either polymorphic form. Metastable form clusters revealed higher values for R_{wp} compared to the stable form at similar cluster size.

modelled when comparing bulk and surface molecules, following cluster relaxation. Additionally, the magnitude of change of the atomic coordinates for the metastable β-form cluster were found to be significantly greater than those for the stable α-form cluster despite the fact that the size of these two clusters was chosen to be similar. This effect could possibly be due to the molecules being more conformationally labile for the metastable β-form when compared to the stable α-form. This result could also possibly be related to the much lower melting point[3] observed for the metastable β-form of 297–299 K compared to that of the stable α-structure of 321 K, *i.e.* a lower energy barrier to melting for the less stable and more conformationally disordered metastable β-polymorphic phase.

4. Conclusions

The work reported presents an integrated examination of the conformational variation in, and energetic stability of, small molecular clusters of the two polymorphic forms of benzophenone in terms of its nucleation and subsequent growth process. Calculated lattice energies were shown to be in good agreement with experimental values with the lattice energy breakdown, demonstrating that the van der Waals intermolecular interactions form the main contribution to the lattice energy (97%) and that the coulombic interactions play only a minor role. Morphological simulation for the crystal shapes of the two polymorphic forms revealed prismatically-shaped crystals for both forms with some deviation with respect to the observed morphologies, these reflecting the effects of surface conformational changes on the crystal habit planes. Analysis of the polymorphic stability as a function of the size of the shaped molecular cluster, built using the observed morphological shapes, revealed very similar energetic stability for the two polymorphic forms at small cluster sizes with the clusters of the stable α-phase possessing greater stability for clusters with sizes exceeding four molecules. These modelling observations were consistent with the lack of stability for the metastable β-form of benzophenone and the observation that it readily transforms into the stable α-form.

The intramolecular clash due to a close H–H contact in the benzophenone molecule was shown to be relieved, to some extent, by optimization of the molecules in the crystal structure, and further relieved for the bulk cluster molecules and the molecules at the surface of the molecular clusters. The torsion angle distribution for molecular clusters of different sizes revealed, as expected, that for both polymorphic forms there was more conformational disorder for the molecules in small compared to larger size clusters. It was also found that molecules of clusters for the metastable β-form had greater variation of the two torsion angles than those in clusters for the stable α-form. Examination of the R_{wp} parameter, used as a measure of the degree of change in the atomic coordinates of the cluster molecules, revealed the magnitude of the change in the atomic coordinates to be significantly greater for the metastable β-form. However, there was no significant variation in the extent of this change on going from the bulk to surface of the molecular clusters which could be due to the lack of a hydrogen-bonded network in the crystal structure of benzophenone.

Acknowledgements

We are grateful to the EPSRC for funding molecular modelling (GR/R/14491 and GR/R/19328) and experimental crystallization research programs (GR/L/43860) at Leeds. One of us (KP) gratefully acknowledges the UK's Overseas Research Students (ORS) Awards Scheme and Tetley and Lupton scholarship schemes and Institute of Particle Science and Engineering at University of Leeds for funding support. This work forms part of the PhD study of one of us (KP).[27]

References

1. B. Grossner, *Z. Kristallogr.*, 1904, **38**, 110.
2. M. J. Burger and M. C. Bloom, *Z. Kristallogr.*, 1937, **96**, 182.
3. H. Kutzke, H. Klapper, R. B. Hammond and K. J. Roberts, *Acta Crystallogr., Sect. B*, 2000, **B56**, 486.
4. G. Lobanova and E. Vul, *Sov. Phys. Crystallogr.*, 1967, **12**, 355.
5. K. J. Roberts, R. Docherty, P. Bennema and L. J. Jetten, *J. Phys. D: Appl. Phys.*, 1993, **26**, B7.
6. K. J. Roberts, J. N. Sherwood, C. S. Yoon and R. Docherty, *Chem. Mater.*, 1994, **6**, 1099.
7. S. Girdwood, *Investigation of Structural Change in Molecular Crystals using High Resolution X-Ray Scattering and Molecular Modelling*, PhD Thesis, University of Strathclyde, UK, 1998.
8. K. Aimi, T. Fujiwara and S. Ando, *J. Mol. Struct.*, 2002, **602–603**, 405.
9. E. Walker, *Modelling the Influence of Growth Environment on the Crystallisation of Organic Molecular Compounds*, PhD Thesis, University of Strathclyde, UK, 1997.
10. R. B. Hammond, K. Pencheva and K. J. Roberts, *Cryst. Growth Des.*, 2006, **6**, 1324.
11. R. B. Hammond, K. Pencheva and K. J. Roberts, *Cryst. Growth Des.*, 2007, in press.
12. R. B. Hammond, K. Pencheva, K. J. Roberts and A. Auffret, *J. Pharm. Sci.*, 2007, in press.
13. R. B. Hammond, K. Pencheva and K. J. Roberts, *J. Phys. Chem. B*, 2005, **109**, 19550.
14. M. Brunsteiner and S. Price, *Cryst. Growth Des.*, 2001, **1**, 447.
15. G. Clydesdale, K. J. Roberts and E. M. Walker, The crystal habit of molecular materials: a structural perspective, in *Molecular Solid State: Syntheses, Structure, Reactions, Applications, vol. 2, Theoretical Aspects and Computer Modelling*, ed. A. Gavezzotti, John Wiley & Sons, Inc., New York, 1996, ch. 7, p. 203.
16. G. Nemethy, M. Pottle and H. Scheraga, *J. Phys. Chem. B*, 1983, **87**, 1883.
17. G. Clydesdale, K. Roberts and R. Docherty, Computational studies of the morphology of molecular crystals through solid-state inter-molecular force calculations using the atom–atom method, in *Controlled Particle, Droplet and Bubble Formation*, ed. D. Wedlock, Butterworth Heinemann, London, 1993, p. 95.
18. MOPAC, version 6.0: Quantum Chemistry Program Exchange Program No. 455. Creative Arts Building 181, Indiana University, Bloomington, IN 47405.
19. G. Clydesdale, R. Docherty and K. J. Roberts, *Comput. Phys. Commun.*, 1991, **64**, 311.
20. R. Docherty, G. Clydesdale, K. J. Roberts and P. Bennema, *J. Phys. D: Appl. Phys.*, 1991, **24**, 89.
21. P. Hartman and P. Bennema, *J. Cryst. Growth*, 1980, **49**, 145.
22. E. Dowty, *Am. Mineral.*, 1980, **65**, 465.
23. A. George, K. Harris, A. Rohl and D. Gay, *J. Mater. Chem.*, 1995, **5**, 133.
24. D. Gay and A. Rohl, *J. Chem. Soc., Faraday Trans.*, 1995, **91**, 925.
25. G. P. Hastie, J. Johnstone, E. M. Walker and K. J. Roberts, *J. Chem. Soc., Perkin Trans. 2*, 1996, 2049.
26. M. Kitamura, *J. Cryst. Growth*, 2002, **237–239**, 2205.
27. K. Pencheva, *Modelling the Solid-State and Surface Properties of Organic Nano-sized Molecular Clusters*, PhD thesis, University of Leeds, 2006.

General Discussion

Professor Harding opened the discussion of Professor Addadi's paper: Are the acidic groups (and in particular the acidic proteins) completely ionised in these systems? Simulations show that the state of ionisation of organic molecules and arrays is of great importance in determining the strength of binding, the crystallisation of amorphous $CaCO_3$ and the morphology of crystals (in particular orientations of crystal growth in polar directions).

Professor Addadi replied: Nothing is known concerning the ionisation state of specific proteins during mineral deposition. However, as the pH is supposed to be basic during deposition of calcium carbonate, the acidic proteins should be completely deprotonated, and neutralized by calcium counterions

Professor Rodger asked: In the description of the nacreous layer you identified a region of highly acidic protein where the aragonite nucleates, and surrounded by a region of sulfate. Is it known what role the sulfate region plays in controlling the aragonite nucleation?

Professor Addadi answered: We don't know for sure. It has been known for a long time, since Crenshaw and Ristedt's experiment in 1976,[1] that there are sulfates at the center of the aragonite polygonal tablets, where the nucleation site was supposed to be located. Back in 1987 we proposed, with great presumption given the state of our knowledge at that time, that the sulfates may contribute to attract the calcium ions, and subsequently the carbonate ions to the nucleation site, where they are organized by the carboxylates into a crystal nucleus. It was suggested that the sulfates concentrate the calcium ions without trapping them in specific ion pairs. This is similar to what occurs in cartilage, where large amounts of ions are concentrated, but not bound, by proteoglycans, thus contributing to the osmotic pressure of cartilage. It appears now that this suggestion may have received further support from the recent experiments, but more surprises may be forthcoming....

1 M. A. Crenshaw and H. Ristedt, in *The Mechanisms of Mineralization in the Invertebrates and Plants*, ed. N. Watanabe and K. M. Wilbur, University of South Carolina Press, Columbia, 1976, p. 355.

Professor Rodger then asked: What is known about the relative sizes of the acidic and sulfate domains within the nacreous layers?

Professor Addadi replied: What we see by light and fluorescence microscopy is a core of carboxylates with a diameter of approximately 1 μm, surrounded by a ring of sulfates with an external diameter of approximately 5–6 μm and an internal diameter of approximately 3 μm.

Professor Caffrey said: I was interested to learn more about the hydrophobic proteins referred to in your talk. How many are there? If they are hydrophobic why do they leach out into water? Are there crystal or NMR 3D structures for any of the matrix proteins with which to rationalize your findings and explain how the proteins function? If not an X-ray structure, for example, then perhaps there are other biophysical studies of lower resolution, such as circular dichroism or infrared spectroscopy, that might be useful in interpreting your results?

Professor Addadi responded: Little is known about the hydrophobic proteins. Two have been sequenced: one has a sequence related to that of spider silk, while the other is gly-rich rather than ala-rich.[1] Their leaching out in water is probably related to the small molecular weight of the fragments. Otherwise, silk-fibroin-like proteins are known to assume at least partial β-sheet conformation, and to form in water hydrogels, which is in agreement with our findings.

1 S. Sudo, T. Fujikawa, T. Nagakura, T. Ohkubo, M. Sakaguchi, K. Tanaka, K. Nakashima and T. Takahasi, *Nature*, 1997, **387**, 563.

Dr Ristic remarked: We have recently done some experiments on crystallisation of calcium carbonate using *in situ* SAXS/WAXS at SRS Daresbury Laboratory. For this purpose, we would mix Na_2CO_3 and $Ca(NO_3)_2$ solutions at initial equimolar reactant concentrations in the range 0.2–0.8 M at 10 °C. Under these conditions the spontaneous precipitation of a white, 'gelatinous' (amorphous) matter would appear instantaneously after the mixing. This phase lasted for a short period of time (a few minutes); the length of that duration depends on the initial concentration of the reactants. Since WAXS could not detect the appearance of any crystalline phase in this interval of time, we examined the evolution of amorphous state using polarised light. After one or two minutes, a few crystalline "spots" would occur in the amorphous phase. Since the cross-polar technique is also limited to some extent in terms of its resolution power, I am just wondering whether small traces of calcite might have been formed, but they are undetectable with the applied technique. With this in mind, I have two questions:

(1) What technique did you use to make sure that you are dealing with a pure amorphous state?

(2) Your amorphous state is likely to be different from ours in terms of structure (polymorphism) and therefore stability, would you like to comment on this?

Professor Addadi replied: The techniques that we use to characterize biogenic and synthetic amorphous calcium carbonate (ACC) deposits are X-ray diffraction, electron diffraction, X-ray absorption spectroscopy, infrared and Raman spectroscopy, as well as polarized light microscopy. References to the use of the different techniques are reported in our paper. Each has its sensitivity and its limitations. We would never presume to declare a specimen 'purely amorphous', and also, because there are many types of ACC, I would be hard put to define what 'purely amorphous' ACC is. The transient ACC precipitate that you observe is certainly highly hydrated. It is probably similar to the precipitate described by Jens Rieger in his paper,[1] which I find gives a fascinating description of the evolution in time of such a synthetic ACC. Your assumption that there are various types of ACC is correct. Beyond the distinction between stable and transient ACC, each of the two classes includes various types of ACC with different 'structures': they may be hydrated or essentially anhydrous, and have different degrees of short-range order. Biogenic transient ACC forms were observed to have short-range order resembling the crystalline structure into which ACC eventually transforms

1 J. Rieger, T. Frechen, G. Cox, W. Heckmann, C. Schmidt and J. Thieme, *Faraday Discuss.*, 2007, **136**, DOI: 10.1039/b701450c.

Professor Kahr asked: Your stories today have evolved largely through microscopy. What are the frontiers of biomineral spectroscopy? What methods of analysis or capabilities not currently available would you wish for so as to further your understanding of these complex composites?

Professor Addadi replied: The stories I've told you today indeed evolved mainly around light, fluorescence and especially electron microscopy, but there are many other techniques that may be, and are, applied to the study of biological mineralisation. Among these are all kind of spectroscopic measurements, X-ray and electron diffraction, X-ray absorption spectroscopy (EXAFS, XANES and micro-XANES), infrared and Raman spectroscopy. What we are only now starting to investigate, which would advance our state of knowledge immensely, are time-resolved measurements, to provide a real time description of the evolution of the mineralized tissue from its inception to the mature product. Observing the final product of mineralisation we miss possible precursors or intermediates that have an overwhelming importance in the evolution of mineralization. Unfortunately, there are not many systems that allow such a follow up. There are some, such as the larval spicules of the sea urchin, larval mollusc shells, the continuously growing teeth of the sea urchin or the teeth in the mollusc radulae. In most of the mineralized tissues, however, the precursors coexist with already mature mineral, which inevitably masks them.

Professor Kahr then asked: What is the timescale that characterizes these dynamic processes? How much time between the deposition of two layers of, say, the nacre?

Professor Addadi responded: The exact timescale will change from species to species and in different seasons. In any case, the order of magnitude for the deposition of each layer is hours.

Professor Bensch asked: Is it not possible to use X-ray scattering data (powder patterns) and transform these data from the reciprocal space to real space to acquire more information about the second nearest, third nearest *etc.* neighbours around Ca^{2+} in the ACC phase ?

Professor Addadi answered: Until a short time ago, the technology available by X-ray was just not good enough to do this. Nowadays it is, and you are right, we should look into it.

Dr Gich asked: What is the relevance of the 3D structure of the organic matrix to the stabilisation of the different 3D polymorphs? In other words, what would happen if you could reproduce the same chemical environment provided by the proteins leading to calcite and aragonite but in a 2D layer?

Professor Addadi responded: There is no doubt that the complex 3D environment contributes as a whole to the stabilisation of the polymorph produced. Calcite is the stable polymorph of calcium carbonate at ambient temperature, and is thus formed as a default in all kinds of environments. Aragonite, however, is not stabilized unless a certain number of requirements are met. Thus, back in 1985 we performed experiments in which we studied how the acidic proteins extracted from nacre, adsorbed on glass or plastic, influence calcium carbonate crystallisation.[1] Oriented nucleation of calcite crystals was induced. In 1996 we performed experiments in which calcium carbonate crystallization was induced inside a 3D chitin framework impregnated with silk hydrogel, on which the acidic proteins were adsorbed, and then aragonite was formed.[2] If either the chitin, silk or proteins were absent, aragonite was not stabilized.

1 L. Addadi and S. Weiner, *Proc. Natl. Acad. Sci. U. S. A.*, 1985, **82**, 4110.
2 G. Falini, S. Albeck, S. Weiner and L. Addadi, *Science*, 1996, **271**, 67.

Professor Vlieg said: The chitin network in the prismatic layer plays a role during the crystallisation (as you showed). Does it also play a role in the mechanical properties? Did you indeed try to break the small calcite needles to test this?

Professor Addadi responded: We did not perform mechanical testing of single prisms. I have, however, no doubt that the chitin network reinforces the crystals against fracture. Calcite single crystals easily cleave along the {104} planes, which are known as 'cleavage planes' for this very reason. The chitin fibers present inside the crystal in a very dense meshwork will interfere with the propagation of cleavage. In the past we have shown that intracrystalline proteins intercalated inside calcite single crystals do have a similar crack-stopping activity, and subsequently reinforce the crystal against fracture. The chitin fibers are much thicker, and are probably active at a larger lengthscale, but I would expect them to have a similar effect.

Dr Rieger asked: You stressed the point that aragonite platelets are nucleated at specific spots on the chitin layers. What then is the role of the rest of the proteins on the chitin layers; how do they interfere with the growth of the aragonite platelets?

Professor Addadi replied: This is a very important question. I won't even try to convince you that we understand the task of each and every protein in this complex system. We do know something concerning the so-called silk-like proteins, which we suggest are in a hydrogel form before mineral deposition, and are squeezed by the growing mineral in the space between the mineral tablets and between the tablet and the chitin layers. These proteins have an inhibiting effect on calcium carbonate crystal nucleation. It is thus conceivable that they prevent aragonite nucleation in the bulk of the hydrogel and at other sites on the chitin, thus indirectly encouraging nucleation at the designated nucleation sites.

Professor Davey asked: In the shells that you describe there is clearly a place where the aragonitic nacre layers are potentially in contact with the calcitic columns. Why does the aragonite not transform to calcite?

Professor Addadi answered: Calcite and aragonite deposition occur at different times in the shell. The shells grow radially. The prismatic layer is deposited first, building the mineral at the external perimeter of the shell. Aragonite is deposited later in the internal side of the shell, coating the already deposited prismatic layer. Different cells of the mollusc, the so-called mantle cells, are active at different sites at the same time, depositing calcite and aragonite. The cells must enter a different program at the different sites, that controls the matrix protein expression and subsequently the deposition of calcite or aragonite following the specific mechanism of each of the two layers

Dr Hare communicated: It seems that even the prismatic layer does not comprise prisms in a mathematically rectilinear sense. At the resolution shown, there appears to be curvature along the length, and a variable cross-section with curvature instead of straight edges. Is a twist, or torsional effect, also to be seen? Would this suggest that even if a model is found (perhaps a set of differential equations after Mazzotti[1]) to generate a linear prism with hexagonal cross-section, it will need further refinement? What then does it mean "to understand" crystal growth in terms of a strategy? Do we think that Vlieg's "gap" (ref. 2, Fig. 2) is destined never to vanish through mathematical modelling alone? Or does use of the term "strategy" suggest that the modeller could be advised to seek heuristic solutions, possibly in terms of events, "CS-encoded" information and information processing in some way pre-programmed into the animate ensemble? In other words, are we really better off asking what the mollusc "knows" that we don't?

1   J. Schöll, C. Lindenberg, L. Vicum, J. Brozio and M. Mazzotti, *Faraday Discuss.*, 2007, **136**, DOI: 10.1039/b616285a.
2   E. Vlieg, M. Deij, D. Kaminski, H. Meekes and W. van Enckevort, *Faraday Discuss.*, 2007, **136**, DOI: 10.1039/b618566p.

Professor Addadi communicated in reply: I believe that the answer to your question is simpler than you expect. As we constantly find out studying biogenic crystal formation, the answers are not simpler or more complicated, they are simply completely different. It is only by entering a different frame of thought that one can place them within an appropriate framework. In this case, growth of the crystals in the prismatic layer proceeds similar to the construction of a bee-hive, *i.e.* each crystal, enclosed in its protein envelope, grows laterally until its growth is limited by the neighbouring prism. This process can and has been modelled mathematically. It simply cannot be modelled as an independent and unlimited growth of a crystal in solution.

Dr Hare communicated: Thank you for your answer, it's interesting. May I follow it up with another question? I am wondering if the biogenic growth depends on the habitat, or responds to a change in it, or does a species always "choose" (or migrate towards) a particular set of conditions that favours its crystal growth?

Professor Addadi communicated in reply: The growth of biogenic crystals does depend on the environment, but not directly as crystal growth from solution. In biologically controlled crystallization all the parameters of crystal growth are regulated by the cells of the specialized organ that produces the skeletal part. This said, environmental parameters such as temperature and impurity concentration do influence biogenic crystal growth. Thus skeletal growth is often seasonal, but this does not have a chemical reason, rather a biological reason. On the other hand, the concentration of ions present in the sea, such as magnesium, inside biogenic crystals may vary with temperature; this may be used as a reporter of the ambient conditions at the time when the crystals were formed. Also, impurities may end up occluded inside biogenic crystals. Occlusion of heavy metal impurities in sea urchin skeletons may be used, for example, as a measure of sea contamination.

Dr Hare then communicated: Might the biological imperative to grow a supportive skeleton or protective shell prove as strong a motivator as the need of an organism for food, or to avoid contamination, *e.g.* at the cellular or molecular level, could the animate ensemble – the specialist part – have an in-built rule that impels it towards a tolerable (or beneficial) calcium concentration range? This would be in contrast with the solution-crystal, which unlike the bio-crystal can have no protein and no encoding telling it to "seek (food, calcium)", and which in an otherwise similar environment might only be impelled by the resolution of conflicting enthalpic and entropic factors.

Dr Schön opened the discussion of Dr Boccaccini's paper: You have verified amorphicity of the samples using XRD. What size limit would you derive from this for your crystallites? Also, when you perform your *in situ* investigations of transformations during heating using ESEM, what is your time resolution? Did you also perform FTIR measurements *in situ*?

Dr Boccaccini replied: The XRD samples were powders prepared by crushing and grinding the porous scaffolds. We consider that the material is XRD amorphous, but we indeed have to say that some crystallites with a size of less than a dozen nanometers could be present, and would not be detected by XRD. Given the relatively slow heating rate, the resolution of the *in situ* ESEM measurements is in fact related to the temperature resolution. The resolution of the heating stage is of the order of 5 °C. This depends on the quality of the contact between the glass particles and the heating stage, which is difficult to quantify. A resolution of 10 °C is reasonable. *In situ* FTIR was not performed.

Dr Ristic asked: Could you describe a thermodynamic pathway of your system which transforms from a crystalline to amorphous state?

Dr Boccaccini responded: This is an energy decreasing pathway. It does seem to conflict with our knowledge that a crystalline state is a lower energy state than its amorphous counterpart of the same composition material. The truth is that the present system is an open one, with an ion exchange mechanism between the system and the surrounding environment being active. Hence the chemical composition of the crystalline phase is very different from that of the amorphous state.

Professor Davey commented: In considering the apparent stability change of the glass you must remember that on immersion into the simulated body fluid (SBF) there is a compositional change.

Dr Ristic said: I agree with Professor Davey's remark, but I expected to see the feasibility of the overall process characterised by the total change of entropy provided that its components could be estimated for all three transformations described in Fig. 14.

Mr Comer asked: Is there a parallel from Nature? Has Nature created materials with holes in, which then get used as support structures for cellular material? It would be a helpful remark in the paper to give examples of naturally occurring substances with this property.

Dr Boccaccini responded: Cancellous bone exhibits a porous structure which we are trying to replicate with our scaffolds.

Professor Addadi asked: Cell activity has an overwhelming importance in the fate of implant materials in the body. Cell activity may cause processes very different from what you see in simulated body fluid. Does the degradation process follow the same chain of events in the presence of cells as it does in simulated body fluid?

Dr Boccaccini replied: Yes. The present scaffolds have been assessed by cell culture *in vitro* (osteoblast-like cells), and XRD analysis showed similar degradation kinetics of the scaffold. Those studies are part of another investigation recently accepted for publication.[1]

1 Q. Z. Chen, A. Efthymiou, V. Salih and A. R. Boccaccini, *J. Biomed. Mater. Res. A*, 2007, DOI: 10.1002/jbmr.a.31512.

Professor Vlieg opened the discussion of Dr Tilocca's paper: A question from a layman's perspective: how sensitive are the results to the specific choice of the potential? In other words, how large are the error bars due to uncertainties in the potential?

Dr Tilocca replied: In general, for these systems the error bars in calculated short- and medium-range structural properties, related to the choice of the potential, are rather small. We have previously found that, compared to rigid-ion potentials, the approximate inclusion of polarization effects in our simulations through a shell-model approach slightly improves the description of the intertetrahedral structure and the local environment surrounding modifier Na and Ca cations, as well as the Q^n distribution of modified silicate glasses, for which our shell-model potential provides a better match with the available experimental data.[1]

1 A. Tilocca, N. de Leeuw and A. N. Cormack, *Phys. Rev. B*, 2006, **73**, 104209.

Dr Schön asked: The experimental density, to which the system is being relaxed, might not be the density one would get if the system is allowed to relax at constant pressure until the end. How would this change the quantities of interest? Similarly, the choice of potential and, in particular, the route of how one reaches the amorphous phase can have strong effects on *e.g.* angular distribution functions or ring-size distributions of the final compound (see *e.g.* ref. 1). Have you checked the robustness of your results in this regard? Amorphous systems exhibit aging phenomena during constant temperature simulations (see *e.g.* ref. 2), and a dependence of the results on cooling rates. Have you looked at how the Q-distribution changes during the simulations as a function of time? Finally, there is the question of when one would call two atoms neighbors in a glass. What definition have you employed?

1 A. Hannemann, J. C. Schön, M. Jansen, H. Putz and T. Lengauer, *Phys. Rev. B*, 2004, **70**, 144201.
2 A. Hannemann, J. C. Schön, M. Jansen and P. Sibani, *J. Phys. Chem. B*, 2005, **109**, 11770.

Dr Tilocca answered:
(a) In our simulations, a full constant-pressure (NPT) cooling run leads to a final density approximately 5% lower than the experimental one.[1] A comparison between the full-NPT structures and the structures obtained using the experimental density did not show significant differences; therefore in our simulations we set the glass density to the experimental value by gradually adjusting the cell volume during the cooling phase.

(b) We have not deemed it necessary to check in detail how the results are affected by the computational procedure used to make the glass, since many previous studies have shown that the structure of melt-derived silicate glasses is adequately reproduced with the standard melt-and-cool by molecular dynamics used in our paper. In particular, earlier work by Kob *et al.*[2] examined the cooling-rate dependence of a number of structural and dynamical properties of amorphous silica, showing that full convergence in properties such as radial, angular and ring size distributions is reached using cooling rates less or equal to the 10 K ps^{-1} value used in our work.

(c) Two atoms are considered neighbours when their distance is shorter than the cut-off distance corresponding to the first minimum in the relevant radial distribution function.

1 A. Tilocca, A. N. Cormack and N. de Leeuw, *Chem. Mater.*, 2007, **19**, 95.
2 K. Vollmayr, W. Kob and K. Binder, *Phys. Rev. B*, 1996, **54**, 15808.

Professor Catlow asked: Could you comment further on the relationship between chain length and stability? Are longer chains less strongly bound to the matrix?

Dr Tilocca answered: The higher bioactivity of BG45 derives from a larger number of both short and long chains, compared to the other glasses examined in our work.
The release of a long, say 6-membered chain, has a stronger impact on the glass dissolution than the release of a shorter trimer, for instance, but at the same time we observe a much larger fraction of shorter chains (Fig.1 in our paper).
Based on the observation that the fraction of Q^1-terminated chains does not seem to depend on the chain length (see Fig.1), we can estimate that the energetic cost needed to set a 6-membered chain free will be half the energy needed to liberate the equivalent *two* trimers; in other words, if the overall effect of the chain release on the partial dissolution of the glass network is taken into account, then releasing longer chains has a more favorable energetic balance.
On the other hand, this effect will be at least partially counterbalanced by the slower migration of longer fragments towards the surface, compared to the shorter chains.

Professor Heyes asked: I was wondering if you think the chain *vs.* ring balance would be the same near the surface as in the bulk? The bioactivity will presumably be sensitive to the structure near the surface. Have you explored this issue?

Dr Tilocca replied: It is certainly possible that the surface of these materials shows somewhat different features with respect to the bulk. For instance, the fraction of highly strained trisiloxane rings, which are often considered active sites involved in the bioactive mechanism,[1] could be higher on the surface than in the bulk. Since the bioactive process starts there, directly modelling the surface represents another important step towards a better understanding of the bioactivity of these materials. The specific bulk structural properties which we discuss in the paper, namely chain and ring distributions, provide a deeper insight into the way in which compositional changes are translated into different bioactivity. These bulk effects, related to the connectivity of the silicate network, are quite general and their interplay with the bioactivity can be discussed independently from the surface properties.[2]

1 L. L. Hench and J. K. West, *Annu. Rev. Mater. Sci.* 1995, **25**, 37.
2 Z. Strnad, *Biomaterials* 1992, **13**, 317.

Professor Kahr opened a general discussion, addressing Dr Tilocca: I am having trouble reconciling why rings should be more highly cross-linked than chains and therefore more difficult to liberate. Is this a consequence of conformational entropy contributions to the free energy? You should be able to calculate the conformational entropy straightaway from discrete structures. Can these numbers be related to experimental solubilities of model compounds?

Dr Tilocca responded: In our analysis a chain fragment is a sequence of several Q^n Si or P atoms with $n = 2$, or 1 for some of the chain terminations. On the other hand, rings can be formed by Q^n Si/P atoms with n equal to or greater than 2.

In other words, rings can be at least partially incorporated in the glass network, whereas chains cannot. Therefore the occurrence of chain fragments as in BG45 greatly enhances the glass solubility, due to the low energetic cost associated with their release. This is essentially an enthalpic effect: in this case, entropy contributes much less to the Gibbs free energy than enthalpy, since dissolution of these materials requires breaking bonds to liberate silicate or phosphate fragments.

Professor Anderson addressed Dr Boccaccini and Dr Tilocca: Your systems would be particularly amenable to study by solid-state NMR. This could give quantitative information about degrees of crystallinity as well as the specific chemical environment. In particular ^{29}Si, ^{31}P, ^{23}Na and ^{17}O could all be used to good effect. Has this been done?

Dr Tilocca replied: ^{29}Si, ^{31}P and ^{23}Na NMR have been used to characterize the chemical environment of these nuclei in bioactive silicate glasses.[1,2] The partial overlap between the broad bands corresponding to different Q^n species tends to complicate the analysis of these signals: for instance, it is hard to quantitatively identify Q^1 species. The main information obtained from ^{29}Si NMR is that compositions close to the 45S5 Bioglass have a predominantly Q^2 structure, whereas the Q^3 fraction increases with increasing silica content, so that glasses containing around 60% SiO_2 are predominantly Q^3, in agreement with our simulations. Another important feature evidenced by NMR is the preferential association of low-n Q sites with Ca ions, which then displace Na ions towards high-n Q sites, as also found in our previous simulations.[3]

1 M. W. G. Lockyer, D. Holland and R. Dupree, *J. Non-Cryst. Solids*, 1995, **188**, 207.
2 I. Elgayar, A. E. Aliev, A. R. Boccaccini and R. G. Hill, *J. Non-Cryst. Solids*, 2005, **351**, 173.
3 A. Tilocca, A. N. Cormack and N. de Leeuw, *Chem. Mater.*, 2007, **19**, 95.

Dr Boccaccini answered: Solid state NMR was not used in this study. However, it is a very good idea. Indeed, NMR should provide additional information to the environment of the different atoms. NMR has been used many times in this context, for example to study crystallisation in $MgO–CaO–SiO_2–P_2O_5$ glass-ceramics.[1]

We should not forget that many other factors have a profound implication on bioactive behaviour of silicate glasses, for example the conditions at the surface, thus the results of the MD simulations should be put in perspective of the experimental capabilities.

1 H.-L. Ren, Y. Yue, C.-H. Ye, L.-P. Guo, J.-H. Lei, *Chem. Phys. Lett.*, 1998, **292**(3), 317–322.

Professor Mazzotti opened the discussion of Professor Vlieg's paper: We have carried out simulations and experiments about the effect of solvent and temperature on the shape of crystals of glycopyrronium bromide grown from solution (the work is the Masters thesis of Lorenzo Codan, an ETH Zurich student, which has been carried out in co-operation with Novartis Pharma, Basel, Switzerland, under the supervision of Dr Gerhard Muhrer). The model used to predict the steady state shape of the crystals is that proposed and applied by Michael Doherty (see Zhang *et al.*, and references therein[1]), which is based on a shape evolution model, where predicted correlations for the relative growth rate of different faces are plugged in. Such relative growth rates are estimated based on molecular properties of the crystal and of the solvent; three different types of growth mechanisms are accounted for (screw dislocation, 2D nucleation, and rough growth). Simulations (single particle) and experiments (ensemble of particles) have been carried out for the seeded growth of glycopyrronium bromide (i) from methanol at 25 °C, (ii) from 1-propanol at 25 °C, and (iii) from 1-propanol at 80 °C. Predicted shapes are very similar and prismatic for the first two cases, whereas platelets are predicted for the third case. Experimental results assessed through SEM images of the final particles are encouragingly consistent with the theory.

1 Y. Zhang, J. P. Sizemore and M. F. Doherty, *AIChE J.*, 2006, **52**, 1906–1915

Professor Vlieg remarked: These results certainly look interesting. If I understand you correctly, you use a kinetic Wulff plot to derive the growth habit. The important issue is how you derive your growth rates. Are they obtained from simulations?

Professor Mazzotti answered: The relative growth rates were calculated using the model developed by Wynn and Doherty.[1] The model is based only on the properties of the pure components involved in the crystallization process, molecular simulations are not needed. The shape is obtained as a function of time during growth by integrating a set of ordinary differential equations. The final steady state shape (at given constant supersaturation) corresponds to that obtained by applying the Wulff construction.

1 D. Winn and M. F. Doherty, *AIChE J.*, 1998, **44**, 2501–2514.

Professor Vlieg said: Attachment energy is indeed one measure for finding the crystal habit, but in our group we are moving away from this because we think the step energy method that I briefly described is more accurate.

Professor Mazzotti replied: The model used in this work accounts for the known and accepted growth mechanisms (screw dislocation, 2D-nucleation and rough growth mechanism) and for the effects on the crystal morphology caused by solvents. The model is based on similar properties as those proposed by Bennema and co-workers.[1] The application of the model requires the knowledge of the properties of the pure components involved in the crystallization process only, namely the crystal structure and the internal energy, the latter obtainable from attachment energy simulations, and the surface free energy of the solvent. Interfacial properties are estimated using classical approaches,[2] whereas simulations or experiments are not necessary.

1   X. Y. Liu, E. S. Boek, W. J. Briels and P. Bennema, *Nature*, 1995, **374**, 342–345; X. Y. Liu and P. Bennema, in *Crystal Growth of Organic Materials*, ed. A.S. Myerson, D.A. Green and P. Meenan, ACS, Washington, 1996.
2   L. A. Girifalco and R. J. Good, *J. Phys. Chem.*, 1957, **61**, 904–909.

Professor Bensch asked: Can you distinguish between the surface of the crystal and the molecules just above the surface with the experimental setup of the scattering experiments? This is a question of the penetration depth of the X-rays. In addition, can you exclude the possibility that the high intensity of the synchroton radiation may disturb the system?

Professor Vlieg answered: By selecting the reflections we measure in reciprocal space, we can indeed distinguish the crystalline from the liquid part at the solid–liquid interface. This requires measuring a full set of crystal-truncation rods, including the specular rod that is most sensitive to the liquid. This selection in reciprocal space is a form of Fourier filtering and therefore the penetration depth of the X-rays is only of secondary importance. We do optimize the penetration depth in order to have an optimal signal-to-background ratio. Radiation damage is something that one always needs to consider. The KDP system described in my paper turns out to be very insensitive to the X-ray beam. We check this by regularly measuring the intensity of a reference reflection. We have, however, also encountered with systems that are radiation sensitive. Brushite was our most extreme system in this respect: in the full beam the intensity of the surface-sensitive reflections decayed in 20 s. Only by attenuating the beam and by having the surface in a humid environment were we able to obtain reliable data in that case.

Dr Ristic commented: It is known that the formation of structural defects such as dislocations and inclusions is caused by surface morphological instabilities and that these defects are quite detrimental to the quality of a crystalline material. In Fig 2 of your paper it says: "Experimental methods always deal with real systems,..." My question is: How much of the crystal growth reality is missing by not taking into account this important and unavoidable phenomenon?

Professor Vlieg replied: Including defects is certainly crucial in order to understand crystal growth. After all, a very common growth mechanism is spiral growth, caused by screw dislocations. In an experiment, such defects are automatically included, but at the moment there is no (simple) theory to predict whether such defects will occur. Some theories for morphology prediction take the growth mechanism into account. In our group we do this, for example in the Monte Carlo simulation programme Monty.[1]

1   S. X. M. Boerrigter, G. P. H. Josten, J. van de Streek, F. F. A. Hollander, J. Los, H. M. Cuppen, P. Bennema and H. Meekes, *J. Phys. Chem. A*, 2004, **180**, 5894.

Professor Anderson said: You mention in your paper that AFM is delivering a much better understanding of crystal growth. Is it possible in your calculations to simulate both crystal habit and surface topology in one self-consistent calculation? In such a manner you will increase the amount of experimental input with which to match your theory.

Professor Vlieg replied: We use different approaches for predicting the crystal habit. For several (industrial) applications, the main goal is finding the macroscopic shape and then the details of the surface topology are of little interest. This is the philosophy of the Hartman–Perdok theory and also of the new approach based on steps that I described in my paper. Different approaches, however, are possible. The work of Julian Gale that we have already mentioned is an excellent example.[1] We also have a kinetic Monte Carlo simulation programme ("Monty") that generates the surface topology. While there are thus promising developments, there is, in my opinion, still plenty of work to do to get all the details right that one can observe using AFM.

1   S. Piana, M. Reyhani and J. D. Gale, *Nature*, 2005, **438**, 70.

Professor Anderson commented: The paper by Julian Gale alluded to seems to provide an excellent approach to the calculation of both crystal morphology and surface topology. In that paper on MD, calculation is used to determine fundamental growth rates which are then fed into a kinetic Monte Carlo simulation.

Professor Vlieg replied: I fully agree with that. This is the type of simulations that can really help to understand what is happening during growth. At the same time, there is room for further connection with experimental data; it would, for example, be very interesting to see how the MD simulations agree with the structural information on the solid–liquid interface that X-ray diffraction can provide. For urea results of this type have not (yet) been obtained.

Professor Heyes said: What timescale were you contemplating in the context of using simulation (molecular) to understand the crystal growth process? Even with today's computers, the accessible time window is still less than, or of order, a few nanoseconds. Therefore, the experimental/model system would need to be chosen carefully for significant growth to be observed.

Professor Vlieg responded: In the first instance I was thinking of simulations that can couple to the type of results from X-ray diffraction that I described, *i.e.* the static, average structure of a solid–liquid interface. In that case no specific time dependence is involved and the simulations have to run long enough to obtain these structural details. Once a (reasonable) agreement between simulations and experiment is established, it becomes of course interesting to consider the dynamics of the interface. It will be highly relevant to observe the various steps that an atom/ion/molecule needs to take in order to get incorporated into the crystal. Experiments are very difficult, and thus reliable simulations can provide valuable insights. A little further into the future, there are prospects of doing time-resolved X-ray diffraction experiments using the X-ray free electron laser (XFEL) that will be constructed in Hamburg and that will enable experiments with a time resolution of 100 fs.

Professor Rodger opened the discussion of Dr Hughes' paper: The difference in the α/γ phase stability on deuteration of glycine raises very interesting challenges for the modelling described in your paper. Is there any evidence about what interactions change on deuteration to effect this change in phase? How is it proposed to incorporate these effects into the modelling, since most modelling methods (including *ab initio* MD) use classical dynamics and hence give no isotopic differences in structural properties?

Professor Roberts added: It was interesting to note the effects of deuteration on the formation of the various polymorphs of α-glycine. Interestingly deuteration of compounds can change phase transformation behaviour of materials, *e.g.* deuteration of the tetragonal hydrogen-bonded solid ADP can give rise to stabilisation of the monoclinic form. The reason for this is not really clear as the inter-atomic interactions should not really change beyond an adjustment in the bond length and associated inter-atomic vibration model. Thus it is not clear at least to me why these effects should be seen in MD simulations.

Dr Hughes responded: Classical MD may indeed find it difficult to simulate these effects as it can only model factors arising from the change in mass and average bond lengths and not those which depend explicitly on quantum effects.

Professor Bensch asked: Is it not an oversimplification to assume that only crystals with a single size are present? Do you agree that it is more realistic that particles with different sizes and shapes coexist?

Dr Hughes replied: It would be very much an oversimplification to make such an assumption. We have fitted the data to models which use single sizes but we do so purely in order to establish a rough estimate of the range of sizes present. As we state in the paper (Section 4.2), "It is expected that, under the conditions of the experiments reported here, the molecular aggregates present during the early stages of the crystallization process should be described by relatively broad distributions of both size and shape, and moreover that the nature of these distributions may also be time-dependent."

Dr Schön asked a number of questions: In the experiment, the critical size of clusters of glycine is greater than what value? How does this compare with the sizes of the simulated clusters? How do the results of the simulations depend on the starting configuration? What is the actual stability of the oligomers ($n = 2, 3,...$) during the simulations? Can you derive lifetimes of the clusters as function of n and temperature? In particular, what is the lifetime of the suspected "building blocks" of this particular modification, the dimers? Could one use the results of your simulations to construct a kinetic Monte Carlo model of the association and chemical reaction of the oligomers until clusters of the critical size have been reached?

Dr Hughes replied: Analysis of lifetimes, stabilities and sizes of clusters observed in the MD simulations is in progress and will be reported subsequently. The simulations could in principle be used to construct a kinetic Monte Carlo analysis, provided the progress of aggregation and dissociation can be expressed in terms of a kinetic scheme involving a number of well defined discrete steps.

Dr Schön then asked: You did not mention the nucleation aspect, just the growth phase. Surely, modelling/measuring nucleation is still an open question?

Dr Hamad responded: Preliminary results of a simulation of an α-glycine nanoparticle in water, show that the particle (formed by 800 glycine molecules) undergoes a large structural reconstruction, but does not dissolve. Therefore, we have an upper estimate of the cluster size that we need to study in order to observe nucleation processes. We are currently simulating larger simulation cells, with more glycine molecules, aiming to observe the formation of the first long-lived clusters.

Professor Catlow commented: Kinetic processes can be effectively modelled using kinetic Monte Carlo (KMC) techniques, provided data on the relevant rate processes for elementary steps can be obtained from modelling or experiment.

Dr Schön responded: I agree with Professor Catlow's comment, in principle. However, the tricky part is often to identify the 'elementary' steps and/or the relevant rate processes on the longer timescales where it is possible to actually model the full process. If we are stuck with (easily identifiable) elementary steps that take place on the below-picosecond timescale, we need extremely long simulations possibly negating the advantage of KMC. In order to address this issue, we might want to proceed via separation of timescale approaches (*e.g.* as we have done in modelling the sol–gel synthesis of $Si_3B_3N_7$, see ref. 1), where the modelling leads to different elementary steps on different timescales.

1 J. C. Schön, A. Hannemann and M. Jansen, *J. Phys. Chem. B*, 2004, **108**, 2210; A. Hannemann, J. C. Schön and M. Jansen, *J. Mater. Chem.*, 2005, **15**, 1167.

Professor Rodger asked: During the discussion it was mentioned that you had calculated hydrogen bond lifetimes to be very short, typically less than 1 ps (though I can't find any mention of lifetimes in the paper, so this may not be such a relevant comment). It is well known that H-bond lifetimes are difficult to calculate, largely because the H-bond geometry is subject to large amplitude vibrations, and it is often difficult to determine when a bond has broken, and when it is simply and excited (but stable) vibration. Two methods for dealing with this can be found in Astley *et al.* and Luzar and Chandler.[1,2]

1 T. Astley, G. G. Birch, M. G. B. Drew and P. M. Rodger, *J. Phys. Chem. A*, 1999, **103**, 5080.
2 A. Luzar and D. Chandler, *Nature*, 1996, **379**, 6560.

Dr Hughes responded: We did not mention lifetimes in the paper but we have since begun to investigate the H-bond lifetimes employing the correlation function used by Luzar and Chandler. Our preliminary results indicate that this leads to apparent lifetimes on the order of 100 to 200 ps.

Professor Catlow commented: The most important result of this paper is that the solution chemistry is not dominated by dimers. This does not, of course, mean that dimers do not play a significant role in crystal growth.

Professor Davey asked: β-glycine is relatively easy to make. Why can't it be made in Cardiff?

Dr Vonk commented: β-glycine can be made using water–ethanol mixtures.

Dr Hughes responded: Various papers cite this method for producing β-glycine but others (*e.g.*, ref. 1 and 2) point out that this method often gives a mixture of α- and β-glycine. These papers give another preparation method but admit that this method is "unstable", sometimes producing the γ polymorph.

1 E. V. Boldyreva, V. A. Drebushchak, T. N. Drebushchak, I. E. Paukov, Y. A. Kovalevskaya and E. S. Shutova, *J. Therm. Anal. Calorim.*, 2003, **73**, 409.
2 V. A. Drebushchak, E. V. Boldyreva, T. N. Drebushchak and E. S. Shutova, *J. Cryst. Growth*, 2002, **241**, 266.

Dr Hare communicated: How is the effect of the deuterium atom on crystal density explained (*c.f.* Table 1)?

Dr Hughes communicated in reply: We do not currently have a clear explanation for this observation. We believe, however, that the contrasting manner in which crystal density changes upon deuteration for the α and γ polymorphs may be a key issue underlying the isotope effects reported in this paper. Obtaining a fundamental understanding of this issue is currently the focus of our ongoing investigations in this area.

Professor Rodger communicated: The data presented in Dr Hughes' paper on H/D effects on the stability of different phases of glycine raises some fundamental questions about isotope effects and the way they should be considered both in interpreting experiment and in modelling. In interpreting neutron scattering data it is usually assumed that isotopic substitutions merely change the relative scattering intensity of the different atoms, but has not intrinsic effect on structure. The shift in thermodynamic stability of the α and γ phase of glycine on deuteration is dramatic example that this is not always the case, yet it is not unique: the phase diagrams of H_2O and D_2O are not the same, and hence one must expect differences in the liquid structure to arise—at least close to freezing—merely because of the difference in isotopes. Similarly, any modelling based on classical equations of motion must give no differences between the equilibrium thermodynamic properties of different isotopes. How important are the differences between H and D in both experimental and theoretical studies? Do we need to include quantum particles into our modelling, or can the H/D effects be adequately described either through quantum perturbations to classical simulations, or by developing effective potentials that are different for H and D?

Professor Kuroda opened the discussion of Professor Roberts' paper: Crystal of benzophenone has been known to exhibit strange polymorphic behaviour; the transformation of the metastable β-form into the stable α-phase occurs very easily, sometime spontaneously, by the application of mechanical load and/or through contact with the stable α-phase. In this paper, the authors calculated that the relative energetic stability of clusters represented for the two forms is substantially different at larger cluster size but becomes very similar at very small cluster size. They also showed conformational disorder depending on the crystal habit surfaces as compared with the bulk. Based on these findings, can the authors explain the unusual polymorphic transformation of the compound? We are interested in the fact that the metastable β-form is achiral whereas the stable α-form is chiral. Benzophenone is a non-chiral molecule. Did the polymorphic transition proceed to produce enantiopure crystals or conglomerates?

Professor Roberts replied: This is a most interesting comment concerning why the two structures have different structural chirality (note the molecule is not chiral) and I am not sure that the reasons for this are all that clear. However, the transformation is monotropic and first order involving the complete destruction of the crystals and so the potential for conglomerate formation is remote given the unstable energetics of the metastable form.

Dr Murray asked: Should we consider very small clusters as perfect crystals? Might relaxation at the surface be important?

Professor Roberts answered: Yes, indeed and particularly for conformational flexible molecules such as benzophenone where in this study we show that there are significant differences between the surface energetics for clusters built from the bulk crystallographic structure and the same following minimisation of the molecular positions and following relaxation of the molecular torsions angles.

Dr Schön asked: Concerning Fig. 8, what are the critical cluster sizes for the system under the various experimental conditions, and how do they fit into this figure?

Professor Roberts replied: Detailed nucleation kinetic studies for benzophenone have not, to my knowledge, been carried out and thus the relation between melt undercooling and critical cluster size is now known. What we know from our modelling studies is that the relative stabilities of the two polymorphic forms are very close at small cluster sizes but rapidly diverge at larger cluster sizes. This results broadly agrees with the experimental observations that crystallisation of the metastable form requires substantial melt undercoolings to a degree that would be consistent with rather small, albeit undefined, cluster sizes.

Dr ter Horst asked: How do your simulations relate to classical nucleation theory (CNT)? Excess energy of clusters relates to surface (Fig. 8) and not cluster size, for instance, a route to interfacial energy. It is interesting to see how interfacial energy develops with cluster size. CNT assumes constant interfacial energy to come to the currently standard nucleation rate equation. In the end you want to determine nucleation rates for the different polymorphs. How will that be done with your method?

Professor Roberts replied: The weakness of the current studies is that we are only considering enthalpic changes and hence not taking into account the entropic component in the surface free energy. The latter effects for small molecular species are probably not as significant a component as, *e.g.* for a high molecular weight polymer and so the assumption is probably reasonable. Given the surface relaxation process can be expected, normally, to be size-dependant, then our method has added-value in that it provides in principle a system-specific capability to provide more reliable interfacial tension data for input to CNT than is currently available. However, much remains to be done to create a robust methodology and in particular the cluster relaxations do need to be carried out within a salvation environment, perhaps *via* the use of molecular dynamics simulations for the minimisation/ relaxation process.

Professor Anwar commented: The paper discussed calculated potential energies. As clusters become smaller, thermal effect became important. This is particularly important for systems with weak interactions as low thermal energies are likely to disrupt the weak interactions. Ultimately, we need to go to free energies. Such calculations both for bulk and surfaces, are now possible.

Professor Roberts replied: Yes, I agree that surface free energy calculations would be the best way forward particularly if these could be carried out in a solvating environment. However, for small molecule systems I am not sure that the entropic contribution would be significant and this would need to be assessed in some careful benchmarking studies. In addition, I do feel that such free energy calculations would be best carried out on clusters, though calculations should be carried out without the use periodic boundary conditions.

Professor Davey commented: Solvent recrystallisation is technologically most important for molecular materials.

Dr Vonk asked: What will happen to crystals in an industrial environment with respect to secondary nucleation, in view of the fact that industrial circumstances are never equilibrium circumstances?

Professor Roberts answered: Yes, this is an interesting point and it is important to note that these studies really are directed towards the primary nucleation case. Industrially, the scaling-up of crystallisation processes to larger scale sizes do tend to involve higher degrees of secondary nucleation as driven by reactor hydrodynamics. However, extension of the work to model secondary nucleation is entirely feasible *via* modelling cluster binding to crystal surfaces and very recent work (see *e.g.* ref. 1) has demonstrated the feasibility of such an approach which would benefit from more further development.

1 R. B. Hammond, K. Pencheva and K. J. Roberts, *Cryst. Growth Des.*, 2007, **7**, 875–884.

Dr Hare communicated: In seeking a more general theory of crystal nucleation and growth — from a solution or a melt (such as S_8 or S_x) — should we attempt to treat a melt as a special case of a solution in which solvation and solvent properties may be exhibited by species that are identical *chemically* to the solute?

Professor Roberts communicated in reply: Yes, this could be a potential way forward. Organic melts, in particular, can exhibit substantial melting point undercooling, *e.g.* molten benzophenone (MP *ca.* 48 °C) can, with care, be stabilised at room temperature. This phenomena perhaps reflects the molecular disordering of this material in the molten state and the energy barrier associated with the concomitant conformational change needed for the molecule to self assemble into a well-ordered crystal structure as described in this paper.

Professor Addadi opened a general discussion by addressing Professor Vlieg: I agree that the most important site to be looked at in order to understand crystal growth processes is the interface between the crystal and the solution. In the crystal you've been addressing, KDP, water is absorbed at the surface and is partially ordered by it, but is not part of the crystal itself. Could it be helpful to look at hydrate crystals where one can in principle follow how water transforms at the interface from an ordered crystal arrangement into a disordered solution state?

Professor Vlieg responded: That is indeed possible. In the case of KDP we found remarkably strong ordering in the first 'liquid' monolayers. Inspired by this, we have also investigated the solid–liquid interface of brushite, $CaHPO_4 \cdot 2H_2O$, a biomineral that contains water layers as part of its crystal structure. We therefore expected especially strong water ordering at the interface, but this turns out not to be the case. The brushite {010} interface consists of two water bilayers, of which only the first one is highly ordered and can be considered as part of the crystal. The second layer, which in the bulk would also be fully crystalline, did not show any in-plane order.[1] We can thus still learn interesting things from experiments.

1 For more details, see J. Arsic, D. Kaminski, P. Poodt and E. Vlieg, *Phys. Rev. B*, 2004, **69**, 245406.

Professor Roberts said: I agree that this would be a useful way to look at the growth process and might provide valuable information concerning how the desolvation process proceeds in solution growth. For example in $CuSO_4 \cdot 5H_2O$ (see *e.g.* ref. 1) the Cu atomic site is octahedrally coordinated to oxygen (via a d^9 Jahn–Teller distortion giving rise to this compound's characteristic blue colour) *via* first shell coordination to 4 water molecules crystallisation and two SO_4^{2-} ions mimicking the solution structure which involves bonding to 6 water molecules crystallisation, *i.e.* surface adsorption simply involves exchanging coordination to two waters with two surface phosphate groups. Some clues can also be gleaned *via* studies of impurity incorporation such as Mn^{3+} ions in KDP (see *e.g.* ref. 2). Here, the Mn^{3+} also binds as an octahedral complex, exchanging two co-ordinations with water in the solution state to surface phosphate groups associated with growth. These though are isolation studies and clearly the underlying science of solution growth would benefit from more detailed study.

1 D. A. H. Cunningham, D. R. Armstrong, G. Clydesdale and K. J. Roberts, *Faraday Discuss.*, 1993, **95**, 347–365.
2 X. Lai, K. J. Roberts, M. J. Bedzyk, P. F. Lyman, L. P. Cardoso and J.-M. Sasaki, *Chem. Mater.*, 2005, **17**, 4053–4061.

Professor Hyne addressed Professor Vlieg and Professor Roberts: Do new investigative technologies such as HR-TEM credibly suggest that the molecular arrangement of species initially bonded onto surface sites from the liquid phase may not in fact retain the same arrangement as they are incorporated into the growing crystal bulk structure and if so how might this be expected to affect the bulk mechanical properties of the solid crystalline material?

Professor Vlieg answered: It is likely that the molecular arrangement of a species will indeed change when going from the solution, to the surface and finally into the bulk. The bulk properties of a crystal will only indirectly be affected by this, because the molecules will have lost their memory on this process, but defect formation and impurity incorporation will be influenced by the growth history and these do change the mechanical properties.

Professor Roberts responded: The existence of surface relaxation/reconstruction effects are well known for many inorganic materials and thus it would not be a surprise if the under-coordinated surfaces atoms changed following adsorption and subsequent incorporation into the bulk crystal lattice. These changes would be expected, in turn, to effect a change in the mechanical properties for the surface regions of the crystals which could impact on physical properties related to inter-particle interactions *etc*.

Dr Schön addressed Dr Hughes, Professor Roberts and Professor Vlieg: Classical crystallographical tools such as XRD have a problem: nucleation is hidden from view. We need to get big enough clusters in the simulations. This might in the end turn out to be much more problematic than the calculations needed to yield input for semi-phenomenological/analytical/simulation models. What is your estimate of the situation in this regard?

Professor Vlieg answered: Understanding nucleation is indeed a main challenge. It appears that simulations are making more progress here than experiments, simply because the size of the critical nuclei make experimental observations nearly impossible. For simplified models, applicable to colloids and proteins in particular, Daan Frenkel and coworkers have been able to simulate the nucleation process, including heterogeneous nucleation.[1,2] A different approach is that of Joop Ter Horst and Dimo Kashchiev, where the growth probability of clusters with different sizes are simulated and used to estimate the nucleation rate.[3] These clusters are modelled using a realistic crystallographic structure and thus the approach can be used to look at differences between polymorphs. See for example ref. 4.

1 P. R. ten Wolde and D. Frenkel, *Science*, 1997, **277**, 1975.
2 A. Cacciuto, S. Auer and D. Frenkel, *Nature*, 2004, **428**, 404.
3 J. H. ter Horst and D. H. Kashchiev, *J. Chem. Phys.*, 2003, **119**, 2241.
4 M. A. Deij, J. H. ter Horst, H. Meekes, P. Jansens and E. Vlieg, *J. Phys. Chem. B*, 2007, **111**, 1523.

Professor Roberts replied: It is certainly true that XRD yields very little concerning cluster structure below 100 nm sizes. Electron diffraction is a better technique in this respect but suffers from problems associated with sample preparation and electron beam damage. X-Ray absorption spectroscopy in favourable circumstances is a good option but is restricted to probing the structures of dense inorganics (see *e.g.* ref. 1) rather than organic molecular crystals. Whilst current cluster modelling has dealt with much smaller cluster sizes of up to 500 molecules, there is scope to increase this and so bridge the gap between current measurements and modelling approaches. Also, mindful of the fact that intermolecular interactions are quite short-range for most organic solids, there is significant further scope for further efficiency gains in terms of modelling CPU overheads *via* the use of interaction potential coarsening approaches which are commonly adopted in mezzo-scale molecular modelling.

1 A. B. Edwards, C. D. Garner and K. J. Roberts, *J. Phys. Chem.*, 1997, **101**, 20–26.

Dr Hughes answered: As stated clearly in our paper, we have employed XRD only to study the final bulk phases obtained from the crystallization process and other techniques have been used to study earlier stages (aggregation and nucleation). Simulations are certainly able to model increasingly large systems and to model nucleation processes directly.

Insights into the crystal growth mechanisms of zeolites from combined experimental imaging and theoretical studies†

Ben Slater,*[a] Tetsu Ohsuna,[bcd] Zheng Liu[be] and Osamu Terasaki*[b]

Received 21st December 2006, Accepted 7th February 2007
First published as an Advance Article on the web 8th May 2007
DOI: 10.1039/b618677g

Detailed investigations into surface mediated crystal growth at zeolite external surfaces are presented. High resolution TEM is able to directly resolve surface and bulk crystallographic features and the unusual surface structural features are interpreted from simulation work. The growth of the double 4 ring is found to be a crucial and rate-determining step in the surface mediated, post-nucleation crystal growth mechanism of zeolite Beta C. Growth of 4 rings is found to be more favourable on fast growing rather than slow growing faces, explaining the relative growth rate of crystal faces in this materials. Similarly, the terminating structures of zeolite Y/Faujasite can be partly explained by considering the condensation of 6 ring and double 6 ring species at the crystal surface. Whilst 4-ring and double 4 rings are known solution species, 6 rings and double 6 rings are not, and hence it is speculated that post-nucleation crystal growth may involve competition between primary building unit and secondary building unit mediated crystal growth mechanisms.

Introduction

One of the grand challenges in materials science is to manufacture tailored microporous materials at will for a specific application. Aluminosilicate zeolites and AlPOs and to some extent mesoporous materials are distinct from most compounds precisely because of their porosity. The porosity or openness of the zeolite (or other framework material), comprised of channel or cage like void space can be more important than the precise structure of the framework itself, when the target use is for separation applications. In catalytic applications, the channel or cage shape can impose stereochemical constraints on both the intermediate and products of chemical reactions occurring within the framework material. Over the last few decades, scores of entirely new materials have been synthesised, swelling the

[a] Davy Faraday Research Laboratory, The Royal Institution of Great Britain, 21 Albemarle Street, London, UK W1S 4BS. E-mail: ben@ri.ac.uk
[b] Structural Chemistry, Arrhenius Laboratory, Stockholm University, SE-10691, Stockholm, Sweden. E-mail: terasaki@struc.su.se
[c] CREST, Japan Science and Technology Agency, Honcho 4-1-8, Kawaguchi-shi, Saitama 332-0012, Japan
[d] Kagami Memorial Laboratory for Materials Science and Technology, Waseda University, Nishiwaseda 2-8-26, Shinjuku-ku, Tokyo 169-0051, Japan
[e] National Institute of Advanced Industrial Science and Technology (AIST), Higashi 1-1-1, Tsukuba 305-8565, Japan

† The HTML version of this article has been enhanced with colour images.

international zeolite database to over 160 materials. However, few, if any of the new materials can be claimed to have been designed purely '*ab-initio*', relying on trial and error chemistry and serendipity. No criticism is implied within the last statement, simply the observation that arguably we have yet to fully understand the elements and variables of the crystal growth mechanism, which prevents us from designing and fabricating more desirable materials. Beyond fabrication of a framework lies the goal of controlling the crystallinity (degree of faulting and defect concentrations) of the product and then controlling the surface termination and the crystal shape. For porous materials in particular, the availability of channels for molecular sieving is a function of crystal morphology and the integrity of those channels—properties that can show acute sensitivity to synthesis parameters.

In attempt to try and form a fundamental understanding of the crystal growth process and hence control the product of the synthesis, highly detailed and systematic work has been undertaken by numerous investigators to try and better understand the composition of the synthesis pot. For example, NMR studies have been performed on silicate solutions to try and establish the identity and populations of oligomeric structures.[1-3] Various studies have shown that secondary building units such as double 4 rings (D4Rs) are present in solution and the fact that these are readily observed in many zeolites has been taken as evidence of a 'secondary building unit' mechanism at work in zeolite growth. Curious then, that 3 rings which are highly populous according to NMR had only been observed mainly in a berylozeolite—*e.g.* framework code OSO, one aluminosilicate (MEI) and a recently identified silicogermanate[4] and that there is little evidence for the existence of double 6 rings (D6Rs) or even single 6 rings (S6Rs) in solution, yet they are present in many natural and synthetic zeolites.

In this work and our related previous studies, the aim of the approach is to examine the terminating surface structure of zeolites for clues about the crystal growth mechanism. In the following few sections we describe how the surface structure of a zeolite can be constructed from known (and in some cases, as yet unknown) solution fragments reacting with crystal faces. Once the crystal has nucleated and there is onset of faceting, the relative growth rates of crystal faces are well defined and are limited by rate-determining steps (terrace nucleation, step evolution and kink annihilation). The surface terminating structures in fact are rate determining steps, evidenced by their consistent structure and longevity, hence a study of the microscopic processes leading to these rate determining steps may potentially give us enough insight into how the path of synthesis may be manipulated to influence the growth rates of faces and hence the morphology of the product. More fundamentally, understanding the mechanism of growth could enable us to disfavour or stop growth of certain zeolite polymorphs and bias the formation of more desirable materials or as yet unsynthesised materials. We present here results pertaining to materials which have distinct secondary building units (SBUs) to try and identify commonalities in the potential role of these species within crystal growth.

Zeolite Beta C

Zeolite Beta C was synthesised only relatively recently but had been postulated by Newsam and Treacy in 1988.[5] The passing of nearly two decades before the material was finally isolated[6] probably indicates a narrow region of phase space in which Beta C can be synthesised. In the reported synthesis it was prepared in purely siliceous form using HF as a mineraliser, which led to the incorporation of ~2 fluoride ions in the unit cell. One presumes that the fluoride is present as a point impurity rather than a defect complex, as there is no discernible evidence from high resolution transmission electron microscopy (HRTEM) of repeated structural defects.

The distinguishing feature of Beta C is that it contains a 3 dimensional interconnecting 12 ring channel system. The relatively large and uniform channels offer the possibility of performing fine chemical separation and catalytic applications. In previous work, we have reported both the synthesis and HRTEM imaging[6] of a single crystal and a combined experimental/modelling study.[7] In Fig. 1a, a low magnification HRTEM image showing several crystals of zeolite Beta C growing off ordinary Beta. Notice that the tips of the crystals vary from sample to sample, indicating very local differences in the growth rates of crystal faces of the same material. In particular, the aspect ratio of the (001) : {110} faces varies from cases in which {001} growth is clearly considerably faster than {110}, but in other instances, the growth rates of the faces are approximately equal (see Fig. 1d). The aim of the previous work was to explain the regular surface terminations shown in Fig. 1b, corresponding to the {100} surface on the right hand side of the figure. To recap briefly, two distinct surface terminations are observed on this face: a D4R surface and a termination where the D4Rs are absent. Using interatomic potential approaches we found three equally thermodynamically stable terminations, the D4R, the single 4 ring (S4R) and no-D4R surface suggesting that from thermodynamic considerations alone, all three terminations could be expected to be visible. Total energy DFT calculations showed that the absence of the intermediate S4R termination could be explained by considering the condensation of 4Rs upon the crystal surface. In essence, the calculations showed the intermediate S4R stage was

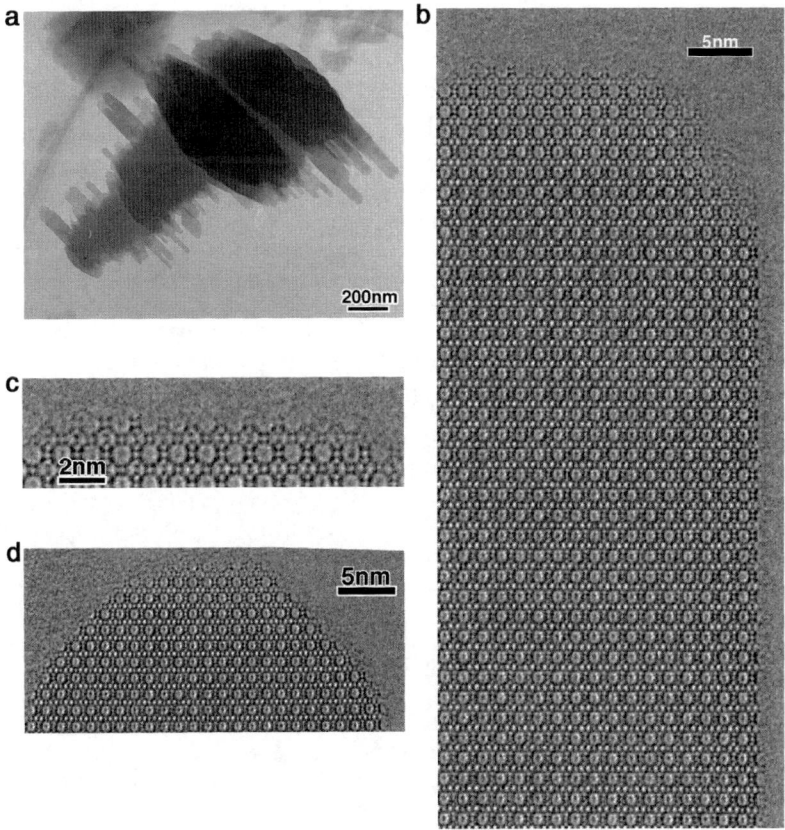

Fig. 1 (a) Low magnification HRTEM image of several Beta C crystals growing off ordinary Beta. (b) High magnification HRTEM image of zeolite Beta C [100] axis. Tip of the crystal exposes the (001) surface, the (010) surface is seen at the right hand side of the crystal. (c) HRTEM image of zeolite Beta C highlighting the (001) surface structure. (d) A view of an another crystal of Beta C taken from Fig. 1a, with a more similar aspect ratio between {001} : {110} crystal faces. Note the well-defined fine structure of the {110} plane.

thermodynamically unstable because the condensation of a S4R onto the surface was found to be slightly endothermic whilst overgrowth by a S4R to give the D4R termination was found to be exothermic by ~30 kJ mol^{-1} and of course the S4R mediated dissolution was also found to be exothermic. In this work we return to this material to examine the faster growing (001) crystal surface, which is also imaged in Fig. 1b and c. All the HRTEM images were taken at the same conditions as reported previously.[6]

The (001) surface has been magnified in Fig. 1c, from which two crystal terminations are apparently evident. On first inspection, the one termination appears to expose the D4R ring, whilst the second termination exposes what appears to be an S4R termination, reminiscent of the absent termination on the {100} surface. In contrast again to the {100} surface, the (001) face does not appear to show a surface structure denuded of a D4R. The crystal structure in the c axis is moderately more complicated than that parallel to the a and b planes and hence in Fig. 2a and b, two orientations of the cell are shown to highlight the structure viewed along [100] and [010]. In Fig. 2a, the double 4 ring is visible, but in Fig. 2b it can be seen that the D4R is separated by bridging silanol groups. To identify the relative stabilities of each face, a 2D slab was created of (001) orientation from a relaxed unit cell, employing the interatomic potential set described previously in Slater *et al*. A total of 6 layers were used in the calculation, where the upper 3 layers are relaxed, whilst the lower 3 layers are held fixed.

Five terminations were considered in all, and most of the relaxed geometries are shown in Fig. 2a–f. The terminations were generated by cutting the crystal parallel to the (001) plane, to generate surfaces with as few terminal bonds as possible. Chemical intuition suggests that terminations that expose the least number of dangling bonds are likely to be the most stable, as shown previously.[8] It has been shown that polar surfaces are inherently unstable unless passivated by, for example, a locally ordered polar solvent that counterbalances the dipole of the surface. Therefore upon cutting the surface, the surface is reconstructed to give an equal number of under-coordinated silicon and oxygen lattice sites at the upper surface (to be relaxed) and at the lower surface. Since zeolites are ubiquitously synthesised under hydrothermal conditions, the plentiful supply of water provides the possibility for water to dissociate at the surface, and hence fully coordinate surface species. In previous work by ourselves,[9] it was shown that under most conditions, and without exception for low index faces, dissociative rather than associative adsorption of water occurs at the surface to generate silanol, geminal and tripodal groups.

In Fig. 2a, the surface has been cut laterally to expose a D4R. In Fig. 2b viewed along a different zone axis, two silanol groups are exposed that appear to bridge the D4Rs. We considered another termination, where one can imagine monomers have condensed on the bridging silanol groups to yield two tripodal O-Si(OH)$_3$ groups (not shown). In fact, Fig. 2a shows a D4R terminated surface absent of tripodal groups. Fig. 2c shows the S4R terminated surface, where an S4R has been removed from the starting structure shown in Fig. 2a. Fig. 2d shows the surface denuded of the D4R. In Fig. 2e, the configuration is identical to that in Fig. 2d but is viewed along an alternative zone axis. Finally in Fig. 2f, the bridging silanol species are removed (compare with Fig. 3a). Table 1 shows the surface energies for five terminating structures after relaxation and correction for the reaction of water with the surface.[10] It is clear that, two terminations are especially stable, those shown in Fig. 2a and c, corresponding to the terminations observed in the TEM image. For clarity, simulated TEM images have been generated and are displayed in Fig. 3a–c, for comparison with observed images and the sampled topologies. Note that the TEM images of the structure corresponding to Fig. 2c and d would be essentially identical apart from slightly enhanced contrast of the upper part of the S4R, a feature which is too subtle to detect *via* TEM. A TEM view from a different axis might permit resolution between which, if either, of the terminations are preferred. Attention is drawn to the fact that the tripodal termination is relatively unfavourable

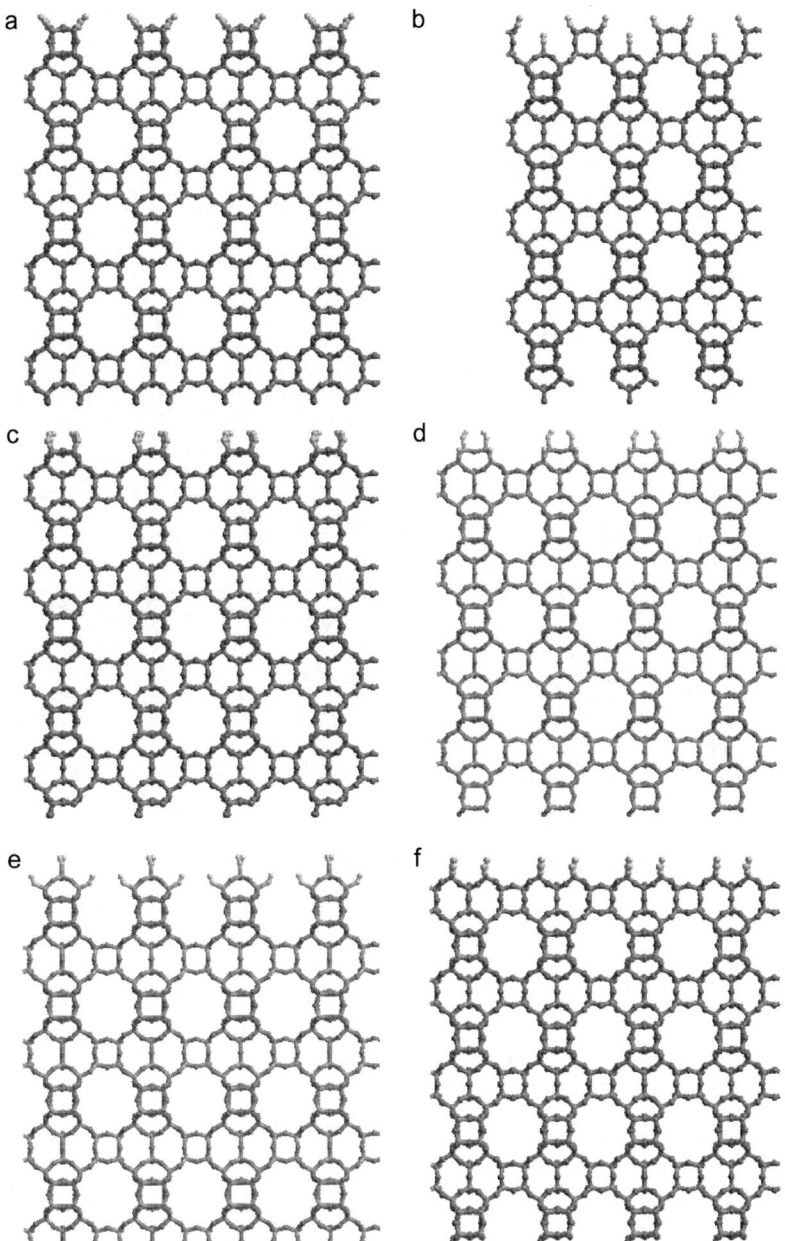

Fig. 2 (a) Relaxed surface of Beta C (001), D4R termination, viewed along [100]. Note the surface is shown in cross section, with the top of the crystal representing the terminating surface. This slab is periodically repeated within the calculation parallel and perpendicular to the plane of the paper. (b) Relaxed structure of (001) Beta C viewed along [010]. (c) Relaxed structure of (001) Beta C, S4R termination, viewed along [100]. (d) Relaxed structure of (001) Beta C, missing D4R and but retaining bridging silanol species, viewed along [100]. (e) Relaxed structure of (001) Beta C, missing D4R and retaining bridging silanol species, viewed along [010]. (f) Relaxed structure of (001) Beta C, missing D4R and bridging silanol species, viewed along [100].

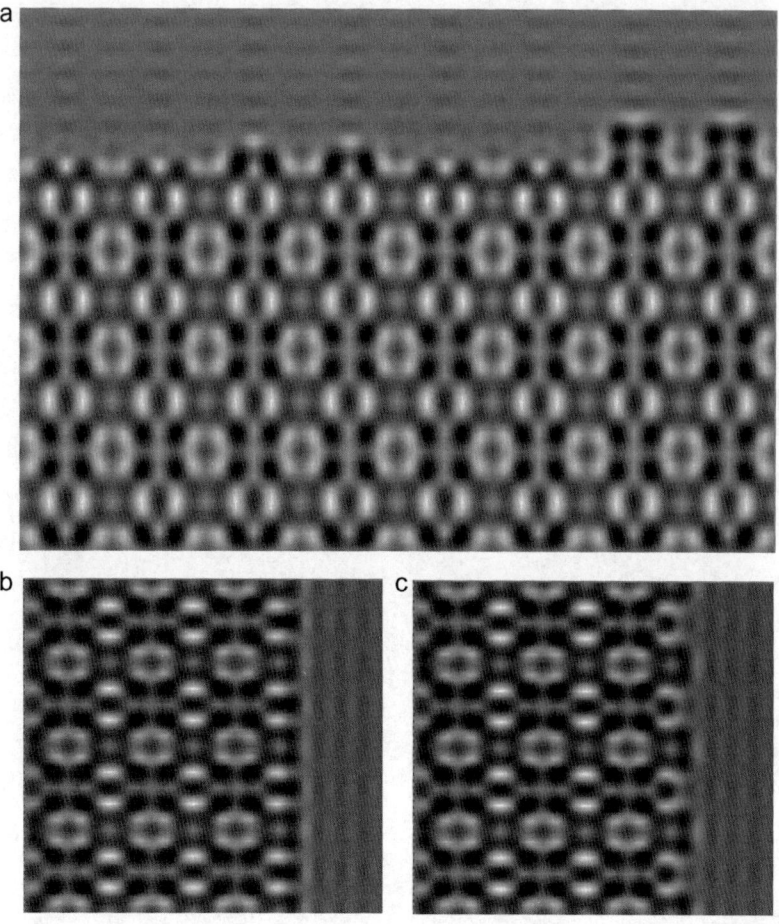

Fig. 3 (a) Simulated HRTEM images of the BEC (001) face viewed from [100]. From left to right, (i) the terminations correspond to removing a D4R and bridging silanol groups, (ii) removal of a D4R retaining the bridging silanol groups, (iii) removal of the bridging silanol groups—as (i) and repeated for contrast for (iv) the D4R termination. (b) The D4R termination shown along the [010] axis with the tripodal silanol groups present. (c) The D4R termination viewed along [010] with the tripodal silanol groups absent.

to all the other terminations considered—the tripodal species are extremely weakly bound to the surface, indeed their dissolution is favourable according to our model and hence we predict terminations 2a/2b, 2c and 2d/2e are the most probable

Table 1 Relaxed surface energies for the (001) and {100} surfaces of zeolite Beta C

Surface structure	Surface energy/J m^{-2}
D4R + tripodal(001) (not shown)	1.190
D4R(001) (Fig. 2a, b)	0.892
S4R(001) (Fig. 2c)	0.832
-D4R(001) + silanol groups (Fig. 2d, e)	0.869
-D4R(001)-silanol groups (Fig. 2f)	1.174
D4R{100}	0.554
S4R{100}	0.534
noD4R{100}	0.557

structures corresponding to those observed in the TEM images. Note also that when the bridging silanol groups are removed (as shown in Fig. 2f), the structure becomes very unstable.

In Fig. 3a, simulated TEM images are displayed which should be compared with the structures apparent in Fig. 1c. Clearly terminations 3a (ii) and 3a (iv) correspond very closely with the terminations seen in Fig. 1c, d and e—which were generated from terminations visualised in Fig. 2a, c and d. For reference, in Fig. 3b and c the {001} surface is projected along [010] to contrast the tripodal termination and silanol terminated surface (corresponding to the structures referred to on the first two entries in Table 1). Although the difference in the TEM images is quite subtle, with the latest generation of TEM machines it ought to be feasible to ascertain which, if any, of the terminations is preferred. According to the calculations presented here, the silanol terminated surface is expected to be observed rather than the tripodal termination.

The S4R structure in projection along the [100] is identical to the surface denuded of D4R and so we cannot resolve the structure unequivocally—but viewed along [010] differences in the structure should be observable. According to the calculations, the stability of the surface when the D4R and bridging silanol groups have been removed is much reduced (1.174 J m^{-2}). On the basis of thermodynamic predictions, ignoring reaction enthalpies of condensing pre-formed species on the surface, the S4R structure is expected to be more long-lived than the D4R free surface with silanol groups although arguably the surface energies are too similar to distinguish.

To examine the microscopic processes in more detail, first principles calculations are required but some insight into the mechanism of growth can be obtained from dissolution studies from classical calculations. In the manner we have explored previously,[8] the configurations that have been sampled form part of a notional crystal growth mechanism involving endo- and exothermic steps. The mechanism can be further investigated by considering a monomer mediated growth pathway as we recently studied for the {100} surface. Starting from the configuration in Fig. 2a until the configuration is regenerated, a monomeric $Si(OH)_4$ silicic acid species was removed from the surface according to the recipe reported previously. Briefly, the crystal is "dissolved" site by site, but the number of SiO_2 species remains approximately constant, because the dissolved $SiOH_4$ is dehydrated and placed at the base of the slab as a SiO_2 unit, whilst a SiO_2 unit is promoted from the fixed region to the lower part of the free region. In this manner, the total energies of each configuration can be compared from step to step allowing for the fact that removal of a $SiOH_4$ unit can break different numbers of bonds depending on the surface site from which it is taken. For example, by analogy, the terrace site breaks one bond, the step site two bonds and the kink site three bonds. The choice of which $SiOH_4$ is removed and hence the pathway is dictated by considering which of the symmetry distinct surface silicon sites is least strongly bound. This last step is achieved by generating each symmetry distinct surface at each step and calculating the relaxed energy of the configuration. The most stable surface relative to the previous step is taken to be the most favourable process and in a stepwise fashion an insight into the microscopic stages involved in dissolution and arguably, growth can be eluted.

In this work, we find that the cost of forming the double 4 ring at the {100} surface involves a maximum endothermic step of approximately ~ 0.55 eV. The analogous process at the (001) surface requires ~ 0.45 eV by comparison, suggesting that the barrier to growth on the (001) surface is slightly lower than that on the {100} surface (assuming that the transition state is very similar in each case). Consistent with the former finding, dissolution of the D4R on the (001) surface is more favourable than on the {100} surface. Given that the surface energy of the {100} face is ~ 0.6 J m^{-2} in comparison to ~ 0.9 J m^{-2} on the (001) face, a coherent picture emerges which can be constructed from this and the other energetics calculated for key stages of the growth and dissolution. Firstly, the surface energy of (001) is higher than that of {100} and the smaller the surface energy (assuming it to be positive) the more stable the surface is predicted to be. Under equilibrium conditions, stable surfaces with low surface energies are expected to be highly pronounced in the crystal habit. Conversely, surfaces with high surface energies are predicted to have very low morphological importance. Hence, surfaces with low surface energies are expected to grow more slowly than surfaces with high surface energies, despite the fact that the surface energy does not have an obvious kinetic component to it. In Fig. 1a, the thin rectangular needle-like crystal clearly favours the {100} surface, and the aspect ratio is very roughly 1 : 10 for (001) : {100}. The equilibrium growth rates based upon the relative surface energies are clearly much lower than the observed growth rates but do give the correct order of growth rates. The microscopic growth rate of the crystal is of course determined by the rate determining steps on these crystal faces. In general, we have noted that fast growing faces have low interlayer spacing and conversely slow growing crystal faces have large interlayer spacing. One is tempted to relate the growth rates to the density of potential nucleation sites—which is high on low d_{hkl} faces and low on high d_{hkl} faces invoking a probabilistic argument. Of course, the sticking probability and dissolution probabilities will be sensitive to pH conditions and naturally the probability of nucleation will also be sensitive to the density of nucleation sites. Zeolites are distinct from many other minerals in that the crystal planes are very inhomogeneous. For a given crystal face an arbitrary cut parallel to the crystal plane can expose very different numbers of dangling bonds to a cut made elsewhere but in the same plane (for example 12 is the minimum and the maximum number of under-coordinated silicon sites is 24 for zeolite Y in the {111} plane). It is tempting to forecast that the layer-by-layer growth rate is then far from uniform, and will feature rapid stages of growth when dangling bond density is high (and the number of available nucleation sits is therefore large) whilst conversely when dangling bond density is low, growth is expected to be slow. The microscopic growth of crystal faces is then expected to consist of fast and slow stages of growth. The relative growth rate of crystal faces will be determined by additional variables such as terrace advance rate, terrace nucleation rate, kink annihilation rate, *etc.*

According to an earlier statement, the surface energy is proportional to the number of dangling bonds per unit area. Hence again, the growth rates of crystal faces can be inferred, but again of course the predicted aspect ratio is underestimated in comparison to the observed crystals in Fig. 1a. However, these calculations only probe end points in the growth process, giving no information on the intermediate terminations and barrier heights that are surely crucial. As noted by Agger *et al.*[11] the aspect ratio and relative growth rates of crystal faces can be used to extract approximate free energy differences—in the case of silicalite, Agger noted that a 4 : 1 aspect ratio can be attributed to free energy differences of ~ 5 kJ mol^{-1} at a typical synthesis temperature of 400 K. Such miniscule energy differences are arguably beyond the capabilities of pair potential models. Nevertheless, recently Piana *et al.*[12] reported a very successful study on the influence of solvent on crystal morphology utilising a molecular dynamics scheme feeding into a kinetic Monte Carlo approach which provides a source of optimism that zeolite growth could be investigated using potential based approaches.

Formation of the D4R is less favourable (more slow) on the {100} face compared with the (001) face and the fact that the D4R is observed on each crystal face implies that this is a rate determining step in the crystal growth process. Additionally, dissolution is more favourable at the fast growing but relatively unstable (001) surface compared to the {100} surface. The insights from interatomic potentials are instructive and consistent with expectation from the observed relative growth rates of crystal faces. However, to confirm the findings, total energy DFT calculations were performed on thin slabs of Beta C using the Quickstep module of the CP2K software.[13] The brevity of this article dictates that limited detail on the Quickstep approach is given. In essence it is a mixed plane-wave, Gaussian basis set approach to the model electron density and orbitals, respectively. Full plane wave calculations for this material are still relatively expensive, since the unit cell contains 96 atoms and is relatively low density (0.04TO$_2$/ unit cell). A single layer of Beta C was used in the following calculations within a cell containing a 9.0 Å vacuum gap. In the case of the (001) surface a maximum of 236 atoms were used. To test the accuracy of the method, the dimerisation (condensation) energy of two monomers was computed at different cut-offs and compared against recent double numeric precision results using the DMol code marketed by Accelrys. Using a DZVP and the BLYP functional, a condensation energy of -14 kJ mol^{-1} was found at a cut-off of 280 Ry compared to -11 kJ mol^{-1} from a recent study.[14] The energy of condensing two S4Rs to give a D4R is calculated to be $+24$ kJ mol^{-1}, but recent work[14] has shown that the entropy term is typically about -80 kJ mol^{-1} at synthesis temperatures for cyclic species, hence the free energy of forming the D4R is very favourable. The relative energies of attaching S4R and D4R ring and monomeric species onto the {100} and (001) faces were calculated. Using these settings, in agreement with the interatomic potential calculations, the most endothermic step in assembly of the D4R was found to be $+0.42$ eV on the {100} face and $+0.31$ eV on the (001) face and is associated with attaching a monomer S4R on the D4R denuded surface (*i.e.* nucleating the D4R on the zeolite terrace.)

In summary, observed surface terminations from TEM show the presence of double 4 rings as rate determining steps on the fast and slow growing crystal faces of Beta C. Microscopic, monomer mediated assembly of the D4R at the crystal face is more favourable for the faster growing (001) face than the {100} surface. DFT calculations show that on the (001) face, the two most stable terminations are the D4R terminated with bridging silanol groups (Fig. 2a, b) and the D4R denuded surface with bridging silanol groups (Fig. 2d, e). The S4R structure (Fig. 2c) is metastable according to DFT, and can be overgrown by attaching a further S4R (-14 kJ mol^{-1}) or the S4R can be dissolved (-16 kJ mol^{-1}). At the classical model level, because the stabilities of solution fragments is not taken into account, the surface energy calculations are extremely close in energy, this technique is unable to distinguish between the two structures.

The very presence of double 4 rings at the crystal surface of this material serves to indicate the importance of double 4 rings in assembling other materials that contain this secondary building unit. It has long been postulated that the double 4 ring may be a key growth unit for many zeolites. The obvious and perhaps most useful model system to further explore this topic is pure silica zeolite A, which is composed entirely from D4Rs[15] and hence is one of the few materials which could be considered to be formed from a single SBU (most zeolites are the sum of small building units added with yet smaller building units). The aluminosilicate form of LTA has been imaged using TEM[16] and also using AFM,[17] revealing a number of surface crystallographic features, but only the {100} faces have been examined and it is the contrast between fast growing faces {110} and {111} that is most revealing in relation to the growth mechanism.

The calculations presented here identify three terminations of the (001) surface that can explain the TEM images observed. Two of the terminations appear identical in the projection along the [100] but are non-equivalent in the [010] plane. DFT calculations that take into account the formation energy of the key growth species S4R and D4R, suggest that the S4R termination may be metastable, but further TEM images are required to confirm this prediction. More generally, comparison of the energetics of attaching species to the slow and fast growing crystal faces show that condensation is more exothermic at the fast growing face. However, the observed aspect ratio cannot be completely explained by the relative energies of reacting D4Rs with the two crystal faces. Our calculations make no attempt to calculate a barrier for condensation, and barrier heights will have a critical effect on the rate of attachment. It is tempting to speculate that the barrier to condensation will be lower on the (001) face because it has a higher density of T sites than {100}. The D4R will therefore be attracted to the (001) surface more strongly than the {100} surface due to van der Waals attraction. However, there are many other factors which may make major contributions to the barrier height which cannot be accounted for in this modelling approach. Nevertheless, the calculations also suggest that both a monomer mediated reaction and the D4R mediated reaction are energetically viable mechanisms of reaction, but the D4R reaction will be more favourable if pre-formed D4Rs exist in solution and they can diffuse through the zeolite–gel interface on a meaningful timescale.

Intergrowths of Faujasite (FAU) and EMT

In trying to understand the fundamental processes of post-nucleation crystal growth, intergrowth materials are particularly interesting as they arise in finely balanced syntheses at intersections in the phase diagram, where small changes in the reagents can influence whether a single or multiple phases are formed (and their yields). In the case of Beta C, it is the relative growth rate of crystal faces that contains information about the mechanism of growth. In the case of the FAU/EMT intergrowth, a widely studied topic within zeolite science,[18–25] investigating why small changes to the synthesis parameters affect the product may aid our understanding of crystal growth mechanism and potentially shed light on the role of PBU and SBUs. Again, Newsam and Treacy[19,26] have reported early important work where they hypothesised the existence of a regular intergrowth between the FAU and EMT materials. The FAU and EMT frameworks are shown in Fig. 4a and b, respectively, where the relationship between the two materials is evident, the stacking sequence being ABCABC in FAU and ABABAB in EMT. EMT contains a mirror plane between the hexagonal sheets whilst FAU contains an inversion centre in the D6R. EMT is more difficult to synthesise than FAU, although using specific crown ether templates it can be prepared under relatively mild conditions. In future work, we will focus on the role of the crown ether in templating FAU and EMT and the intergrowth material but in this work we report preliminary data on a hypothetical intergrowth and some speculations on templates for synthesising intergrowths with a more detailed study

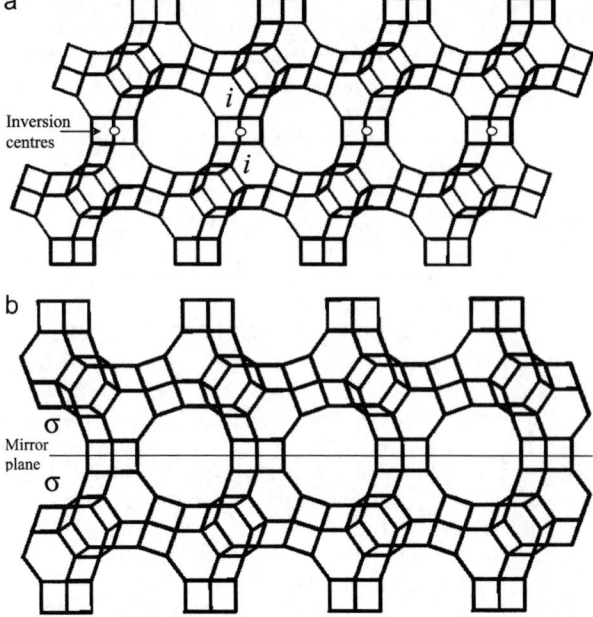

Fig. 4 (a) The structure of the FAU lattice, where the silicon atoms correspond to vertices. (b) The structure of the EMT lattice, where the silicon atoms correspond to vertices.

of the reaction of S6Rs and D6Rs at the zeolite Y surface for direct comparison with Beta C.

The as yet hypothetical intergrowth of FAU and EMT proposed by Newsam and Treacy is the sandwich of alternating *single* FAU and EMT layers with a period of 1 to give the stacking sequence ABCABABCAB. Whilst FAU/EMT intergrowths have been successfully synthesised by numerous groups, none have succeeded in creating the perfectly periodic structure—only randomly periodic structures, typically on the 10–20 nm length scale.[27,28] We note that many groups have been able to create a sandwich of FAU and EMT, where FAU overgrows a seed of EMT and *vice versa* but the films of FAU and EMT contain multiple layers.[24] Fig. 5 shows an image of a typical FAU/EMT intergrowth and note that intergrowth exclusively occurs in the [001] plane of EMT. The HRTEM image of FAU/EMT was taken with the electron beam incidence parallel to [110] of FAU and [100] of EMT by JEM-4000EX at 400 kV.

Firstly, we examine the stability of the intergrowth to ascertain whether the faulted material is particularly unstable. The stability of FAU, EMT and the intergrowth can be obtained from relatively simple interatomic potential lattice energy calculations using the potential set described previously with a cut-off of 12 Å. The work of Henson *et al.*[29] has demonstrated the remarkable accuracy of the Sanders and Catlow forcefield[30] in comparison to calorimetric data and indeed the mean error in geometry and energy relative to quartz is comparable to *ab-initio* methods. In Table 2, the absolute total energy of pure silica forms of FAU, EMT and the as yet unsynthesised hypothetical regular intergrowth of FAU/EMT. The all silica form of LTA, recently synthesised by Corma *et al.*[15] is also given as a guide to the thermal stability and accessibility of large pore silica zeolites. The regular intergrowth lies almost exactly halfway between FAU and EMT indicating that the marriage of these two phases incurs no special energy penalty although FAU rich intergrowths will always be expected to be more stable than EMT rich intergrowths. In the literature intergrowth concentrations have never reached close to 50 : 50, yet according to the computed thermal stabilities, the intergrowth ought to be accessible (ignoring the influence of Si : Al ratio and the extra-framework cation). To our knowledge the structural properties of the regular intergrowth have not been explored, hence we now focus on the interface between the EMT and FAU layers. To aid the discussion, the void volumes of EMT, FAU and FAU/EMT are shown in Fig. 6a, b, c, computed using probe molecules with a diameter of 6 Å. In FAU the supercages are connected with tetrahedral symmetry, whilst in EMT the key

Fig. 5 HRTEM image of FAU/EMT intergrowths.

Table 2 Total energies of all-silica zeolite phases

Zeolite phase	Total energy/eV
FAU	−24 671.6601
EMT	−24 671.1903
FAU/EMT	−24 671.4203
LTA	−24 672.9381

difference from FAU is that there are two cage types, a hypercage and a hypocage (following the nomenclature from Martens *et al.*) which are both elliptical—the hypercage being the larger of the two and having a greater capacity than the FAU cage. However, the connecting channels through 12 rings between supercages are actually slightly narrower than in FAU (9.81 Å *versus* 9.90 Å). Whilst the more voluminous cage of EMT permits larger molecules to form than in FAU, only small molecules narrow enough to diffuse through the connecting channels may be recovered. In comparison to FAU, the smaller channels of EMT either restrict the diffusion of products or may block diffusion completely. Further details of the structure will be given in a future publication.

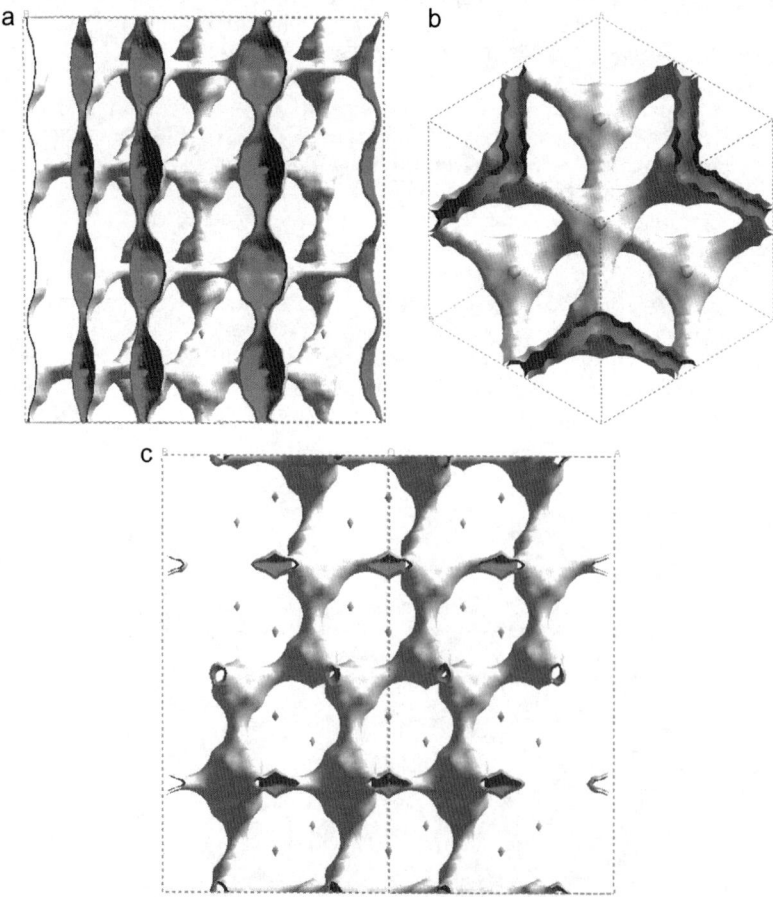

Fig. 6 (a) Accessible volume of EMT viewed along the [110] axis. (b) Accessible void space in FAU viewed along the [111] axis. (c) Accessible volume of EMT/FAU regular intergrowths viewed along the [110] axis.

The regular intergrowth displays both of the characteristic FAU supercage and the EMT cages, but in the *c* axis the channel connecting the two cages is slightly larger than both the FAU and EMT channels. If the intergrowth were to be employed as an oriented membrane, the layered double cage system may have some interesting catalytic applications. Fractionally larger products will be able to diffuse from the FAU and EMT cages along the *c*-axis through other FAU and EMT cages than those permitted to diffuse in the FAU or EMT parent materials. The hybrid of cages is reminiscent of hierarchical membranes, where the porosity of the material has a gradient as a function of the thickness of the film—open at the exterior but extremely narrow pores and hence highly selective at the crystal interior.

Whilst this material may have some desirable properties it remains an interesting hypothetical curiosity until it can be successfully synthesised. Since the crown ether used to form the EMT end-member is known to be a space-filling agent, that can occupy both the hypo- and hypercage, a different agent will be needed to try and enforce regular intergrowth. Looking closely at the structure, one can hypothesise that a dumbbell shaped template may aid regular stacking of FAU/EMT. One end should space fill the FAU supercage and the other larger end would fit EMT better than FAU. Recently Bonilla et al.[31] have reported systematic work to explore changes in the morphology of MFI using templates with different end groups including TPA, and dimer and trimer forms based on TPA with varying chain lengths between the TPA groups to span 1, 2 or 3 channels. In this instance, we wish the template to selectively link between FAU and EMT *cages*, and hence we consider the absorption of a hypothetical diquaternary ammonium template with one end exposing TPA and the other end displaying TBA connected by 5 carbon atoms—as shown in Fig. 7. The chain length is long enough to permit the bulky end groups to span two cages without excessive hindrance.

Using a fixed framework and considering only van der Waals repulsion between the template and framework, the relative absorption energies were calculated in a $2 \times 2 \times 1$ intergrowth cell, using anneal dynamics over 100 ps (using a 1 fs timestep). The temperature interval ranged from 300 to 750 K with 50 fs temperature ramping stages. We found that the lowest absorption point was found for the template with the TBA end to be absorbed in the EMT layer and TPA end absorbed in the FAU layer across the intergrowth layer. The next most stable configuration was found for the template absorbed TBA in FAU and TPA in EMT, across the interface (with a relative absorption energy of $\sim +5$ kJ mol^{-1}). The third most stable conformation was that in which the template is absorbed entirely within the EMT layer, parallel to the (110) plane ($\sim +7$ kJ mol^{-1}). The least favourable location for the template was within the FAU layer ($\sim +27$ kJ mol^{-1}). These very primitive calculations do suggest that an appropriately tailored template may enhance the growth rate of

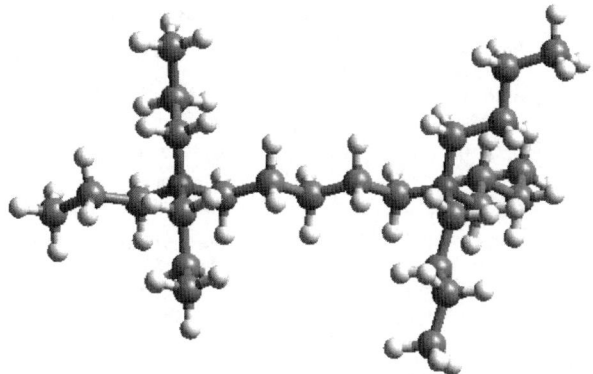

Fig. 7 The diquaternary ammonium template with propyl and butyl end groups.

EMT, but more importantly, it may enhance the FAU/EMT intergrowth rate relative to FAU. Given that the crown ether template concentration is far from excess in reported intergrowth studies, it may be that addition of a template which shows the selectivity suggested by these calculations may well enhance the frequency of intergrowths. It is hoped that these indicative calculations will stimulate experimental studies within the community.

In this final section, we focus on FAU surface structures noting that we have previously reported the surface structure of the (111) FAU surface[32] but have reinvestigated this using different software (GULP[33]). TEM has shown the existence of two terminations, on the {111} surface, a D6R terminated termination and D6R denuded surface.[28] A S6R terminated surface which is created by removing a S6R from the D6R is found to have an identical surface energy to that of the other two known surface terminations according to this interatomic potential model approach. To elucidate this apparent anomaly, first principle slab calculations were carried out on FAU to analyse the reaction condensation energies of attaching single and double 6 rings that are assumed to be present in solution. Again using the quickstep code, a single slab of FAU expressing the D6R termination was relaxed using the same recipe as that used for Beta C. The initial surface cell contained 396 atoms; after relaxation this geometry was edited so that the surface expressed S6Rs, and finally no D6R. To compare the surface reaction enthalpies, the energy of condensing two S6Rs to give a D6R was calculated within the same surface cell utilised for the surface relaxations. The total enthalpy of reaction for this process was found to be $+64$ kJ mol^{-1}, which is rather endothermic. Nevertheless, the entropy ($T\Delta S$) term associated with forming this structure at typical reaction temperatures is $\sim +80$ kJ mol^{-1}, implying that the free energy of formation of the D6R is very favourable. In a similar manner to the Beta C study, we considered the two-stage reaction of S6Rs on to the D6R denuded crystal surface and the one step condensation of the D6R upon the surface. The two-stage process involves a very endothermic step of $+67$ kJ mol^{-1} followed by a less endothermic step of $+40$ kJ mol^{-1}. The one step process is found to be endothermic by $+49$ kJ mol^{-1}. Under reaction conditions (high T) and taking into account aqueous solvation effects (which have found to systematically make the total energy of formation of chain and ring species more exothermic[14]), we therefore expect that both reactions are feasible. In the two-step reaction process, the intermediate structure is thermodynamically unstable with respect to further condensation and hence the S6R terminating structure is not expected to be seen. The D6R one-step reaction would also be expected to occur and hence both reactions apparently explain the absence of the intermediate termination. To assess why the S6R terminated surface is disfavoured, the relaxed 6 ring upon the crystal face was extracted, and fully hydroxylated and then relaxed within the surface cell, keeping both the silicon and oxygen atoms fixed. The conformation adopted at the crystal face was found to be $+16$ kJ mol^{-1} more unfavourable than that found for the gas phase fully relaxed S6R and hence the reaction pathway is less favourable.

It is worth noting two qualifying points relating to these results: (1) it has been found that the entropy term can be the dominant term in the free energy of formation of species but that the entropy term may not be the same for the S6R and D6R species (because of their obvious disparity in conformational freedom—presumably, the S6R will have a smaller $T\Delta S$ term than D6R, which would tend to favour the D6R one step process). (2) the systematic deprotonation energies of different species are non-linear quantities and moiety specific and recent work from Trinh et al.[34] has shown that the barrier heights associated with condensing charged species (as would be found under high pH synthesis conditions) are considerably more favourable than neutral species. We have calculated that deprotonation of the crystal surface is typically at least 10–20 kJ mol^{-1} more favourable than removing the first proton from an oligomeric solution species and hence the reaction mechanism would undoubtedly feature a charged surface silanol with a neutral or charged oligomer. A final and most critical point relates to the existence of S6Rs and

D6Rs in solution. In brief, there is no data which proves the existence of such species within the reaction media and hence the mechanism we have explored may be purely hypothetical. We will return to this topic in the Discussion.

Discussion

We have examined the surface structure of two zeolites, whose characteristic SBU are the D4R and D6R within the Beta C and FAU/EMT frameworks, respectively. HREM images show surface terminations in which the secondary building unit is either present and complete, or completely absent. First principles calculations show that the formation of a terminating structure exposing a D4R or D6R can proceed *via* at least two mechanisms. In the first mechanism, a S4R or S6R attaches to the crystal surface, followed by attachment of another ring *i.e.* a two-step mechanism. The second mechanism is a one-step reaction in which a D4R or D6R ring attaches directly to the surface. Overall, both mechanisms are similarly viable (taking into account the entropy contribution), but the two-step mechanism proceeds by a relatively unfavourable step followed by a very favourable step. The intermediate termination is therefore thermodynamically unstable, due to the unfavourable conformation of the ring at the surface and hence such terminating structures are not expected to be seen. On the crystal faces examined here, the intermediate terminations are not observed, hence providing an explanation for the observed surface structure. The one-step mechanism involving condensation with a DR also accounts for the observed terminations and the energetics suggest this mechanism is viable. However, both mechanisms rely upon an adequate concentration of ring species and therein lies a major source of contention. D4Rs and S4Rs have been observed in NMR experiments, but to the best of our knowledge S6Rs and D6Rs have not yet been isolated (possibly their concentration is too small to detect). However, it is not beyond the realms of possibility that the solution conditions which have thus far been examined for species, suppress the formation of S6R and D6R species, since pH, dilution and cation concentration are all thought to influence oligomer populations. One the one hand, for Beta C for at least one part of the crystal growth mechanism (layer by layer growth of the crystal terrace), the SBU can participate in crystal growth, which is apparently at odds with the assertions of Knight and Kinrade[2,3,35,36] who argue that oligomeric units are simply spectator species in crystal growth, and that primary building units facilitate growth. A very recent study[37] strongly supports their view and has presented evidence that at the crystallisation stage of growth, there are no pronounced concentrations of oligomers which deplete on precipitation. Furthermore, in our own recent work on analcime, we show that the measured step heights, which coincide with the height of a T site can only be explained by a monomer mediated growth.

In the case of zeolite Y, since the S6R and D6R have not been seen in solution, one presumes they cannot be intrinsic to the growth mechanism of this zeolite. One speculation is that growth mechanisms involving primary *and* secondary building units may be important for forming certain zeolites, whilst in other materials such as analcime, the layer-by-layer growth mechanism must proceed *via* primary building units.[38] If secondary building units do indeed participate at all in crystal growth, the kinetics of forming these species could vary in very local regions within solution, hence the growth rates of individual faces would be strongly influenced by local populations of oligomers, leading to disparate morphologies. The results described here pertain to inferred pathways in the post nucleation crystal growth stage, when the crystal surfaces are well formed, whereas the solution characterisation experiments relate to pre-nucleation or germinal stages of crystallisation. The greater importance of thermodynamic rather than kinetic factors during the latter stages of crystal growth could imply distinct growth mechanisms during these regimes.

Finally, invoking the SBU mediated growth mechanism, we note that at the crystal–gel interface, it seems unless there is a very large concentration of D4Rs, diffusion limited growth would occur caused by the low density of D4Rs in a viscous medium occasionally reaching the nucleation sites on the surface. Because the conditions at the interface are very distinct from the conditions probed by NMR and mass spectroscopic studies, we cannot discount the possibility that the gel interface is abnormally rich in oligomers not found in clear solution conditions sampled by various spectroscopic techniques.[35,37] The work presented here highlights the need for more detailed investigations of the evolution of oligomeric silicate species throughout the crystallisation process, to unravel whether they are active or inactive participants in any regime of crystal assembly.

Acknowledgements

BS would like to acknowledge EPSRC for the award of computing time on the HPC(x) facility and additionally for the provision of in house computing facilities. BS would like to thank Dr Joost VandeVondele for invaluable guidance on the Quickstep code and Dr Miguel Mora-Fonz and Dr Dewi Lewis for data on the stability of silicate species. A part of this work is financially supported by VR, Sweden and JST, Japan to TO, ZL and OT.

References

1 M. Haouas and F. Taulelle, *J. Phys. Chem. B*, 2006, **110**, 3007–3014.
2 S. D. Kinrade, C. T. G. Knight, D. L. Pole and R. T. Syvitski, *Inorg. Chem.*, 1998, **37**, 4272–4277.
3 C. T. G. Knight, R. T. Syvitski and S. D. Kinrade, in *Zeolites: a Refined Tool for Designing Catalytic Sites*, Elsevier, Amsterdam, 1995, pp. 483–488.
4 A. Corma, M. J. Diaz-Cabanas, J. L. Jorda, C. Martinez and M. Moliner, *Nature*, 2006, **443**, 842–845.
5 J. M. Newsam, M. M. J. Treacy, W. T. Koetsier and C. B. Degruyter, *Proc. R. Soc. London, Ser. A*, 1988, **420**, 375.
6 Z. Liu, T. Ohsuna, O. Terasaki, M. A. Camblor, M. J. Diaz-Cabanas and K. Hiraga, *J. Am. Chem. Soc.*, 2001, **123**, 5370–5371.
7 B. Slater, C. Richard, M. A. Catlow, Z. Liu, T. Ohsuna, O. Terasaki and M. A. Camblor, *Angew. Chem., Int. Ed.*, 2002, **41**, 1235.
8 M. E. Chiu, B. Slater and J. D. Gale, *Angew. Chem., Int. Ed.*, 2005, **44**, 1213–1217.
9 M. Mistry, *PhD Thesis*,, University of London, London, 2005.
10 S. C. Parker, N. H. de Leeuw and S. E. Redfern, *Faraday Discuss.*, 1999, **114**, 381–393.
11 J. R. Agger, N. Hanif, C. S. Cundy, A. P. Wade, S. Dennison, P. A. Rawlinson and M. W. Anderson, *J. Am. Chem. Soc.*, 2003, **125**, 830–839.
12 S. Piana and J. D. Gale, *J. Am. Chem. Soc.*, 2005, **127**, 1975–1982.
13 J. VandeVondele, M. Krack, F. Mohamed, M. Parrinello, T. Chassaing and J. Hutter, *Comput. Phys. Commun.*, 2005, **167**, 103–128.
14 M. J. Mora-Fonz, C. R. A. Catlow and D. W. Lewis, *Angew. Chem., Int. Ed.*, 2005, **44**, 3082–3086.
15 A. Corma, F. Rey, J. Rius, M. J. Sabater and S. Valencia, *Nature*, 2004, **431**, 287–290.
16 T. Wakihara, Y. Sasaki, H. Kato, Y. Ikuhara and T. Okubo, *Phys. Chem. Chem. Phys.*, 2005, **7**, 3416–3418.
17 J. R. Agger, N. Pervaiz, A. K. Cheetham and M. W. Anderson, *J. Am. Chem. Soc.*, 1998, **120**, 10754–10759.
18 F. Dougnier, J. Patarin, J. L. Guth and D. Anglerot, *Zeolites*, 1992, **12**, 160–166.
19 M. M. J. Treacy, D. E. W. Vaughan, K. G. Strohmaier and J. M. Newsam, *Proc. R. Soc. London, Ser. A*, 1996, **452**, 813–840.
20 T. Ohsuna, O. Terasaki, D. Watanabe, M. W. Anderson and S. W. Carr, *Chem. Mater.*, 1994, **6**, 2201–2204.
21 V. Alfredsson, T. Ohsuna, O. Terasaki and J. O. Bovin, *Angew. Chem., Int. Ed.*, 1993, **32**, 1210–1213.
22 A. M. Goossens, B. H. Wouters, P. J. Grobet, V. Buschmann, L. Fiermans and J. A. Martens, *Eur. J. Inorg. Chem.*, 2001, 1167–1181.
23 N. Hanif, M. W. Anderson, V. Alfredsson and O. Terasaki, *Phys. Chem. Chem. Phys.*, 2000, **2**, 3349–3357.
24 A. M. Goossens, B. H. Wouters, V. Buschmann and J. A. Martens, *Adv. Mater.*, 1999, **11**, 561.

25   E. J. P. Feijen, J. A. Martens and P. A. Jacobs, *J. Chem. Soc., Faraday Trans.*, 1996, **92**, 3281–3285.
26   J. M. Newsam, M. M. J. Treacy, D. E. W. Vaughan, K. G. Strohmaier and W. J. Mortier, *J. Chem. Soc., Chem. Commun.*, 1989, 493–495.
27   T. Ohsuna, O. Terasaki, V. Alfredsson, J. O. Bovin, D. Watanabe, S. W. Carr and M. W. Anderson, *Proc. R. Soc. London, Ser. A*, 1996, **452**, 715–740.
28   O. Terasaki, T. Ohsuna, V. Alfredsson, J. O. Bovin, D. Watanabe, S. W. Carr and M. W. Anderson, *Chem. Mater.*, 1993, **5**, 452–458.
29   N. J. Henson, A. K. Cheetham and J. D. Gale, *Chem. Mater.*, 1994, **6**, 1647–1650.
30   M. J. Sanders, M. Leslie and C. R. A. Catlow, *J. Chem. Soc., Chem. Commun.*, 1984, 1271–1273.
31   G. Bonilla, I. Diaz, M. Tsapatsis, H. K. Jeong, Y. Lee and D. G. Vlachos, *Chem. Mater.*, 2004, **16**, 5697–5705.
32   L. Whitmore, B. Slater and C. R. A. Catlow, *Phys. Chem. Chem. Phys.*, 2000, **2**, 5354–5356.
33   J. D. Gale and A. L. Rohl, *Mol. Simulation*, 2003, **29**, 291–341.
34   T. T. Trinh, A. P. J. Jansen and R. A. van Santen, *J. Phys. Chem. B*, 2006, **110**, 23099–23106.
35   C. T. G. Knight, J. P. Wang and S. D. Kinrade, *Phys. Chem. Chem. Phys.*, 2006, **8**, 3099–3103.
36   C. T. G. Knight and S. D. Kinrade, *J. Phys. Chem. B*, 2002, **106**, 3329–3332.
37   S. A. Pelster, W. Schrader and F. Schuth, *J. Am. Chem. Soc.*, 2006, **128**, 4310–4317.
38   J. R. Agger, M. Shoaee, M. Mistry and B. Slater, *J. Cryst. Growth*, 2006, **294**, 78–82.

Crystal growth in nanoporous framework materials†

Michael W. Anderson,* Jonathan R. Agger, L. Itzel Meza, Chin B. Chong and Colin S. Cundy

Received 5th December 2006, Accepted 30th January 2007
First published as an Advance Article on the web 10th April 2007
DOI: 10.1039/b617782b

Future applications of nanoporous materials will be in opto-electronic devices, magnetic and chemical sensors, shape-selective and bio-catalysis, structural materials and nuclear waste management. Crucially, in all such applications, an understanding of crystal growth to the same depth as has been achieved in semiconductor technology is needed. Therefore, defects, intergrowths, dopants and isomorphous substitution must be controlled, and crystal habit and size (*e.g.* single crystal films) must be fabricated with precision. These goals elude the community because of lack of understanding of crystal growth processes. Modern microscopy techniques including AFM, ultra-high resolution SEM and HREM coupled with theoretical calculations are beginning to reveal the details of these growth processes yielding the important thermodynamic data crucial to effect synthetic control such as: controlling defects; controlling intergrowths; introducing chirality; modifying surface access; altering diffusion pathways; controlling crystal habit; synthesising templated materials cheaply in order to render them economically viable; controlling crystal size for instance as single crystal films. In this paper we will discuss recent results including: the details of surface alteration processes in nanoporous materials, measured *in situ*, under different chemical environments and the ability to switch processes on and off by the control of growth conditions. Further we illustrate an approach to theoretically model the crystal growth in such complex systems which ultimately delivers activation energies for fundamental growth processes.

Introduction

Crystal growth in nanoporous framework materials is a subject that has received considerable attention over many decades.[1,2] This interest has been fueled by the enormous industrial significance of this class of material in the areas of petrochemicals, ion-exchange and gas separation. Indeed much of this work has been performed within an industrial setting with the goal normally to synthesise new crystalline phases with desired pore architecture and functionality. Crystal size and morphology are important for many of these applications.[3] For instance, for

Centre for Nanoporous Materials, School of Chemistry, The University of Manchester, Oxford Road, Manchester, UK M13 9PL. E-mail: m.anderson@manchester.ac.uk; Fax: +44 161 306 4551; Tel: +44 161 306 4519

† We would like to acknowledge the assistance of R. J. Plaisted for help with the continuous flow experiments. Also to EPSRC and ExxonMobil for funding.

application in catalysis it is often important to limit crystal size in order to reduce diffusion path-lengths from the crystal periphery to the internal void space. For similar reasons, when the pore structure is anisotropic, such as in a material with unidimensional porosity, it is important to reduce the crystal size along the direction of these pores. As many zeolites with unidimensional pores have a natural tendency to grow with a long crystal dimension parallel to the pores, this requires a modification of the crystal morphology or aspect ratio.[4] As with most synthetic work these desired effects can often be achieved using an informed trial-and-error approach coupled with high-throughput synthetic methods.[5] However, the phase space is enormous (incorporating both inorganic and organic components) and even high-throughput methods are hampered by the mechanical complexity of developing a robotic approach for every synthetic and characterisation step associated with this kind of work. Nevertheless, it could be argued that the subject had reached a reasonable degree of maturity by the early 1990s despite quite a range of new zeolitic and related phases being discovered since this time.[6]

New impetus to the study of crystal growth in many fields, including nanoporous framework materials, was ignited with the advent of scanning probe microscopy methods.[7] This opened up a considerable new level of detail, with possibilities for monitoring growth processes at the molecular scale, *in situ*, thereby extracting detailed growth mechanisms and related thermodynamic constants. In practice realisation of this is far from straightforward: first, because the experimental procedures are delicate; second, because the theoretical models to simulate experimental results are not in place and must be developed from scratch.

In this paper we discuss results from atomic force microscopy, AFM, applied to zeolite A and the siliceous analogue of zeolite ZSM-5 known as silicalite. The experimental methodology involves a combination of *ex situ* and *in situ* AFM. The former approach is the most flexible allowing crystals to be grown under conditions which are not suitable for the microscope but optimised to answer specific questions. Following crystallisation under controlled conditions the crystals are transferred to the atomic force microscope. The *in situ* measurements by contrast need to be designed to be compatible with the atomic force microscope. For zeolite preparations desirable conditions for growth would often necessitate growth from a gel phase. This is incompatible with the microscope as the laser cannot penetrate the cloudy suspension. For this reason we restrict ourselves in this paper to discussions of dissolution experiments under clear basic conditions. A further important requirement for the study of crystal growth or dissolution is that the kinetics must be slow enough to be captured at a frame rate of *ca.* 1–2 min. Such a slow rate means that supersaturation must be adjusted close to equilibrium conditions and consequently the thermodynamics will be that associated with near equilibrium.

In order to interpret the AFM results it is necessary to simulate surface topography.[8] In this paper we demonstrate a method which is able to simulate crystal growth in a nanoporous framework material from which fundamental growth rates can be extracted. In order to maximise the experimental input to the simulations we adopt an approach which endeavours to simulate both crystal habit and surface topography using one set of growth parameters. At present the method is hard-wired to simulate growth in cubic systems but the same general approach could be used to simulate crystal growth in any crystal system.

Setting the scene

The fundamental importance of the nanoporous class of materials which include zeolites is the nature and accessibility of the internal pore structure which imparts very high surface area and also high selectivity for molecular recognition. For zeolites the accessibility and integrity of this pore structure is governed by the precise nature of the crystal. Fig. 1 shows three examples of crystals of zeotype materials. The upper image shows small 5 µm crystals of the siliceous microporous material

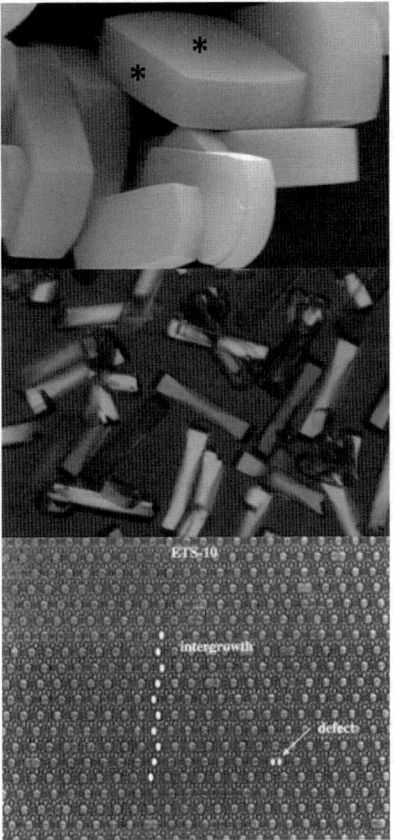

Fig. 1 Upper, SEM shows 5 μm near perfect crystals of silicalite, tunnels exit on facets marked with *. Middle, large 280 μm long silicalite crystals exhibiting faulted structure. Lower, HREM of microporous titanosilicate ETS-10 showing details of defects and intergrowth, pore size ca. 7 Å.

silicalite. The crystals have been grown at a comparatively low temperature of 130 °C and the morphology is indicative of crystals with low defect concentration and no intergrowths. In other words the crystal form is as near perfect as might be expected. The pore structure of silicalite is two dimensional with tunnels exiting the crystal surface on the two flat facets marked with an asterisk. The aspect ratio of the crystal defines the overall tunnel length which is consequently different in different directions. The middle image shows the same crystal structure, silicalite, grown at a much higher temperature, 180 °C, which produces much larger crystals, ca. 180 μm in length and modified aspect ratio and overall habit. Most strikingly in the optical micrograph of these crystals is the birefringent *hour-glass* effect which is due to internal faulting. This faulting will directly affect the diffusion of guest gas molecules through the pore network. Finally, the lower high-resolution electron micrograph of the microporous titanosilicate ETS-10 shows that the *normal* structure contains both defects and intergrowths. The defects are very well defined structural imperfections, in this case double pores, and the intergrowth is a random connection of the pore structure which is built up of layers from the bottom of the image to the top. The defects will affect porosity and thereby selectivity of the crystal but also the thermal and hydrothermal stability of the material. The intergrowth changes the entire structure, one possible ordered intergrowth of ETS-10 exhibits a spiral pore in one direction.[9] All the features discussed in these examples should be controllable if it is possible to understand and control crystal growth. Crystal aspect ratio becomes particularly important for structures with unidimensional pores (such as zeolite L[4]) where typically the crystals form needles with the pores along the long needle direction. For catalytic purposes this is undesirable as short diffusion path-lengths are preferred in order to increase accessibility to the internal pore space. Another very important factor which must be considered in the synthesis of microporous structures is the cost of any organic template (or structure-directing agent, SDA). Such molecules are crucial for the synthesis of the vast majority of zeotype materials and preclude the preparation for commercial purposes on the basis of cost. Reducing the amounts, or elimination of the SDA, is a highly desirable goal to attain. Another aspect which is of interest is the size of the crystal—most microporous materials grow as micron-sized crystals. This is ideal for catalysis where the smaller the crystal usually the more active the catalyst. However, for opto-electronic device technology large crystals are preferred. Similarly single crystal films would be of great interest for separation applications. All these desires elude the community at present because our understanding of the fundamental growth processes is insufficient. Consequently, the goal of our work on crystal growth in nanoporous materials is to be able to ultimately exert control over the following important parameters:

- Defects
- Intergrowths
- Introduction of chirality
- Surface access
- Diffusion pathways
- Crystal habit (*e.g.* short *c*-axis)
- Template (especially for cheaper synthesis)
- Crystallite size (*e.g.* single crystal films)

The strategy that we adopt along with collaborators in Stockholm, Lund, Versailles, Royal Institution, UCL and Bath[10] to tackle these problems is interdisciplinary but relies heavily on microscopy techniques in particular: *in situ* and *ex situ* atomic force microscopy; very high-resolution scanning electron microscopy; high-resolution electron microscopy. This is coupled with theoretical calculations of surface energies and growth pathways and simulations of surface topologies and crystal habit. These theoretical methods lead to rate constants for fundamental growth processes. Finally we are utilising NMR and mass spectrometry methods to look at solution speciation within the mother liquor from which the crystals grow.[11] This contribution details results from recent AFM measurements coupled with simulation of the experimental data.

Experimental

The zeolite A crystals were synthesized from a gel with molar composition SiO_2 : 2.23 Na_2O : 5.18 TEA : 0.89 Al_2O_3 : 246 H_2O. Two solutions were prepared as follows: solution 1, 0.85 g aluminum wire (BDH) was dissolved in an alkaline solution of 3.14 g NaOH (BDH) dissolved in 39.39 mL deionised water, the final mixture was filtered to remove impurities; solution 2 was prepared by adding 3.65 g tetraethyl orthosilicate (TEOS from Aldrich) to a solution of 13.57 g triethanolamine (TEA from Acros organics) and 39.39 g water. Both solutions were mixed with vigorous stirring and then aged for 30 min in polypropylene bottles.

Atomic force micrographs were recorded on a Digital Instruments MultiMode™ SPM with Nanoscope III controller using AFM Contact Mode with 0.58 N m^{-1} force constant silicon nitride tips at scan rates of 3 Hz. A syringe pump (Razel™ model A-99 from SEMAT) was used for the experiments conducted under continuous flow. All experiments were carried out at room temperature, 25 °C. Errors in terrace heights were determined by measuring at least 50 separate heights.

1.6 μm long, un-twinned, lozenge-shaped silicalite seed crystals were synthesised as detailed elsewhere.[12] Seed crystals at a concentration of 9.5 g L^{-1} in their spent mother liquor were placed in a specially modified 1 L autoclaved reactor stirred at 300 rpm. The slurry was heated to 130 °C over a period of 30 min and then left to equilibrate for 3 h, after which nutrient feed was switched on at an initial rate of 90 mL h^{-1}. The nutrient feed comprised silica sol, tetrapropylammonium bromide and sodium hydroxide with the following composition [Si] = 0.48 M, Si : OH = 8.72 and Si : TPA = 10.00. Samples were removed at regular intervals in order to maintain an effectively constant working volume in the reactor. In the absence of secondary nucleation, the assumption of effective stirring to create homogeneous crystal distribution enabled calculation and monitoring of the required nutrient feed to produce constant linear crystal growth rates of 0.4 μm h^{-1}. During the course of the reaction the nutrient feed was disconnected for periods of *ca.* 16 h allowing gel depletion akin to the latter stages of a normal batch synthesis. Quoted reaction times refer to the total period for which the nutrient feed was switched on.

The simulation of micrographs was performed using a computer model written in Fortran. Graphical output from the program is produced using Mathematica.

Zeolite A dissolution

Understanding precisely which structural units attach to the crystal surface from solution is a difficult problem to solve. Techniques, such as NMR and mass spectrometry[11,13] may be applied to monitor solution speciation, however, it is usually difficult to differentiate spectator species from active species in solution. One technique which has been successfully adopted to determine the species which attach to the solid for nanoporous aluminophosphates[14] is to look at the reverse dissolution reaction. Under favourable conditions the rate constants for competitive processes result in high concentrations of the active species being formed by the dissolution process whereas high concentrations of spectator species are formed during the growth process. With a similar purpose in mind we have monitored the dissolution of zeolite A using *in situ* AFM.

Fig. 2 shows a series of images showing the dissolution of the (100) facet of zeolite A under 0.5 M NaOH solution recorded every three minutes. The schematic on the right depicts the changes in the surface topology during this treatment. The 1.2 ± 0.1 nm high square terraces dissolve initially *via* terrace retreat. The terrace height corresponds to half a unit cell consisting of one sodalite cage and a double 4-ring (see Fig. 3). Interestingly the straight edges of the terraces dissolve more slowly than the holes in the terraces which open up preferentially. This is in accordance with our previous work on growth of zeolite A which showed that growth at kink sites is about 15 times faster than growth at edge sites.[8] The curved terrace edges present in the holes will show a high kink site density compared to the straight terrace edges and, consequently, it appears that dissolution is also favoured at kink sites over edge sites. The height of the receding terrace edge is only 0.9 ± 0.1 nm rather than the full 1.2 ± 0.1 nm. Indeed the remaining 0.3 ± 0.1 nm can be seen as an undisturbed terrace edge which is depicted as a dotted line in the schematic in Fig. 2. Consequently, there is a second structure 0.3 ± 0.1 nm in height which is more stable than the 0.9 ± 0.1 nm structure. The final dissolution of the top terrace occurs as the terrace breaks up into smaller squares which are rather uniform and *ca.* 90 × 90 × 9 nm in size. This phenomenon can be seen even more clearly in Fig. 4. After the 0.9 ± 0.1 nm high terrace is removed the dissolution repeats and another terrace is removed which is 0.9 ± 0.1 nm high. This means that dissolution of the intervening 0.3 ± 0.1 nm high terrace is not observed by AFM. The most likely explanation for this is that the 0.3 ± 0.1 nm high terrace dissolves orthogonal to the surface. This can be understood by careful consideration of the structure of zeolite A. Zeolite A is composed of sodalite cages (truncated octahedra) connected to each other by double 4-rings in a cubic manner. The most likely division of the unit cell into structural units corresponding to the terrace heights observed by AFM is that the 0.9 ± 0.1 nm terrace corresponds to a sodalite cage and the 0.3 ± 0.1 nm terrace corresponds to

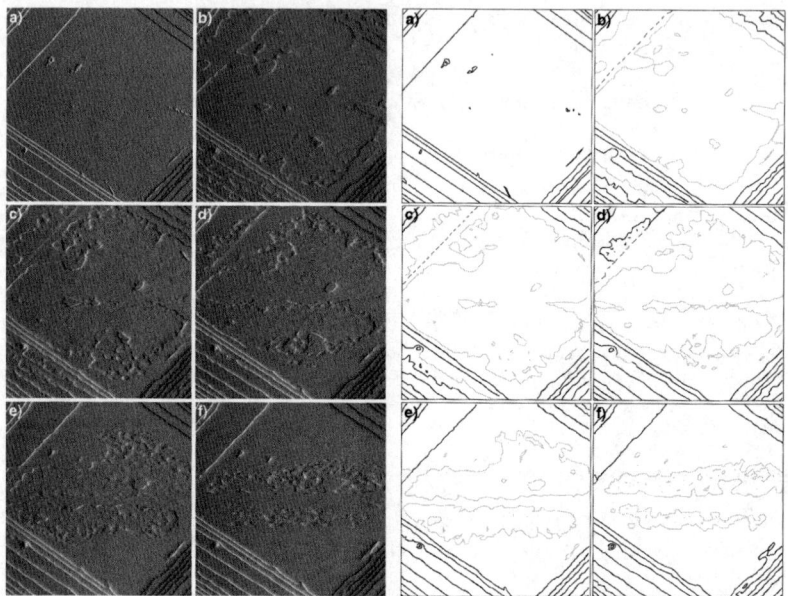

Fig. 2 Left: 4.3 × 4.3 µm² deflection AFM images of a zeolite A crystal under NaOH 0.5 M at 0, 31, 33, 36, 38 and 47 min. Right: schematic representation showing the dissolution process observed in the left AFM images. The black lines correspond to step heights of 1.2 nm, grey areas represent layers with step heights *ca.* 0.9 nm and dashed lines correspond to step heights of *ca.* 0.3 nm.

the addition of a single 4-ring. As the sodalite cages are connected to each other parallel to the (110) facet it is not surprising that they dissolve in a correlated fashion, *i.e.* by terrace retreat. One sodalite cage must be dissolved to expose the next sodalite cage. However, the single 4-rings are not connected to each other and consequently they should dissolve in an uncorrelated manner. This will be impossible to distinguish by AFM as the single 4-rings have a lateral dimension of only

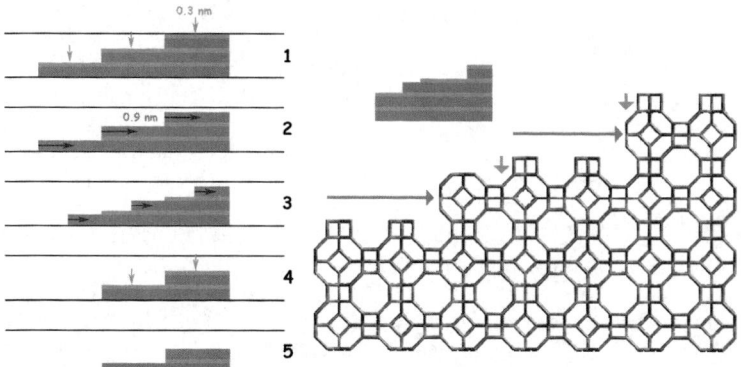

Fig. 3 Schematic illustration of the dissolution process for zeolite A under alkaline conditions. Uncorrelated removal of a single 4-ring capping the double 4-ring shown on the surface of the structure (right, vertical arrow) amounts to 0.32 nm. Correlated removal of a β-cage cage, except the 4-ring base, amounts to 0.92 nm. Layers are successively removed, 4-rings orthogonal to surface and β-cage cages parallel to surface *via* terrace retreat. The process is either 1–2–3–4 repeat or 2–3–4–5 repeat depending on the initial exterior surface of the crystal. Note how the top surfaces of 1 and 4 or 2 and 5 are only displaced orthogonal to the surface by 1.2 nm and therefore indistiguishable by AFM.

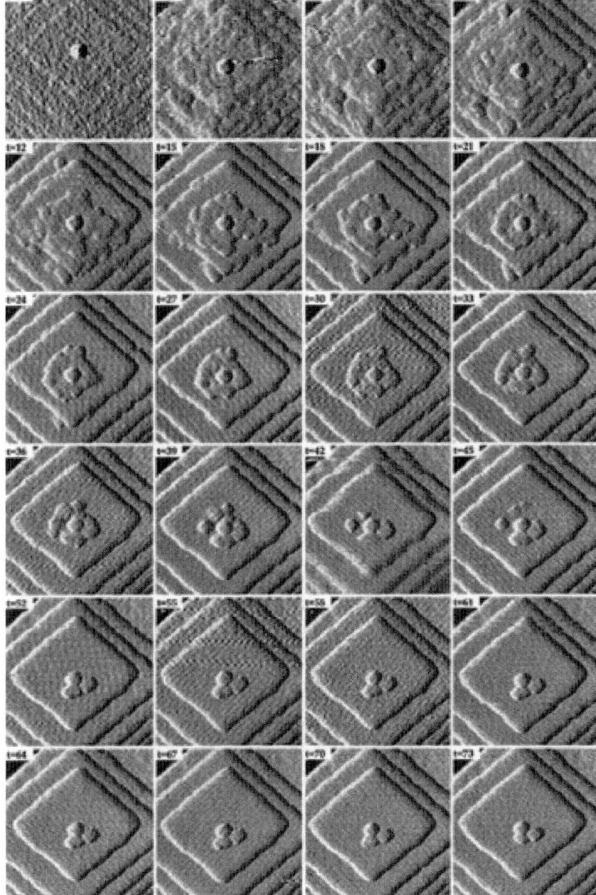

Fig. 4 Series of 1 × 1 μm² deflection AFM micrographs of a zeolite A crystal under a static solution of mother liquor diluted to 67%. Images from 0–73 min.

ca. 0.3 nm and the lateral resolution of the AFM is typically *ca.* 3 nm. These two structural terminations for zeolite A are consistent with the theoretical work of Slater *et al.*[15] who calculate that these are the energetically preferred surface structures.

Another feature of this mechanism of dissolution in zeolite A is that periodically the terrace configuration returns to the same lateral topology (compare schematic 1 and 4 or schematic 2 and 5). This is a result of the difference in kinetics of dissolution of the two processes. The dissolution of the single 4-rings is sufficiently slow to arrest the terrace retreat of the sodalite cages essentially at the same position as the preceding terrace edge. This is observed experimentally and explained well by our model of zeolite dissolution.

A similar dissolution process is observed when zeolite A is exposed to an undersaturated mother liquor solution, diluted to 67%, as seen in Fig. 4. Particularly apparent in this series of images is the final break up of the retreating 0.9 ± 0.1 nm high terrace into small, rather uniform squares each with *ca.* 90 × 90 nm lateral dimension. This rather strange phenomenon suggests that at this critical size the free energy of these nano-slabs is a minimum and therefore stabilised. The final dissolution of these units is then rapid but not by terrace retreat. Theoretical modelling of this phenomena would be useful to understand the interplay of both entropic and enthalpic contributions to this stabilisation.

Fig. 5 Deflection AFM images of a zeolite A crystal (a) 4.1 × 4.1 μm^2 under a static solution of mother liquor diluted to 50%, (b) 4.0 × 4.0 μm^2 under a continuous flow solution of mother liquor diluted to 10%, (c) 5.0 × 5.0 μm^2 under a static solution of mother liquor 100%.

Defects and crystal growth in zeolite A

Crystals of zeolite A often have a tendency to exhibit pyramids of terraces all seemingly emanating from the same point on the crystal surface. Dissolution of zeolite A also reveals another interesting feature. Owing to the preferential dissolution of defects in the material it is revealed that there is a defect at the centre of each pyramid, see Fig. 5. The exact nature of this defect is not clear and is observed in the AFM simply as a depression in the surface. Because the AFM tip is unable to penetrate deep into such a small hole it is impossible to ascertain the full depth of the defect, however, intuitively it is expected that it will penetrate at least the depth of the exposed pyramid. Such line defects would be extremely difficult to observe using electron microscopy owing to the rather unstable nature of zeolite A in the electron beam, however, it would be of interest to try to locate these on siliceous forms of zeolite A which should be substantially more beam stable. Electron microscopy might give a better understanding of the fine structural nature at such defects. Our work shows, however, that AFM is a very straightforward technique for locating defects which emanate at the crystal surface and that selective dissolution at the defect sites is a useful strategy to expose otherwise topographically hidden defects.

As well as defect initiated layer growth in zeolite A in some cases we also observe spiral growth from screw dislocations. Similar spiral growth has been reported before for the natural zeolite heulandite.[16] The nature of possible screw dislocations in zeolites has been discussed and modelled by Slater *et al.*[17] Owing to the large unit cell parameter for a zeolite, 2.4 nm for zeolite A, it is necessary for a screw dislocation to spiral around a mesoscopic void or substantially twisted structural conformation. This is rather different to a screw dislocation in a metallic crystal which simply requires a simple displacement of atoms close to the dislocation line. Slater *et al.*[17] showed how it was possible to twist the structure in such a way that essentially no mesoscopic void is created but there is a knock on effect to the structural twisting for several unit cells away from the dislocation line. The AFM shown in Fig. 6 shows that the screw dislocation is often accommodated around a more substantial mesoscopic void *ca.* 60 nm diameter, equivalent to *ca.* 25 unit cells. This should certainly be sufficient for relaxation of the structure. We have also observed spiral growth where any mesoscopic void present is below the lateral resolution of the AFM tip and so we can certainly not rule out the possibility of Slater-type screw dislocations. The predominant growth mechanism in zeolite A is, however, layer-by-layer growth and growth at screw dislocations can be viewed as an important but subsidiary mechanism.

Supersaturation and controlling layer growth

The key to control over defects, intergrowths, crystal habit *etc.* is to be able to control fundamental growth processes independently of each other. This can either be achieved by enhancing or retarding certain growth processes or by facilitating

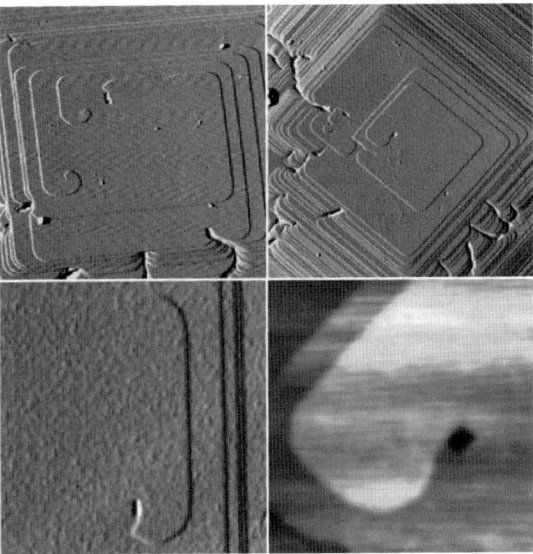

Fig. 6 Spirals observed on the {100} faces of zeolite A: top images AFM deflection images with scan sizes of (left) 4 × 4 μm, (right) 6 × 6 μm, lower images zoomed.

completely new avenues for crystal growth. In this work we demonstrate how supersaturation can be used to effectively control the nucleation of events *versus* the spreading of surface terraces. The principle of this strategy is that different fundamental processes have different activation energies. By changing supersaturation, or temperature, it is possible to change the rates of these processes. However, because the rates vary in a non-linear fashion, and the response to changes in supersaturation is dependent upon activation energy it is possible to alter relative rates for different processes.

A simplistic view of the growth of a zeolite crystal is depicted in Fig. 7a and b. The figure shows how the crystal size and growth rate varies as a function of supersaturation. At the beginning of a crystallisation a silica source is dissolved in a basic medium. During this period the supersaturation level is low and the crystals remain small. However, as the supersaturation level grows so does the crystal growth rate until a steady-state is reached at constant supersaturation and constant growth rate. During this period the crystal length increases linearly with time. Finally the silica nutrient is exhausted and the saturation level drops and the crystals eventually cease to grow. Crystals removed from a growth medium at the end of a synthesis would exhibit surface features consistent with growth under low supersaturation conditions. In order to control the supersaturation conditions we have grown crystals under conditions of constant supersaturation using a continuous flow autoclave. Nutrient is continually fed in to the vessel at the same rate as the nutrient is being depleted by crystal growth.

The model system we have chosen for study is the siliceous microporous material, silicalite. Silicalite has a two dimensional medium pore system and the orthorhombic crystal symmetry is manifested in crystals with two flat facets and a circular facet, see Fig. 7c. This morphology is consistent with crystals which have low levels of defects and faulting. The crystal growth was conducted in a manner whereby the supersaturation was maintained at a high level, area shaded in Fig. 7a, and then the level allowed to drop, area shaded in Fig. 7b. This was repeated twice such that the supersaturation level was allowed to drop after 9.6 and 16.4 h. Fig. 7c shows scanning electron micrographs, reproduced to scale, of the crystals removed after various periods of time. The crystals grow in a highly linear fashion while the

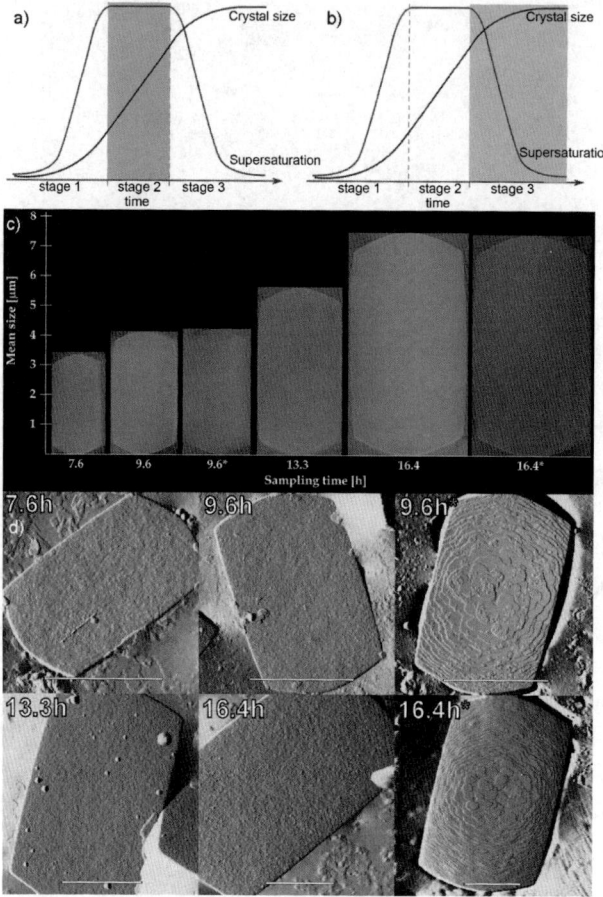

Fig. 7 Parts (a) and (b) show a schematic representation of the supersaturation and crystal length changes during crystal growth; (c) series of scanning electron micrographs of silicalite crystal grown under continuous nutrient feed conditions, times marked with an asterisk show when the nutrient feed was switched off; (d) series of AFM images of the (010) facet of silicalite.

supersaturation is high and the crystals almost cease growing when the supersaturation level is allowed to drop. Fig. 7d shows a series of AFM images of the (010) surface of this set of crystals. What is most noticeable is the substantial difference in the surface topology when the supersaturation drops, times 9.6* and 16.4* h. In these images 1 nm high terraces are clearly distinguished and the concentration of nucleation events is fairly low. In contrast, in the AFM images recorded when the supersaturation level is high, the crystal surface has a high density of surface nucleation events without clearly established terraces. These images indicate that when the supersaturation level is dropped the first process to be limited is surface nucleation, the event which will have the highest activation energy. However, surface spreading continues and the surface roughness grows out. This effect can be seen more clearly in the following section where we have simulated the surface topology *via* crystal growth models.

With silicalite growth conditions have been established whereby it is possible to selectively turn off the process with the highest activation energy while maintaining growth *via* less energy expensive processes. If this is a general phenomena then it offers an excellent prospect for controlling intergrowth and defect populations which are often the result of the competitive layer nucleation and layer spreading processes. Such is the case for the important intergrowth system hexagonal and cubic zeolite Y (structure codes EMT/FAU).[18] It may also be illustrated with reference to the microporous titanosilicate system ETS-10 which was referred to in the Introduction.[9] ETS-10 is a system which is built from layers and contains both well defined defects as well as an intergrowth structure resulting from a random stacking of these layers. The layers are themselves composed of rods of titanate chains surrounded by a silica network, Fig. 8. However, if we initially consider the layer growth as depicted on the high-resolution electron micrograph in Fig. 8 then the different colour circles show the pore structure in the different layers. The layer depicted at the bottom of the micrograph in white has three layers above. These are a result of multiple nucleation such that the layer does not spread fast enough to ensure complete coverage. If the multiple nuclei are displaced relative to each other then at the boundary where the nuclei meet a defect is created which is manifested as a double pore. This mechanism then repeats. In order to decrease the level of defects it would be necessary to decrease the nucleation rate relative to the spreading rate, in a similar manner to that already achieved in the silicalite system. If it were possible to achieve the opposite such that the rate of nucleation was increased relative to the spreading rate then the concentration of double-pore defects could in principle be increased.

Controlling the intergrowth structure requires controlling the stacking sequence which will require making one polymorph more energetically favoured. This will probably require a template either to force a particular polymorph or poison a particular polymorph.

The situation in ETS-10 is further complicated because the crystal actually grows from a succession of rods rather than layers and this results in another type of defect

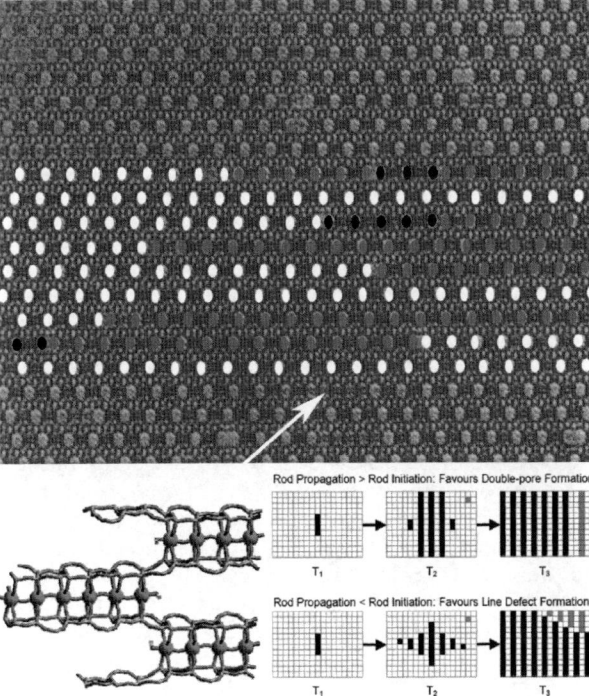

Fig. 8 Top image shows a high-resolution electron micrograph with pore arrangement white, grey or black to illustrate layer nucleation and spreading to incorporate defects; below show rod structure in ETS-10 and how the rods nucleate defects.

illustrated by the white arrow in Fig. 8. Here the pores appear to be blocked and is a result of interruption in the rod structure as shown in Fig. 8. The density of such defects is very high, indeed too high to be caused as a result of a random stacking of rods resulting from the relative nucleation and spreading rates of the rods. Consequently it is necessary to invoke a further possible scenario that the rod ends (the black rods in Fig. 8) act as nucleation points for the displaced rods (grey rods in Fig. 8). This last point is made both to illustrate the complexity of crystal growth in nanoporous materials, but also to illustrate that we are able to disentangle much of this complexity by combining, in particular, modern microscopic techniques.

Simulation of crystal habit and surface topology in nanoporous materials

Ultimately, the atomic force micrographs contain important information not only about growth mechanism but also about growth rates and activation energies for growth. However, extracting such information is far from straightforward. Our approach to at least partially tackle this problem is to try to simulate the surface topology *via* computer models. Similar approaches have been adopted in the past for other systems such as water[19] and urea.[20] An added complication with nanoporous materials is that it is still not clear from which solution species the crystals grow. Basic silicate solutions exhibit a plethora of oligomeric silicate species, the concentration and nature of which vary according to temperature, pH and the presence of organic structure directing agents.[13] Many of these species will be spectators and the true nature of the active species is still unclear. Consequently, we have chosen not to consider the primary growth unit as the unit for our modelling. Instead we have chosen closed cage structures which we know from the surface topology (AFM and HREM) are preferred surface structures. As the structures are, at least, metastable these structures can be considered rate determining and consequently it is not necessary to consider smaller, more fundamental, growth units. This simplifies the modelling considerably. However, even with this simplification in parameterisation the growth model will still have far more independent variables than parameters to which the model can be tested. In order to increase the level of input to the model from experiment we perform a three-dimensional simulation whereby crystal habit is simulated concurrently with surface topology. The input to the model are the probabilities for growth at structurally different types of sites. These probabilities equate with the fundamental rates in a model such as the one used by Gale *et al.* in the simulation of the crystal growth of urea.[20]

A hard-wired model has been developed for cubic zeolite systems such as zeolite A or sodalite, however, we are in the process of developing a general program, based on the same philosophy which can be applied to any crystal symmetry and connectivity network. Previous two-dimensional models have successfully predicted the topology and density of surface terraces and from this information derived the relative rates of surface nucleation *versus* terrace spreading.[8] Those models were based upon a first nearest neighbour difference in attachment energies. The three-dimensional model, however, invokes a second-nearest neighbour difference in attachment energies which is necessary in order to simulate facets with indices higher than (100). A few examples of the type of growth sites considered in our modelling program are illustrated in Fig. 9. For example, growth site 2_8 has two nearest neighbours and 8 second nearest neighbours. Such a site is not necessarily unique and there will be topologically distinct sites with the same coordination, however, for this model they are considered to be degenerate. Fig. 9 shows some of the sites considered on both the (100) and (110) facets. The model will effectively simulate facets up to the (111) index. There are some sixty site types in this description which illustrates the problem of undetermination. As a consequence we apply some strict interrelationships between the probabilities for growth

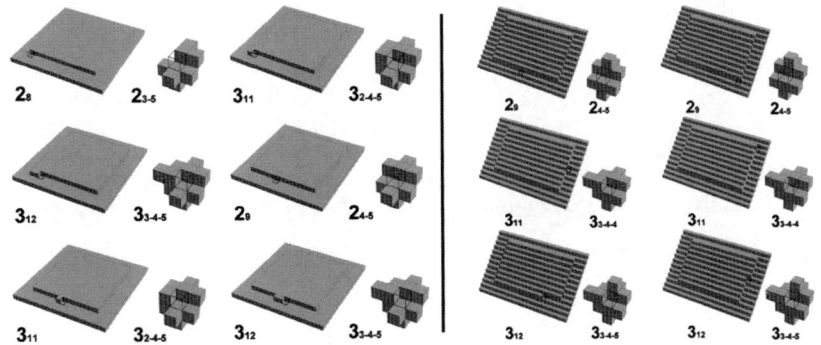

Fig. 9 A sample of the site types used in the modelling of zeolite A growth. Main number is the first-nearest-neighbour connectivity and the subscript the second-nearest-neighbour connectivity.

at the various sites. First the primary consideration for growth probability is the first nearest neighbour coordination. The higher the coordination the more probable the growth. Second sites with a first nearest neighbour coordination greater than three are assigned a probability which is so high, relatively, that they are now rate determining and can effectively be eliminated from consideration. Also, a number of sites with low second nearest neighbour coordination are only important in the early stages of crystal growth and are non-effectual in the later important stages of crystal growth. Finally, the probabilities for growth are arranged in sequence according to second nearest neighbour coordination. With all these restrictions the problem becomes determinable.

Fig. 10 shows the result of one simulation of a 0.26 μm crystal. The series incorporates a change in probabilities to simulate a dramatic decrease in the rate of surface nucleation relative to surface spreading. Although the simulation is essentially for zeolite A this is the situation which was encountered in the previous section for silicalite. Two consequences are immediately apparent. First the facetting of the crystal becomes pronounced. Second the multiple nucleation on the crystal surface ceases and the speeding of these nuclei result in a surface with a low population of well-defined square terraces. This is very similar to the result observed for silicalite and shown in Fig. 7. Terracing is also observed on both the (100) and (110) crystal surfaces, the latter similar to that observed in the high-resolution scanning electron micrographs reported by Sacco et al.[21] A much more detailed description of these results and calculations of activation energies for fundamental processes in zeolite A will be reported elsewhere.[22]

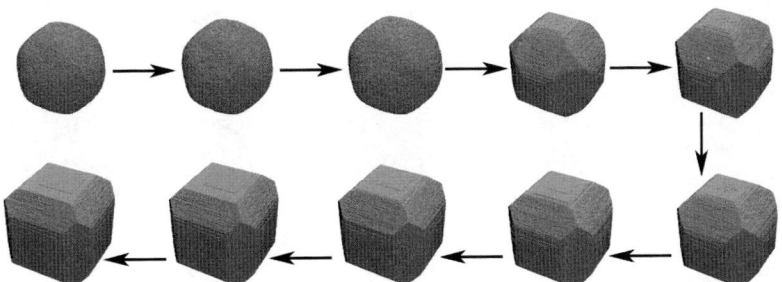

Fig. 10 10 000 000 iterations to grow a 0.26 μm zeolite crystal. The probabilities are switched to show the effect of suppression of surface nucleation.

Conclusions

In conclusion we have shown that atomic force microscopy, recorded both *ex situ* and *in situ* is a valuable tool to understand fundamental growth processes in nanoporous materials. When coupled with computer simulation it is possible to determine the underlying mechanisms and ultimately activation energies for these processes. By altering crystal growth conditions it is possible to switch processes on and off relative to one another and this is demonstrated for the nanoporous silicate silicalite.

References

1. C. S. Cundy and P. A. Cox, *Chem. Rev.*, 2003, **103**, 663.
2. C. S. Cundy and P. A. Cox, *Microporous Mesoporous Mater.*, 2005, **82**, 1.
3. M. E. Davis, *Nature*, 2002, **417**, 813.
4. O. Larlus and V. P. Valtchev, *Chem. Mater.*, 2004, **16**, 3381.
5. D. E. Akporiaye, I. M. Dahl, A. Karlsson and R. Wendelbo, *Angew. Chem., Int. Ed.*, 1998, **37**, 609.
6. W. M. Meier, in *Atlas of Zeolite Framework Types*, ed. Ch. Baerlocher, W. M. Meier and D. H. Olson, Elsevier, Amsterdam, 5th edn, 2001.
7. G. Binnig, C. F. Quate and C. Gerber, *Phys. Rev. Lett.*, 1986, **56**, 930.
8. J. R. Agger, N. Pervaiz, A. K. Cheetham and M. W. Anderson, *J. Am. Chem. Soc.*, 1998, **120**, 10754.
9. M. W. Anderson, O. Terasaki, T. Ohsuna, P. J. O'Malley, A. Philippou, S. P. MacKay, A. Ferreira, J. Rocha and S. Lidin, *Philos. Mag. B*, 1995, **71**, 813.
10. Collaboration funded by EPSRC.
11. P. Bussian, F. Sobott, B. Brutschy, W. Schrader and F. Schüth, *Angew. Chem., Int. Ed.*, 2000, **39**, 3901.
12. C. S. Cundy, M. S. Henty and R. J. Plaisted, *Zeolites*, 1995, **15**, 353.
13. A. V. McCormick, A. T. Bell and C. J. Radke, *Zeolites*, 1987, **7**, 183.
14. C. Serre, C. Lorentz, F. Taulelle and G. Ferey, *Zeolites*, 2003, **15**, 2328.
15. B. Slater, J. O. Titiloye, F. M. Higgins and S. C. Parker, *Curr. Opin. Solid State Mater. Sci.*, 2001, **5**, 417.
16. L. Scandella, N. Kruse and R. Prins, *Surf. Sci.*, 1993, **281**, L331.
17. A. M. Walker, B. Slater, J. D. Gale and K. Wright, *Nat. Mater.*, 2004, **3**, 715.
18. J. M. Newsam, M. M. J. Treacy, D. E. W. Vaughan, K. G. Strohmaier and W. J. Mortier, *J. Chem. Soc., Chem. Commun.*, 1989, 493.
19. B. Wathen, M. Kuiper, V. Walker and Z. Jia, *J. Am. Chem. Soc.*, 2003, **125**, 729.
20. S. Piana, M. Reyhani and J. D. Gale, *Nature*, 2005, **438**, 70.
21. S. Bazzana, S. Dumrul, J. Warzywoda, L. Hsiao, L. Klass, M. Knapp, J. A. Rains, E. M. Stein, M. J. Sullivan, C. M. West, J. Y. Woo and A. Sacco, Jr, *Stud. Surf. Sci. Catal.*, 2002, **142A**, 117.
22. C. Chong, PhD Thesis, University of Manchester, 2007C. B. Chong, J. R. Agger and M. W. Anderson, in preparation.

PAPER

New insights into the formation of microporous materials by *in situ* scattering techniques

Gopinathan Sankar,*[ab] Tatsuya Okubo,[c] Wei Fan[c] and Florian Meneau[d]

Received 3rd January 2007, Accepted 15th February 2007
First published as an Advance Article on the web 24th May 2007
DOI: 10.1039/b700090c

The formation of zeolite A (LTA) in the presence of tetramethylammonium cations and CoAlPO-5 (AFI structure) in the presence of tetraethyl ammonium hydroxide was studied using time-resolved, *in situ* energy dispersive X-ray diffraction (EDXRD) and small angle and wide angle X-ray scattering (SAXS/WAXS) techniques. The *in situ* SAXS measurements show the formation of homogeneous precursors, *ca.* 10 nm in size, prior to the crystallization of LTA, and consumed during the crystallization. The crystal size is estimated by fitting the SAXS patterns with an equation for a cubic particle, and it is revealed that the final crystal size of the LTA depends on the synthesis temperature. The crystallisation of CoAlPO-5 occurs through the formation of poly-dispersed particles with an average size of the precursor particle of *ca.* 50 nm.

Introduction

Microporous aluminosilicates and aluminophosphates and many other metal ion substituted variants are part of the family of zeotype materials whose intricate pores and channels are in the size range of 0.3–2 nm, and have widely been used in a variety of applications which include catalysis, adsorbents and ion-exchangers.[1,2] Interest in the preparation of novel microporous materials continues to dominate the field, and several studies were directed towards the understanding of the mechanism involved in the production of microporous materials.[3–6] Despite these efforts, due to the complexity of the zeolite formation process, its mechanism is not completely understood and it has been recognised that it is necessary to understand the crystallisation mechanism in detail to synthesize microporous materials by rational design. Several attempts have been made in the past to unravel the complex mechanism of the zeolite formation, both computationally and experimentally (employing both *in situ* and *ex situ* methods).[7–9] For example, NMR,[10] dynamic light scattering and small angle X-ray scattering (SAXS) techniques[11,12] were employed to investigate the formation of zeolite A; the synthesis of nanosized silicalite-1 (MFI) from a clear solution with tetrapropylammonium (TPA$^+$) cation as a structure directing agent (SDA) has been studied using *in situ* small-angle and

[a] *Daresbury Laboratory, Warrington, Cheshire, UK WA4 4AD*
[b] *Davy Faraday Research Laboratory, The Royal Institution of Great Britain, 21 Albemarle Street, London, UK W1S 4BS*
[c] *Department of Chemical System Engineering, The University of Tokyo, 7-3-1 Hongo, Bunkyo-ku, Tokyo 113-8656, Japan*
[d] *SWING-Synchrotron SOLEIL, L'Orme des Merisiers, BP 48, Saint-Aubin, 91192, Gif sur Yvette, Cedex, France*

wide-angle X-ray scattering (SAXS and WAXS) techniques.[11–15] Mintova et al.[16–18] studied the synthesis of nanosized aluminosilicate LTA from a colloidal solution with TMA^+ and showed the steps of the generation and densification of precursor particles to nanosized LTA by high resolution transmission electron microscopy (HRTEM). More recently,[30] we reported the study of formation of zeolite A from clear solution and the study showed that, depending on the content of sodium hydroxide, 5 nm particles were observed to form prior to the crystallisation of zeolite A.

Another system of considerable interest is the formation of microporous aluminophosphate materials.[19] Both large pore AlPO-5 (IZA code AFI) and small pore AlPO-34 (IZA code CHA) based systems were extensively studied due to their potential applications in catalysis.[20,21] Earlier studies using energy dispersive X-ray diffraction methods clearly showed that, in the presence of cobalt ions, small pore materials, either AlPO-18 or AlPO-34, are formed[22–25] unless the pH of the starting gel is maintained around 6. Thus it is of considerable interest to investigate the formation of the large pore AFI material in the presence of cobalt, in particular the changes that take place prior to the formation of crystalline material.

To study the formation of crystalline materials, in particular prior to the formation of ordered structure, it is necessary to use techniques such as SAXS or X-ray absorption spectroscopy. These techniques, when combined with X-ray diffraction (combined SAXS/WAXS or EXAFS/WAXS or SAXS/WAXS/EXAFS) become even more powerful for the study of materials crystallising from a clear solution. However, X-ray absorption spectroscopy is limited to the study of materials containing elements that have absorption energies higher than 5 keV; more recently, Al K-edge XAS studies have been performed,[26] but these were restricted to a single technique, since it is difficult to combine with XRD technique at low energies. Thus the combined SAXS/WAXS and XAS/XRD methods have been used during the last 10 years for the study of crystallisation of microporous materials.[27–29] Here, we report a time-resolved *in situ* study of the nucleation and crystal growth of nanosized aluminosilicate LTA using the combined SAXS/WAXS and energy dispersive X-ray diffraction (EDXRD) techniques and formation of CoAlPO-5 using SAXS/WAXS technique.

Experimental

A solution containing $11.25SiO_2 : 1.8Al_2O_3 : 13.4(TMA)_2O : 1.2NaOH : 700H_2O$ was used for the preparation of zeolite A, similar to the one reported by Mintova et al.[18] Tetramethylammonium silicate was used here, since the colloidal silica will interfere with the SAXS measurements of zeolite A crystal.[30] A solution containing the appropriate molar ratio was prepared and aged for 2 h prior to introduction into a specially designed *in situ* cell. This *in situ* cell was placed in an electrically pre-heated (to a specific temperature, 100 °C) oven. *In situ* SAXS/WAXS measurements were started as soon as the cell containing the gel was introduced and they were performed at station 6.2 of the synchrotron radiation source at Daresbury Laboratory, UK, which operates at 2 GeV with a typical current between 120 and 220 mA. Sample to detector distances of 3.5 m and 1.3 m were used in this work and the data were recorded using a wavelength of 1.4000 Å. High-quality SAXS and WAXS patterns were collected simultaneously every 2 min using RAPID detectors.[27,31] The scattering from water was subtracted as the background. The scattering pattern of a standard sample (wet rat tail collagen) and the diffraction pattern of NaA zeolite were used to calibrate the patterns of SAXS and WAXS, respectively. SAXS data were analysed using the XOTOKO program available at Daresbury Laboratory. EDXRD patterns were recorded at station 16.4 of Daresbury Laboratory. In a typical experiment, the reaction mixture was introduced in to a stainless steel autoclave and it was placed inside the oven just prior to the start of the measurements. EDXRD patterns were recorded every two minutes during the hydrothermal reaction.

Gel containing aluminium hydroxide, phosphoric acid, cobalt acetate, water and tetraethyl ammonium hydroxide (in the molar ratio of 0.96 : 1.0 : 0.04 : 25 : 0.8, respectively) was prepared to investigate the crystallisation process associated with the formation of crystalline microporous CoAlPO-5 material.[22] SAXS/WAXS data were collected during the hydrothermal treatment of this gel at 160 °C, using a setup identical to that employed for the study of zeolite A crystallisation, described above. EDXRD patterns were recorded at station 16.4 of Daresbury Laboratory[24,28,32] and the data were recorded every two minutes during the hydrothermal reaction.

Results and discussion

First we discuss the energy dispersive X-ray diffraction data recorded during the crystallisation of zeolite A followed by the SAXS and WAXS data and then discuss the results of the crystallisation of CoAlPO-5 material.

Zeolite A

In Fig. 1(a) we show a structural model of zeolite A and in (b) we show a typical EDXRD pattern recorded during the crystallisation of zeolite A, at 100 °C. A structureless, broad hump (circled in Fig. 1) appears during the initial stages of the crystallisation, which is due to the presence of non-crystalline material. After several minutes of hydrothermal reaction, diffraction peaks appear and all the reflections seem to grow at similar rate indicating uniform growth. In Fig. 1(c) we plotted the variation of the normalised intensities of (200), (220), (222), (420) reflections with time. It is clear from Fig. 1(c) that there are no differences in the increase in intensity of all the reflections with time, indicating uniform growth of zeolite A. The full width at half maximum (FWHM) obtained from the line-profile fitting of the (220) reflection is plotted with time to determine the variation in particle size during the growth process. As one would expect, the decrease in FWHM from *ca.* 1.7 to 1.3 keV

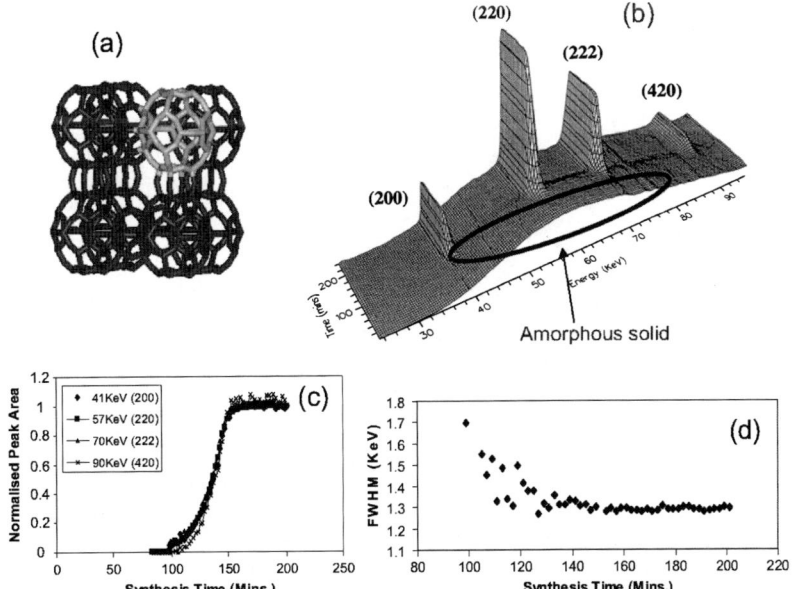

Fig. 1 A typical zeolite A structure is shown in (a); the sodalite cage is highlighted here. In (b) we show a typical EDXRD pattern of zeolite A, recorded during the hydrothermal synthesis at 100 °C. In (c) we show the variation of the area of the peaks corresponding to (200), 220), (222) and (420) reflections, with time, which suggests uniform growth. The variation in the full width at half maximum (FWHM) of the (220) reflection with time is shown in (d).

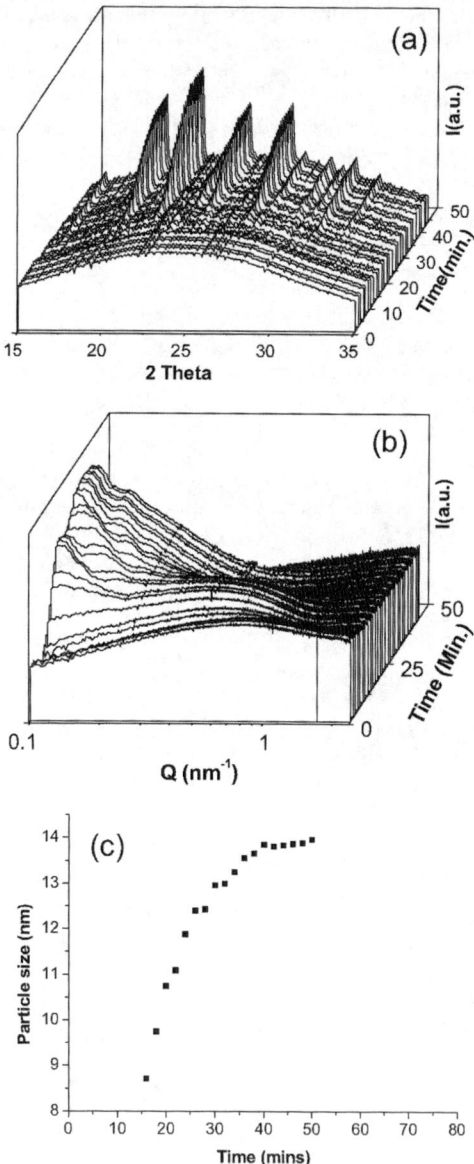

Fig. 2 *In situ* time-resolved WAXS in (a), SAXS in (b), recorded during the hydrothermal synthesis of zeolite A, at 100 °C, are shown here along with the variation in particle size, estimated from the broad hump in (b), with time is shown in (c).

suggest that particles are small and grows for about 20 min after the initiation of the crystallisation processes and remain unaltered after 120 min of the hydrothermal reaction. Although the information derived from EDXRD technique enabled us to determine the kinetics of crystallisation and activation energy of the process,[22] it was not possible to determine the nature of particles formed just prior to the crystallisation of zeolite A. Hence we investigated this system, in detail, using SAXS/WAXS technique.

Time-resolved stacked WAXS and SAXS patterns are shown in Fig. 2(a) and (b), respectively. (The SAXS data shown here are processed to remove the background by subtracting the SAXS pattern of water, similar to the method used by de Moor et al.[15]). It is clear from Fig. 2(b) that a broad hump appears in the SAXS pattern after heating the clear solution for about 16 min. This hump spreads over a Q range of ca. 0.7 and 0.4 nm^{-1} which corresponds to ca. 5 and 15 nm with centre of the hump at ca. 10 nm (calculated using the equation $d = 2\pi/Q$, where Q is $4\pi\sin(\theta)/\lambda$).[13–15,33] The position of the maximum point of the hump moves to a lower Q value with time; the hump centred around ca. $Q = 0.79$ nm^{-1} ($d = 8$ nm), measured after 16 min of the reaction moves to ca. $Q = 0.48$ nm^{-1} ($d = 13$ nm) at 32 min. Particle size estimated from the position of the broad hump clearly shows that it has increased from 8 to 13 nm over a period of 16 min. WAXS patterns recorded simultaneously during the SAXS measurements did not show any Bragg reflections, indicating that the crystallization has not taken place, which is similar to the one observed by EDXRD data. After 42 min of the hydrothermal reaction, although the intensity of the broad hump appears to have decreased slightly, its position ($Q = 0.44$ nm^{-1} ($d = 15$ nm)) remained close to the earlier observation made at ca. 32 min, suggesting that the particle size has not significantly changed. More interestingly, an oscillatory pattern appears in the low Q region of the SAXS patterns recorded after 32 min, which also coincides with the appearance of reflections in the WAXS pattern.

We also carried out a series of experiments using a short sample to detector distance (1.3 m). The advantage of using a short sample detector distance is that the accessible Q range is ca. 0.025–0.50 Å^{-1} (which correspond to ca. 8–225 Å) compared to a large sample detector distance (Q range is ca. 0.011–0.262Å^{-1} which corresponds to ca. 25–575 Å). Hence it is possible to collect scattering from both the non-crystalline and crystalline part of the system using the same detector when a short sample detector distance is used; the first diffraction peak for the majority of the nanoporous materials appears below a d-spacing of 10 Å. A typical stacked plot, recorded during the crystallisation of zeolite A from a clear solution is shown in Fig. 3(a). Here we can see clearly several changes in the small angle region, in particular formation of a broad hump similar to the one observed using a long sample to detector distance, described above. A Bragg peak related to crystalline zeolite A starts to appear at high Q region (marked as X in Fig. 3(a)) after about 35 min of crystallisation. In order to show that the changes in the SAXS is not an artefact of the experimental setup, we also collected SAXS/WAXS patterns for solutions containing only silicate (Fig. 3(b)) or aluminate (Fig. 3(c)); the alumina or silica source, respectively, were not included in the solution. (The data were recorded under identical conditions to that of the solution containing ingredients that produces zeolite A.) Both these patterns did not show any change in the scattering, which include the presence of a hump in the initial stages, that corresponds to the formation of specific particles prior to crystallisation and the typical oscillatory behaviour observed during the formation of zeolite crystals. More importantly, there were no Bragg peaks associated with the formation zeolite A or any other phase in both SAXS and WAXS patterns.

We analysed both SAXS (data collected using the long detector camera distance) and WAXS data, in detail, to determine the size and shape of the particles formed prior and during the crystallisation of zeolite A. We calculated the particle size based on the centre position of the hump (shown in Fig. 2(b)) and plotted the variation in the particle size, with time, in Fig. 2(c). As expected the particle size continues to increase during the initial stages of the hydrothermal reaction and attains a maximum. At the same time, an oscillatory pattern, dominant in the low Q region (see Fig. 2(b)), appears in the SAXS pattern which not only grow in intensity, but also moves towards the low Q region which is typical for a growing particle. More importantly this oscillatory pattern clearly indicates the presence of homogeneously dispersed particles in the solution. The appearance of the oscillatory pattern in the low Q region also coincides with the emergence of Bragg reflections corresponding to zeolite A crystals (see WAXS data of Fig. 2(a) and Fig. 3(a)). The oscillations in the

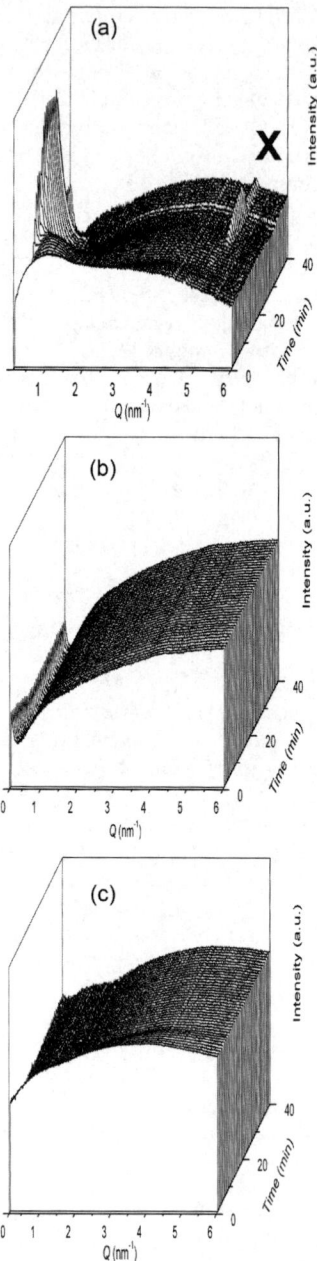

Fig. 3 Time-resolved SAXS patterns recorded during the hydrothermal reaction of (a) for a typical chemical composition for zeolite A (Na = 0.45 mol), (b) chemical constituents used for zeolite A synthesis without silica source and (c) chemical composition used for zeolite A synthesis without alumina source. This study clearly shows that the oscillations and hump seen in the zeolite A crystallisation (in (a)) is not an artefact of the experiment.

low Q region ($Q < 0.3$ nm^{-1}) in SAXS in combination with the diffraction peaks present in the WAXS (as well as in the SAXS data when a short detector–camera distance was used) suggest the formation of well-defined morphologies of crystalline zeolite A with narrow size distribution. We made use of the well-defined oscillatory behaviour of the SAXS data and calculated the particle size by fitting the data using an equation corresponding to cuboidal particles.[30] Once again the SAXS data support the XRD finding (particle size was estimated using the Scherrer equation) that at the onset of crystallisation, 50 nm particles exist which grow with time. We estimated the final particle size of *ca.* 120 nm based on the analysis of SAXS data, which is similar to the one estimated from WAXS data, based on the Scherrer equation, employing the data collected after 120 min; SEM measurements also estimate the particle size to be about 100 nm. It is very difficult to determine the initial size of the particle at which both the oscillatory part in the SAXS and the first diffraction peak appears in the WAXS data, since the signal to noise is not sufficiently large for accurate analysis. However, it appears from the WAXS data that approximately 50 nm particles are present at this stage.

CoAlPO-5

The typical structure of AFI is shown in Fig. 4(a) and the EDXRD pattern recorded, *in situ*, during the crystallisation of CoAlPO-5 (Co content was 4%, the pH of the starting gel was maintained at 6 and the synthesis temperature was *ca.* 163 °C) is shown in Fig. 4(b). The first reflection was analysed by fitting the peak and the area

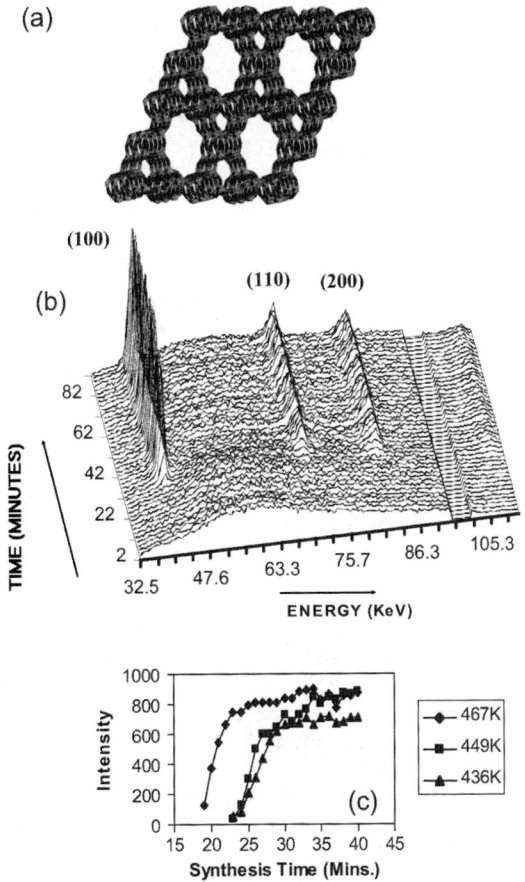

Fig. 4 Structure of AlPO-5 (AFI, International Zeolite Association code) is shown in (a). A typical time-resolved EDXRD data recorded during the hydrothermal reaction of a CoAlPO-5 gel at 160 °C is shown in (b). In (c) we show the variation in the area under the peak of reflection (100), with time, recorded at three different temperatures.

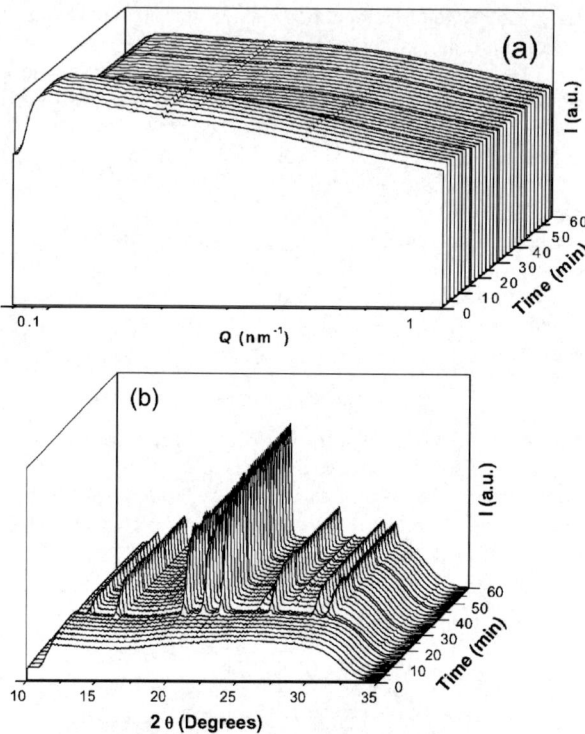

Fig. 5 The time-resolved SAXS (in (a)) and WAXS (in (b)), recorded during the formation of CoAlPO-5, under hydrothermal conditions at *ca.* 160 °C.

under the curve determined is plotted against time in Fig. 4(c). The EDXRD pattern shown in Fig. 4(b) shows a structureless pattern in the data recorded during the initial stages of the hydrothermal reaction and after *ca.* 20 min the reflections representing pure AFI phase starts to appear in the EDXRD patterns. A plot of the area under the (100) reflection with time, shown in Fig. 3(c) indicates the growth of the CoAlPO-5 material during the hydrothermal reaction. Hydrothermal reactions conducted at various temperatures (see the growth curve obtained from reactions conducted at three different temperatures in Fig. 4(c)) shows that at higher temperatures the reaction proceeds faster, and the induction period is decreased to a lower value, typically seen for the kinetically driven processes. Earlier studies[25] using combined XRD/XAS (at the Co K-edge) revealed that the Co(II) ions present in octahedral coordination are converted to tetrahedral coordination just prior to the crystallisation of CoAlPO-5 material.

We performed *in situ* SAXS/WAXS studies of the crystallisation of microporous aluminophosphate, using a gel mixture similar to the one used for EDXRD and combined XRD/XAS studies and the stacked plot of the SAXS and WAXS patterns are shown in Fig. 5. The SAXS pattern in Fig. 5(a) did not show well-defined characteristics as seen for zeolite A. Although the SAXS data (see Fig. 5(a)) appears to be featureless, we utilised the change in slope of the SAXS pattern of the data recorded simultaneously when the WAXS patterns showed indications of the appearance of reflections corresponding to AFI structure and the particle size estimated from the analysis indicated that approximately 50 nm particles are formed just prior to the crystallisation of AlPO-5 structure.[34,35] The WAXS data (see Fig. 4(c)) clearly showed the presence of only phase pure AlPO-5[25] material under the reaction conditions used in this work.

In summary, the *in situ* SAXS/WAXS investigation indicates the formation of zeolite A precursors with a particle size in the range of *ca.* 10 nm before the onset of crystallisation. The occurrence of oscillations in the low Q region of the SAXS data suggest that a homogeneous nucleation leading to LTA crystals with a narrow size distribution are formed under these synthetic conditions. However, similar homogeneous crystallisation is not observed in the formation of the CoAlPO-5 system.

Acknowledgements

GS thanks the Royal Society for a UK-Japan visiting fellowship. GS also thanks CLRC, Daresbury Laboratory for provision of beam time and facilities and Dr C. Martin and Dr D. Taylor for useful discussions. WF also acknowledges a Grant for 21st Century COE Program "Human-Friendly Materials based on Chemistry" from the Ministry of Education, Culture, Sports, Science, and Technology of Japan.

References

1. R. M. Barrer, *Hydrothermal Chemistry of Zeolites*, Academic Press, London, 1982.
2. D. W. Breck, *Zeolite Molecular Sieves*, Wiley, New York, 1974.
3. C. S. Cundy and P. A. Cox, *Chem. Rev.*, 2003, **103**, 663.
4. C. S. Cundy and P. A. Cox, *Microporous Mesoporous Mater.*, 2005, **82**, 1.
5. F. Schuth, *Curr. Opin. Solid State Mater. Sci.*, 2001, **5**, 389.
6. D. P. Serrano and R. van Grieken, *J. Mater. Chem.*, 2001, **11**, 2391.
7. C. R. A. Catlow, D. S. Coombes and J. C. G. Pereira, *Chem. Mater.*, 1998, **10**, 3249.
8. M. J. Mora-Fonz, C. R. A. Catlow and D. W. Lewis, *Angew. Chem., Int. Ed.*, 2005, **44**, 3082.
9. S. A. Ojo, L. Whitmore, B. Slater and C. R. A. Catlow, *Solid State Sci.*, 2001, **3**, 821.
10. J. M. Shi, M. W. Anderson and S. W. Carr, *Chem. Mater.*, 1996, **8**, 369.
11. P. S. Singh and J. W. White, *Phys. Chem. Chem. Phys.*, 1999, **1**, 4131.
12. M. Smaihi, O. Barida and V. Valtchev, *Eur. J. Inorg. Chem.*, 2003, **1**, 4370.
13. P. de Moor, T. P. M. Beelen, B. U. Komanschek, L. W. Beck, P. Wagner, M. E. Davis and R. A. van Santen, *Chem. Eur. J.*, 1999, **5**, 2083.
14. P. de Moor, T. P. M. Beelen, R. A. van Santen, L. W. Beck and M. E. Davis, *J. Phys. Chem. B*, 2000, **104**, 7600.
15. P. de Moor, T. P. M. Beelen and R. A. vanSanten, *Microporous Mater.*, 1997, **9**, 117.
16. S. Mintova, N. H. Olson and T. Bein, *Angew. Chem., Int. Ed.*, 1999, **38**, 3201.
17. S. Mintova, N. H. Olson, J. Senker and T. Bein, *Angew. Chem., Int. Ed.*, 2002, **41**, 2558.
18. S. Mintova, N. H. Olson, V. Valtchev and T. Bein, *Science*, 1999, **283**, 958.
19. P. Norby, A. N. Christensen and J. C. Hanson, *Inorg. Chem.*, 1999, **38**, 1216.
20. J. M. Thomas, *Angew. Chem., Int. Ed. Engl.*, 1994, **33**, 913.
21. J. M. Thomas, R. Raja, G. Sankar and R. G. Bell, *Acc. Chem. Res.*, 2001, **34**, 191.
22. A. T. Davies, G. Sankar, C. R. A. Catlow and S. M. Clark, *J. Phys. Chem. B*, 1997, **101**, 10115.
23. G. Muncaster, A. T. Davies, G. Sankar, C. R. A. Catlow, J. M. Thomas, S. L. Colston, P. Barnes, R. I. Walton and D. O'Hare, *Phys. Chem. Chem. Phys.*, 2000, **2**, 3523.
24. F. Rey, G. Sankar, J. M. Thomas, P. A. Barrett, D. W. Lewis, C. R. A. Catlow, S. M. Clark and G. N. Greaves, *Chem. Mater.*, 1995, **7**, 1435.
25. G. Sankar, J. M. Thomas, F. Rey and G. N. Greaves, *J. Chem. Soc., Chem. Commun.*, 1995, 2549.
26. A. M. Beale, A. M. J. van der Eerden, D. Grandjean, A. V. Petukhov, A. D. Smith and B. M. Weckhuysen, *Chem. Commun.*, 2006, 4410.
27. W. Bras, G. E. Derbyshire, A. J. Ryan, G. R. Mant, A. Felton, R. A. Lewis, C. J. Hall and G. N. Greaves, *Nucl. Instrum. Methods Phys. Res., Sect. A*, 1993, **326**, 587.
28. G. Sankar and J. M. Thomas, *Top. Catal.*, 1999, **8**, 1.
29. G. Sankar, J. M. Thomas and C. R. A. Catlow, *Top. Catal.*, 2000, **10**, 255.
30. F. Wei, M. Obrien, M. Ogura, M. Sanchez-Sanchez, C. Martin, F. Meneau, K. Kurumada, G. Sankar and T. Okubo, *Phys. Chem. Chem. Phys.*, 2006, **8**, 1335.
31. W. I. Helsby, A. Berry, P. A. Buksh, C. J. Hall and R. A. Lewis, *Nucl. Instrum. Methods Phys. Res., Sect. A*, 2003, **510**, 138.
32. R. J. Francis, S. J. Price, J. S. O. Evans, S. Obrien, D. Ohare and S. M. Clark, *Chem. Mater.*, 1996, **8**, 2102.
33. P. de Moor, T. P. M. Beelen, R. A. van Santen, K. Tsuji and M. E. Davis, *Chem. Mater.*, 1999, **11**, 36.

34 A. M. Beale, A. M. J. van der Eerden, S. D. M. Jacques, O. Leynaud, M. G. O'Brien, F. Meneau, S. Nikitenko, W. Bras and B. M. Weckhuysen, *J. Am. Chem. Soc.*, 2006, **128**, 12386.
35 D. Grandjean, A. M. Beale, A. V. Petukhov and B. M. Weckhuysen, *J. Am. Chem. Soc.*, 2005, **127**, 14454.

Cocrystal architecture and properties: design and building of chiral and racemic structures by solid–solid reactions†‡

Tomislav Friščić and William Jones*

Received 9th November 2006, Accepted 5th February 2007
First published as an Advance Article on the web 16th April 2007
DOI: 10.1039/b616399h

The concept of the so-called "supramolecular synthon" has been employed to construct chiral and centrosymmetric cocrystals with predictable short-range order. The reactants included nicotinamide, mandelic acid and ibuprofen. In order to maximize the efficiency of cocrystal synthesis, the solids were constructed using the liquid-assisted grinding approach. The predictability of the cocrystal architecture was further employed to study in detail the effects of chirality upon physical properties of the synthesized materials, especially the melting point. The combined results of crystallographic and thermochemical studies enable, at least partially, the rationalisation of cocrystal thermal behaviour with respect to the corresponding cocrystal formers.

Introduction

Interest in organic solid-state materials has steadily increased.[1,2] The past ten years, however, have witnessed a significant increase in research in the field, mainly as the result of the advances achieved using supramolecular and crystal engineering approaches.[3] An overview of recent papers[4] indicates that current state-of-the-art design of molecular solids is achieved through the construction of multi-component molecular solids, or cocrystals.[5] Indeed, cocrystals have proven particularly successful as functional materials with applications in pharmaceuticals,[6] molecular electronics,[7] optical applications,[8] and synthetic organic chemistry.[9] The principal reasons for the successful use of cocrystals are the discovery of reliable supramolecular synthons for cocrystal design, as well as modularity, an emergent property of multicomponent molecular solids.[10] Modularity provides an opportunity to fine-tune the structural and physical properties of a solid without significantly affecting chemical properties (*e.g.* when dissolved).[11,12]

Opportunities and challenges provided by cocrystals have inspired both the exploration of optimum supramolecular synthons, as well as the development of synthetic methodologies for producing cocrystals. Whereas the first topic, *i.e.* addressing the synthesis and identification of supramolecular synthons for cocrystal

Pfizer Institute for Pharmaceutical Materials Science, Chemistry Department, University of Cambridge, Lensfield Road, Cambridge, UK CB2 1EW. E-mail: wj10@cam.ac.uk; Tel: +44 1223 336468

† The HTML version of this article has been enhanced with colour images.
‡ Electronic supplementary information (ESI) available: Fig. S1–S15: X-ray powder diffraction patterns, DSC thermograms and hot-stage microscopy images of cocrystals. See DOI: 10.1039/b616399h

synthesis, has been pursued with great intensity by the research groups of Aaköry,[13] Zaworotko,[4] Desiraju[14] and Nangia,[15] our own research has focused on the second topic, *i.e.* methods for cocrystal synthesis. In that context we have been, along with others,[16] exploring mechanochemical approaches (*i.e.* grinding) as a means of constructing molecular cocrystals.[17] Specifically, we have been developing the technique of liquid-assisted grinding, a rapid and efficient method to conduct supramolecular synthesis in the solid state. In our laboratory, the potential of liquid-assisted grinding has established the approach as a general way to discover and construct cocrystals. For that reason, we find it appropriate to first provide a brief overview of the application of liquid-assisted grinding to construct cocrystals and identify specific advantages over conventional solution cocrystallisation methods.[18] In the light of the ongoing discussion on the definition of the term cocrystal, we have used the term herein to describe all multi-component crystalline molecular solids, including the ones that would otherwise be classified as solvates, clathrates or inclusion compounds.[5]

Neat and liquid-assisted grinding

In its simplest approach, a cocrystal can be formed by the neat grinding of two (or more) components together. The addition of a small amount of a liquid phase can, however, significantly accelerate the formation of cocrystals, and grinding methods involving a catalytic amount of a liquid have been termed "kneading",[19] as well "solvent-drop" grinding.[20] The mechanism by which an accelerating effect is produced is still not fully understood. In some cases it has been interpreted as a purely physical effect, *i.e.* the liquid acting as a lubricant for the process.[21] In other cases, however, effects related to the solubility of the materials and the chemical nature of the liquid phase have been suggested. Indeed, the importance of the chemical nature of the liquid is apparent in the solvent-dependent selective formation of different polymorphs of cocrystals of caffeine and glutaric acid.[22] In many cases, however, the actual role of the liquid phase during grinding has not been established with certainty, and it is possible that it may vary from case to case. For that reason we consider it appropriate to encompass all possible mechanisms for using a liquid phase to accelerate cocrystal formation under the common name liquid-assisted grinding.[23] Defined in that way, liquid-assisted grinding describes the use of a liquid to accelerate (or enable) a cocrysallisation reaction between two solids, regardless of whether the liquid acts largely through physical effects (*e.g.* lubrication, enhanced dissolution of the reactants), or if specific chemical properties of the liquid phase are more relevant (*e.g.* formation of an intermediate solvate of a reactant or molecules of the added liquid becoming constituents of the final cocrystal). From that perspective, solvent-drop grinding is a strategy of liquid-assisted grinding that exploits the solubility of solid reactants in the liquid phase.[24]

Discovery of new cocrystals

That cocrystallisation by grinding can provide a cocrystal of different composition than cocrystallisation from solution has been demonstrated in the case of cocrystals involving caffeine and acetic acid. Cocrystallisation from acetic acid resulted in a solid made up of finite assemblies comprising a molecule of caffeine and two molecules of acetic acid. Identical cocrystals were obtained by grinding caffeine with an excess of the acid. However, upon grinding equimolar amounts of caffeine and acetic acid, a solid containing the two components in a 1 : 1 molar ratio was obtained. Repeated attempts to produce the 1 : 1 material from solution failed, however, and structure solution from X-ray powder diffraction data became necessary, revealing two-component 1 : 1 molecular assemblies.[25]

Changing the reaction scale can also affect the nature of the cocrystal formed by grinding: small amounts of caffeine and trifluoroacetic acid yield an orthorhombic form of the 1 : 1 cocrystal, whereas a triclinic polymorph is obtained upon increasing the reaction scale.[25]

Construction of ternary solids and efficiency of liquid-assisted grinding

In addition to forming binary cocrystals, we have recently reported liquid-assisted grinding as a means to construct ternary solids. In particular, we have compared the efficiencies of neat grinding, liquid-assisted grinding and cocrystallisation from solution in screening inclusion compounds of a two-component lattice host.[23] The results demonstrated liquid-assisted grinding is more efficient than cocrystallisation from solution, providing a ternary solid with 20 out of 30 investigated guest molecules, compared to four obtained from solution. The effectiveness of ternary cocrystal formation was independent on the state of aggregation of the guest (*i.e.* whether the guest was a liquid or a solid).

Exploring the effect of chirality: chiral *vs.* racemic cocrystal formers

Although cocrystals are readily amenable to synthesis by design, the relationship between cocrystal structure and physical properties remains unexplored. Indeed, most approaches to functional solids *via* cocrystals are based on screening, rather than an understanding of the cocrystal structure–property relationship.[26,27] In order to explore such relationships and develop strategies to tailor physical properties of cocrystals, we have begun to investigate the influence of chirality on the reactivity of potential cocrystal formers and on the thermal properties of resulting cocrystals.

Our first experiments involved the cocrystallisation of model active pharmaceutical ingredients (APIs) caffeine and theophylline with either chiral or racemic forms of tartaric acid.[28] The results showed that theophylline and caffeine exhibited different reactivity towards different "forms" of tartaric acid. Theophylline readily provided cocrystals with both chiral and racemic tartaric acid by liquid-assisted grinding. Crystal structure analysis revealed that racemic cocrystals consist of hydrogen-bonded tapes based on ring-like assemblies of theophylline and tartaric acid, whilst in chiral cocrystals the tapes are based on hydrogen-bonded helices (Fig. 1).§

For caffeine, liquid-assisted grinding provided cocrystals with *R*- or *S*-tartaric acid, but not with *RS*-tartaric acid. Different behaviour of chiral and racemic tartaric acids towards caffeine motivated us to further explore the effect of chirality on cocrystal formation and physical properties. To facilitate our study, we focused on cocrystal components that were likely to interact through a set of robust supramolecular synthons, providing racemic and chiral solids having similar architectures. In order to make use of model APIs for these studies, we selected

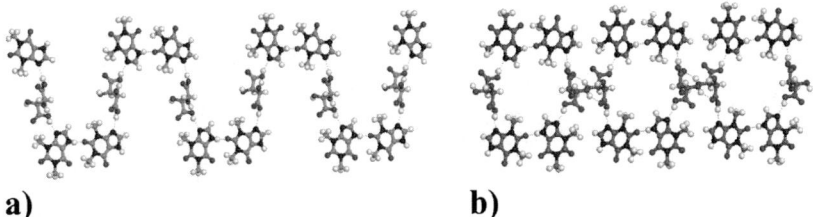

a)   b)

Fig. 1 A ball-and-stick representation of a fragment of the theophylline cocrystal with: (a) *R*- or *S*-tartaric acid and (b) *RS*-tartaric acid.

§ We use the description "chiral" and "racemic" to indicate that the solid is either enantiomerically pure or contains both enaniomers.

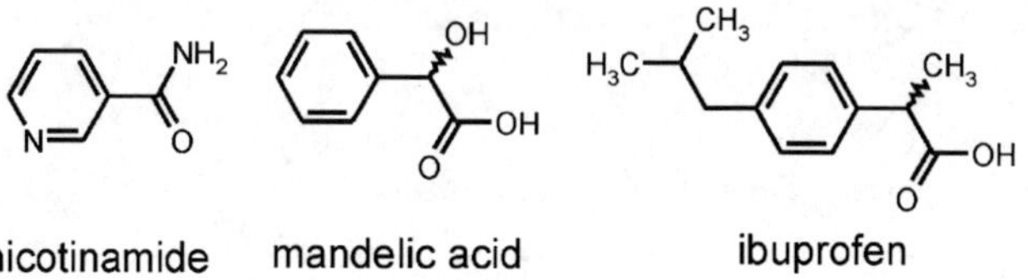

Scheme 1 Schematic representation of nicotinamide, mandelic acid and ibuprofen.

nicotinamide (vitamin D) and pairs of chiral and racemic monocarboxylic acids: R/S- and RS-mandelic acid, as well as S- and RS-ibuprofen (Scheme 1), as suitable cocrystal components.¶

An overview of the Crystal Structure Database revealed that cocrystals involving nicotinamide and a monocarboxylic acid generally comprise four-membered assemblies involving two nicotinamide and two acid molecules. The central part of each assembly is a nicotinamide dimer, held together by an amide–amide supramolecular synthon of two N–H···O hydrogen bonds (Scheme 2a). The carboxylic acids are held at the peripheral part of the assembly through the carboxylic acid–pyridine supramolecular synthon consisting of an O–H···N and a C–H···O hydrogen bond between carboxylic acid and pyridine groups that are coplanar (Scheme 2b). This arrangement is anticipated from the hydrogen bonding hierarchy formulated by Etter.[29] A similar design was exploited with great success by Aakeröy and coworkers to construct cocrystals of isonicotinamide with benzoic acids.[30] Thus, we anticipated nicotinamide would assemble with mandelic acids in a 1 : 1 stoichiometric ratio to form chiral and racemic cocrystals **1a** and **1b**, respectively, and with S- and RS-ibuprofen to form chiral and racemic cocrystals **2a** and **2b**, respectively. All cocrystals were expected to contain four-membered assemblies with nicotinamide in the center and acid molecules on the periphery (Scheme 3).

As our first entry into investigating the physical properties of chiral and racemic cocrystals, we decided to focus on thermal behavior. In that context, studies by Bond and Nangia indicated that the melting points of a series of cocrystals can reflect the thermal behaviour of parent cocrystal formers.[31] Our interest was to investigate a possible relationship between thermal properties of cocrystals built on a similar architecture but involving either enantiomerically pure or racemic cocrystal formers. We have found mandelic acid and ibuprofen to be especially interesting cocrystal formers for that purpose, since they exhibit different relationships between thermal properties of chiral and racemic forms. The melting point and density are higher for racemic ibuprofen than for the pure enantiomer, in accordance with the empirical Wallach's rule.[32] The two forms of mandelic acid contradict the rule, with the melting point and density being higher in the case of the enantiomerically pure form.

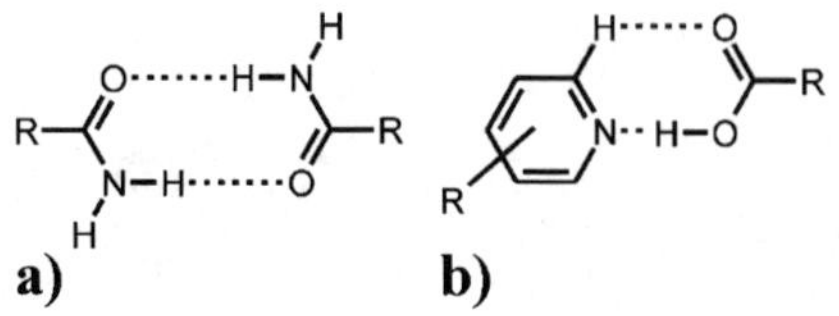

Scheme 2 Schematic representation of: (a) an amide–amide synthon and (b) pyridine–carboxylic acid synthon. Hydrogen bonds are shown as dotted lines.

¶ Results of cocrystallisation experiments involving nicotinamide and D/S- or RS-tartaric acid from solution and via liquid-assisted grinding will be published at a later date. Attempts to cocrystallise caffeine or theophylline with either mandelic acids or ibuprofen have, so far, failed.

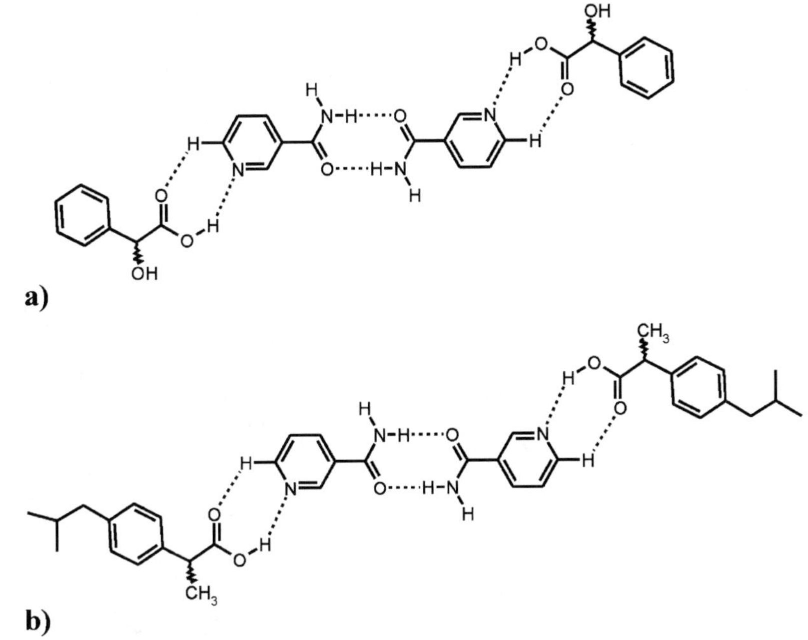

Scheme 3 Structure of molecular assemblies expected to form in the cocrystal of nicotinamide with: (a) mandelic acid and (b) ibuprofen.

Materials and methods

Synthesis of cocrystals

Nicotinamide, *R*-, *S*- and *RS*-mandelic acids, *S*- and *RS*-ibuprofen and nitromethane were commercially available from Sigma-Aldrich and were used without further purification. All liquid grinding experiments were performed using a Retsch MM200 grinder mill operating at a frequency of 30 Hz. For the synthesis of cocrystals, 0.20 g of an equimolar mixture of nicotinamide and the cocrystal former were placed in a 25 mL volume stainless steel grinding jar, along with 5 drops of nitromethane. The samples were then ground over a period of 30 min using two stainless steel grinding balls of 7 mm diameter. The samples prepared were then dried in air.

All cocrystals were also obtained by cocrystallisation of equimolar amounts of the appropriate two components from solutions in either nitromethane, acetonitrile or ethyl methyl ketone.

X-Ray diffraction experiments

X-Ray powder diffraction patterns were obtained using a Philips X'Pert Pro diffractometer equipped with an X'celerator RTMS detector, using Ni-filtered Cu-Kα radiation. Single crystal X-ray diffraction data for **1a** were collected on a Nonius Kappa CCD diffractometer equipped with a graphite monochromator and an Oxford Cryosystems cryostream, using Mo-Kα radiation.

Differential scanning calorimetry (DSC) and hot-stage microscopy

Thermal calorimetric measurements were performed on samples with a weight in the range of 5–12 mg using a Mettler DSC30 instrument. The samples were placed in 40 μL aluminium crucibles and measurements were taken in the temperature range

30–150 °C, using a heating rate of 5 K min^{-1} in a dynamic atmosphere of nitrogen gas. Hot-stage microscopy experiments were performed on a Mettler FP84HT TA microscopy cell connected to the Mettler FP90 central processor unit. The samples were placed in open glass crucibles and their thermal behaviour was monitored in the temperature range 30–150 °C at a heating rate of 10 K min^{-1}.

X-Ray powder diffraction patterns of all reactants and products are provided as ESI,‡ along with selected hot-stage microscopy images and DSC thermograms.

CCDC reference number 626647 for **1a**.

For crystallographic data in CIF or other electronic format see DOI: 10.1039/b616399h

Experimental results

Liquid-assisted grinding of equimolar quantities of nicotinamide with either *R*- or *S*-mandelic acid and nitromethane produced a solid with a X-ray powder diffraction (XRPD) pattern different from that of the solid reactants. The product powder pattern was identical for both *R*- and *S*-mandelic acid as the reactant, suggesting (as expected) that the two product crystals were enantiomeric. The XRPD pattern of the product of liquid-assisted grinding of nicotinamide with *RS*-mandelic acid and nitromethane also contained no evidence for unreacted material. It was also different than the pattern obtained using the chiral mandelic acid (Fig. 2). Cocrystals with *R*- or *S*-mandelic acid could be readily obtained from acetonitrile solution and were identical to those obtained by liquid-assisted grinding. Cocrystals of nicotinamide and the racemic acid could be obtained from acetonitrile solution only after prolonged standing or rapidly upon addition of a few seeds of the solid obtained by liquid-assisted grinding. In both cases the resulting cocrystals exhibited the same XRPD pattern as the material obtained *via* grinding. As established by ^{1}H NMR spectroscopy, each cocrystal contained equimolar amounts of nicotinamide and mandelic acid. The results are consistent with the products of liquid-assisted grinding being the anticipated cocrystals **1a** and **1b**.

In a similar way it was established that liquid-assisted grinding of nicotinamide with either *S*- or *RS*-ibuprofen produces new crystalline solids (Fig. 3). As evidenced by XRPD, the products obtained by grinding were identical to those obtained from solution. Moreover, ^{1}H NMR spectroscopy indicated the cocrystals were composed

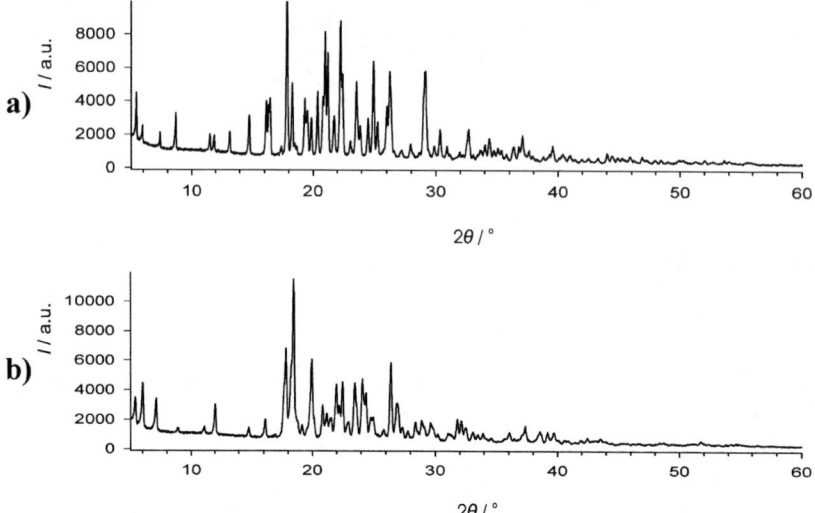

Fig. 2 X-Ray powder diffraction patterns of cocrystals: (a) **1a** and (b) **1b**, obtained by liquid-assisted grinding.

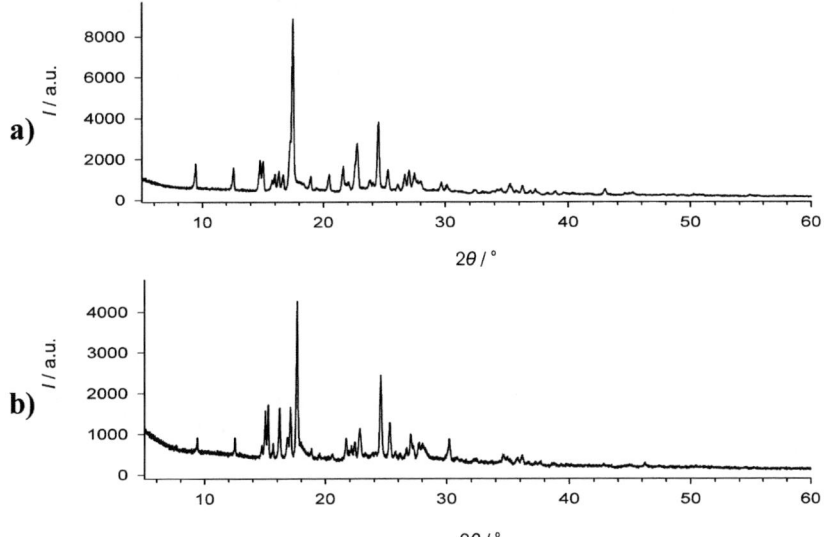

Fig. 3 X-Ray powder diffraction patterns of cocrystals: (a) **2a** and (b) **2b**, obtained by liquid-assisted grinding.

of nicotinamide and ibuprofen in a 1 : 1 stoichiometric ratio. Again, the results are consistent with the formation of anticipated cocrystals **2a** and **2b** both by grinding as well as solution cocrystallisation.

Thermal analysis

For all cocrystals, DSC thermograms were characterised by a sharp endothermic signal that was interpreted as the melting point of the solid. The interpretation was confirmed using hot-stage microscopy. The melting points of the cocrystals, along with the melting points of precursor materials are given in Table 1.

Crystal structure analysis

In order to verify the formation of the anticipated four-membered molecular assemblies, we undertook a single crystal X-ray diffraction study of **1a**. Needle-shaped single crystals of **1a** of sufficient quality for the X-ray diffraction experiment were obtained by slow cooling and evaporation of a solution of an equimolar mixture (0.100 g) of nicotinamide and *R*-mandelic acid in acetonitrile (2 mL). The most relevant crystallographic data for **1a** are summarised in Table 2. Crystal structure determination revealed that the asymmetric unit of the cocrystal consists of the expected molecular assembly comprising two nicotinamide and two mandelic acid molecules, see Fig. 4.

Table 1 Melting points of reactants and prepared cocrystals

Reactant	$T_{melting}/°C$	Cocrystal	$T_{melting}/°C$
R/S-mandelic acid	130	**1a**	89
RS-mandelic acid	117	**1b**	76
S-ibuprofen	50	**2a**	82
RS-ibuprofen	76	**2b**	91
Nicotinamide	150		

Table 2 General and crystallographic data for **1a**

Empirical formula	$C_{14}H_{14}N_2O_4$	Z	8
Formula weight	274.3	$\rho_{calc}/g\ cm^{-3}$	1.384
Temperature/K	150(2)	μ/mm^{-1}	0.103
Wavelength/Å	0.71073	$F(000)$	1152
Crystal system	Monoclinic	Crystal size/mm^3	0.46 × 0.07 × 0.07
Space group	C2	θ range/°	3.77 to 27.43
Unit cell dimensions/Å, °	a = 32.6557(9)	Data/restraints/parameters	5773/1/361
	b = 5.475(1)	S	1.12
	c = 14.9264(5)	R_1, wR_2 for $I > 2\sigma(I)$	0.055, 0.128
	β = 99.400(1)	R_1, wR_2 for all data	0.070, 0.138
$V/\text{Å}^3$	2632.9(5)	$\Delta/e\ \text{Å}^{-3}$, min, max	−0.295, 0.293

The assemblies in the crystal link by way of intermolecular N–H⋯O hydrogen bonds, forming chains that propagate parallel to the crystallographic *b*-direction (Fig. 5).

In contrast to our expectations, only one carboxylic acid moiety in each assembly is coplanar with a nearby pyridine group, while the second one is significantly twisted out of the plane of the pyridine ring (respective angles between planes defined by six non-hydrogen atoms of the pyridine ring and the three non-hydrogen atoms of the carboxylic acid functionality: 11.3(4)° and 34.2(3)°). Consequently, only one carboxylic acid group forms the expected pyridine—carboxylic acid supramolecular synthon through a pair of O–H⋯N and C–H⋯O bonds (O⋯N and C⋯O separations: 2.60 and 3.31 Å, respectively). The second carboxylic acid group is held to a neighbouring pyridine moiety through only a single O–H⋯N hydrogen bond (O⋯N separation 2.60 Å).

The failure of one mandelic acid molecule in the assembly to form the carboxylic acid–pyridine synthon can be related to the behaviour of the alcohol moiety in the acid. Notably, the hydroxyl group forms an intramolecular O–H⋯O hydrogen bond with an oxygen atom of the carboxylic acid group (O⋯O separation: 2.67 Å).

The ability of the carboxylic acid oxygen atom to act as a hydrogen bond acceptor is thus diminished, hindering the formation of a C–H⋯O bond. In the case of the mandelic acid unit that forms the C–H⋯O bond, the hydroxyl group is involved in an intermolecular O–H⋯O hydrogen bond (O⋯O separation: 2.68 Å). Such O–H⋯O bonds act together with intermolecular N–H⋯O bonds (N⋯O distances 2.97 and 3.00 Å), to link neighbouring molecular chains into layers. The layers are parallel to the {100} set of crystallographic planes. The core of each layer is populated mostly by polar groups, whereas hydrophobic aromatic residues decorate the layer surface (Fig. 6).

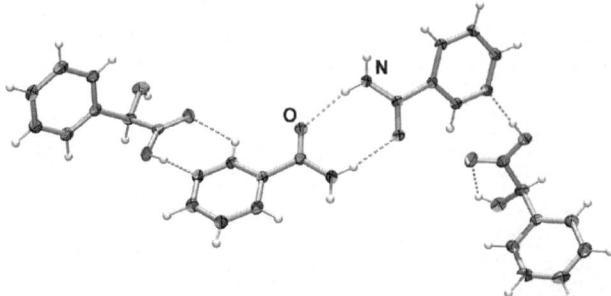

Fig. 4 ORTEP representation of the asymmetric unit of the **1a** cocrystal. Non-hydrogen atoms are shown as ellipsoids at 30% probability level.

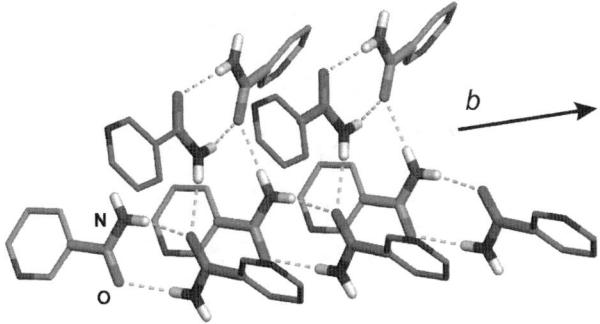

Fig. 5 A wire frame representation of molecular chains in the cocrystal **1a**. For clarity, hydrogen atoms bonded to carbon atoms, and mandelic acid molecules are omitted.

Discussion

Liquid-assisted grinding has provided cocrystals of nicotinamide with all the examined cocrystal formers. Consequently, in the cases studied, nicotinamide does not appear to discriminate between chiral and racemic cocrystal formers. The products obtained *via* liquid-assisted grinding were identical to products obtained by cocrystallisation from solution. In contrast to cocrystallisation from solution, however, the products obtained by grinding did not require any purification procedures (*i.e.* filtering) and, according to the X-ray powder diffraction patterns, the reaction yield was quantitative. Crystal structure analysis of **1a** demonstrated that short-range ordering of molecules in a cocrystal of nicotinamide and a carboxylic acid can be tentatively predicted by considering possible supramolecular synthons. Such an ability to design the architecture of a cocrystal simplified the study of the effects of chirality on the physical properties of the cocrystals. Whilst only compound **1a** could at present be investigated by single crystal X-ray analysis (attempts to determine the crystal structures of **2a** and **2b** *via* single crystal X-ray diffraction have so far failed, owing to the pronounced needle-like habit of the cocrystals), we note that the similarity of XRPD patterns between **1a** and **1b**, and between **2a** and **2b**, suggests some structural similarity within the members of each pair.

Chirality affects the melting point of the ibuprofen cocrystals in a way different to those of mandelic acid as a cocrystal former. Specifically, the melting point of the racemic cocrystal **2b** is higher than the melting point of the corresponding enantiomerically pure cocrystal **2a**. In the case of mandelic acid cocrystals the situation is reversed, *i.e.* the melting point of the chiral cocrystal **1a** is higher than the melting point of the racemic cocrystal **1b**. As evident from Table 1, differences in the melting points of chiral and racemic cocrystals agree well with the behaviour of the pure cocrystal formers. Such a result is reminiscent of the reports by Bond, and Nangia,[31]

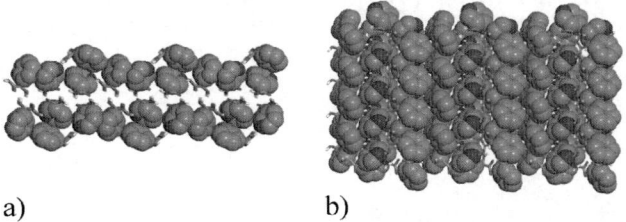

a)   b)

Fig. 6 Two views of molecular layers in **1a**: (a) cross-section and (b) a view perpendicular to the layer surface, exhibiting the predominance of polar groups within and aromatic moieties on the surface of each layer. Aromatic residues are displayed using a space-filling model.

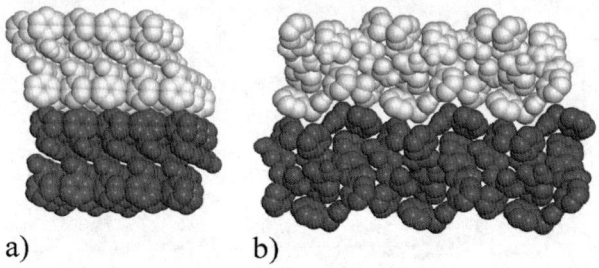

Fig. 7 Two adjacent layers in the crystal structure of: (a) *R*-mandelic acid and (b) cocrystal **1a**. For clarity, hydrogen atoms not bonded to oxygen are omitted.

that the alternation in melting points of a homologous series of alkanecarboxylic acids is also exhibited in the series of corresponding cocrystals. Although the few results presented herein do not allow a broad generalisation, they nevertheless suggest that a simple relationship may exist between the thermal properties of cocrystals and the parent cocrystal formers.

It is also interesting to consider the melting points of the cocrystals with respect to those of the pure cocrystal formers. Here again the cocrystals of ibuprofen follow a different trend to those of the cocrystals of mandelic acid. The melting points of **1a** and **1b** are lower than the melting point of either *R*- or *RS*-mandelic acid. The opposite is true for the melting points of **2a**, **2b** and ibuprofen. We believe that such behaviour can be rationalised through a comparison of the solid-state structures of the cocrystal and the corresponding cocrystal former. Our rationalisation is based on the assumption that all the cocrystals studied are based on molecular assemblies similar to the ones that constitute **1a**. Following that assumption, cocrystals of ibuprofen would be expected to have a higher melting point than the cocrystal former, since the isolated carboxylic acid dimers in crystalline ibuprofen would be replaced by a network of N–H···O hydrogen bonds in the cocrystal. In the case of chiral mandelic acid, molecules in the free acid, as well as in the corresponding cocrystal **1a**, are connected into hydrogen-bonded layers. In *R*-mandelic acid the layers are held together *via* O–H···O and C–H···O hydrogen bonds, and in **1a** the molecules in each layer are connected by O–H···O, O–H···N and N–H···O hydrogen bonds. Consequently, lowering of the melting point of the cocrystal with respect to the cocrystal former is probably not a result of changes to the hydrogen-bonding framework. We find that the analysis of the packing of layers in *R*-mandelic acid and **1a** provides a plausible explanation of the thermal behaviour. In layers of the crystalline acid, phenyl moieties of each molecule are approximately perpendicular to the plane of the layer. Consequently, the adjacent layers are partially interdigitated, reminiscent of a molecular version of a Velcro™ fastener. Such interdigitation is expected to make the parallel sliding of layers difficult. In contrast, the surface of each layer **1a** is covered by flat aromatic moieties (Fig. 7). As a result, adjacent layers in **1a** are expected to slide more readily than in *R*-mandelic acid, thereby possibly facilitating melting. Based on structural similarity evident from the similarity of the XRPD patterns, a similar explanation is expected to be valid for the cocrystal **1b**.

Conclusion

We have applied a synthon-based approach to construct cocrystals that would have similar architectures, but different symmetry properties. The difference in symmetry is introduced by utilising an enantiomerically pure or a racemic cocrystal former. We demonstrate that the thermal properties of cocrystals change with the symmetry of the cocrystal former. The melting points of pairs of chiral and racemic cocrystals, based on a similar supramolecular architecture, appear to follow the same trend as the melting points of the pure cocrystal. Notably, we have recognised a pair of racemic and chiral cocrystals that contradicts the Wallach's rule as far as melting points are concerned. We consider the observation is promising in the context of engineering cocrystals with desired thermal stability, a possibility that is particularly attractive for pharmaceutical materials. Indeed, the results presented in this contribution also demonstrate that cocrystallisation can increase the melting of a low-melting point drug ibuprofen. However, the change in thermal stability upon cocrystallisation is difficult to predict and we show that both changes in hydrogen bonding pattern, as well as changes to molecular close packing are relevant. To achieve a better understanding of how the thermal behaviour of a cocrystal emerges from thermal properties of cocrystal components, we are now pursuing structural characterisation of nicotinamide cocrystals with RS-mandelic acid, S- and RS-ibuprofen, as well as other cocrystal formers that exist in chiral and racemic forms.

Acknowledgements

We thank Dr Neil Feeder and Dr Pete Marshall for useful discussions. The Pfizer Institute for Pharmaceutical Science is acknowledged for funding. Dr John E. Davies is acknowledged for providing single crystal X-ray diffraction data.

References

1 (a) G. M. J. Schmidt, *Pure Appl. Chem.*, 1971, **27**, 647; (b) M. C. Etter, *Acc. Chem. Res.*, 1990, **23**, 120.
2 (a) M. A. Garcia-Garibay, *Proc. Natl. Acad. Sci. USA*, 2005, **102**, 10771; (b) A. Troisi, G. Orlandi and J. E. Anthony, *Chem. Mater.*, 2005, **17**, 5024.
3 J. D. Wuest, *Chem. Commun.*, 2005, 5830.
4 (a) P. Vishweshwar, J. A. McMahon, M. L. Peterson, M. B. Hickey, T. R. Shattock and M. J. Zaworotko, *Chem. Commun.*, 2005, **12**, 4601; (b) B. K. Saha, A. Nangia and M. Jaskolski, *CrystEngComm*, 2005, **7**, 355.
5 (a) C. B. Aakeroy, J. Desper and J. F. Urbina, *Chem. Commun.*, 2005, 2820; (b) G. R. Desiraju, *CrystEngComm*, 2003, **5**, 466; (c) J. D. Dunitz, *CrystEngComm*, 2003, **5**, 506.
6 P. Vishweshwar, J. A. McMahon and J. A. Bis, *J. Pharm. Sci.*, 2006, **95**, 499.
7 A. N. Sokolov, T. Friščić and L. R. MacGillivray, *J. Am. Chem. Soc.*, 2006, **35**, 3523.
8 M. D. Hollingsworth, *Science*, 2002, **295**, 2410.
9 (a) L. R. MacGillivray, G. S. Papaefstathiou, T. Friščić, D. B. Varshney and T. D. Hamilton, *Top. Curr. Chem.*, 2005, **248**, 201; (b) J. H. Kim, S. M. Hubig, S. V. Lindeman and J. K. Kochi, *J. Am. Chem. Soc.*, 2001, **123**, 87.
10 T. Friščić and L. R. MacGillivray, *Croat. Chem. Acta*, 2006, **79**, 327.
11 T. Friščić and L. R. MacGillivray, *Chem. Commun.*, 2003, 3523.
12 A. V. Trask, W. D. S. Motherwell and W. Jones, *Cryst. Growth Des.*, 2005, **5**, 1013.
13 C. B. Aakeröy, A. M. Beatty and B. A. Helfrich, *Angew. Chem., Int. Ed.*, 2001, **40**, 3240.
14 G. R. Desiraju, *Angew. Chem., Int. Ed. Engl.*, 1995, **34**, 2311.
15 B. K. Saha, A. Nangia and M. Jaskólski, *CrystEngComm*, 2005, **7**, 355.
16 (a) D. Braga and F. Grepioni, *Angew. Chem., Int. Ed.*, 2004, **43**, 4002; (b) F. Toda, K. Tanaka and A. Sekikawa, *J. Chem. Soc., Chem. Commun.*, 1987, 279–280; (c) E. J. Cheung, S. J. Kitchin, K. D. M. Harris, Y. Imai, N. Tajima and R. Kuroda, *J. Am. Chem. Soc.*, 2003, **125**, 14658–14659.
17 A. V. Trask and W. Jones, *Top. Curr. Chem.*, 2005, **35**, 3523.
18 For an overview of the use of grinding as a synthetic method, see: (a) D. Braga, S. L. Giaffreda, F. Grepioni, A. Pettersen, L. Maini, M. Curzi and M. Polito, *Dalton Trans.*, 2006, (12), 1249; (b) F. Toda, *Chem. Rev.*, 2000, **100**, 1025.
19 D. Braga and F. Grepioni, *Angew. Chem., Int. Ed.*, 2004, **43**, 4002.
20 N. Shan, F. Toda and W. Jones, *Chem. Commun.*, 2002, 2372.
21 D. Braga and F. Grepioni, *Chem. Commun.*, 2005, 3635.
22 A. V. Trask and W. Jones, *Top. Curr. Chem.*, 2005, **254**, 41.
23 T. Friščić, A. V. Trask, W. Jones and W. D. S. Motherwell, *Angew. Chem., Int. Ed.*, 2006, **45**, 7546.
24 N. Rodríguez-Hornedo, S. J. Nehm, K. F. Seefeldt, Y. Pagán-Torres and C. J. Falkiewicz, *Mol. Pharm.*, 2006, **3**, 362–367.

25 A. V. Trask, J. van de Streek, W. D. S. Motherwell and W. Jones, *Cryst. Growth Des.*, 2005, **5**, 2233.
26 T. Friščić and L. R. MacGillivray, *Chem. Commun.*, 2003, 1306.
27 A. V. Trask, W. Jones and W. D. S. Motherwell, *Cryst. Growth Des.*, 2005, **5**, 1013.
28 T. Friščić, L. Fábián, J. C. Burley, W. Jones and W. D. S. Motherwell, *Chem. Commun.*, 2006, 5009.
29 M. C. Etter, *J. Phys. Chem.*, 1991, **95**, 4601.
30 (*a*) C. B. Aakeroy, A. M. Beatty and B. A. Helfrich, *J. Am. Chem. Soc.*, 2002, **124**, 14425; (*b*) C. B. Aakeroy, J. Desper and B. A. Helfrich, *CrystEngComm*, 2004, **6**, 19.
31 (*a*) A. D. Bond, *CrystEngComm*, 2006, **8**, 333; (*b*) L. S. Reddy, A. Nangia and V. M. Lynch, *Cryst. Growth Des.*, 2004, **4**, 89; (*c*) A. D. Bond, *Chem. Commun.*, 2003, **12**, 250.
32 C. P. Brock, W. B. Schweizer and J. D. Dunitz, *J. Am. Chem. Soc.*, 1991, **113**, 9811.

PAPER | www.rsc.org/faraday_d | Faraday Discussions

The nucleation of inosine: the impact of solution chemistry on the appearance of polymorphic and hydrated crystal forms†

Renato A. Chiarella,‡[a] Amy L. Gillon,§[a] Rebecca C. Burton,[a] Roger J. Davey,[a] Ghazala Sadiq,[a] Anthony Auffret,[b] Marina Cioffi[c] and Christopher A. Hunter[c]

Received 6th November 2006, Accepted 2nd February 2007
First published as an Advance Article on the web 12th April 2007
DOI: 10.1039/b616164m

This contribution concerns the issue of crystal nucleation in the polymorphic and hydrate forming system inosine–water. A combination of computational and experimental tools have been used to explore the relationship between solution phase inosine species and the structural synthons as found in its crystal structures. It is evident that the initial nucleation of a metastable polymorph at temperatures above 10 °C is directed by dimeric self-association as revealed through proton NMR. At lower temperatures a dihydrate structure becomes the most stable solid phase and in this region of the phase diagram this is the only form that appears even though the solution species remain unchanged. This can only be rationalised in terms of a combination of water binding to the solution dimers and the thermodynamic stability of the hydrate crystal structure.

Introduction

The phenomenon of crystal nucleation is central to the isolation and purification of molecular materials. The process of nucleation is characterised by a number of well known macroscopic phenomena including the existence of a metastable limit, which must be exceeded for nucleation to proceed, a characteristic 'induction' time required for the appearance of the first crystals and the occurrence of both solvated and polymorphic structures of a single molecule. While the kinetic aspects of these features may be described by use of well known kinetic formalisms,[1] originating in the work of Volmer and others, this approach does not provide molecular level insight into the mode of self-assembly of the species involved.

In recent studies aimed at understanding structural aspects of the nucleation process two things are becoming evident,[2,3] firstly that the use of crystal structure data alone is insufficient to understand and design control strategies for the

[a] *Colloids Crystals and Interfaces Group, School of Chemical Engineering and Analytical Sciences, University of Manchester, P.O. Box 88, Manchester, UK M60 1QD. E-mail: roger.davey@manchester.ac.uk*
[b] *Pharmaceutical R&D, Pfizer Global R&D, Sandwich, Kent, UK CT13 9NJ*
[c] *Centre for Chemical Biology, Krebs Institute for Biomolecular Science, Department of Chemistry, University of Sheffield, Sheffield, UK S3 7HF*

† The HTML version of this article has been enhanced with colour images.
‡ Current address: Transform Pharmaceuticals, 29 Hartwell Ave., Lexington, MA, USA.
§ Current address: Astra Zeneca, Charnwood, Loughborough, Notts, UK.

nucleation process and hence secondly that more knowledge and characterisation of solution structure is needed in order to make a step change in our understanding.

In the light of this we have been interested in exploring the application of molecular spectroscopies—vibrational, electronic and NMR—together with solubility and crystallographic data of various compounds, to improve our knowledge of molecular self-association and solvation in supersaturated solutions. Thus, for example in studies using the measured concentration dependence of proton NMR chemical shift in sulfamerizine solutions Spitaleri et al.,[4] have noted a direct correlation between the structure of dimers in the solution phase and those in the resulting crystal. The use of FTIR to explore the relationship between growth and structural synthons in carboxylic acids has been reported by Davey et al.[5] and more recently, the early use of molecular dynamics to study tetrolic acid solutions[6] has been extended to simulations of 5-fluorouracil in aqueous and nitromethane solutions.[7] A clear outcome of this work is that while there appears, in many cases, to be a relationship between the self assembled dimers in solution and the assemblies that appear in the crystalline phases this is by no means a universal characteristic. Mandelic acid is an example where this is apparently not the case[5] and where it appears that molecular rearrangement in the clusters is essential to yield the observed crystal packing.

In this current work we set out to extend these studies beyond polymorphic systems to address the question of 'if and how' solution chemistry drives the nucleation of a solvated structure.[8] The issue of solvate formation in general has been the subject of considerable previous work[9] but much of this has been from a structural perspective aimed at rationalisation of the role of solvent molecules within crystal structures. Perhaps the most well known conclusion from this is the idea that solvents tend to be used to address any mismatch between the number of hydrogen bonded donors and acceptors in a crystallising molecule.[10]

Here we use a combination of crystal structure analysis, in situ crystallisation experiments, solubility data, proton NMR and molecular modelling in order to progress towards an understanding of the origin of the crystalline dihydrate that appears in the inosine–water system.

Inosine

Inosine is a nucleoside (Fig. 1), a purine derivative antimetabolite, containing both a polar furan ring and an apolar aromatic group. It is an important component of ribonucleic acids, with the hydrophilic–hydrophobic balance of the molecule contributing to their conformational stability. Previous vapour pressure and proton NMR studies[11,12] of the aqueous solution chemistry of this and other purine nucleosides (e.g. purine, adenosine and their derivatives), has led to the consistent

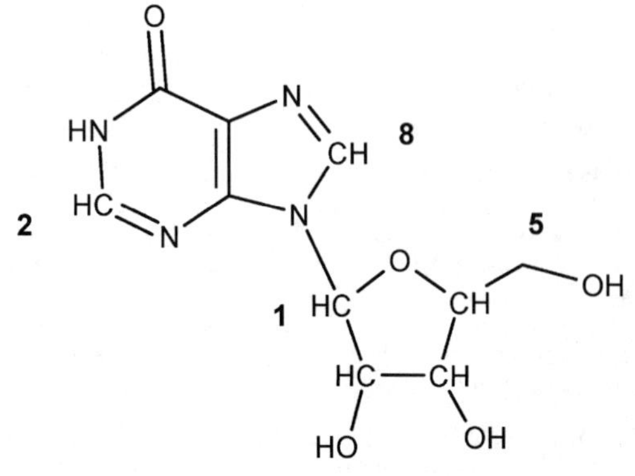

Fig. 1 The inosine molecule with protons numbered as per proton NMR data.

Table 1 Inosine crystallographic data[13–15]

Phase	α	β	Dihydrate
Chemical formula	$C_{10}H_{12}N_4O_5$	$C_{10}H_{12}N_4O_5$	$C_{10}H_{12}N_4O_5 \cdot 2H_2O$
RMM, g mol^{-1}	286.23	286.23	322.23
Crystal system	Orthorhombic	Orthorhombic	Monoclinic
Space group	$P2_12_12_1$	$P2_1$	$P2_1$
Unit cell			
a/Å	13.261	4.818	17.573
b/Å	21.285	10.450	11.278
c/Å	8.097	10.970	6.654
β/°	90	90.72	98.23
Z	8	2	4
Density/kg m^{-3}	1576	1613	1542

view that such molecules aggregate *via* vertical stacking of the bases rather than *via* H-bonding.

Inosine is known to have two anhydrous polymorphic crystal forms, the orthorhombic α form[13] and the monoclinic β form[14] together with a dihydrate.[15] The crystallographic details of these phases are summarised in Table 1. We have previously described the differences in the H-bonded nature of these packings[16] as well as the solid state dehydration of the dihydrate. Fig. 2 shows the difference in aromatic stacking arrangements of dimers taken from each structure: it is evident that α and β dimers minimize and maximize respectively the steric interactions between the furan rings while the dihydrate adopts an intermediate position. The three forms have distinct pXRD patterns which can be used to distinguish them. For example, inosine dihydrate can easily be recognised by its peak at 2θ of 5°; α inosine has two small peaks at 8° and a large peak at 11°, while β inosine has only one large peak at 12°.

Solubility measurements have been reported in the literature[17,18] and are reproduced in the form of the solubility diagram, Fig. 3. The least soluble (*i.e.* most stable) form at room temperature and above is the β form, while below 10 °C the most stable form in aqueous solutions becomes the dihydrate. The α polymorph is related monotropically to the β form. The measured solubility of β inosine in water at 25 °C is 0.0014 mol/mol. The heat of fusion, ΔH_f, has been measured in this work to be 48.77 kJ mol^{-1} with a melting temperature, T_m of 493.5 K. These values may be used along with the van't Hoff relationship:[19]

$$\ln x_{\text{ideal}} = -\Delta H_f(T_m - T)/RT_mT + \Delta C_p(T_m - T)/RT$$

to calculate x_{ideal}, the ideal solubility of β inosine, at 25 °C. Because its melting point is significantly above 25 °C the second term in this equation becomes significant and

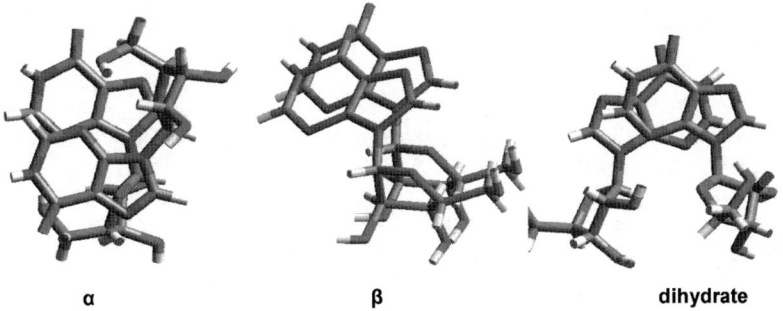

Fig. 2 Dimer stacking arrangement in the α, β and dehydrate structures of inosine.

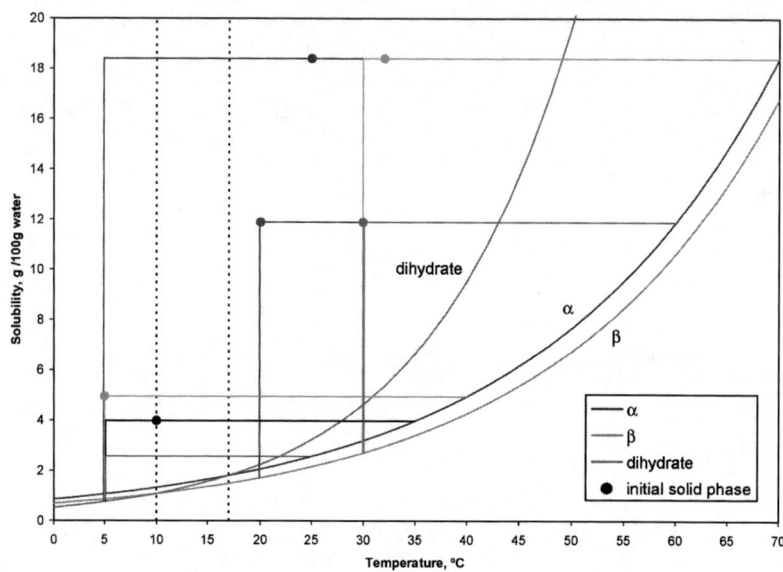

Fig. 3 Inosine solubility in water[18] and crystallisation trajectories (see Table 3 for respective experiments).

the difference in heat capacity (ΔC_p) between crystalline inosine and supercooled liquid inosine at 25 °C must be known. Although we were unable to measure this due to decomposition of liquid inosine above its melting point, it is evident that, for any value of ΔC_p greater that 15 J mol^{-1} K^{-1}, the ideal solubility will exceed the measured value. Given that typical values for this quantity appear to be well in excess of this—for paracetamol values of 74, 77 and 99 J mol^{-1} K^{-1} have been estimated[20] while for inosine itself the value of ΔC_p between crystal and saturated aqueous and ethanolic solutions have been measured[21] to be 122 and 155 J mol^{-1} K^{-1}—it is evident that saturated aqueous solutions of inosine show positive deviations from ideality, a feature consistent with molecular aggregation. Taking a value of ΔC_p of 120 J mol^{-1} K^{-1}, for example, suggests an ideal solubility of 0.04 mol/mol. Interestingly, in DMSO the measured solubility of β inosine at 25 °C is 0.09 mol/mol, an order of magnitude greater than in water suggesting a negative deviation resulting from much stronger solvation in this more hydrophobic solvent.

Experimental

All chemicals were purchased from Aldrich. (−)-Inosine was used without any further purification, D$_2$O was opened immediately before use. The pKs of inosine are 1.2, 8.9; 12.4[22] and the pH of a saturated aqueous solution is 4.0 indicating that in all of the experiments the uncharged species predominates.

Crystallisation experiments

The major objective of the crystallisation experiments described here was to determine which of the three forms nucleated first under given conditions of temperature and supersaturation. This required an experimental design in which the chosen conditions could be achieved as quickly as possible and in which the identity of the initial crystal form could be rapidly established. A combination of methods was employed.

In a series of small scale (20 ml) experiments a temperature jump method was developed in which an agitated jacketed vessel was connected to two water baths. A solution of known composition could then be prepared at an elevated temperature and by switching to the second water bath could be brought to a lower, crystallisation temperature within 1 min. Upon crystallisation, a sample of the first solids were removed, filtered and immediately subjected to pXRD (Bruker AXS D8) analysis. Subsequent transformations were recorded by allowing the experiments to equilibrate for up to two days.

A second series of experiments were performed on station 16.4 at the SRS Daresbury Laboratory, UK using energy dispersive X-rays. Here a small (70 ml), Teflon, batch crystalliser described previously[23] was used and a diffraction pattern of its contents could be recorded every minute. In these experiments cooling of solutions took longer (up to 30 min) but the first phase and its temperature of appearance could be identified *in situ* thereby removing the possibility of changes taking place during sample characterisation. Time constraints meant that these experiments were terminated after about 45 min—no data were recorded on the subsequent transformation of the crystallised phases.

Fig. 3 shows some typical crystallisation trajectories. The starting solutions were prepared from α inosine and had compositions corresponding to saturated solutions at the quoted starting temperature of the experiment. Supersaturations were calculated using the relationship:

$$\sigma = \ln \{x/x_s\}$$

in which x is the initial solution mole fraction and x_s the saturation mole fraction at the crystallisation temperature. For the dihydrate these compositions were expressed in terms of moles of inosine dihydrate.[24]

While the majority of this work was performed in aqueous solutions, the results were extended *via* a limited number of experiments in DMSO and acetone–water mixtures. In this latter, mixed solvent system the transition temperature between the dihydrate and β polymorph can be depressed as low as 1 °C.[25]

Crystals from these experiments were observed using optical microscopy (Zeiss Axioplan 2) and SEM (Philips XL30). Heats and temperatures of fusion of β inosine were recorded on a Mettler Toledo DSC 30.

A limited number of evaporative crystallisation experiments were carried out in water–DMSO mixtures in order to grow large crystals. These were used to determine the fast growth direction of each form from single crystal diffraction (Nonius SMART diffractometer).

All NMR solution spectra were recorded on a 400 MHz Bruker DPX Spectrometer using residual solvent as an internal standard. Spectra were analysed using *XWin-NMR* and MestRec Software, and the chemical shifts are quoted in ppm on the δ scale. ^{1}H NMR dilution experiments were carried out at 7 and 25 °C by preparing 3 ml stock solutions of the inosine in D$_2$O at concentrations of 0.033 and 0.237 M, respectively. Aliquots of the stock solution were added successively to a sample of D$_2$O, and ^{1}H NMR spectra were recorded as a function of concentration. In addition, the previously published data of Broom *et al.*[11] measured at 25 °C were reanalysed using the same software. In this way we have recorded data above and below 10 °C, the dihydrate/β transition temperature in water. The data were fitted to dimerisation and non-cooperative linear polymerisation isotherms with purpose written software, *NMRDil_Dimer* or *NMRDil_Agg.*,[26] which use a Simplex procedure to fit the experimental data and determine the optimum solutions for the association constant, and the limiting bound and free chemical shifts. Thus, for example, *NMRDil_Dimer* fits the data to a dimerisation isotherm by solving the following equations

$$[AA] = \frac{1 + 4K_d[A]_0 - \sqrt{\{1 + 8K_d[A]_0\}}}{8K_d} \quad (1)$$

$$[A] = [A]_0 - 2[AA] \qquad (2)$$

$$\delta_{obs} = \frac{2[AA]}{[A]_0}\delta_d + \frac{[A]}{[A]_0}\delta_f \qquad (3)$$

in which $[A]_0$ is the total concentration, $[A]$ is the concentration of unbound free species, $[AA]$ is the concentration of dimer, K_d is the dimerisation constant and δ_f and δ_d are the chemical shifts of the free and limiting bound dimer, respectively.

NMRDil_Agg fits the data to a non-cooperative linear oligomerisation/polymerisation isotherm by solving the following equations.

$$[Agg] = [A]_0 \left\{ 1 - \frac{2}{1 + \sqrt{\{1 + 4K[A]_0\}}} \right\} \qquad (4)$$

$$[A] = [A]_0 - [Agg] \qquad (5)$$

$$\delta_{obs} = \frac{2[Agg]}{[A]_0}\delta_b + \frac{[A]}{[A]_0}\delta_f \qquad (6)$$

where $[A]_0$ is the total concentration, $[A]$ is the concentration of sites which are unbound (this is the sum of the free species and the ends of the aggregate which are not bound), $[Agg]$ is the concentration of sites involved in intermolecular interactions in the aggregate, K is the association constant for chain extension of the aggregate and δ_f and δ_b are the chemical shifts of the free and limiting bound sites in the aggregate, respectively.

Signals for H1, H2 and H8 (defined as in Fig. 1) were followed in each experiment, and the values quoted for the association constant is the weighted average based on the observed changes in chemical shift. The data were then refitted using this average association constant to obtain a self-consistent set of free and bound chemical shifts. The method used to determine and visualise three-dimensional structures from the measured complexation-induced changes in the chemical shifts (CIS), *SHIFTY*, has been described in detail elsewhere.[26–28] CIS values are calculated by building the molecules in XED 2.8 using standard bond lengths and angles. The anisotropy parameters used for the base were taken from *ab initio* calculations. The representation used treats the five-membered ring as aromatic with a ring current factor of 1.03 and the six-membered ring as non-aromatic and represented simply using the bond contributions of the carbon–carbon double bonds and amide.[29] A genetic algorithm was used to optimise the structure of the dimeric complex, so that the calculated CIS values matched the experimental values as closely as possible. The conformational search was divided into five steps, each with population size of 750 and 2000 generations. In the first step, the intermolecular distance was set to 10 Å, the range of allowed rotations of one molecule relative to the other was set to 360°. Intramolecular torsions were allowed to change within the full range of 360° for both molecules. In subsequent steps, each search space parameter was reduced to half of its previous value. The self-association constant (K_f) was varied by a factor of 10, to scale the experimental values to allow for errors in the isotherm curve fitting. To restrict the search space, VdW clashes were penalised at distances of less than 3 Å for intermolecular clashes and 2 Å for intramolecular clashes for non-hydrogen atoms. 20 different calculations, starting from random geometries, were carried out. All the solutions showed a root mean square difference between the experimental and calculated chemical shifts (RMSD) of less than 0.002 ppm. The scaling factor applied to the self-association constant was close to one in all cases with an average value of 1.2, suggesting that the experimental value is reasonably accurate. The final structure predicted using *SHIFTY* may be compared with the actual crystal structures of the associated polymorph.

Visualisation of crystal structures and lattice energy calculations were performed in CERIUS^2.[30] Two scripts were used, linking several functions available within CERIUS^2 in an automated fashion. Both used the molecular mechanics Drieding 2.21 MM force field, but varied in the method employed (Gasteiger or QEq) for the assignment of atomic point-charges. Geometry optimisation (AM1) and heats of formation (PM3) of the three inosine dimers (single point energies) were calculated in MOPAC.[31]

Results

Crystallisation experiments

The crystallisation of inosine from aqueous solutions always resulted in powder samples comprising crystals of a few microns in size, with all three forms having a needle like morphology. Examples of these morphologies as revealed by SEM are shown in Fig. 4. The needle axes (fast growth directions) were determined from larger crystals to be c, a and c for α, β and dihydrate, respectively. The results of the crystallisation experiments are reported in Tables 2 and 3 which give the identification of the initial phase and the extent of supersaturation of each phase at the end point temperature. For the energy dispersive data, the temperature and supersaturation at which the first phase appeared is reported and for the laboratory data the crystallisation induction times are given together with the phase present after ageing of the slurry. Considering the laboratory data of Table 2 a number of factors are apparent. For the three crystallisation experiments carried out at 50 and 60 °C the solutions are supersaturated only with respect to the α and β polymorphs and the consistent initial appearance of the α form followed by its transformation to the

Fig. 4 SEM images of (a) α inosine, (b) β inosine and (c) inosine dihydrate crystals grown from aqueous solutions.

Table 2 Crash cooling crystallisation of inosine (Cooling time < 1 min)

Exp.	Temp. drop/°C	Supersaturation σ_α, σ_β, $\sigma_{dihydrate}$	Start of cryst./ min	Solid phases		
				Initial	1 day	2 days
1a	70–50	2.4, 2.7, —	8	α	β	
2a	70–60	1.5, 1.7, —	3	α	α	α + β
3a	60–50	1.5, 1.8, —	13	α		α + β
4a	60–30	3.7, 4.4, 2.5	2	α + β		
5a	60–30	3.7, 4.4, 2.5	8	α + dh	α	β
6a	55–30	3.0, 3.5, 2.1	11	α	α + β	
7a	15–10	1.2, 1.5, 1.1	10	dh		
8a	15–5	1.5, 1.9, 2.2	8	dh	dh	
9a	15–2	1.8, 2.2, 2.7	3	dh		

β polymorph is as expected and follows Ostwald's Rule.[32] Comparing experiments 1a and 3a, both at 50 °C, shows, as expected, the induction time for the appearance of α to increase with a fall in supersaturation from 2.4 to 1.5. Comparison of experiments 2a and 3a show how for constant supersaturation the induction time falls with increasing temperature. For experiments 4a, 5a and 6a all performed at 30 °C the outcome is not so consistent. All three solutions are supersaturated with respect to all phases and the initial solids to form all contain α, on one occasion combined with some β and on another with dihydrate. Left to equilibrate it appears that these systems do evolve to the stable β phase.

At the lower temperatures (experiments 7a, 8a, 9a) at 10 °C and below, where the dihydrate is the most stable phase, the outcome is again very consistent, but not as might be expected on the basis of Ostwald's Rule. Surprisingly the most stable dihydrate phase now becomes the first to appear with no evidence for the existence of either of the polymorphs. As the temperature falls from 10 to 3 °C the induction times decrease from 10 to 3 min indicating the nucleation rate of dihydrate to increase with falling temperature and rising supersaturation.

Results from the SRS *in situ* studies, seen in Table 3, broadly confirm these outcomes. Some representative time resolved diffraction patterns are seen in Fig. 5a–d. Experiments 2b and 3b crystallised at 10 and 5 °C, respectively, both in the region of dihydrate stability. In agreement with the laboratory experiments, the dihydrate was the first and only form to appear. Experiment 4b crystallised at 25 °C where the metastable dihydrate appeared despite it having the lowest supersaturation, while experiments 5b, 6b and 7b at 20 and 30 °C all yielded α. These outcomes seem again

Table 3 Crystallisation of inosine monitored by energy dispersive X-ray diffraction. Experiment 1b did not yield sufficient solid to be detected (<2% solids by weight)

Exp.	Temp. drop/°C	Mean cooling rate/°C min^{-1}	End point supersaturation σ_α, σ_β, $\sigma_{dihydrate}$	Initial solid phase (σ_α, σ_β, $\sigma_{dihydrate}$)
1b	25–5	1.1	2.4, 3.0, 3.4	None
2b	35–5	2.1	3.7, 4.6, 5.3	Dihydrate at 10 °C (3.0, 3.7, 3.7)
3b	40–5	2.9	4.7, 5.8, 6.6	Dihydrate at 5 °C (4.7, 5.8, 6.6)
4b	70–5	1.8	17.5, 21.7, 24.7	Dihydrate at 25 °C (7.3, 8.7, 5.8)
5b	60–20	2.0	6.0, 7.2, 5.5	α at 20 °C (6.0, 7.2, 5.5)
6b	60–30	1.9	4.4, 5.2, 3.0	α at 30 °C (4.3, 5.1, 3.0)
7b	70–30	2.2	5.9, 7.0, 4.1	α at 32 °C (5.4, 6.3, 3.5)

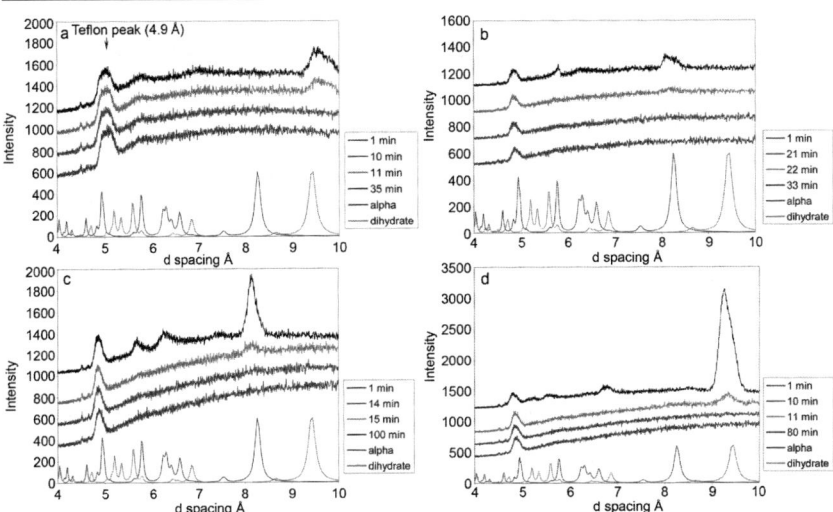

Fig. 5 (a) *In situ* crash cooling crystalisation (35–5 °C) of inosine in pure water (experiment 2b). (b) *In situ* crash cooling crystallisation (60–20 °C) of inosine in pure water (experiment 5b). (c) *In situ* crash cooling crystallisation (70–30 °C) of inosine in pure water (experiment 7b). (d) *In situ* crash cooling crystallisation (70–5 °C) of inosine in pure water (experiment 4b).

to be consistent with the somewhat mixed results obtained in the laboratory for this region of the phase diagram.

In the water–acetone mixtures 3 laboratory experiments were performed. At 80 wt% water, where the transition temperature is approximately 7 °C[25] the dihydrate was the first phase to appear at 0 °C and at 20 °C it was β; at 35 wt% water the transition temperature is about 3 °C[25] and crystallisation at 0 °C again gave the dihydrate directly while at 5 °C a mixture of dihydrate and α resulted. Whilst paucity of data does not allow calculation of supersaturations for these experiments they do at least appear to confirm an important result that crystallisation experiments performed below the β to dihydrate transition temperature always result in direct nucleation of the stable hydrated phase with no evidence of the metastable polymorphs.

In contrast to the results obtained in aqueous systems, cooling crystallisation from DMSO always gave the β form directly.

Taking the totality of the data measured in aqueous solutions enables the construction of Fig. 6 showing the appearance domains of the forms, divided into the Ostwald region in which α consistently appears first, the mixed region in which mixtures of forms were seen and the non-Ostwald region in which crystallisation proceeds directly to the dihydrate. It is interesting to note that in aqueous solutions there appears to be no region which favours the consistent, direct nucleation of the β form.

Solution structure determination using solution NMR spectroscopy

Table 4 gives the self-association constants (M^{-1}) and limiting $\Delta\delta$ values (ppm) for aggregation of inosine in D_2O derived from the measured chemical shift changes for the protons H1, H2 and H8. It was evident that the data could be fitted equally well to dimerisation and to polymerisation isotherms, because the fully bound end of the isotherm is not experimentally accessible, and in the early stages of polymerisation, dimerisation is the dominant process. All of the protons gave similar self-association constants, which indicates that these changes in chemical shift are consistent with a simple two state equilibrium process that affects the whole molecule in the same way.

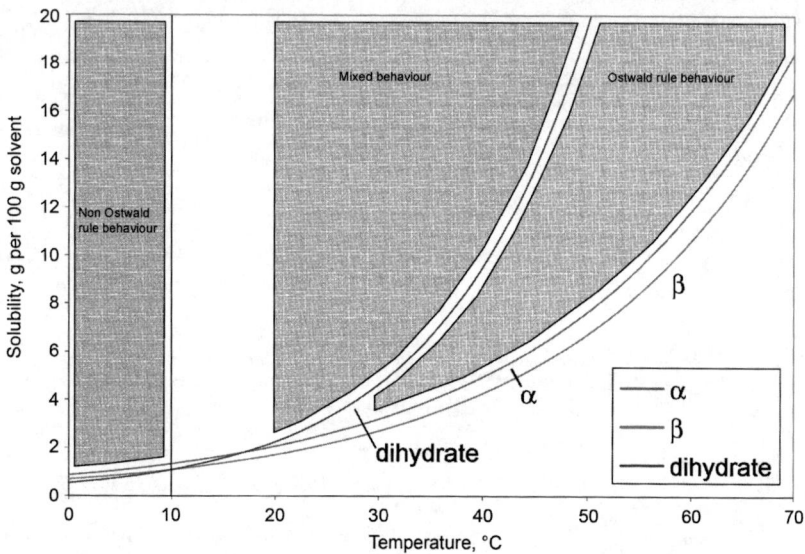

Fig. 6 Nucleation domains interpolated from crash cooling crystallisation of inosine in pure water at different supersaturations and temperatures.

The self association binding constant decreases from 9 to 7 M^{-1} in going from 7 to 25 °C. The fact that only half of the binding isotherm can be experimentally sampled means that there is potentially a significant error in these values and indeed values of 2.69 m^{-1} and 1.7 m^{-1} reported previously[12,33] based on osmotic coefficient data are somewhat lower (note that the units differ which would lead to a further 2% reduction of m^{-1} compared to M^{-1}). However, it is only the absolute CIS values that are affected by this error. The relative CIS for different protons in the molecule are accurately determined by the experiments, so they contain information that can be used in the structure determination. Thus in using *SHIFTY* the self-association constant was allowed to be an additional variable in the structure optimisation process to allow for errors in the self-association constant discussed above.[4] This variable scales the absolute values of the experimental CIS values without changing the relative values. A large number of solutions were obtained from the *SHIFTY*

Table 4 Self-association constants and limiting $\Delta\delta$ values for aggregation of inosine in D$_2$O

Temp./ °C	$\Delta\delta$ (ppm)			Maximum % bound	Binding constant/M^{-1}
	H1	H2	H8		
25	−0.195	−0.176	−0.143	66	7 ± 1
7	−0.048	−0.051	−0.041	36	9 ± 1

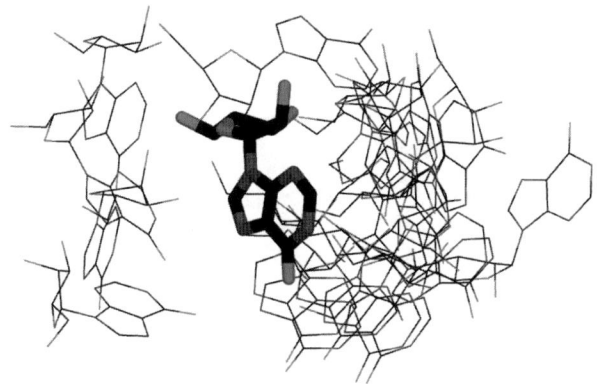

Fig. 7 Visualisation of an inosine dimer—an overlay of 20 *SHIFTY* runs.

structure optimisation, and Fig. 7 shows that the structure of the dimer complex is not well-defined by the experimental data. This may reflect the fact that the complex genuinely samples a large number of conformations, or that there are too few experimental observables (3 chemical shift changes) to unambiguously define the structure of a single highly populated complex.

Although *SHIFTY* does not accurately define a single structure for the inosine dimer in solution, it can be used to assess the likelihood that a specific dimer structure of interest is populated based on the experimental chemical shift data. Thus we calculated the patterns of CIS values that would be expected for formation of specific dimer motifs and compared these with the corresponding experimental data. In this case, the structures of dimers taken from the crystal structures of the different inosine phases (Fig. 2) were examined and Table 5 lists their calculated CIS values. Table 6 then compares the relative experimental and calculated CIS values and from these data it is evident that only the α dimer gives rise to large negative CIS values for all three protons. The other dimers show quite different behaviour, which is not consistent with the experimental findings in solution. Thus it seems reasonable to infer that the formation of the stacked α dimers is responsible for the observed concentration-dependent chemical shifts in aqueous solutions of inosine. Interestingly, Broom *et al.*[11] arrived at a very similar conclusion based on qualitative arguments.

MOPAC calculations

Only for the α dimer was the heat of formation compared to the two single molecules stabilising (-2.6 kcal mol^{-1}). Fig. 8 on the other hand shows the impact of allowing the dimers, as taken from their crystal structures, to adopt geometry optimised comformations. It is clear that both the β and the dihydrate dimers as they appear in the crystal structure are inherently unstable and optimise to dimers with potential for H-bonding between adjacent furan rings, a situation definitely not consistent with the NMR data. On the other hand the α and dihydrate dimer including its four water molecules do retain the stacking motif and remain essentially intact. The calculated

Table 5 Calculated $\Delta\delta$ values for crystal structure dimers of inosine in ppm

Dimer structure	H1	H2	H8
α	−0.34	−0.27	−0.12
β	−0.53	−0.09	−0.16
Dihydrate	−0.09	−0.67	−0.00

Table 6 Relative Δδ values for dimerisation of inosine

Proton	Experimental 25 °C	Experimental 7 °C	Calculated α	Calculated β	Calculated dihydrate
H1	−1.0	−0.9	−1.0	−1.0	−0.1
H2	−0.9	−1.0	−0.8	−0.2	−1.0
H8	−0.7	−0.8	−0.4	−0.3	0.0

lattice energies of the forms are −31.0, −34.2 and −43.2 kcal mol^{-1} for the α, β and dihydrate, respectively. These values are in agreement with the relative stabilities as reflected in the solubility data.

Discussion and conclusions

The results of NMR studies are consistent with all previous data on aqueous solutions of inosine in illustrating that molecules are self-associated in molecular stacks. In utilising the NMR data to gain a molecular level visualisation of the likely nature of such stacks it is clear that dimerisation yields a species which is related to the dimer as it appears in the α crystal structure. This is consistent with both the single point energy and geometry optimisation which demonstrate the relative stability of this dimer compared to the β or dihydrate dimer.

In terms of crystallisation this suggests a mechanism in which α-like dimers aggregate to form nuclei having the α structure. As Fig. 9a shows, the fast growing c axis of this structure can then grow by addition of further dimers from the solution. These dimers are linked to those below by weak van der Waals interactions (3.7 Å separates the aromatic rings) and to those in the neighbouring stacks by two –C=O···H–O–, two –N–H···O< and one –O–H···O< hydrogen bonds. Hence according to this idea both the nucleation and growth of the α polymorph will be favoured over the β form since this is unable to make use of the available solution dimers either to create nuclei or, also shown in Fig. 9b, to facilitate its growth. This

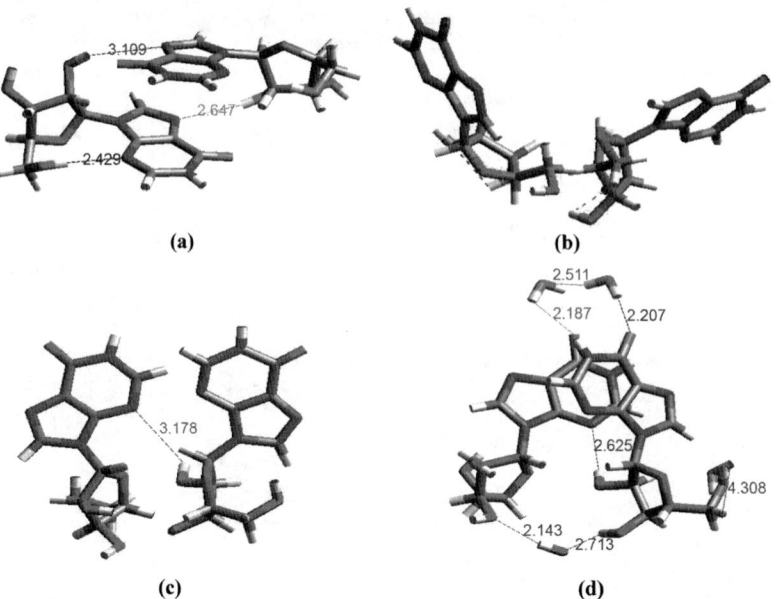

Fig. 8 The impact of geometry optimisation on the (a) α dimer, (b) β dimer, (c) dihydrate dimer without water molecules, (d) dihydrate dimer with water molecules.

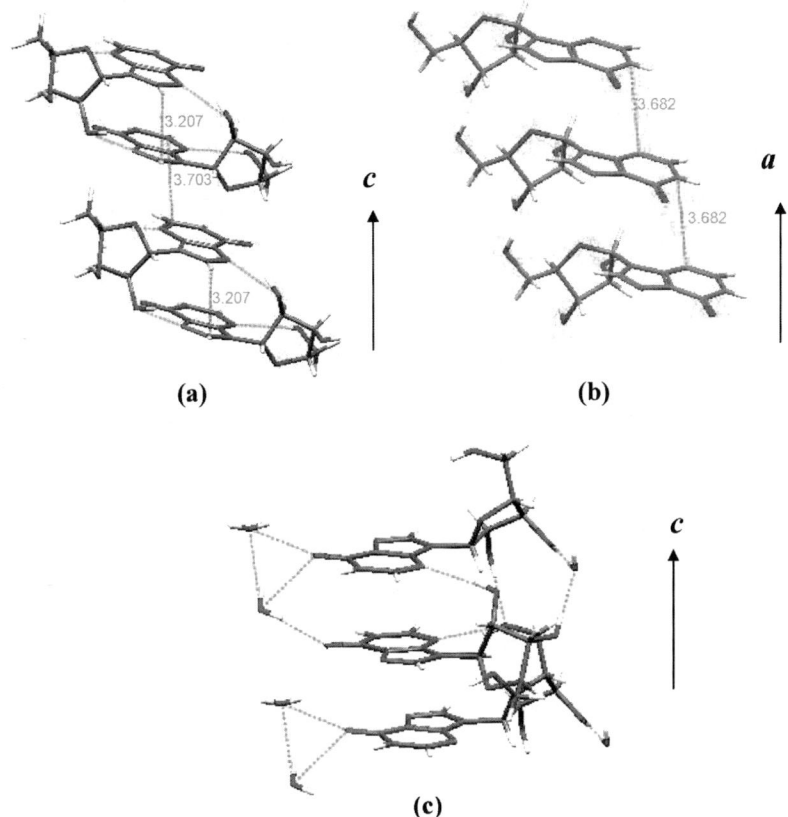

Fig. 9 The fast growth directions of (a) α inosine (b) β inosine and (c) inosine dihydrate.

mechanism is thus totally consistent with the observed nucleation of the α form at temperatures above 30 °C in the Ostwald region of Fig. 6 and would appear to explain why the direct nucleation of the β form is not observed. Of course, as indicated by the solubility data and the calculated lattice energies the β form is the more stable polymorph and so in this region of the phase diagram β will be the ultimate product, as is indeed observed. Expressed more simply in language of reaction chemistry, α is the kinetic and β the thermodynamic product. This overall picture is further consistent with the outcome of DMSO crystallisations. Here, β is seen to nucleate directly and we may rationalise this as being a consequence of the change in solution chemistry which results in strong solvation of inosine in DMSO, favouring monomeric rather than dimeric species. These monomers could clearly join the fast growing a-axis of the β structure without problem hence leading to the direct appearance of the stable polymorph in DMSO.

As the temperature falls below 10 °C the NMR data reveal no apparent change in the nature of the solution dimers and yet the outcome of the crystallisation experiment switches spontaneously to the dihydrate. Thus, as a result of falling temperature, it appears that the system moves from a situation in which the molecular interactions present in solution are transferred directly to the resulting crystal to one in which the molecular packing in the crystal is distinct from that in solution. Clearly the formation of dimers in solution exposes the hydroxyl rich furan moieties and carbonyl groups to solvation by water. As the temperature falls it may be expected that the binding of water to the dimers and the subsequent desolvation free energy necessary to create an anhydrous form may become important factors in determining the crystallisation outcome: *i.e.* the more difficult it becomes to remove the water the greater the possibility that a hydrate crystal will form.[34] Indeed the benefit of the water as a stabilising factor is highlighted by the geometry optimisation (Fig. 8) showing how water can play a key role in maintaining the integrity of the dihydrate dimer. Thus it might be imagined that the clusters formed below the transition temperature comprise both α-like inosine dimers together with water of solvation. These clusters must then undergo rearrangement to create nuclei comprising new dihydrate dimers stabilised by bound water molecules. Like nucleation, growth along the fast growing c axis of this structure is not facilitated by the solution dimers since they have the wrong configuration, instead it must rely on the monomer population for its growth units. Interestingly however, as seen in Fig. 9c, in addition to the aromatic ring interactions (separation 3.357 Å) growth can also make use of one –O–H···N<, one –OH···OH– and three hydrogen bonds to water all binding a single molecule to the layer below and two –N···H–N< and one carbonyl–water hydrogen bonded interactions binding it to adjacent stacks. In this sense one can imagine that once the dihydrate nuclei form their growth is actually more favourable than that of potential α nuclei. Therefore in this case the outcome of the dihydrate represents both the kinetic and thermodynamic product, the latter being favoured due both to its preferential growth and its the overwhelming stability compared to the α and β polymorphs.

Acknowledgements

R. J. D., A. L. G., R. A. C. and G. S. would like to thank Pfizer Global R&D, Sandwich U. K. for funding of this work.

References

1 R. J. Davey and John Garside, *From Molecules to Crystallisers—an Introduction to Crystallisation*, Oxford University Press, Oxford, 2002.
2 R. J. Davey, K. Allen, N. Blagden, W. I. Cross, H. F. Lieberman, M. J. Quale, S. Righini, L. Seton and G. J. T. Tiddy, *CrystEngComm*, 2002, **4**, 257–264.
3 I. Weissbuch, M. Lahav and M. L. Leiserowitz, *Cryst. Growth Des.*, 2003, **3**, 125–150.
4 A. Spitaleri, C. A. Hunter, J. F. McCabe, M. J. Packer and S. L. Cockroft, *CrystEngComm*, 2004, **6**, 489–493.
5 R. J. Davey, G. Dent, R. K. Mughal and S. Parveen, *Cryst. Growth Des.*, 2006, **6**, 1788–1796.
6 A. Gavezzotti, *CrystEngComm*, 2002, **4**, 343–347.
7 S. Hamad, C. Moon, C. R. A. Catlow, A. T. Hulme and S. L. Price, *J. Phys. Chem. B*, 2006, **110**, 3323–3329.
8 S. Parveen, R. J. Davey, G. Dent and R. G. Pritchard, *Chem. Commun.*, 2005, 1531–1533.
9 A. L. Gillon, N. Feeder, R. J. Davey and R. Storey, *Cryst. Growth Des.*, 2003, **3**, 663–673.
10 G. R. Desiraju, *J. Chem. Soc., Chem. Commun.*, 1991, 426–428.
11 A. D. Broom, M. P. Schweizer and P. O. P. Ts'o, *J. Am. Chem. Soc.*, 1967, **89**, 3612–3622.
12 M. P. Schweizer, S. I. Chan and P. O. P. Ts'o, *J. Am. Chem. Soc.*, 1965, **87**, 5241.
13 A. R. I. Munns and P. Tollin, *Acta Crystallogr., Sect. B*, 1970, **26**, 1101–1113.
14 E. Subramanian, *Cryst. Struct. Commun.*, 1979, **8**, 777–85.
15 U. Thewalt, C. E. Bugg and R. E. Marsh, *Acta Crystallogr., Sect. B*, 1970, **26**, 1089–1091.
16 A. L. Gillon, R. J. Davey, R. Storey, N. Feeder, G. Nichols, G. Dent and D. C. Apperley, *J. Phys. Chem. B*, 2005, **109**, 5341–5347.
17 Y. B. Tewari, R. Klein, M. D. Vaudin and R. N. Goldberg, *J. Chem. Thermodyn.*, 2003, **35**, 1681–1702.
18 Y. Suzuki, *Bull. Chem. Soc. Jpn.*, 1974, **47**, 2549–2550.
19 P. Atkins and J. de Paula, *Physical Chemistry*, Oxford University Press, Oxford, 7th edn, 2002, p. 178.
20 M. Sacchetti, *J. Therm. Anal. Calorim.*, 2001, **63**, 345–350.
21 J. Stern and D. R. Oliver, *J. Chem. Eng. Data*, 1980, **25**, 221–223.
22 J. J. Christensen, J. H. Rytting and R. M. Izatt, *Biochemistry*, 1970, **9**, 4907–4913.
23 N. Blagden, R. J. Davey, M. Song, M. Quayle, S. Clark, D. Taylor and A. Nield, *Cryst. Growth Des.*, 2003, **3**, 197–201.
24 O. Sohnel and J. W. Mullin, *Chem. Eng. Sci.*, 1978, **33**, 1535–1538.

25 R. Chiarella, PhD thesis, School of Chemical Engineering and Analytical Sciences, University of Manchester, UK, 2005.
26 M. Gardner, A. J. Guerin, C. A. Hunter, U. Michelsen and C. Rotger, *New J. Chem.*, 1999, **23**, 309–313.
27 C. A. Hunter and M. J. Packer, *Chem. Eur. J.*, 1999, **5**, 1891–1897.
28 C. A. Hunter, M. R. Low, M. J. Packer, S. E. Spey, J. G. Vinter, M. O. Vysotsky and C. Zonta, *Angew. Chem., Int. Ed.*, 2001, **40**, 2678–2685.
29 C. Zonta and O. De Lucchi, *Eur. J. Org. Chem.*, 2006, **2**, 449–452.
30 CERIUS[2], Accelrys, Cambridge, UK.
31 J. J. Stewart, MOPAC 6.0, ed 1990.
32 W. Ostwald, *Z. Phys. Chem. (Leipzig)*, 1897, **22**, 289–330. See also ref. 1.
33 M. M. Magar and R. F. Steiner, *Biochim. Biophys. Acta*, 1970, **224**, 80–87.
34 R. Banerjee, P. M. Bhaatt, M. T. Kirchner and G. R. Desiraju, *Angew. Chem., Int. Ed.*, 2005, **44**, 2515–2520.

PAPER

Membrane protein crystallization in lipidic mesophases. A mechanism study using X-ray microdiffraction

Vadim Cherezov[a] and Martin Caffrey[b]

Received 13th December 2006, Accepted 2nd February 2007
First published as an Advance Article on the web 10th April 2007
DOI: 10.1039/b618173b

The membrane structural biologist seeks to understand how membrane proteins function at a molecular level. One of the most direct ways of accomplishing this requires knowing the structure of the protein, ideally at atomic resolution. To date, this can only be done by the method of macromolecular crystallography. Integral to the method is the need for three-dimensional crystals of diffraction quality and their production represents a major rate-limiting step in the overall process of structure determination. The *in meso* method is a novel approach for crystallizing membrane proteins. It makes use of lipidic mesophases, the cubic phase in particular. A mechanism for how the method works has been proposed. In this study, we set out to test one aspect of the hypothesis which posits that the protein migrates from the bulk mesophase reservoir to the face of the crystal by way of a lamellar conduit. Using a sub-micrometer-sized X-ray beam the interface between a growing membrane protein crystal and the bulk cubic phase was interrogated with micrometer spatial resolution. Characteristic diffraction from the lamellar phase was observed at the interface as expected. This result supports the proposal that the protein uses a lamellar portal on its way from the bulk mesophase up and into the face of the crystal.

1. Introduction

One of the primary impasses on the route that eventually leads to membrane protein structure through to activity and function is found at the crystal production stage. Diffraction quality crystals, with which structure is determined, are particularly difficult to prepare currently when a membrane source is used. This difficulty reflects our limited ability to manipulate proteins with hydrophobic/amphipathic surfaces that are normally enveloped with membrane lipid. More often than not, the protein gets trapped as an intractable aggregate in its watery course from membrane to crystal. As a result, access to the structure and thus function of tens of thousands of membrane proteins is limited. In contrast, a veritable cornucopia of soluble proteins have offered up their structure and valuable insight into function, reflecting the relative ease with which they are crystallized. There exists therefore an enormous need for new ways of producing crystals of membrane proteins. One such promising approach makes use of lipidic liquid crystalline phases or mesophases.[1] To date, this

[a] *Department of Molecular Biology, The Scripps Research Institute, La Jolla CA 92037, USA*
[b] *Department of Chemical and Environmental Sciences, and the Materials and Surface Science Institute, University of Limerick, Limerick, Ireland*

Fig. 1 Cartoon representation of the events proposed to take place during the crystallization of an integral membrane protein from the lipidic cubic mesophase. The process begins with the protein reconstituted into the highly curved bilayers of the 'bicontinuous' cubic phase (bottom left hand corner of the figure). Added 'precipitants' shift the equilibrium away from stability in the cubic membrane. This leads to phase separation wherein protein molecules diffuse from the continuous bilayered reservoir of the cubic phase by way of a sheet-like or lamellar portal (left upper mid-section of figure) to lock into the lattice of the advancing crystal face (right upper mid-section of figure). Salt (positive and negative signs) facilitates crystallization by charge screening. Co-crystallization of the protein and its native lipid is shown in this illustration. As much as possible, the dimensions of the lipid (light brown oval with tail), detergent (pink oval with tail), native membrane lipid (purple oval with tails), protein (blue; outer membrane Vitamin B_{12} transporter, BtuB; PDB code 1NQE), bilayer and aqueous channels (purple) have been drawn to scale. The lipid bilayer is approximately 40 Å thick.

so-called *in meso* method is responsible for 48 of the 563 crystal structures of membrane proteins in the Protein Data Bank (http://www.mpdb.ul.ie).[2]

Our working hypothesis concerning the mechanism of membrane protein crystallization from the lipidic mesophase is that crystals grow from a local lamellar phase that is contiguous between the crystal and the bulk cubic phase where the protein is uniformly dispersed initially (Fig. 1).[3] The hypothesis is based on experience gained with growing crystals of the membrane protein bacteriorhodopsin (bR), a light-driven proton pump.

There are two reports in the literature that address the *in meso* growth of membrane protein crystals by way of a lamellar conduit. The first of these involved freeze–fracture electron microscopic (EM) examinations of microcrystal of the acetylcholine receptor-α-bungarotoxin complex grown from within a lipid mesophase.[4] EM images showed highly ordered domains of the complex next to lipid lamellae consistent with our working hypothesis. In the second study, atomic force microscopy was used to characterize the *in meso* crystallogenesis of bR.[5] The authors of that work state that evidence has been obtained for the existence of a lamellar conduit between the protein crystal and the bulk cubic phase.

In the current paper we report on a study designed to test the hypothesis by using a sub-micrometer-sized X-ray beam to profile the immediate environment of the bR crystal. Measurements were made at the Advanced Photon Synchrotron Source and a zone plate was used to focus the X-ray beam (1.83 keV) to a spot with a diameter of 400 nm. bR crystals were grown in a specially designed 25 μm thick cell with 100 nm silicon nitride windows. Low-angle diffraction data were recorded with the focused beam while the sample was translated from the bathing cubic phase up and into the bR crystal. The cubic and lamellar mesophases have signature diffraction patterns that were observed to come and go in a predicable manner as the beam interrogated different parts of the sample. The lamellar phase which extends for 1–2 μm from the growing bR crystal may serve as the postulated portal for proteins to pass from the bulk cubic phase to be ratcheted into position at the crystal surface.

Also described in this paper (section 3.9) are the results of an array of important control measurements performed to rule out artifacts that might arise due, in part, to radiation damage.

2. Experimental

2.1. Materials

Monoolein (1-oleoyl-rac-glycerol, lot M239-029-L, 356.54 g mol^{-1}), monopalmitolein (1-palmitoyl-rac-glycerol, lot M219-A22-E, 328.5 g mol^{-1}) and cholesterol (lot CH-800-N22-K, 386.66 g mol^{-1}) were purchased from Nu Chek Prep, Inc (Elysian, MN). 2,3-Dihydroxypropyl (7Z)-9,10-dibromooctadecanoate (bromo-monoacylglycerol, lot 180BR-10) was a gift from Avanti Polar Lipids (Alabama, AL). n-Octyl-β-D-glucoside (OG, Anagrade lot OG14, 292.4 g mol^{-1}) was obtained from Anatrace Inc. (Maumee, OH). Chicken egg white lysozyme (lot 71K7032) and salts of the highest quality available were purchased from Sigma-Aldrich (St. Louis, MO). Silver behenate powder was a gift from T. Blanton (Kodak, Rochester, NY).

Bacteriorhodopsin (bR) was solubilized with OG detergent from the purple membrane isolated from *Halobacterium salinarum* (strain S9) using established protocols.[6] Water (resistivity > 18 MΩ cm) was purified by using a Milli-Q Water System (Millipore Corporation, Bedford, MA) consisting of a carbon filter cartridge, two ion exchange filter cartridges and an organic removal cartridge.

2.2. Microfocus diffraction

Experiments were performed at the Advanced Photon Source (Argonne, IL) on beamline 2-ID-B. Complete details of the experimental setup have been described.[7] Briefly, soft X-rays (1830 eV, 6.755 Å) were focused by a Fresnel zone plate (ZP) to a spot with a minimum diameter of ∼400 nm at a focal distance of 11.4 mm from the ZP. The diameter of the beam on the sample was changed by moving the ZP out of focus in the direction of or away from the sample. A 20 μm diameter pinhole was placed 8.6 mm downstream from the ZP to serve as an order sorting aperture (OSA) removing higher diffraction orders from the ZP. A central beamstop, 30 μm in diameter, was glued directly to the exit window of the beampipe upstream of the ZP to block the direct beam from passing through the ZP and the OSA.

A two-circle goniometer (Huber, Model 424) was used to position the sample and the detector in the X-ray beam. The sample cell was mounted on a small manual goniometer head (Huber, Model 1002) fixed to a motorized X–Y translation stage which, in turn, was attached to the inner (theta) circle of the two-circle goniometer. The X–Y stage allowed for sample translations in the direction perpendicular to the beam with a minimum step size of 150 nm. The detector was mounted on a linear Z stage attached to the outer circle (2-theta arm) of the two-circle goniometer. The Z stage was used to move the detector parallel to the X-ray beam and to set the camera (sample-to-detector) distance.

Diffraction patterns were recorded with a liquid nitrogen-cooled CCD detector (Brandeis; 1024 × 1024 pixels; pixel size, 24 μm × 24 μm). The sample-to-detector distance was ∼110 mm which was calibrated using silver behenate (d_{001} = 58.4 Å).[8]

To avoid high intensity scatter from the OSA, the detector was repositioned in the beam by adjusting the 2-theta setting. Temperature inside the closed hutch where the sample was housed ranged from 24 to 26 °C and was monitored throughout the study. The hutch was in an air-conditioned room held at ~24 °C. No other temperature control of the sample was implemented.

Diffraction patterns were radially integrated in batch mode using the Fit2D program.[9] Diffraction peaks in intensity *versus* 2-theta plots were fit using PeakFit (SPSS Inc., Chicago, IL).

2.3. Sample microcell

Details of the design and construction of the wet sample microcell have been reported.[7] The cell was made of two silicon chips with 1 mm × 1 mm and 100 nm thick, low-stress silicon nitride windows that were custom fabricated by Silson Ltd (Northampton, England). One of the chips had a 1 mm wide by 25 μm thick spacer frame made from a photoresist polymer SU-8 around its perimeter that confined the sample to a 10.5 × 3 mm^2 area. The sample was sandwiched between the two chips so that the 25 μm spacer defined the thickness of the sample. The maximum sample volume of the microcell was 0.7 μL. Chips were glued to a copper support frame on a 3 mm diameter pin that fit directly into the X-ray goniometer head.

2.4. Samples preparation

All samples of lipid mesophases were prepared using a home-built syringe mixer,[10] as described.[7] Accurate dispensing of the mesophase into the sample cell was done by means of a 50-step repetitive dispenser (Model PB-600, Hamilton Co., Reno, NV) attached to a 10 μL gas-tight syringe (Hamilton Co., Reno, NV), as described.[11] Thus, 200 nL of the mesophase was dispensed onto the front face of the chip with the spacer at a location just outside the window area. On top of the mesophase was placed up to a maximum of 500 nL liquid solution, typically the precipitant. A second chip without the spacer was used to gently squeeze the lipidic mesophase so that it filled the area between the two silicon nitride windows. After the cell was closed, the edges were sealed hermetically with 2-ton epoxy (Devcon, Danver, MA). Specific sample compositions and preparations used in the study are listed below.

2.4.1. Bacteriorhodopsin (bR) crystals.
To grow bR crystals inside the microcells, 60% (w/w) lipid (monoolein, monopalmitolein, or 10 mol% bromo-monoacyl-glycerol in monoolein) was mixed with 40% (w/w) of a 15 mg bR per mL solution in 25 mM Na–K phosphate buffer pH 5.6 using the syringe mixer. Due to the relatively high OG detergent concentration in the bR samples the lipid/protein solution mixture produces a mixed phase system consisting of the cubic *Pn3m* and L$_\alpha$ phases.[12] Microcells were loaded with 200 nL of this mixture together with 500 nL of precipitant (2.1–2.4 M Na–K phosphate, pH 5.6) and the samples were incubated at 20 °C. Crystallization was initiated 5–7 days before diffraction measurements were performed. By this time the crystals had grown to ~20–25 μm in the longest dimension and were randomly oriented inside the microcell (Fig. 2(a)–(c)).

2.4.2. Lysozyme crystals.
Monoolein was mixed with 50 mg mL^{-1} lysozyme solution in 0.1 M sodium acetate pH 4.8 buffer in a 3/2 weight ratio which spontaneously formed the cubic phase. Microcells were loaded with 200 nL of the lysozyme-laden mesophase and the bolus was overlain with 500 nL of 7–10% (w/v) NaCl in 0.1 M sodium acetate pH 4.8 as the precipitant. Crystals, grown within the cubic phase for 3–5 days at 20 °C, reached ~100 μm in the maximum dimension (Fig. 2(d)). Two types of lysozyme crystals were observed under these conditions; 3-dimensional tetragonal crystals and needle clusters. The phase behavior in the vicinity of both crystal types was examined in this study.

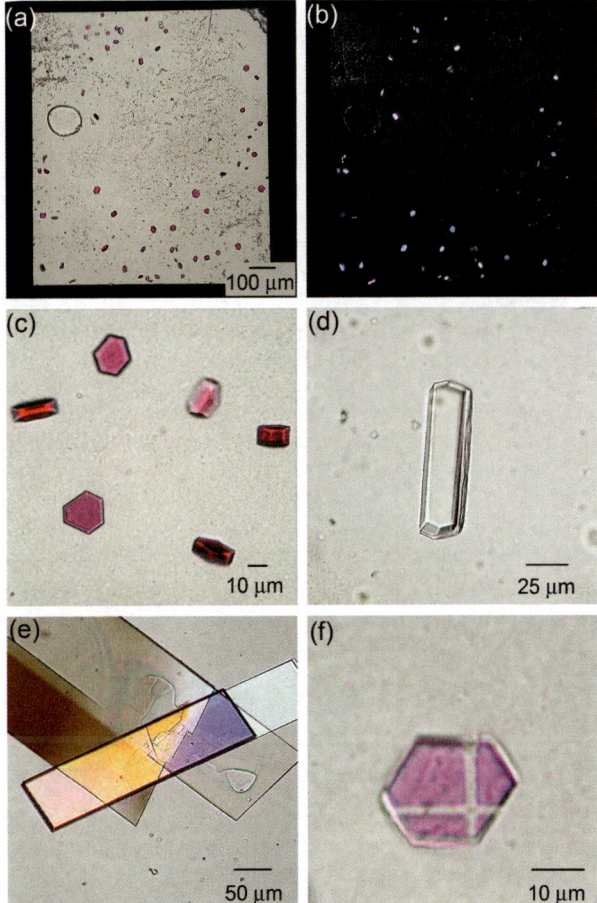

Fig. 2 Photomicrographs of crystals grown by the *in meso* method in microcells at 20 °C. (a) bR crystals growing *in meso* in the microcell. In this view, the entire window of the microcell is shown. (b) The same view as in (a) but taken between crossed polarizers to highlight crystal birefringence and the non-birefringent nature of the hosting cubic mesophase (dark background). At certain orientations the bR crystals appear as bluish birefringent objects. Clusters of such dots in the upper left hand corner likely originate from bR microcrystals that are not visible under normal light in (a). (c) A zoomed in view of several bR crystals growing *in meso* in a microcell. Crystals typically reach a size of 20–25 µm and are randomly oriented with respect to the microcell windows. (d) A crystal of lysozyme growing *in meso* in a microcell. (e) Plate-like crystals of cholesterol growing *in meso* in a microcell. The crystals were usually quite large and were aligned parallel to the microcell windows. Often crystals grew to fill the depth (25 µm) of the microcell. (f) X-Ray damage footprints left by the focused 400 nm beam after stepwise scans along orthogonal directions across a bR crystal. This picture illustrates the level of accuracy attained in positioning the sample in the X-ray beam. In this instance, scan lines were supposed to cross at the centre of the crystal. The mismatch of ~6 µm represents the error in beam position determination. Such tracks of radiation damage were used during analysis to improve absolute positional accuracy.

2.4.3. Cholesterol crystals. Cholesterol and MO at a 1 : 3 mole ratio were co-dissolved in methanol. Solvent was removed under a stream of argon initially and subsequently under vacuum (20 mTorr) for 24 h at room temperature (~20 °C). The dry cholesterol/MO was then homogenized with water in a 3/2 weight ratio using the lipid mixer. The mixture formed a transparent and homogeneous cubic phase. Crystallization of cholesterol was initiated by the addition of 500 nL 0.3 M Na–K phosphate, pH 5.6. Crystals grew as large plates oriented parallel to the cell windows (Fig. 2(e)) and reached a maximum size within a few days at 20 °C. Often crystals were as thick as the depth of the cell.

2.4.4. Control samples. There were four control sample types prepared for use in this study as follows: (I) Samples of monoolein in water were prepared at a 3/2 ratio by weight, as described above. 200 nL of the corresponding mesophase was loaded into a microcell on top of which 500 nL water was added to guarantee excess water conditions.[13] (II) Control samples of monoolein–OG detergent were prepared by combining monoolein and a 0.4 M OG solution at a 3/2 ratio by weight. 200 nL of the mesophase was added to the microcell followed by 500 nL 0.4 M OG solution. (III) Control samples of monoolein–salt solution were prepared by homogenizing lipid and 4 M Na–K phosphate, pH 5.6, solution at a weight ratio of 3/2. The microcell was loaded with 200 nL of the mesophase followed by 500 nL 4 M Na–K phosphate buffer, pH 5.6. (IV) Samples that mimicked bR crystallization conditions but that did not contain protein were made by homogenizing monoolein with 0.3 M OG in 25 mM Na–K phosphate buffer, pH 5.6, at a weight ratio of 3/2. The microcell was loaded with 200 nL of the mesophase and 500 nL 2.2 M Na–K phosphate buffer, pH 5.6.

3. Results and discussion

3.1. Target crystals

Crystals of a membrane protein (bR) (Fig. 2(a)–(c)), a soluble protein (lysozyme) (Fig. 2(d)) and a lipid (cholesterol) (Fig. 2(e)) were successfully grown in the cubic phase of hydrated monoolein in microcells specially designed for use in this microfocus diffraction study.[7] Under standard conditions, the bR crystals ranged in size from 20–25 μm and the crystals were randomly oriented inside the cell (Fig. 2(c)). Lysozyme produced either three-dimensional tetragonal crystals up to 100 μm in maximum dimension or needle clusters (Fig. 2(d)). Cholesterol crystals grew as plates several hundreds of micrometres long that often filled the depth (25 μm) of the microcell (Fig. 2(e)).

The experimental objective of this study was to scan the interfacial region of the crystals within the bulk mesophase. This was done with either a sub-micrometre-sized focused beam having a diameter of ~400 nm or a defocused beam with a diameter of 5 μm. Typical scanning patterns consisted of 1 μm steps in the X-direction (parallel to the synchrotron orbit) and either 2 or 5 μm steps in the Y-direction (perpendicular to the synchrotron orbit) in the case of the focused beam, and 5 μm steps in the X-direction and 10 μm steps in the Y-direction in the case of the defocused beam.

Accuracy in positioning the crystals with respect to the beam was estimated to be within ~10 μm.[7] The X-ray damage footprint left by the beam on the crystals was used as a 'landmark' to relate scanning paths to microphotographs of the crystals. This enabled improved accuracy in defining beam location with respect to the crystal after data collection was complete (Fig. 2(f)).

3.2. Lamellar 'signatures'

Mesophases have unique small-angle X-ray diffraction patterns. With enough reflections a pattern can be indexed and the phase identified and structurally characterized (see, for example, ref. 14). In this study, we wished to determine the identity of the phase that acts as a conduit between the bulk cubic phase and the crystal. However, the conditions under which these microdiffraction measurements were necessarily made place constraints on the number of reflections that can be recorded at any one time. In most cases, this was limited to a first or second order.

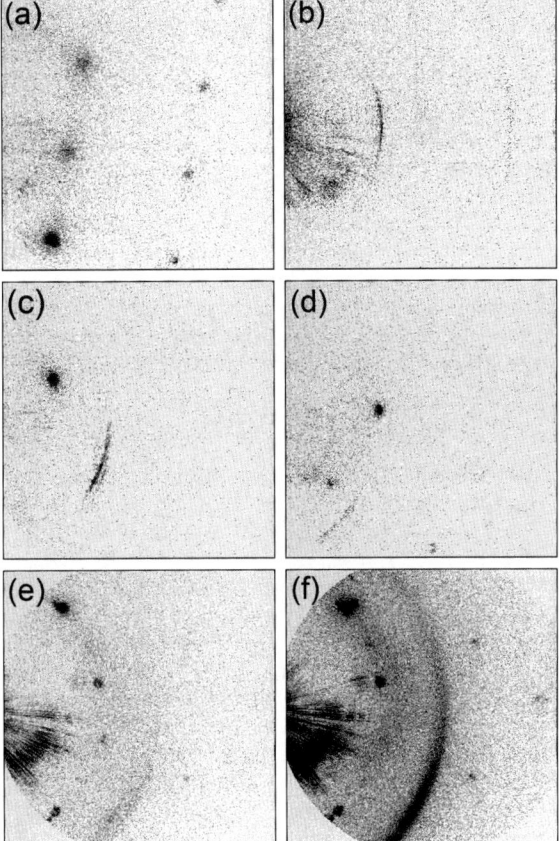

Fig. 3 Microdiffraction patterns recorded with lipidic mesophases and next to bR crystals growing *in meso*. (a) Pattern from a fully hydrated monoolein sample in the cubic *Pn3m* phase recorded with a focused beam. Exposure, 1 s. (b) Pattern from a sample of monoolein in excess 0.4 M OG detergent in the L_α phase recorded with a focused beam. Exposure, 1 s. (c) Pattern recorded with a focused beam next to a bR crystal grown in a mesophase prepared with 10% bromo-monoacylglycerol in monoolein. Exposure, 1 s. (d) Pattern recorded with a defocused 5 μm beam in the vicinity of bR microcrystals grown in a mesophase prepared with monoolein. Exposure, 1 s. (e) Pattern recorded with a focused beam next to a bR crystal grown in a mesophase prepared with monoolein. Exposure, 1 s. (f) As in (e) with an exposure of 10 s.

Accordingly, we sought to exploit other characteristics of the assorted phases to assist in their identification. Under the prevailing experimental conditions the bulk cubic phase always gave rise to discrete Bragg reflections or spots (Fig. 3(a)). These arise because the domain size of the cubic phase exceeds the diameter of the beam (≤5 μm) and the thickness of the sample-holding microcell (25 μm). Accordingly, a spotty pattern plus the corresponding d-spacings of the discrete reflections were used to identify the cubic phase.

In contrast, the lamellar (L_α) phase tended to produce powder diffraction rings or arcs (Fig. 3(b)). Such patterns were never observed with the cubic phase in the microcells. Accordingly, powder-like diffraction along with a d-spacing range that is characteristic of the phase were used as hallmarks or signatures of the lamellar phase.

Our working hypothesis posits that a lamellar phase acts as a conduit between the bulk cubic phase and the crystal. Thus, diffraction characteristic of the lamellar phase was looked for in the vicinity of crystals growing in the cubic phase housed in the sample microcell. Indeed, such signatures were seen (Fig. 3(c)–(f)). However, the lamellar diffraction was usually weak and was not always present. For this reason, we considered it necessary to perform a survey involving many crystals and thousands of diffraction patterns with a view to judging the frequency with which the lamellar signature was seen and to determine if this correlated in anyway with the orientation of the crystal in the cell.

A total of 84 hexagonal plate-like crystals (see, for example, Fig. 2(c) and (f) and 4(a)–(d)) were included in the survey and all of the crystals were in excess of 10 μm in maximum dimension. Both the focused and defocused beams were used to probe crystal interfaces which was done using step sizes of 1 μm and 5 μm, respectively. The results of the survey are as follows. Crystals that were oriented with the plane of the hexagonal plate perpendicular to the microcell windows had the lamellar signature in all of the nine crystals that were examined, corresponding to a hit score of 100%. In the case of crystals that were tilted, with the plate plane at an angle of greater than 45° to the plane of the window, the hit score dropped to 40% (16 out of 40 crystals). It dropped even further to 11% (4 out of 35 crystals) when the tilt angle went below 45°.

These observations are consistent with our working hypothesis. Thus, proteins are in a sheet-like arrangement in the crystal with the sheet planes parallel to the plane of the hexagonal plate. The expectation is that the bilayers of the lamellar phase in the conduit are continuous and coplaner with the sheets of protein in the crystal (Fig. 1). Thus, since diffraction will occur when the X-ray beam and the plane of the lamellae are parallel, the hit score observed is as expected.

The lamellar phase signature diffraction was usually seen within ~ 10 μm of a crystal (Fig. 4(a) and (d)). However, in some cases the signature was observed but in the absence of a nearby macroscopic crystal. Careful examination of such regions in the sample using crossed polarized microscopy consistently revealed the presence of dot-like birefringence (Fig. 4(e) and (f)). The frequency of these observations was as follows: 5% of regions with high birefringence had the lamellar signature compared to <0.5% for zones of low birefringence (analysis based on ~ 1000 patterns). The dot-like birefringence very likely arises from sub-micrometre-sized crystals and is consistent with the hypothesis for a lamellar conduit.

The region in which a continuous lamellar signature was observed depended on the size of the crystal. Thus, for example, with small and sub-micron-sized crystals its maximum extent was 1–2 μm. For big crystals however the zones were generally larger but their extent did not exceed the size of the crystal by more than 1–2 μm in any direction (Fig. 5(a) and (b)).

3.3. Lattice parameter of the lamellar 'signature' and the cubic phase

The lattice parameter of a mesophase depends on a variety of factors including temperature and sample composition. In this study, we found that the d-spacing of the lamellar signature, which is related to the lattice parameter of the mesophase, also depended on the size of the crystal and that of the X-ray beam. Thus, with large crystals the d-spacing observed with the 400 nm and 5 μm beams was 50–54 Å and 53–54 Å, respectively. In contrast, with small (<10 μm) and sub-micron-sized crystals the corresponding values were 42–43 Å with the focused 400 nm beam, 45–48 Å with a 1 μm beam, and 51–54 Å for the defocused 5 μm beam (Fig. 6(a)). This dramatic change in d-spacing with beam size is likely attributable to X-radiation damage as described below (section 3.4 Radiation damage).

The lattice parameter of the cubic $Pn3m$ phase surrounding scanned crystals varied in the range from 85–97 Å depending on crystallization conditions. The domain size of the cubic phase was typically 50 μm. However, in some samples domains in excess of 200 μm were observed. Further, the cubic phase in crystallization samples displayed some disorder as evidenced by an excess of diffuse scatter in the region between the (110) and (111) reflections, as noted.[7]

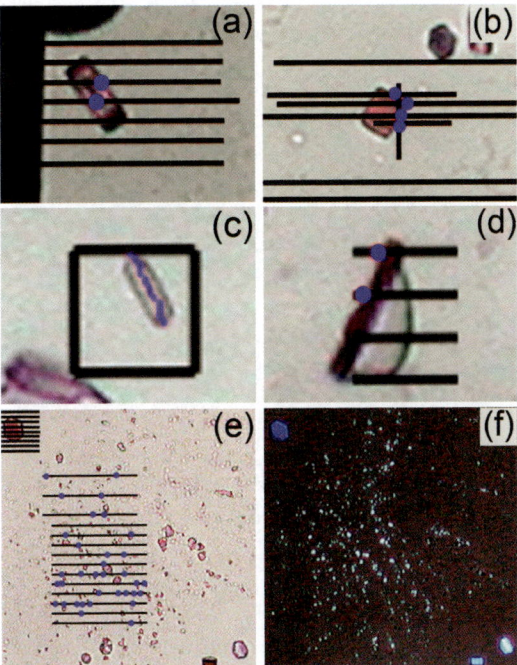

Fig. 4 The occurrence of the lamellar signature in stepwise scans across bR crystals and microcrystals growing *in meso*. The horizontal and vertical lines represent the regions of the sample interrogated by the X-ray beam in sequential 1 μm steps. Solid blue circles identify locations where the lamellar signature was observed. (a)–(d) bR crystals oriented perpendicular to the microcell windows and to the X-ray beam. The entire area within the box in (c) was scanned with the focused beam in steps of 2 μm. (e), (f) bR microcrystals. The photomicrograph in (f) was taken using crossed polarizers. Dot-like birefringence likely corresponds to sub-micrometer-sized crystals.

3.4. Radiation damage

Damage to lipidic mesophases by synchrotron X-radiation has been well documented.[15–19] How a particular mesophase responds to a given X-ray exposure depends on the lipid, sample composition including hydration, and temperature. The response also depends on the X-ray dose, dose rate and wavelength.[19]

In the current study, we observed that the d-spacing of the lamellar signature changed depending on the size of the beam used to make the measurement, particularly in the vicinity of small and sub-micron-sized crystals (Fig. 6(a); section 3.3). We attribute this effect to radiation damage. To begin with, the delivered X-ray dose rate depended inversely on beam size. Thus, the 400 nm and 5 μm diameter beams deliver dose rates of 1.4 Grad s^{-1} and 9 Mrad s^{-1}, respectively. Further, the rate at which the lamellar signature disappeared upon continuous irradiation depended inversely on beam size. Thus, the lamellar ring or arc intensity faded in matters of seconds in the focused 400 nm beam. In contrast, with the defocused 5 μm beam reflection intensity did not change significantly in the first 2 min after which there was a slow loss of intensity. Because intensity did not alter dramatically with time in the latter case, it was possible to track the change in the d-spacing of the lamellar signature throughout the exposure. The initial value was 54 Å and it dropped to 43 Å over a period of about 3 min. It then stabilized in the 42–43 Å range. When mapped to equivalent accumulated dose (compare Fig. 6(a) and (b)), these data support the view that the different d-spacings recorded for the different beam sizes arise due to X-ray damage.

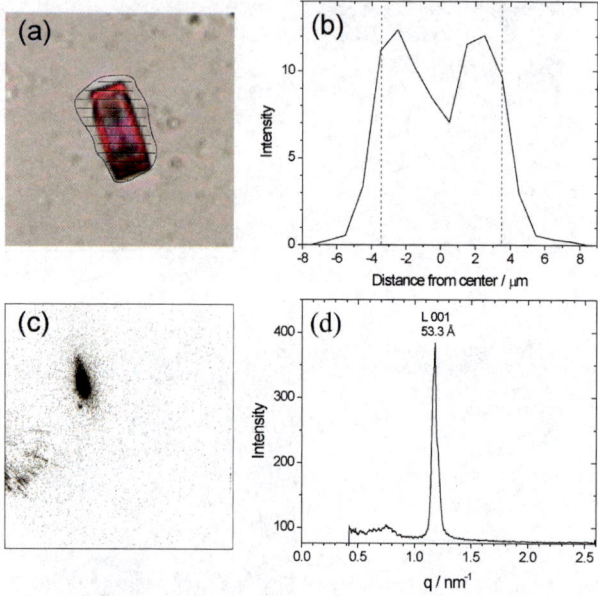

Fig. 5 The lamellar signature and its coincidence with a crystal of bR growing *in meso*. (a) Photomicrograph of the target bR crystal oriented perpendicular to the microcell windows and to the X-ray beam. The entire 50 μm × 50 μm area of the microcell shown was scanned using a focused 400 nm beam with sequential 1 μm steps in the *X*-(horizontal) direction and sequential 2 μm steps in the *Y*-(vertical) direction. Exposure time at each location was 1 s. The black horizontal lines mark where the lamellar signature was observed. (b) The intensity of the lamellar signature along one of the horizontal scan lines in (a). The vertical dashed lines represent the edges of the bR crystal. (c) An example of a diffraction pattern recorded during data collection in (a) that includes the lamellar signature. (d) An intensity *versus* scattering vector plot produced by radially integrating the diffraction pattern in (c). The d-spacing and Miller index of the lamellar reflection are indicated.

Only a limited number of studies on radiation damage to lipidic mesophases have been reported. The above damage-induced rapid and substantial drop in d-spacing, for what we are assuming to be the lamellar phase, has not been documented before.

In most cases, the lamellar signature appears initially as diffraction arcs reflecting a high degree of orientation, as expected. With exposure time and accumulated dose

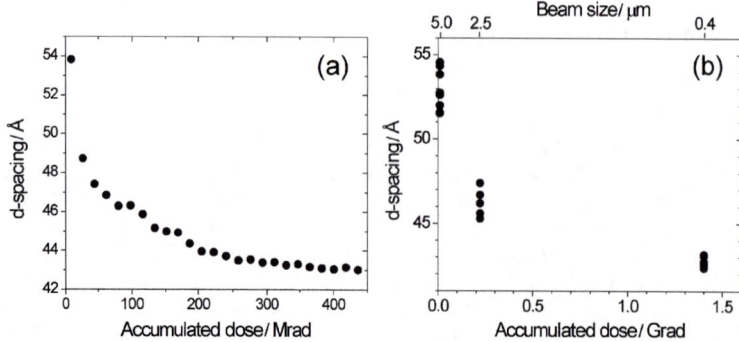

Fig. 6 d-Spacing of the lamellar signature and its sensitivity to beam size and accumulated X-ray dose. (b) Diffraction patterns were recorded in the vicinity of bR crystals growing *in meso* using beams of the indicated sizes and giving rise to the corresponding accumulated dose. In all cases, the exposure time was 1 s. (a) Patterns were recorded during continuous irradiation at a fixed location in the sample using a defocused 5 μm diameter beam and 1 s exposures.

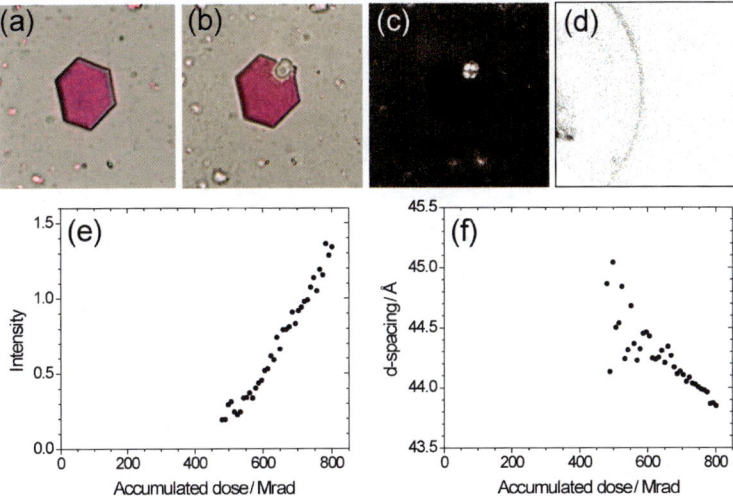

Fig. 7 X-Ray damage-induced expression of the lamellar signature next to a bR crystal growing *in meso*. (a) The target crystal was oriented parallel to the window of the microcell and to the X-ray beam. (b) The same crystal as in (a) following one hundred exposures of 1 s duration to the defocused 5 μm beam. (c) As in (b) but viewed between crossed polarizers. The site of damage has associated with it the 'extinction cross' texture characteristic of the lamellar phase. (d) Powder diffraction ring recorded at the site of damage in (b) characteristic of an unoriented lamellar phase. (e) Diffracted intensity in the lamellar signature during X-irradiation that produced the damage shown in (b)–(d). (f) d-Spacing of the lamellar signature during X-irradiation that produced the damage shown in (b)–(d). Note that in (e) and (f) there is no trace of the lamellar signature in the diffraction pattern during the early stages of the exposure.

the arcs evolve into powder rings. Presumably therefore damage disrupts the oriented multilayers and induces the formation of structures reminiscent of multi-lamellar vesicles. The results shown in Fig. 7 demonstrate this effect convincingly. The target crystal was oriented parallel to the microcell windows and to the X-ray beam (Fig. 7(a)). Thus, the lamellar signature was not expected to be seen at the crystal/mesophase interface because the lamellae and the beam are orthogonal. Indeed, this was found to be the case (Fig. 7(e) and (f)). However, with time in the beam and thus accumulated dose, the lamellar signature appeared, grew in intensity and developed into a well-defined powder ring (Fig. 7(d) and (e)). At the end of the experiment damage to the crystal was clearly visible (Fig. 7(b)). When viewed between crossed polarizers the damaged zone showed the characteristic extinction cross texture (Fig. 7(c)) of the lamellar phase. What likely happened here is as noted above. Damage induced the lamellar conduit to lose its natural alignment with the crystal and to form what amount to multilamellar vesicles with their characteristic birefringence and powder diffraction. One other possibility not discounted would have the products of crystal damage inducing lamellar phase formation.

3.5. Higher order diffraction

Thus far, we have relied on a single diffraction ring or arc as the hallmark of the lamellar phase. In a separate study, we have shown that under conditions of measurement with the microfocused beam only the lamellar and inverted hexagonal phases produce such diffraction.[7] But a single reflection does not allow for unambiguous phase identification. For proper indexing, higher order reflections are needed. Every effort was made to collect such data in the current study with a view to firming up phase identity. Thus, the detector was repositioned such that data in the vicinity of the (002) reflection of the lamellar phase and the (11) and (20)

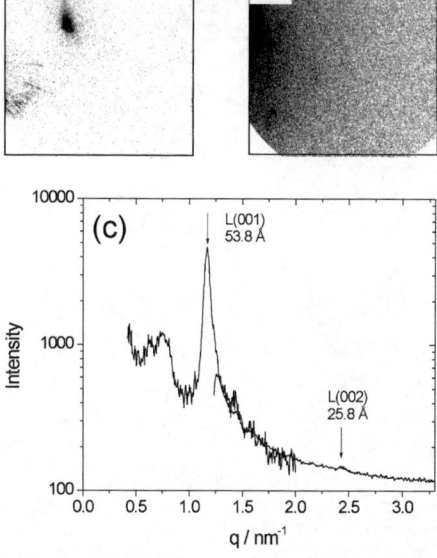

Fig. 8 Higher order diffraction from the lamellar signature recorded next to a bR crystal growing *in meso*. (a) The presumed (001) lamellar reflection recorded with a detector offset of 8° and with an exposure time of 1 s. (b) The presumed (002) lamellar reflection recorded with a detector offset of 14° and with an exposure time of 120 s. (c) An intensity *versus* scattering vector plot produced by radially integrating and combining the diffraction patterns in (a) and (b). The d-spacing and Miller index of the lamellar reflections are indicated. All measurements were made with the defocused 5 μm beam.

reflections of the hexagonal phase could be recorded. Further, the exposure time was increased to 100–200 s and the beam was defocused to 5 μm to minimize and slow radiation damage. Typically, the (002) reflection from the lamellar phase of hydrated monoolein is very weak.[13] Accordingly, additional measurements were made with monopalmitolein and a bromo-monoacylglycerol–monoolein (1 : 9 by mol) mix. Both sample types form the lamellar and cubic *Pn3m* phases and in both cases the (002) reflection of the lamellar phase is more pronounced. However, of the many crystal-containing samples tested in the course of this study only one showed a trace of the (002) reflection (Fig. 8). In all other cases convincing higher order diffraction was not observed. It is important to appreciate that a failure to see this higher order diffraction does not negate the existence of the lamellar phase.

3.6. Birefringence at the interface

A polarized light micrograph of bR crystals in the cubic phase showed 'extra' birefringence in the form of a halo and that spread for ∼5 μm around the crystal (see Fig. 3 in ref. 20). This was interpreted as evidence in support of the existence of the proposed lamellar conduit. However, the photomicrograph was taken through a thick-walled and highly curved glass tube in which the crystals grew where light is distorted and where scattering from defects in the cubic phase is possible. We have taken advantage of the microcells used in this study, which have extremely thin and flat windows, to record high quality images of the crystals where optical artifacts are minimized. Photomicrographs of plate-like crystals oriented perpendicularly and parallel to the window plane taken with and without crossed polarizers are presented in Fig. 9. These show no evidence of 'extra' birefringence around the crystals to within the resolution of the measurement (∼1 μm). This agrees with the

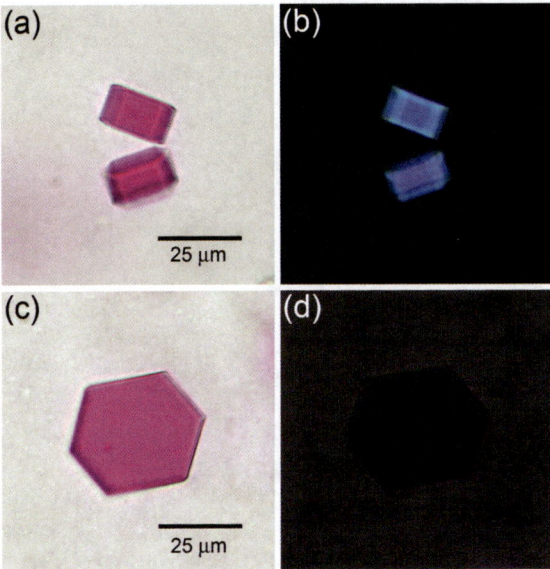

Fig. 9 Photomicrographs of bR crystals growing *in meso* oriented edge-on (a, b) and parallel to the microcell windows (c, d) recorded in normal light (a, c) and between crossed polarizers (b, d). The cubic phase is optically isotropic and non-birefringent (dark background in b, d). Crystals are birefringent when oriented parallel to the direction of the illuminating light (b) and lack birefringence when oriented perpendicular to that light. In (d) the light intensity and exposure were increased to produce a silhouette of the crystal. No 'extra' birefringence outside the crystals is apparent in either orientation (b, d).

microdiffraction data presented above indicating that the lamellar conduit is likely oriented and that it does not extend for more than 1–2 µm beyond the crystal.

3.7. Relative orientation of the lipidic cubic phase

The proposal for a lamellar conduit between the bulk cubic phase and the face of the crystal suggests that the two mesophases are contiguous. It was of interest therefore to determine if there was any preferential relative orientation of the lamellar and cubic phases. The data in Fig. 10 would suggest that this is so. In that figure is shown a series of patterns recorded as the sample was moved in the X-ray beam from the bulk cubic phase on one side of a crystal, through the crystal and back into the bulk cubic phase on the other side of the crystal. The crystal was oriented with its hexagonal plane perpendicular to the cell windows. The (200) reflection from the cubic phase was observed in the first frame which arises from the bulk mesophase on one side of the crystal (Fig. 10(a)). In the second frame (Fig. 10(b)) the intensity of the (200) reflection dropped off dramatically and the (001) lamellar signature appeared right next to it. At the mid-point in the scan the lamellar signature dominated and the cubic phase reflection was barely detectable (Fig. 10(c)). Upon continued movement of the sample through the beam the lamellar signature disappeared and the (200) reflection from the cubic phase returned (Fig. 10(d) and (e)).

The near coincidence of the two reflections (Fig. 10(b)) indicates that the layer spacing giving rise to the (001) reflection of the lamellar signature and to the (200) reflection of the cubic phase is very similar. Further, the reflecting planes in the two phases are close to being parallel. In a separate scan, the same observations were made but with the (111) reflection of the cubic *Pn3m* phase in place of the (200) reflection (data not shown).

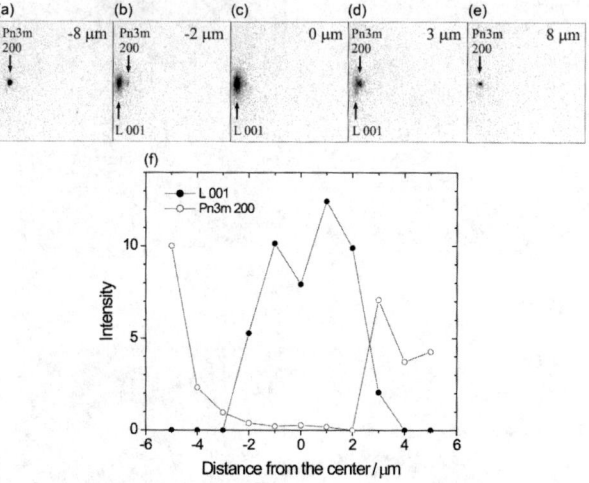

Fig. 10 Small-angle diffraction recorded while a bR crystal growing *in meso* was translated horizontally through a focused 400 nm X-ray beam. (a)–(e) A series of diffraction patterns were recorded at the indicated locations in the sample. Location is referenced with respect to the centre of the ~4 μm wide crystal which was oriented edge-on as in Fig. 5(a). (f) Diffracted intensity associated with the cubic (200) and lamellar (001) reflections as a function of position relative to the centre of the crystal. Exposure time per pattern was 1 s.

These data are also relevant to the lamellar–cubic phase transition mechanism (see ref. 21, for example).

3.8. Displacing the mesophase from the crystal surface

At one point during the course of this study temperature control in the experimental hutch at the synchrotron was lost and the temperature of the sample rose by a few (~5 °C) degrees. The sample was subsequently withdrawn and immediately examined microscopically. As revealed in Fig. 11 the temperature rise triggered the formation of small droplets throughout the sample. Some droplets were free in the bulk cubic phase. However, most were associated with bR crystals and almost all of the crystals had a droplet on at least one face.

The droplets are very likely excess aqueous phase that spontaneously forms with increasing temperature. The corresponding temperature–composition phase diagram for the monoolein–water system[13] supports this conclusion because the excess water boundary of the cubic phase in that diagram shifts to lower hydration levels with heating. What is particularly interesting about the behaviour recorded in Fig. 11 is that the droplets appeared to be associated with the large flat faces of the crystal with very high frequency. It is seldom that droplets are attached to the other crystal faces. Since the droplets develop presumably by prying the mesophase off the crystal, their noted spatial distribution suggests that the flat hexagonal face of the crystal provides the weakest link. This is consistent with the view that the lamellar phase in the conduit contains bilayer sheets that are coplanar with the layers of protein in the crystal (Fig. 1).

3.9. Control measurements

Given the weakness of the lamellar signature intensity encountered in this study and the relatively low frequency with which it was observed, it was considered important that the proper control studies be performed to lessen the likelihood of artifacts and of misinterpreting the data. Artifacts might arise due to radiation damage, as noted,

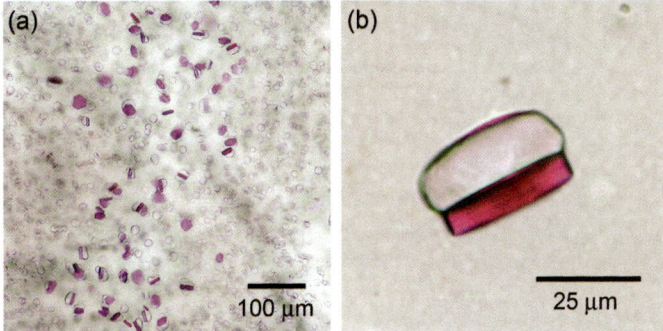

Fig. 11 Droplet formation and mesophase displacement from bR crystals growing *in meso* induced by a rise in sample temperature. (a) Droplets appear distributed throughout the bulk cubic phase and attached to bR crystals. (b) A zoomed-in view of a crystal in (a) showing 'preferential' association of a droplet with one of the faces of a bR crystal.

or to local inhomogeneities developing in the samples. The control measurements performed and the information they provided are described below.

3.9.1. Monoolein in excess water. This sample was prepared as a reference cubic *Pn3m* phase which is stable at room temperature under conditions of excess water.[13] It is the most common phase observed during the *in meso* crystallization of membrane proteins. It was necessary to characterize this phase and its behaviour in the microcell under the prevailing conditions. The lattice parameter recorded for this sample was 103 Å which is within the range of values reported in the literature for the bulk cubic *Pn3m* phase.[13] The average distance over which a continuous cubic *Pn3m* diffraction pattern was recorded in the microcell was 10–15 μm. This then corresponds to the average domain size of the cubic phase. It was also noted that the diffracted intensity and lattice parameter of the cubic *Pn3m* phase were very stable with respect to radiation exposure.

3.9.2. Monoolein and detergent. Our test membrane protein bR brings with it into the lipidic mesophase a certain amount of OG detergent used to solubilize it from the purple membrane of its host cell. A typical bR sample contains 0.3–0.6 M OG.[12] The detergent can alter mesophase behavior and we have established separately that at high enough concentrations the cubic phase transforms to the lamellar phase. For this control measurement therefore OG was combined with hydrated monoolein at a level that mimics the conditions that prevail during crystallization trials with bR. Here, the sample was found to exist in the lamellar (L_α) phase and the d-spacing of its (001) reflection recorded in the microdiffraction cell was 48 Å. Because the intensity of the second order (002) reflection from this phase was weak, signal-to-noise was improved by summing intensities over several diffraction patterns (Fig. 3(b)). The lamellar phase was sensitive to radiation damage and this was evidenced by a drop in the diffracted intensity of the (001) reflection (2 times in 100 s, ∼ 10 times in 800 s with the focused beam) and a very small reduction in d-spacing (by no more than 0.5–1 Å in 800 s).

3.9.3. MO and salt. High concentrations of salt induce formation of the inverted hexagonal (H_{II}) phase in hydrated monoolein at room temperature.[14,22] bR crystallization *in meso* is initiated in the presence of 2–2.5 M Na–K phosphate which ordinarily is not enough to trigger H_{II} phase formation in hydrated monoolein. However, as part of this control study we felt it important to produce the H_{II} phase and to examine its behaviour in the microcells under prevailing conditions. To this end, a sample of monoolein with 4 M Na–K phosphate, pH 5.6, was prepared which formed the H_{II} phase as required. The diffraction pattern of the H_{II} phase was powder-like, as was observed with the lamellar phase. The higher order (11) and (20) reflections were just barely detectable. The H_{II} phase was not particularly sensitive to radiation damage as was observed with the cubic phase.

3.9.4. MO, detergent and salt. We have argued that it might be possible for the lamellar signature described above to arise artifactually from microinhomogeneities that develop in the sample during the course of the measurements but that are not related to crystal formation. To test this hypothesis control samples were prepared which included everything in the usual crystallization mix (lipid, detergent, and salt) except the protein. Further, the concentration of detergent and salt was adjusted to match crystallization conditions and samples were prepared, treated and scanned in the same way as the protein-containing samples. Of the 3812 patterns recorded with these control, protein-free samples not one included the lamellar signature. This result supports the view that the lamellar signature is associated with the presence of protein and, by extension, with crystal growth.

3.9.5. Lysozyme crystals. It has been shown that the cubic phase supports the crystallization of water-soluble proteins.[3,23] A mechanism whereby this occurs has been proposed. Here, the cubic phase is considered to act as an inert gel to slow diffusion in a way that supports crystal growth. However, the proposed mechanism does not involve a lamellar conduit which distinguishes it from that for membrane proteins. An appropriate control measurement therefore was to look for the lamellar signature in the vicinity of a water-soluble protein crystal grown under *in meso* conditions. This was done using the water-soluble enzyme lysozyme and two different crystal forms of the protein were investigated. A total of 4649 diffraction patterns were recorded in the course of 106 scans performed on four tetragonal crystals and seven needle clusters. In none of the patterns was there a trace of the lamellar signature. This result is consistent with our working hypothesis that the lamellar conduit is integral to the mechanism of membrane protein crystal growth by the *in meso* method. It essentially rules out the possibility that the lamellar signature arises at the interface between any crystal and the surrounding cubic phase in which it grows.

3.9.2. Cholesterol crystals. We have shown that it is possible to crystallize cholesterol, which is itself a highly apolar lipid, from within the cubic phase.[7] Prior to crystallogenesis, cholesterol presumably resides in the lipid bilayer of the cubic phase and its path into the crystal may be analogous to that of a membrane protein. If so, then the lamellar signature might be expected to show up at the surface of cholesterol crystals. To explore this possibility cholesterol crystals were grown *in meso* in the microdiffraction cells (Fig. 2(e)). Unfortunately, profiling the crystal interface in search of the lamellar signature was not possible because the hexagonal phase formed almost immediately upon exposing these samples to the X-ray beam. This profound sensitivity to radiation damage was not anticipated.

In the course of these measurements it was noted that the cholesterol-laden cubic phase had an unusually small domain size in the 1–2 μm range. It was expected therefore that the crystals produced in such a system would be equally small if the source cholesterol were to come from a single domain. However, crystals up to 0.2 mm in size grew readily. This finding suggests that, in the case of cholesterol at least, movement of the sterol *between* domains and up to the crystal face is possible. Whether this reflects bilayer contiguity between domains or some other form of inter-domain transport is not known.

4. Conclusions

Crystallization of membrane proteins by the *in meso* method has been proposed to involve a lamellar phase-like conduit between the crystal and the bulk lipidic cubic phase. The hypothesis has been tested in the current study by using small-angle X-ray diffraction in combination with a synchrotron-generated microfocused beam to probe the crystal/mesophase interface with micrometre resolution. A hermetically-sealed sample holder with X-ray transmitting windows was built in which crystals could be grown and interrogated by means of the microfocused beam. Signature diffraction from the lamellar phase was observed at the interface between the cubic phase and the surface of suitably oriented crystals consistent with the working hypothesis. Signature diffraction was not observed in the bulk cubic phase unless accompanied by birefringence suggesting the presence of microcrystals. Nor indeed was it seen next to crystals of a water-soluble protein grown *in meso* as expected. The lamellar signature was found to extend to no more than about 2 µm from the edge of the crystal suggesting that the conduit is of this approximate dimension. Under certain conditions the mesophase pulled away from the crystal surface. It did so in a way that is consistent with our working model for how the lamellar conduit is tethered to the crystal.

Radiation damage was a problem that had to be dealt with in this study. It required many control measurements and the implementation of special sample handling and data collection procedures to minimize its effects.

Acknowledgements

The authors thank J. Clogston, J. Lyons, I. McNulty, Y. Misquitta, D. Patterson, K. Riedl, and O. Slattery for their invaluable help during data collection on the 2-ID-B beamline at the Advanced Photon Source, Argonne National Laboratory. Supported by Science Foundation Ireland (02-IN1-B266), the National Institutes of Health (GM61070, GM075915), and the National Science Foundation (IIS0308078). Use of the Advanced Photon Source was supported by the US Department of Energy, Office of Science, Office of Basic Energy Sciences, under Contract No. DE-AC02-06CH11357.

References

1. E. M. Landau and J. P. Rosenbusch, *Proc. Natl. Acad. Sci. U. S. A.*, 1996, **93**, 14532.
2. P. Raman, V. Cherezov and M. Caffrey, *Cell. Mol. Life Sci.*, 2006, **63**, 36.
3. M. Caffrey, *J. Struct. Biol.*, 2003, **142**, 108.
4. Y. Paas, J. Cartaud, M. Recouvreur, R. Grailhe, V. Dufresne, E. Pebay-Peyroula, E. M. Landau and J. P. Changeux, *Proc. Natl. Acad. Sci. U. S. A.*, 2003, **100**, 11309.
5. Y. Qutub, I. Reviakine, C. Maxwell, J. Navarro, E. M. Landau and P. G. Vekilov, *J. Mol. Biol.*, 2004, **343**, 1243.
6. N. A. Dencher and M. P. Heyn, *Methods Enzymol.*, 1982, **88**, 5.
7. V. Cherezov, D. Patterson, I. McNulty and M. Caffrey, 2007, manuscript in preparation.
8. T. N. Blanton, T. C. Huang, H. Toraya, C. R. Hubbard, S. B. Robie, D. Louër, H. E. Göbel, G. Will, R. Gilles and T. Raftery, *Powder Diffr.*, 1995, **10**, 91.
9. A. P. Hammersley, S. O. Svensson, M. Hanfland, A. N. Fitch and D. Hausermann, *High Pressure Res.*, 1996, **14**, 235.
10. A. Cheng, B. Hummel, H. Qiu and M. Caffrey, *Chem. Phys. Lipids*, 1998, **95**, 11.
11. V. Cherezov and M. Caffrey, *J. Appl. Crystallogr.*, 2003, **36**, 1372.
12. Y. Misquitta and M. Caffrey, *Biophys. J.*, 2003, **85**, 3084.
13. H. Qiu and M. Caffrey, *Biomaterials*, 2000, **21**, 223.
14. M. Caffrey, *Biochemistry*, 1987, **26**, 6349.
15. M. Caffrey, *Nucl. Instrum. Methods*, 1984, **222**, 329.
16. A. C. Cheng, J. L. Hogan and M. Caffrey, *J. Mol. Biol.*, 1993, **229**, 291.
17. A. Cheng and M. Caffrey, *Biophys. J.*, 1996, **70**, 2212.
18. V. Cherezov, A. Cheng, J. M. Petit, O. Diat and M. Caffrey, *Cell. Mol. Biol.*, 2000, **46**, 1133.
19. V. Cherezov, K. M. Riedl and M. Caffrey, *J. Synchrotron Radiat.*, 2002, **9**, 333.

20 P. Nollert, H. Qiu, M. Caffrey, J. P. Rosenbusch and E. M. Landau, *FEBS Lett.*, 2001, **504**, 179.
21 V. Cherezov, D. P. Siegel, W. Shaw, S. W. Burgess and M. Caffrey, *J. Membr. Biol.*, 2003, **195**, 165.
22 V. Cherezov, H. Fersi and M. Caffrey, *Biophys. J.*, 2001, **81**, 225.
23 E. M. Landau, G. Rummel, J. P. Rosenbusch and S. W. Cowan-Jacob, *J. Phys. Chem. B*, 1997, **101**, 1935.

General Discussion

Professor Roberts opened the discussion of Professor Terasaki's paper: The TEM studies show the formation of surface processes and understanding this with respect to AFM resolved step system would be interesting. Specifically, it would be helpful if the relative rates of the kinetic steps associated with in-plane step (8-member) and isolated surface adsorbed (4 member) in terms of step structure.

In addition, the potential to use selected area electron diffraction, mindful of its superior capability for examining crystallographic order, might be useful to examine the crystallographic structure and symmetry of the surface regions and potentially the ordered area close to the surface. Could these techniques be applied to studies of cluster crystallography post-nucleation?

Dr Slater replied: Rates will come from barrier calculations that we are looking at the moment—we hope to shed light on this. My guess is that the surface will not be resilient enough to withstand probing from the electron beam but if it can be done, the information would be absolutely invaluable. With respect to cluster crystallography, these structures are probably too fragile to use such a technique.

Professor Catlow asked: Can you comment on the long running issue of whether growth occurs by the addition of monomers or oligomers. Do your modelling studies provide any insight into this key issue?

Dr Slater answered: From the experimental evidence perspective, the recent work of Pelster *et al.*[1] does provide convincing evidence that secondary building units (SBUs) do not play a pivotal role in the critical nucleation stage of growth—however, as with a number of NMR studies, the conditions under which the measurements are taken are far removed from typical syntheses, being clear solution rather than gel phase.

Our calculations relate to the crystallisation stage of growth, after the critical nuclei have formed, within a distinct regime from the pre-nucleation stage. The indications from our work (including the paper presented here) suggests both the primary and secondary building units may participate in growth.

The degree of participation of each moiety will be dependent on its concentration in solution and the associated barrier height of reacting a given unit with a given crystal face or another unit within solution. However, the distribution of species in solution will be expected to be very dependent upon structure directing agent, pH, alkali source and other synthetic variables. The morphology of many zeolites (and hence the relative growth rates of different crystal faces) can be dramatically affected by changes to the synthetic mixture,[2] implying that either the mechanism or distribution of solution species is distinct from other conditions. Notwithstanding all of these factors, in general, our calculations suggest that surface crystal growth for a given crystal face consists of a number of stages which typically involve fast and slow steps. Slow steps involve processes such as terrace nucleation whilst fast processes include steps such as ring closure. My feeling is that *e.g.* pre-formed rings may play an important role in slow processes, whereas monomers participate in fast processes. Similarly, comparing fast and slow growing crystal faces, my feeling is that monomers are more important in assembling fast growing crystal faces than SBUs but SBUs have a role in overcoming rate determining steps in the growth of slow-growing crystal faces. By coupling the kind of work on solution species reported by Mora-Fonz *et al.*,[3] with study of reaction barriers at different crystal faces we hope to examine whether these intuitions are valid or not. The main obstacle to the work is building an understanding of the true nature of the solution under reaction conditions, but it is hoped that modelling can be used to examine extrema of conditions, to investigate how mechanism is affected by variations in the mother liquor.

1 S. A. Pelster, W. Schrader and F. Schüth, *J. Am. Chem. Soc.*, 2006, **128**, 4310–4317.
2 See, for example, S. D. Loades, S. W. Carr, D. H. Gay and A. L. Rohl, *J. Chem. Soc., Chem. Commun.*, 1994, 1369–1370.
3 M. J. Mora-Fonz, C. R. A. Catlow and D. W. Lewis, *Angew. Chem., Int. Ed.*, 2005, **44**, 3082–3086.

Professor Unwin addressed Professor Terasaki, Professor Anderson and Dr Slater: Stunning though the HRTEM data are, is there any danger in interpreting processes at the solid/liquid interface, and identifying characteristic growth units, from surface structure measurements after transferring crystals to high vacuum?

Professor Anderson responded: Of course it is important to recognize that the electron microscopy are recorded under high vacuum which is very different to the synthetic conditions. Nevertheless, comparing electron microscopy results with atomic force microscopy results recorded under solution we are able to map out very similar structures thereby reinforcing the belief that the electron microscopy reveals the important surface structural features.

Dr Slater replied: We can say from the calculations that reconstructions are not thermodynamically favourable and in the Chiu *et al.* paper we did establish that the most stable terminating structures of zeolites under aqueous solvent are the same as those *in vacuo*. These two factors tend to suggest that no processes will occur as a result of extracting the crystal from the mother liquor that will alter the structure. There are microscopic changes due to relaxation effects of the surface in different conditions but no change in the actual molecular architecture. The chief danger (from my perspective) of interpreting TEM data is that being a projection technique, detail can be obscured or ambiguous—in a sense this is highlighted by the microscopic mechanistic studies of the (001) surface reported in our paper showing how several terminating structures can give the same projected image when viewed along a particular axis.

Dr Slater then said to Professor Anderson: Following on from the question about whether the surface structure is a function of the conditions in which it is formed, it is noteworthy that the step heights you observe *in vacuo* are the same as those found under hydrous conditions, suggesting a common and regular surface termination irrespective of the medium in which the AFM measurement is made. This is in line with theoretical results which suggest the surface energy and hence thermodynamic stability is optimal for particular surface terminations.

Professor Rodger addressed Professor Terasaki and Dr Slater: In discussing the growth of a double 6-ring on the surface, the possibility of assembling it from three single 4-rings was dismissed as unlikely due to the improbability of a 4-ring adding edge-on. However, edge-on addition of a double 4-ring is much more likely, which could then rearrange on the surface to begin to form the double 6-ring. What is the relative importance of the addition of pre-formed clusters from solution and interconversion of clusters at the interface?

Dr Slater replied: The proposal of double 4-ring edge-on addition is a very interesting one and we intend to address this issue very soon. The questions raised are absolutely fundamental; quite simply we don't have a definitive answer to the question of the participation of oligomers and their interconversion at the interface.

Again this is a topic we will investigate in the future. I have made some inferences and conjectures in my earlier response to Professor Catlow that are relevant here.

Dr Schön said: My concern is the issue of timescales and whether the processes can be assumed to have equilibrium aspects. What kind of timescales can you deduce from your calculations regarding the activation of the various processes you compare? Note that if the ranking/sequence of moves is based on the relaxed energies of the states with and without the $SiOH_4$ unit, then this implies that we are considering the system in full equilibrium regarding the various states considered. Thus, very long times are involved, and we have to be concerned about re-attachment of the unit somewhere else at the surface during such a time. (Unless you can argue that each dissolved unit is removed right away from the system by some additional process.) For shorter timescales re-attachment might be irrelevant, but then we would expect barriers to control the dissolution process, and we cannot use energy gains to choose where and how units are removed.

Dr Slater answered: Certainly reattachment is a distinct possibility, not least since the solution is unlikely to be uniform, showing variations in pH and distribution of oligomers in proximity to the surface. The calculations are static lattice and make no attempt to probe barriers. Barriers to dissolution and growth are clearly central to the microscopic dynamical processes. Although we don't probe barriers directly, my feeling is that enthalpy is a fair guide to the relative importance of processes; we have found that the enthalpy of reaction is proportional to the number of bonds that are formed or broken—and I think it's reasonable to expect that barriers are also likely to be proportional to the number of bonds that are being formed or broken—barriers to attachment at kinks are expected to be lower than those at terraces. It would be quite wrong to try and infer something about the timescales of the processes from these calculations but the relative importance of processes, reflected by enthalpy differences, I think are likely to be correct. We are working towards building a model that is capable of verifying these assertions by calculating barrier heights.

Dr Schön opened the discussion of Professor Anderson's paper: In what sense is the growth of the terraces underdetermined? Where do the probabilities come from—computation of surface energies?

Professor Anderson replied: In order to model the crystal growth, both in terms of crystal habit and surface topology, it is necessary to define the structural environment with a certain level of accuracy. The required level necessitates a description of many more environments than experimental parameters to which the model is fitted. In this manner the problem is under-determined. However, this does not preclude us from finding a solution to the problem which is more or less unique. We apply a number of intelligent constraints to the problem. For instance, kink site growth is more probable than edge-site growth. Also, all sites with coordination higher than that of a kink site are not rate determining and consequently are given a rate which is too high to affect the result. In this manner we reduce the indeterminacy of the problem to a manageable level.

Dr Schön communicated: Does the amount of defects inside the zeolite have any influence on the envisioned applications for the systems investigated?

Professor Anderson communicated in reply: Defects are often the seat of reactivity in a zeolite. Consequently, they may well have an effect on, for instance, the catalytic activity. Further, they may well affect the stability of the material which is often a prime consideration for commercial application.

Professor Roberts said: It is interesting to observe the hollow, internally facetted, cores to the screw dislocation step structures shown in Fig. 6. Are these consistent with Frank's (1958) model who first predicted hollow core for dislocations with large Burgers vectors? How do these tie in with the pore structure of the zeolite and is the void diameter consistent with this or that proposed by Frank?[1] Also, it would be interesting to compare the void shape with that of the step/terraces that the dislocations give rise to, *i.e.* tiny minor surface step alignments seen for 8-sided step/terrace but not in the 4-sided hollow core.

1 F. C. Frank, *Acta Crystallogr.*, 1951, 4.

Professor Anderson responded: One must be very careful to interpret the shape of holes in a surface imaged with atomic force microscopy. Depending upon the depth of the whole and the lateral dimension the shape of the whole is often just a reflection of the shape of the tip. I would therefore not wish to draw too many conclusions from this.

Dr Slater commented: We recently explored possible structures for the dislocations in zeolite A—as is cited in Professor Anderson's paper. We examined the preferred site for the dislocation core and the effect of Burgers vector length and found, unsurprisingly, that the core is stabilised by locating it in the centre of the void in the 8-membered ring of LTA. We also found that a vector of length $1/2a$ is preferred over a which is probably attributable to the rigidity of the α cage of zeolite A which has a length equal to a.

Because of the large void volume over which strain can be imagined to be dissipated and the wide variation in T–O–T angles that can be accommodated in zeolites with little energetic penalty, we found that the energetic cost of forming dislocations was comparable with those formed in more strongly bonded ionic materials. The structure we hypothesised is clearly quite different from that imaged by Anderson since the imaged dislocation has a core void area apparently on the mesoscale rather than the nanometre scale defect we simulated. Our work suggested that transport mechanisms into the zeolite would be strongly affected by the presence of nanoscale dislocations and the observation of mesoscale screw dislocations suggests that these defects will have a non-negligible effect on *e.g.* the molecular sieving properties of zeolite films and membranes.

In relation to the question about the importance of screw dislocations in zeolite growth, in previous work, Agger *et al.* has seen that dislocations on this particular zeolite are only seen at low levels of supersaturation implying two distinct crystal growth mechanisms; a 2D terrace–ledge–kink model at high supersaturation and screw dislocation growth in the mode of Burton–Cabrera–Frank at low supersaturation. Whether this is a generic feature of zeolite growth I think remains an open question.

Professor Roberts responded: These observations are most interesting, particularly the observations of a half lattice vector Burgers vector. Does this imply, thus, that this is a partial dislocation and by inference one stabilised by the presence of a stacking fault. Is there any experimental evidence for this?

Dr Slater communicated: Actually this is a full dislocation, though it would be very interesting to compare the energy of two partials with the structure and energy we have found thus far. There isn't any evidence to shed any light on the full *versus* partial dislocation question unfortunately. We have found that many stacking faults are relatively low energy—faults can be easily identified from TEM for a host of different materials so their relationship with dislocations is certainly something that deserves some attention.

Professor Hodnett asked: What are the factors which determine rates of dissolution? Is it possible to measure dissolution rates from the various images and associated z-axis measurements shown in Fig. 4, and correlate these rates with the various features which appear in the figure, such as the number of steps and kink sites in an image and the instantaneous dissolution rate?

Professor Anderson answered: In principle atomic force microscopy recorded *in situ* will give a direct measure of the dissolution rates. This is, of course, with the proviso that the dissolution rates occur on a timescale compatible with atomic force microscopy. In order to correlate these rates with specific fundamental processes it will be necessary to model the topology observed. Without such modelling it will be impossible to extract the contributions from each process. However, by putting all this together it will be possible to vary parameters such as pH and temperature. This has not been done in a systematic fashion but is very much the type of experiment we will be doing in the near future.

Professor Unwin said: I have a follow up question on Fig. 4. The initial images in the series look rather different from those elsewhere in the paper and appear to show a 'cleaning' of the surface following the deposition of material. The initial surface structure, before addition of solution, is not shown. What does it look like?

Professor Anderson answered: The initial surface, which is not shown, is extremely clean, so the first image in Fig. 4 is simply the result of changes occurring before the first image can be recorded.

Professor Rodger addressed Professor Anderson and Professor Addadi: The discussion has considered the templating of nanoporous framework materials (Anderson's paper) and that seen in biomineralisation (Addadi's paper) and concluded that these were fundamentally different processes. However, a much more fruitful area for comparison is likely to be the mesoporous systems, where silicates can form in the aqueous part of surfactant stabilised mesophases. The role of the asprich protein in the prismatic layer of mollusc shells could provide useful insights into how to crystallise the silicates from the mesophases as large single crystals.

Professor Anderson replied: The crystallisation of zeolites and the formation mechanisms of mesoporous materials is fundamentally different. The main difference arises from the fact that the zeolite is crystalline whereas the mesoporous material is an ordered amorphous structure. The mesoporous porous architecture is formed as a soft solid, possibly in a similar way to some biological systems. The zeolite, on the other hand, is a crystallisation from solution. Parallels can no doubt be drawn between biological systems and either of these silicate mechanisms, however, it is important to recognize many fundamental differences as well. For instance, the silica materials, whether zeolite or mesoporous, are normally synthesised in either highly alkaline or highly acidic media. This is very different to the typical biological synthesis conditions.

Dr Schön opened the discussion of Professor Sankar's paper: What is the time-resolution of the SAXS/WAXS measurements? How typical is the timescale of 16 min for the growth of particles? Can the results be transferred to other systems? Does the growth rate change as function of time? Can the experiments be extended to demonstrate the existence of typical classical growth laws in this system?

Professor Sankar answered: The time resolution used in most of these experiments are either 60 s or 120 s, which is worked out based on how good the diffraction intensities are. The timescale of 16 min for the growth particle is not typical and cannot be transferred to other systems or the same system at different temperature.

For example, the kinetically controlled process can slow down at lower temperatures.

The growth appears to change only with temperature and not with time under the current synthesis conditions.

It is possible to extend the experiments to derive rate constants and activation energy using classical equations.

Professor Roberts said: It is very interesting to note the application of the analysis of 'amorphous' scattering features in your SR XRD patterns. This is not described in the paper and it would be interesting to reflect on the utility of this approach with respect to providing complementary data to support the SAXS/WAXS data analysis. Also, the value of the high k-range afforded by hard SR could be stressed.

Professor Sankar replied: We did not include the results of the amorphous scattering data in the paper, since we performed the diffraction studies using samples extracted periodically from the synthesising system. In this paper, we only concentrated on *in situ* studies of nucleation and growth process. The reason for doing such an *ex situ* experiment is that water is a major component of the synthesis medium and the O–O scattering from water appears to dominate the data and suppress many of the interesting features from the solid aluminosilicate. Furthermore, it is difficult to obtain a good quality and reliable background with an experiment conducted only with water in the reaction vessel and subtract it from the actual experiment to remove the contribution from water, since the amount of water present in the aluminosilicate containing system will differ and subtracting the background obtained from a water only system is likely introduce error into the analysis.

This amorphous diffraction method is highly relevant for supporting the SAXS/WAXS data, in particular for obtaining the structure of non-crystalline particles formed prior to crystallisation. Using very short wavelengths of *ca.* 0.2 Å it is now possible obtain high-quality amorphous diffraction data up to about 30 Å^{-1}. Such short wavelengths can be achieved routinely at third generation synchrotron sources, for example at SPRING8, ESRF and APS.

Professor Hodnett asked: What fraction of the total solid phases do you detect using the WAXS and SAXS techniques? Can calibration standards be incorporated into the experiments?

Professor Sankar answered: It is not straightforward to measure what fraction of the total solid phase is detected during the experiment, the main reason being that the amount of material in the beam is likely to fluctuate due to turbulence created by water. It is also difficult to incorporate a standard within the experimental chamber since it may promote nucleation. Our aim is to conduct this experiment in a clear solution in the absence of any particles to follow the nucleation and growth of nanoporous materials.

Professor Catlow asked: Could you give more details on the activation energies which are referred to in you paper? How accurate do you consider these to be?

Professor Sankar replied: The activation energies are derived based on the conventional Avrami–Erofeev equation. It is difficult to say how accurate they are, since the temperature is measured outside the reaction chamber. Although independent experiments are performed to calibrate the temperature within the vessel, the fluctuations that arise due to the presence of ingredients are difficult to assess. However, we are in the process of exploring other types of external temperature measuring devices based on infrared techniques to measure and calibrate the actual temperatures in the reaction vessel which may allow us to determine activation energies much more accurately.

Professor Catlow opened a general discussion by addressing Professor Terasaki, Professor Anderson and Professor Sankar: Could you comment on how definitive the Reverse Monte Carlo (RMC) approach is in developing models for the amorphous precursor structures?

Professor Sankar answered: The RMC technique is routinely used in analysing EXAFS, diffraction data and also many other data sets obtained from a variety of techniques. It may be difficult to obtain a unique structural solution from this method and also it is limited to systems containing only a few atoms. However, this method appears to be useful for analysing amorphous diffraction data, in particular for determining the structure of disordered materials. In our system we introduced constraints, in particular that Al–O–Al bonds should be avoided (Lowenstein's rule), for RMC calculations.

Professor Catlow said: Lewis and Mora-Foraz have reported a range of simulation studies relating to speciation in gel chemistry. Perhaps they could comment on their relevance to the present discussion?

Dr Lewis replied: In our recent work[1] and forthcoming papers, we have considered the thermodynamics of the assembly of small oligomers that have been identified in silicate solutions which result in zeolite formation. In doing so we have attempted to firstly develop a method that reproduces the experimental behaviour of silicates at high pH (for example the deprotonation energies and dimerisation energy) and secondly to map possible self-assembly routes of those species present—up to units containing 12 silicon atoms. We have been able to demonstrate how many of the species suggested as key building or dissolution units from experiments (such as those presented in this session) are thermodynamically stable and are therefore likely to be abundant in solution. Furthermore, they allow us to compare with the simulation of surfaces (such as those presented by Slater here) to attempt to deduce likely surface growth pathways. We believe the methods we have are robust and will in the near future allow us to provide further insight into the assembly of nucleation centres and also the mechanisms of growth.

1 M. J. Mora-Fonz, C. R. A. Catlow and D. W. Lewis, *Angew. Chem., Int. Ed.*, **44**, 3082.

Professor Vlieg addressed Professor Anderson, Professor Terasaki and Dr Slater: These appear to be very well-defined systems, with very few defects in some cases. Given this, can you say something about the growth mechanism, *e.g.* by combining simulations with the experimental observations? Can you see a special role for, *e.g.*, the corners of the crystals?

Professor Anderson answered: By combining simulations with experimental observations we are able to extract fundamental growth rates from the atomic force micrographs. It is not so much the corners of the crystals that are important as the edges growing terraces at the surface of the crystals. We are able to determine, for instance, relative rates of kink site growth to edge site growth to nucleation at a crystal surface. This shows that cage structures are very important in the growth mechanism. Because of the diversity of the structures of zeolites there is a large variety of different growth mechanisms. However, a unifying parameter seems to be the importance of closed cage structures.

Dr Slater answered: In these calculations we have focused on terrace nucleation events but do sample kink/corner structures. Corners, which imply kinks and steps have more binding points than terraces and so we find the energy of reacting monomers at these sites is higher than terraces. So the mechanism we infer is consistent with 2D, layer by layer growth, as has been suggested, in particular, from the AFM work of Anderson *et al.* The focus of this work is on addition to terrace-like sites because these steps in the growth process is rate determining.

Professor Catlow addressed Professor Terasaki, Professor Anderson and Dr Slater: Could you comment further on the role of screw dislocations in the growth of zeolite A and the extent to which prediction of modelling studies have been verified experimentally?

Professor Anderson responded: Most of the crystals that we have looked at to date show a layer-by-layer growth mechanism rather than spiral growth. Nonetheless, spiral growth is observed not only in zeolites but also in metal–organic framework materials. Consequently, we should not ignore this mechanism even if it is not the predominant mechanism. In terms of the comparison between experiment and theory we have an excellent mapping in particular in the case of dissolution of zeolite A. The structures we observe at the surface are in direct accordance with the energies calculated for the surface structures.

Dr Slater responded: Screw dislocations have been found on different crystal faces of zeolite A and certainly must play a role in some part of the growth process. Anderson and Aggers work on MFI tends to suggest that dislocations are likely to play a role at the end of the crystallisation process, when supersaturation is low rather than the initial phase. Depending on the synthesis conditions, it's possible the dislocations may have an important role in the growth of zeolite A. Experimental verification of dislocation structures has not been reported for any nanoporous material to my knowledge—I think this a major challenge. There has been work on materials like silicon, metals and olivine which looks very promising but I imagine it will be some time before the structure we suggested from simulation can be validated. In my view, these and other linear defects and stacking faults do merit more experimental investigation, partly because they have a role in crystal growth and partly because they impact profoundly on transport processes—their presence is often ignored in membranes but they can have density of 1 per 5–10 nm (in *e.g.* MFI/MEL).

Professor Vlieg addressed Professor Anderson, Professor Sankar, Professor Terasaki and Dr Slater: It appears that the surface structure of the various facets of the zeolites is an important issue. Has anybody tried to use surface X-ray diffraction to address this issue? You need high-quality single crystals with a well-defined surface, but the size could be less than 100 μm with current X-ray beams at synchrotrons. At the ESRF (Grenoble, France), one surface diffraction beamline has a beam of 40 μm diameter. In the future, beams of 50 nm are expected.

Professor Sankar replied: It is correct that the surface structure of the various facets of the zeolites is an important issue and they may play an important role in the growth processes. To my knowledge the surface diffraction of micron sized zeolite crystals have not been studied yet. We are aware of the developments at ESRF and also at DIAMOND the new synchrotron source, UK and it is now possible to obtain measurements using micro-focus beam. We hope to perform such experiments in the future on micron sized zeolites and zeo-type systems.

Professor Anderson answered: This has not been done to my knowledge.

Dr Slater responded: To my knowledge, no studies have been reported along these lines. I think there are a number of technical issues that would need to be overcome—the particulates are rather fragile to the beam, possibly the surface might be degraded by X-ray techniques. Because the surfaces are often very stepped and hence some of the surface features have a very low incidence, I wonder how much information might be extracted from very weak reflections. Certainly, if these challenges can be met, the technique would be invaluable.

Professor Terasaki answered: I don't think anybody has tried to study surface fine structures of zeolites by "surface X-ray diffraction". We can sometimes synthesise zeolite single crystals larger than 100 mm, therefore we will be able to use the technique utilising the small X-ray beam. Reflection high energy electron diffraction (RHEED) may be another diffraction approach to study the surface structures.

Professor Mazzotti addressed Professor Addadi and Professor Anderson: I wonder what are analogies and differences between zeolites and calcite prisms in shells discussed by Professor Addadi in her Introductory Lecture. I see an analogy in that a zeolite crystal is a single crystal intermeshed with a 3D connected network of nanopores, whereas the calcite prism is a single crystal intermeshed with a 3D network of fibers.

Professor Addadi responded: This is an interesting line of thought. I do believe, however, that the two systems are very different: besides the difference in nanopores meshwork relative to fiber network, there are also clear differences in the lengthscale of the network and especially in its order. The nanopores in zeolites are an intrinsic part of the structure, while the fibers in the calcite crystals are an incorporated network with no structural relation to the calcite itself.

Professor Anderson replied: I think the principal difference between the zeolite system and the calcite system is that the zeolite naturally has nanopores which accommodate the templating agents. They consequently become an integral part of the structure residing at crystallographic locations within the unit cell. For the calcite system a foreign molecule, probably organic in nature, will need to be accommodated by a dense phase crystal lattice. This lattice will have to envelop to foreign species.

Professor Addadi added: In nature, the mother-of-pearl layers of mollusc shells are composed of aragonite crystal platelets, and nucleation of the crystals occurs on acidic proteins adsorbed on chitin fibers. The correlation between the chitin fibers, the proteins and the nucleated crystals may be akin to the template-induced nucleation that Professor Anderson was mentioning. This is very different from the prismatic layer of the same shells, where the chitin meshwork is occluded inside the crystal structure, but is also not related crystallographically to the crystals inside which it is occluded.

Professor Anderson said: In zeolites the template acts in a very specific way to direct the final structure of the crystal. Because a zeolite is nanoporous there is an opportunity for the pores to be filled with a structure-directing agent such as a small organic amine. These amines both act to nucleate growth at a crystal surface and also prevent an desired crystal from forming. From our work we believe that the template keys into the growing surface, fitting into half pockets on the surface, before becoming subsequently clathrated as the next layer of the structure grows. This is slightly different to action of proteins observed in mollusc shells.

Dr ter Horst opened the discussion of Professor Jones' paper: Can co-crystallisation of chiral compounds be used to separate enantiomers?

Professor Jones replied: This is an interesting question and one that needs more detailed examination. In principle, cocrystallisation may well be employed for enantiomer separation in a manner similar to formation of salts with optically active acids or bases. One might imagine for example that using a chiral cocrystal former one might produce a cocrystal with one component with very different solubility to the pure chiral reactant and thereby create a process for creating a separation procedure.

Professor Kuroda said: We are also working on cocrystal formation of organic compounds by solid-grinding. Charge-transfer complexes are formed with benzoquinone (BQ) and 1,1'-bis-β-naphthol (BN) derivatives. In the case of *rac*-BN and BQ, different crystals of different stoichiometry are obtained from solid-grinding and from conventional solution crystallization, where different chiral discrimination operates. In the crystals obtained from solid-grinding, a racemic pair of BN (*i.e.*, aR and aS) sandwiches a BQ molecule, whereas in the crystals formed in solution, two homochiral BN (aR and aR, or aS and aS) sandwich a BQ molecule. Interestingly, chiral BN and BQ crystals do not form CT cocrystals. In your case of caffeine and tartaric acid, you have obtained cocrystals with chiral tartaric acid but never with racemic tartaric acid, from solution nor solid-grinding. Even solid-grinding of caffeine-*R*-tartaric acid crystals and caffeine-*S*-tartaric acid crystals did not form caffeine-*R*,*S*-tartaric acid crystals. Is this correct?

Then, is there any case that cocrystals are obtained only in the solid-grinding but not from solution crystallization?

This is a leading question to the seeding experiment you quoted in the paper, which says that cocrystals of nicotinamide and the racemic acid could be obtained from solution only after a prolonged standing or rapidly upon addition of a few seeds of the solid obtained by liquid-assisted grinding. What is the seed doing then to encourage crystallization? Are they dissolved? Does any particular crystal face play a role?

Professor Jones answered: Professor Kuroda describes some very interesting observations. The case of the cocrystal of theophylline and L- or D-tartaric acid that we present is a case where we have not been successful in constructing the cocrystal from solution, but have been successful by liquid-assisted grinding. We have other cases where the product of grinding can be used to seed a solution to obtain material that previously we had not obtained by solution methods. One must always qualify these sort of observations, of course, by recognizing that continued efforts at solution growth (*e.g.* using a variety of solvents, growth parameters *etc.*) might eventually lead to successful formation of the "grinding product". For this reason we would perhaps emphasize the role of grinding and liquid assisted grinding as a means for screening for alternative forms rather than claiming that grinding will be the only route to their being obtained. Clearly, a knowledge of the corresponding phase diagram will always be important!

Professor Davey asked: Does grinding have to be used to form the cocrystal or could some less energy intensive process be used—'beating' the powder together for example?

Professor Jones answered: Mechanochemical cocrystallisation can be induced in different ways for different systems, and it is quite possible that in some cases beating may be applicable or even the method of choice. It is also likely that "agitation" of a suspension of solids may be a strategy. It is clear that until we fully understand the mechanisms operating in each case we remain in a sort of "trial and error" situation. Objectives include producing new structures, varying stoichiometry and possibly also to screen for hydrate/solvate formation.

Professor Kuroda said: Following up the answer to the question by Professor Davy, the necessity of grinding depends on the compounds.

We have worked on benzoquinone (BQ)–biphenol (BP) cocrystal formation. BQ and 2,2′-BP form charge-transfer cocrystals without grinding. Placing the two crystals side by side is good enough for the formation. The melting point (T_m) of 2,2′-BP is 109 °C. Surprisingly, 4,4′-BP whose T_m is as high as 283 °C, also formed CT cocrystals with BQ without grinding, although it took much longer than the case of 2,2′-BP. In contrast, *rac*-1,2′-bis-β-binaphthol (BN) never produced cocrystals with BQ without grinding. We have done qualitative work on the BN–BQ system, which shows co-grinding below a threshold never produce cocrystals even for a long period of grinding, however, grinding over a threshold produces cocrystals after 15 min. This indicates the importance of lattice energy. The T_m of *rac*-BN is 93 °C, much lower than 4,4′-BP, and the reason for the 4,4′-BP–BQ cocrystal formation without co-grinding is due to the high stability of the product. The T_m of the products are 89, 147 and 93 °C for 2,2′-BP–BQ, 4,4′-BP–BQ and BN–BQ, respectively.[1]

1 R. Kuroda, K. Higashiguchi, S. Hasebe and Y. Imai, *CrystEngComm*, 2004, **6**, 464.

Professor Litster asked: You use a mechanochemical process (mortar and pestle) to produce cocrystals. Is this also proposed as a scaleable process for production of cocrystals in, for example, pharmaceutical applications? Have people studied the energy input in grinding related to conversion in these processes?

Professor Jones responded: In general the procedure used in liquid-assisted grinding is somewhat similar to that used in wet granulation. Since the wet granulation can be a large-scale industrial process, we believe that liquid-assisted grinding should be amenable to scale-up. With regard to energy input, we direct you to Professor Kuroda's comment, which indicates that some mechanochemical reactions exhibit a well-defined threshold to be initiated.

Professor Roberts commented: There is a significant industrial issue relating towards trying to prepare pharmaceuticals with very high purity reflecting the improved processing properties that purer compounds have. It is a common observation in solid-form selection that there is a paradox: hydrated materials can have monomeric structures and hence higher purity but have poorer solid-state stability. In contrast anhydrous structures can exhibit the reverse behaviour. Cocrystals however have superior stability properties and hence it may be attractive to use cocrystal formation to effect high product purity even when the cocrystal form may not be the desired final form. Analogous examples come from clathrate chemistry: *i.e.* purification of *n*-alkanes to remove iso-alkanes *via* forming a clathrate cocrystal with urea. Note in this specific case cocrystal formation does not require hydrogen bonds.

Professor Jones said: We agree entirely with this important observation. Indeed this is one of the possibilities that we have noted regarding the formation of an anhydrous cocrystal starting from hydrated caffeine and citric acid.[1]

1 S. Karki, T. Friščić, W. Jones and W. D. S. Motherwell, *Molecular Pharmaceutics*, 2007, **4**, 347–354.

Professor Hodnett asked: Could you tell us a little more about the physical forms of the starting crystals and your cocrystals? Are there changes in particle size and habit as you grind? Is crystal hardness a factor in forming crystals by grinding? What is the role of glass transition temperatures in this system? Can you indicate the fraction of your sample which is analysable by X-ray diffraction; is there a large amorphous component present as a mixed phase or as separate phases?

Professor Jones replied: The particle size is generally reduced upon grinding, and materials obtained by liquid-assisted grinding generally exhibit a larger particle size than the ones obtained by neat grinding (if one can assume a direct relationship between, for example, line widths and particle size *via* PXRD). Recent results also indicate that whilst neat grinding can result in samples with *ca.* 30% amorphous content (see the recent paper by Nguyen *et al.*[1] where a possible implication of an amorphous intermediate phase is suggested), liquid-assisted grinding appears to lead to complete and faster conversion. These results have been obtained by combined use of X-ray powder diffraction and terahertz spectroscopy.

Glass transition temperatures may affect the products of neat grinding experiments, and this has been documented.[2]

Clearly the existence of an amorphous component will affect (short-term) solubility and this may in some cases be an explanation for the effects associated with liquid assisted grinding.

1 K. L. Nguyen, T. Friščić, G. M. Day, L. F. Gladden and W. Jones, *Nat. Mater.*, 2007, **6**, 206–207.
2 A. Jayasankar, A. Somwangtharanoj, Z. J. Shao and N. Rodriguez-Hornedo, *Pharm. Res.*, 2006, **23**, 2381–2392.

Dr Schön asked: How about the existence of small nucleated crystals inside a common amorphous phase? Or do we see the inter-diffusion between nano-crystals of the two original crystal-types (which only appear as an amorphous phase due to the missing peaks in the XRD)?

Professor Jones replied: Current evidence suggests that cocrystallisation using mechanochemical methods may well involve an amorphous phase, with a well defined glass transition temperature, see the important paper by Jayasankar *et al.*[1] However, in some systems molecular diffusion may be more significant than the formation of an amorphous phase, see Kuroda *et al.*[2] This is an area that requires further investigation.

1 A. Jayasankar, A. Somwangtharanoj, Z. J. Shao and N. Rodriguez-Hornedo, *Pharm. Res.*, 2006, **23**, 2381–2392.
2 R. Kuroda, K. Higashiguchi, S. Hasebe and Y. Imai, *CrystEngComm*, 2004, **6**, 463–468.

Dr Schön communicated: Quite generally, in order to analyze whether the generation of the bi-crystals takes place due to the local heating (and thus very high local diffusivity) associated with the grinding process of grinding both original crystals together or whether the effect is more due to the surface–surface contacts of very small particles, one could try as an alternative synthesis route to first grind the crystals individually, let them cool off and then intermix them.

Professor Jones communicated in reply: We have considered this type of experiment. However, one would also need to take into account the possible formation of an amorphous phase during grinding of each reactant. This would present a difficulty in trying to determine the importance of molecular diffusion *vs.* large contact area. Also perhaps related to this might be the interesting observation of Nakamatsu *et al.*[1]

1 S. Nakamatsu, S. Toyota, W. Jones and F. Toda, *Chem. Commun.*, 2005, 3808–3810.

Professor Caffrey said: It is possible to quantify the relative amounts of disordered and ordered phases using small-angle X-ray scattering. We used this approach in a study of phase transitions in hydrated phospholipid mesophases. It is acknowledged that there are large errors associated with quantifying the disordered phase.

Professor Jones replied: It is clear that we need to develop techniques to assist in determining the ratio of crystalline to amorphous amounts in a sample. I feel that we also need to develop strategies for determining the local packing in amorphous organic solids and ascertaining how different are the packing arrangements in the amorphous and crystalline domains. We have attempted to use computational methods to begin to address this issue. For example, we have recently described our attempts to create models of the amorphous state and use these to predict glass transition temperatures.[1]

1 See, for example, A. Simperler, A. Kornherr, R. Chopra, P. A. Bonnet, W. Jones, W. D. S. Motherwell and G. Zifferer, *J. Phys. Chem. B*, 2006, **110**, 19678–19684.

Professor Kahr opened the discussion of Professor Davey's paper: Your solution structures were refined by modelling the concentration dependence of chemical shift on selected protons. Undoubtedly, the extent to which you can define these structures is critical to all subsequent judgments. The excitonic character of your model of the inosine-dimer predicts several things that could easily be established with electronic spectroscopy. The energy of the absorption should be blue shifted, and the circular dichroism band should be bisignate. Do you observe these things?

Professor Davey answered: I can only reiterate that UV spectra of dilute solutions showed no deviation from Beer Lambert behaviour.

Professor Addadi asked: The silvery shine of fish skin is due to stacks of crystals of anhydrous guanine that form photonic crystals. *In vitro* anhydrous guanine can be grown from DMSO. If there is water in the DMSO, anhydrous guanine does not crystallize. Does this indicate that *in vivo* there is no water inside the crystal chambers where the crystals are formed?

Professor Davey answered: The formation of a hydrated crystal from solution will require a certain critical water activity. thus having no water is not an essential requirement, rather the water must be in a environment in which its activity is held below the critical value.

Professor Addadi then asked: Application of all theories of theoretical morphology prediction yield crystal morphologies elongated along the direction of molecular stacking. Yet, the biogenic crystals are thin plates with the plate face perpendicular to the stacking direction. Can you suggest any explanation of how this may occur?

Professor Davey replied: I can offer nothing specific as an answer. Presumably some form of 'additive' molecule is involved.

Dr Hughes asked:
(1) The paper mentions a study from the 1960s which used vapour pressure to determine that inosine molecules form π-stacked dimers in solution. For the case of glycine, many statements in the literature claim that it forms doubly-hydrogen bonded dimers in solution but this is not in agreement with the freezing-point-depression measurements from the 1930s. I therefore wondered whether freezing-point depression had been measured on inosine to back up the stacking hypothesis.

(2) Again harking back to the glycine case, we have observed an isotope effect on polymorphism and I was hence concerned that the D_2O used in the NMR measurements might be affecting the polymorphism.

Professor Davey replied: I am not aware of any freezing point depression data. We did not perform any crystallisation experiments from D_2O and so I cannot comment on any affects on the polymorphism of inosine.

Dr Schön said: The speed of nucleation of the hydrated form appears to depend on the temperature in such a way that in the cooler system (or the one with higher supersaturation—this need not result in equivalent influences on the nucleation and growth process) the speed is higher. In order to understand whether this is reasonable, it would be important to know whether we are dealing with homogeneous or heterogeneous nucleation. Depending on the details of the nucleation + growth model, both the observed trend and its opposite could be possible.

Also, experimentally observed time evolutions often represent a melange of several nucleation and growth processes, which can exhibit opposing trends, leading to difficulties in interpreting the parameter dependence of the measurements. However, for simulations, it is crucial to know what types of nucleation processes are present. If we only want to consider the later stages of the growth process, then the original cause of the formation of a critical cluster and the initiation of the growth might not be relevant anymore.

Professor Mazzotti asked: With reference to Table 2, experiments 7a, 8a and 9a end up at different temperatures but the same supersaturation. Could the different induction times (*i.e.* 10, 8 and 3 min) be due to the different cooling rates in the three cases, *i.e.* 15 to 10 °C, 15 to 5 °C and 15 to 2 °C, respectively? Looking at runs 4a and 5a, where conditions were exactly the same, one notes that induction times vary between 2 and 8 min. This possibly indicates that we were at the limit of the precision of the induction time measurement.

Professor Davey answered: I agree.

Professor Mazzotti communicated: Beside vapor pressure measurements, there are other ways of measuring dimerisation equilibrium constants. One possibility is to look at non linear effects in the UV absorbance; another one is to look at non linear effects in time response of the polarimeter (in the case of chiral compounds only).[1]

1 R. Baciocchi, G. Zenoni, M. Valentini, M. Mazzotti and M. Morbidelli, *J. Phys. Chem. A*, 2002, **106**, 10461–10469.

Professor Davey communicated in reply: We did use UV to measure the extinction coefficients of inosine in water at 25 and 7 °C. We could only access a small concentration range but found no non-linear behaviour.

Dr Vonk asked: Water has an odd behaviour at temperatures near the melting point: *e.g.* density drops with decreasing temperature. Could the structure of water give an explanation for the decrease of the induction time?

Professor Davey responded: I am not sure. Certainly it will be much more difficult to dehydrate inosine at 7 than at 25 °C.

Dr Hare communicated: With regard to Table 1, does the β form have monoclinic symmetry? In Fig. 5 and 6, what is meant by "pure water"? Does it exist, other than in a model or the gas phase?

Professor Davey communicated in reply: Thank you for pointing out the error in Table 1. The β form has the space group $P2_1$, monoclinic. Pure water in this case refers to water that is distilled and deionised.

Professor Sankar opened the discussion of Professor Caffrey's paper: Please would you comment on the X-ray damage and its effect on your measurement which you mentioned in your presentation. Also, please would you comment on the existence of the cubic and lamellar phases present in your system.

Professor Caffrey responded: X-Ray damage was a major issue in this study. How it was monitored, quantified and controlled for is dealt with very thoroughly in section 3.4 of the paper. Concerning the existence of the cubic and lamellar phases in the system, our working hypothesis for how *in meso* crystallization comes about posits the existence of a lamellar phase-like conduit between the bulk, reservoir cubic phase and the growing face of the crystal (Fig. 1). The data presented in this paper provides convincing support for the hypothesis.

Professor Roberts said: A major challenge lies in having confidence that the structure and function of the protein is not affected *via* its extraction from its natural membrane environment where it may not be present in the degree of aggregation that is present in the native environment. Additionally, presumably the nature of the lipid environment should impact on protein function. How well does the cubic phase replicate the membrane environment and what is the influence of charge on the surfactant and its lipid self organisation? Also, how can 2D XRD studies be used to probe structural order within the natural 2D membrane structures. Might these not provide complementary information to give greater confidence to the 3D PX studies?

Professor Caffrey replied: Recall that *in meso* crystallization is performed using monoacylglycerol as the hosting lipid. Alas, this is not a typical membrane lipid. The possibility exists therefore that it might induce an unnatural conformation in the reconstituted target protein. We have tested this possibility directly with the vitamin B12 transporting protein, BtuB. BtuB is a 22-stranded beta-barrel integral membrane receptor whose structure was solved to 1.95 Å using *in meso*-grown crystals.[1] The protein was reconstituted into the cubic phase as is done in preparation for crystallization trials. The affinity of the reconstituted protein for vitamin B12 was quantified using the change in its intrinsic fluorescence upon binding. The affinity constant was found to be the same as that observed for the protein in its native outer membrane of *Escherichia coli*.[1] This gives us confidence that the bilayer of the cubic phase provides an environment that supports native-like structure and function. Separate studies on rhodopsin in the lipidic cubic phase corroborate this view.[2]

As to the effect of charge, the hosting monoacylglycerol lipid presents two hydroxyl groups and an ester linkage at the membrane/aqueous interface. Neither functionalities engage in protonic equilibrium in the pH range normally encountered and thus, charge is not considered an issue as far as the lipid component is concerned. However, charge is likely to play a part in the nucleation and growth phases of the protein crystallization process. Regarding 2D crystallography, indeed, this methodology has helped enormously in the field of membrane protein structure and function determination. The first membrane protein to be crystallized and to have its structure solved by the *in meso* method was bacteriorhodopsin. Bacteriorhodopsin resides in its native bacterial membrane as so-called purple patches that consist of 2D crystalline arrays of the protein and lipid. 2D electron crystallographic studies of samples such as this provided a reasonably high-resolution structure of the protein.[3] This, in turn, was used to solve the close to atomic-resolution structure of bacteriorhodopsin crystallized *in meso*.[4] The structures are very similar.

1 V. Cherezov, E. Yamashita, W. Liu, M. Zhalnina, W. A. Cramer and M. Caffrey, *J. Mol. Biol.*, 2006, **364**, 716–734.
2 J. Navarro, E. M. Landau and K. Fahmy, *Biopolymers*, 2002, **67**, 167–177.
3 R. Henderson and P. N. T. Unwin, *Nature*, 1975, **257**, 28–32.

4   E. Pebay-Peroula, G. Rummel, J. P. Rosenbusch and E. M. Landau, *Science*, 1997, **277**, 1676–1681.

Professor Caffrey then addressed Professor Terasaki and Dr Slater: I am curious about the possibility of using Professor Terasaki's very powerful and revealing TEM technology to explore the surface of membrane protein crystals growing in or from a lipidic mesophase as described in my talk. I foresee a problem however in that the materials Professor Terasaki works on are solid and hard whilst those I am interested in are soft and fragile.

Dr Slater responded: My guess is that scattering would be a big problem leading to very low resolution and that protein could be beam sensitive, I imagine the prospects of using this technique are not too high.

Professor Terasaki answered: Certainly two of the main problems for the study are (i) membrane protein crystals are very electron beam sensitive and (ii) the system is soft and contains water.

Sometimes we don't need very high resolution TEM image but it is challenging to combine electron low-dose technique and the highly sensitive recording system for the 1st problem, and to apply the cryo-TEM approach for the 2nd problem.

Professor Davey addressed Professor Caffrey: Does the addition of protein affect the phase diagram of the surfactant system?

Professor Caffrey responded: By surfactant I assume you are referring to monoacylglycerol. The concentration of protein used in the crystallization trials is typically low and is not expected to directly influence the phase behavior of the *in meso* system. However, if enough protein were to be added it surely would have an effect. Indeed, we have seen this with the pore-forming, gramicidin.[1] There is another important component that accompanies the protein, which is a little more worrying, and that is the detergent. It rides along with the protein during the extraction and purification process and it ends up in the mix at the reconstitution stage. Detergents can profoundly influence mesophase behavior. In this regard therefore we have studied quite extensively the effect that detergents have on the phase behavior of the hosting lipid–water system.[2] What we find is that the cubic phase is remarkably forgiving in that it can accommodate/incorporate a relatively high level of detergent (and indeed other additives) without undergoing a potentially deleterious phase transition.[2,3] Practically, the first measurement we do with any new protein preparation destined to undergo *in meso* crystallization trials is an evaluation of how it impacts on mesophase behavior. This is done using small-angle X-ray diffraction.[2b]

1   W. Liu and M. Caffrey, *J. Struct. Biol.*, 2005, **150**, 23–40.
2   (a) X. Ai and M. Caffrey, *Biophys. J.*, 2000, 79, 394–405; (b) Y. Misquitta and M. Caffrey, *Biophys. J.*, 2003, **85**, 3084–3096.
3   (a) V. Cherezov, F. Hannan and M. Caffrey, *Biophys. J.*, 2001, **81**, 225–242; (b) V. Cherezov, J. Clogston, Y. Misquitta, W. Abdel-Gawad and M. Caffrey, *Biophys. J.*, 2002, **83**, 3393–3407.

Professor Anderson commented: With regard to complex phase transformations of lipid phases, there is a lot to be learnt from the mesoporous silica phases which essentially trap the mesostructures. These can then subsequently be studied by electron microscopy.

Professor Caffrey answered: The study of mesoporous silica phases by electron microscopy is very well developed. However, I am not sanguine about the prospect of using this approach to investigate the current *in meso* crystallogenesis system. The templating process requires the addition of a variety of materials many of which are likely to perturb the system under investigation. This is particularly true given the high concentrations of additives needed and the often harsh chemical conditions required for polymerization. The products of these reactions can also be destabilizing as in the case of tetramethoxysilane where methanol is produced.

Professor Rodger asked: To what extent can the crystallisation be considered as the crystallisation of a supersaturated solution, and if so, how important is the nature of the protein in determining the free energy of the cubic, lamellar and crystal phases? Do similar phase transitions occur for other macromolecules dissolved in a cubic mesophase?

Professor Caffrey responded: Crystallization happens typically from a supersaturated liquid crystalline dispersion that incorporates the target membrane protein. An assortment of precipitants (salts, small and large organic molecules) are usually added to create the condition of supersaturation which, in the ideal case, leads to phase separation where one of the phases takes the form of a well-ordered, diffraction-quality membrane protein crystal. The free energy of the various meso- and crystalline phases in the system will surely depend on the identity and concentration of the added protein. The focus, of course, is on the protein component and on setting up the conditions that induce it to phase separate as a crystalline solid. Most of the time however, these conditions are not met and the protein remains dispersed in the mesophase or it phase separates as a useless precipitate. I am not aware of studies that have been performed on the crystallization in the lipid cubic phase of macromolecules other than proteins.

PAPER

Test of Cairns-Smith's 'crystals-as-genes' hypothesis†

Theresa Bullard, John Freudenthal, Serine Avagyan and Bart Kahr*

Received 13th November 2006, Accepted 29th January 2007
First published as an Advance Article on the web 20th April 2007
DOI: 10.1039/b616612c

One aspect of the multifaceted proposal by A. G. Cairns-Smith, that imperfect crystals have the capacity to act as primitive genes by transferring the disposition of their imperfections from one crystal to another, is investigated. Rather than examining clay minerals, the most likely crystalline genes in the theories of Cairns-Smith, an experiment was designed in a model crystalline system unrelated to the composition of the prebiotic earth but suited to a well-defined test. Plates of potassium hydrogen phthalate riddled with dislocations were studied in order to ascertain whether, according to Cairns-Smith, parallel screw dislocations could serve as an information store with cores akin to punches in an old computer card. Evidence of screw dislocations was obtained from their associated growth hillocks through differential interference contrast microscopy, atomic force microscopy, and luminescence labeling of hillocks in conjunction with confocal laser scanning microscopy. The dispositions of growth active hillocks were quantified by fractal analysis. 'Mother' crystals were cleaved and inheritance was evaluated by the corresponding patterns of luminescence developed in their 'daughters' after continued growth in the presence of fluorophores. Luminescence microscopy proves to be a versatile tool for studying the dynamics of growth active hillocks. In the aggregate, this work speaks to the need for molecular mechanisms of dislocation nucleation.

1. Introduction

In *Genetic Takeover and the Mineral Origins of Life* (GT),[1,2] Cairns-Smith (CS) proposed that non-nucleic acid genetic systems were an essential part of the early stages of life's evolution. Convinced that the first organisms must have been constructed from entities that, unlike RNA, were abundant on the prebiotic earth, CS argued that imperfect clay crystals were the first genes. He envisioned prebiotic genetics with information stored in patterns of crystal defects. The mutual disposition of the imperfections was presumed to be replicated, or transferred from one crystal to another, *via* fragmentation and epitaxy (Fig. 1).[3] Certain 'phenotypes' offered by the mutability of disordered clay crystals may have reproduced faster than others giving rise to natural selection. In this scenario, a nucleic acid-like genetic

Department of Chemistry, University of Washington, Box 351700, Seattle WA 98195-1700. E-mail: E-mail: kahr@chem.washington.edu; Fax: +1-206-685-8665; Tel: +1-206-616-8195

† We are grateful to the US National Science Foundation, the Royalty Research Foundation of the University of Washington, and the University of Washington Center for Nanotechnology for support of this work.

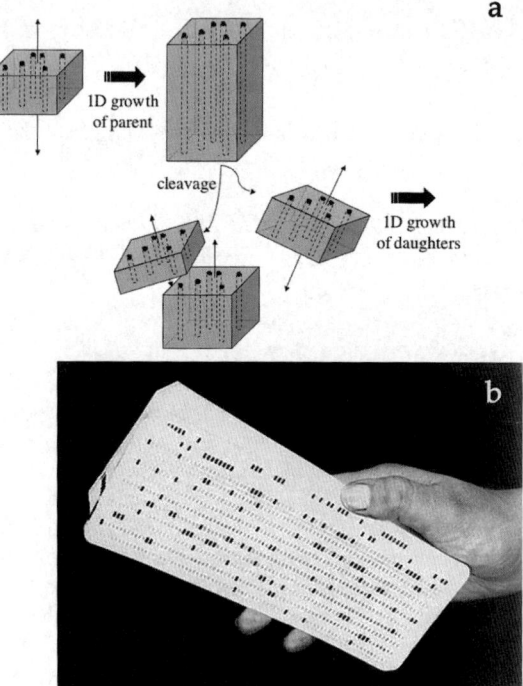

Fig. 1 (a) Genetic 'punched-card' crystal. Information is stored as a pattern of parallel dislocations, akin to an old punched computer card, that can be replicated through cleavage and subsequent crystal growth. (b) IBM punched computer cards. Courtesy of International Business Machines Corporation. Unauthorized use not permitted.

material would have taken over from the primary clay genes that were then discarded as mere scaffolds.

What exactly were these clay genes? CS explored a number of ideas and emphasized the information stored in the relative disposition of polytype twins in clays within and between layers as well as the distribution of cations common in disordered silicate minerals. CS also considered the possibility of screw dislocations as carriers of information as follows:

"A screw dislocation is an example of a defect that is often replicated through crystal growth. This kind of defect is faithfully copied, while other kinds, arising from initial misplacement of adding units, *etc.* can be put right by local reversals. A crystal containing many parallel screw dislocation lines might conceivably replicate information in the form of a particular disposition of these lines. This would be an example of a two-dimensional information store, like a stack of identical punched cards with their lines of superimposed holes analogous to the dislocation lines" (Fig. 1).[1]

Herein, we will test this latter proposition.

CS's ideas have captured a place in the origin of life community and are widely cited, but acknowledgments of their virtues are almost always accompanied by the caveat that they lack verification by experiment.[4] Orgel said that CS's "postulate of an inorganic life form has failed to gather any experimental support. The idea lives on in the limbo of uninvestigated hypotheses."[5] According to Dyson, "There is no experimental evidence to support the statement that clay can act...as a replicator with enough specificity to serve as a basis for life. Cairns-Smith asserts that the chemical specificity of clay is adequate for these purposes. The experiments to prove him right or wrong have not been done."[6] It is remarkable to be faced with an idea so attractive that it is invariably credited in reviews on the science of the origin of life, despite the fact that no one has been able to generate an experiment to test it for decades.

Clays were abundant on the prebiotic earth and could have served as suitable genes and scaffolding for the formation of early life by virtue of their imperfections and mutability. However, when it comes to experimental investigation of CS's proposal, researchers beginning with clays are at a disadvantage because, as crystals go, they are difficult to grow reproducibly, sizably, and with well-defined habits. CS's ideas must first be realized in chemically well-defined systems that are best suited to illustrations of his concept, irrespective of the likelihood that such crystals were part of the prebiotic milieu.

Herein, we report a long overdue experimental investigation of the CS 'crystals-as-genes' hypothesis. This research endeavors to evaluate one proposed mechanism whereby crystals act as a primitive source of transferable information. It includes:
• The introduction of a method for the identification of growth hillocks in crystals based on luminescence labeling.[7]
• The mapping of growth hillocks in a model crystal system and the evaluation of the associated information content.
• The 'reproduction' of crystals by cleavage and subsequent growth to demonstrate the transfer of information between successive generations.
• The evaluation of the frequency of mutations generated during reproduction.

2. Background and technical approach

2A Dyeing crystals

We have made a systematic study of the process of dyeing crystals[7] and have shown that simple molecular and ionic crystals including sulfates, phosphates, and carboxylates can orient and overgrow a wide variety of colored or luminescent molecules. One of the most remarkable crystals in its ability to orient and overgrow dye molecules is potassium hydrogen phthalate (conventionally abbreviated KAP for potassium acid phthalate, $C_6H_4 \cdot COOH \cdot COO^- K^+$, space group $Pca2_1$).[8–11] Selective adsorption of the dye by the crystal host arises as a result of different crystallographic faces having different affinities for the dyes, trapping them differentially in polyhedral sub-volumes called growth sectors. This is known as inter-growth sectoral chemical zoning. Impurities may also inhomogeneously deposit *within* a single growth sector, a process known as intra-growth sectoral zoning.[12] Polygonization of hillocks partitions crystal faces into symmetry independent vicinal slopes that express distinct affinities for additives (Fig. 2a). The c direction of KAP is polar; the steps that propagate in the $+c$ and $-c$ directions are different in structure, therefore intrasectoral zoning may result. Indeed, KAP displays intrasectoral zoning in the (010) growth sectors (Fig. 2d) when grown in the presence of a variety of dyes (more than 100; see for example Scheme 1).[13] Luminescent labels that bind to the widely spaced fast steps in preference to the more closely spaced slow steps will reveal themselves as bright chevrons. The correspondence between hillocks and dye is evidenced by reflected light differential interference contrast (DIC) and fluorescence microscopies (Fig. 2c, d).

Growth active hillocks can begin and end at any point during the growth of real crystals, so they do not always make it to the surface to be observable by DIC or other surface probes. The fluorescent dyes bind to the crystal throughout the growth process. Even after subsequent overgrowth, the luminophores create a 'fossil-record' of hillock evolution in patterns of light that can be 'dug-out' by successive cleavage or with a confocal luminescence microscope.

We used confocal laser scanning microscopy (CLSM)[14] to image as-grown KAP crystals doped with either **1**, **2**, or **3** at approximately 1 molar part dye per 100 000 parts KAP, within the fast slopes of the {010} growth hillocks. Crystals were grown by spontaneous nucleation through slow evaporation in 30 °C water baths. After

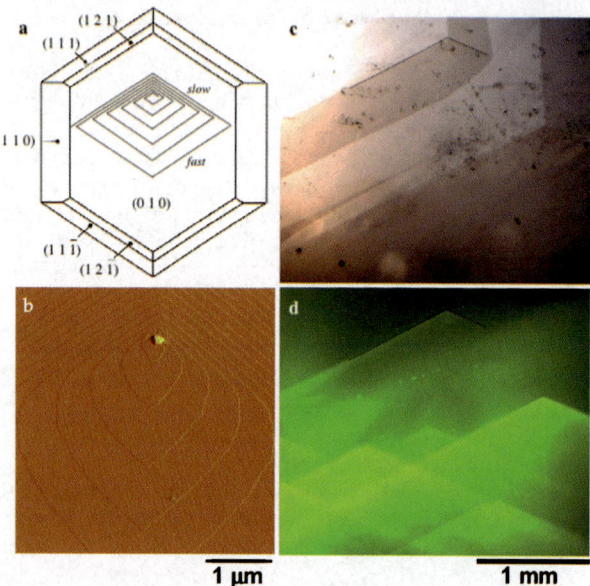

Fig. 2 KAP hillocks. (a) Idealized KAP crystal habit and absolute indices. The orientation of the fast ($+c$ direction) and slow ($-c$ direction) growing slopes of a hillock is shown on (010) in a comparison of images made by (b) *ex situ* atomic force microscopy, (c) reflected light differential interference contrast microscopy, and (d) fluorescence microscopy of KAP dyed with **1** at 10^{-5} M. (c) and (d) represent the same areas.

2–4 days, crystals were harvested and the solution was rapidly removed using a high-pressure nitrogen jet[15] to preserve the surface features.

By scanning from the bottom to the top of a crystal, the number and position of the hillocks that form throughout the (010) growth sector were recorded. Fig. 3 shows successive optical slices of KAP dyed with **1** in which the development of the (010) growth sector can be seen in the evolving pattern of luminescence.[16] The optical sections are thin enough such that the z-coordinate can roughly be taken as time of growth. By mapping hillock positions in each optical section, and comparing them to the positions in previous and subsequent sections, a sequence of activated hillocks was obtained. In addition to the growth sequence, we ascertained the stability of the defect pattern by recording the birth and death of growth-active hillocks in the (010) sector. In one of the crystals imaged[17] with a thickness of 870 µm (not shown) we found 189 hillocks, of which 54 propagated to the surface. The lifetimes of the hillocks varied from ~20 µm (seen in only two sections) to ~520 µm (52 optical sections). The average hillock lifetime was ~130 µm, and the maximum hillock density was 2.2 hillocks mm^{-2}. Each new growth layer is ~14 Å high, the KAP b lattice constant. Thus even in a 20 µm section of KAP, there are more than 13 000 replicated layers. Fig. 4 shows an example with a greater density of dyed KAP hillocks.

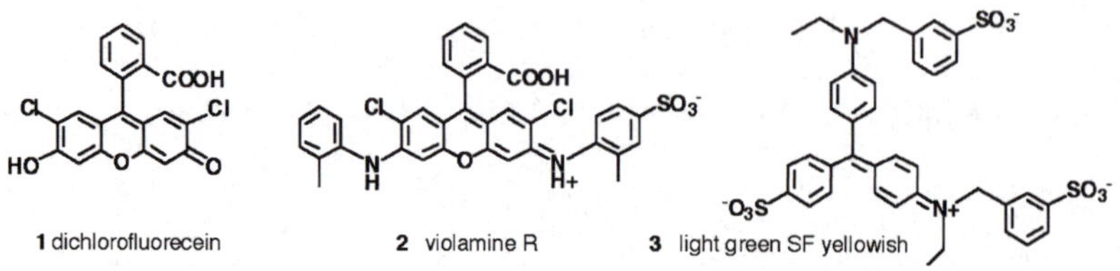

Scheme 1

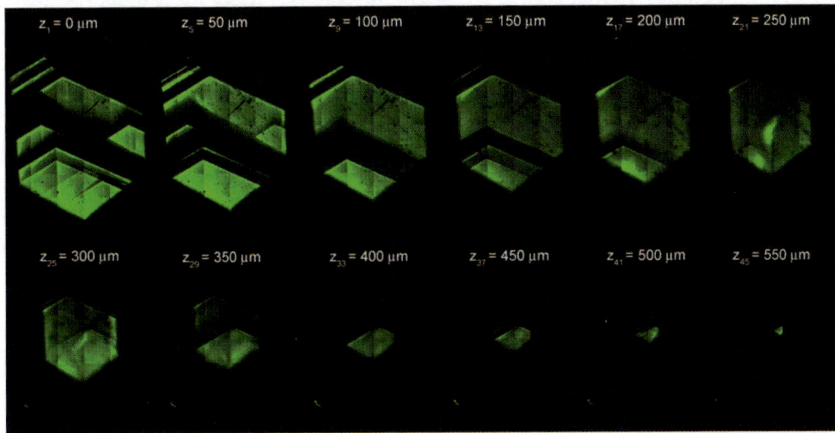

Fig. 3 Confocal slices of KAP with **1**. Sequence read from top left (top of crystal) to lower right (bottom of crystal), depth in the z-series is indicated. Step size is 12.5 μm. Every fourth optical slice is shown. Luminescence develops on the fast slopes of (010) growth hillocks, with vertices marking the dislocation cores. Crystal size = 6.3 mm wide × 8.3 mm long × 0.7 mm thick.

The points corresponding to the hillock generating dislocations were very sharp even at the highest magnifications. In most cases, they maintained the same xy coordinates through successive optical slices. Both of these observations are consistent with pure screw dislocations emergent on the (010) faces. Dislocations with considerable edge character would undoubtedly move in the xy (or crystallographic (010)) plane between optical z-slices. In this way, the vertices of the luminescent chevrons would be diffuse. Were dislocations to have emerged at an angle of just 20° with respect to (010), the vertex would spread across ∼4 μm in one 20 μm optical slice. As many hillocks grow continually for hundreds of microns, this xy motion would easily be observable by convential fluorescence microscopy. We will revisit the nature of the dislocations in the Discussion.

2B Hillock fractal dimensionality as genetic information

CS proposed the transfer of information in the patterns of hillocks through crystal growth. But, what is the nature of this information, and how can it be characterized? Fractal analysis can be used to quantify the information stored in the pattern of growth hillocks. Fractal geometry extends the classical concept of dimensional

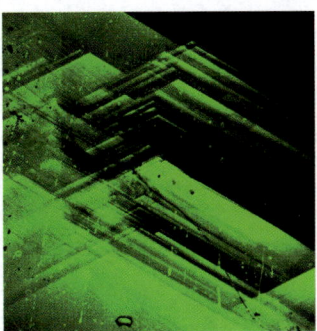

Fig. 4 Confocal image of KAP with **1** showing details of luminescence that has developed in the fast slopes of the (010) growth hillock. Vertices mark dislocation cores. Image size = 0.9 mm × 0.9 mm.

analysis to include a fractional (non-integer) number that describes the structure of complex systems in space–time. There are a variety of methods for computing fractal scaling, many of which rely upon covering a given structure with circles (or boxes) of varying sizes, r, counting the number, $N(r)$, of circles (boxes) it takes to enclose the object, and plotting the results on a log–log scale to draw out a power law, $N(r) \sim r^D$, where D is the fractal dimension of the object. The following discussion follows closely that of Viscek.[18]

A more practical approach when dealing with scale-limited random fractals, abundant in naturally occurring patterns, is to use a pair-correlation function. Scale-limited random fractals possess statistical fluctuations, and thus lack the symmetry found in deterministic fractals generated through iterative functions.[18] The correlation function is as follows:

$$c(\vec{r}) = \frac{1}{V} \sum_{\vec{r}'} \rho(\vec{r}+\vec{r}')\rho(\vec{r}') \qquad (1)$$

Here, $c(r)$ gives the expectation value that two points separated by distance r both belong to the structure. In other words, if a particle is at position r', this equation gives the probability of finding another particle at $r + r'$. The local density, ρ, is equal to one, if the point belongs to the object, and zero otherwise. V is the volume, area, or length the object occupies. Thus, $c(r)$ is a measure of how the object's local density distribution scales with distance relative to any given point in the object. An important component of fractal behavior is that the basic structural pattern (information) re-emerges at varying magnifications. This feature is referred to as scale-invariance or self-similarity. If the correlation function remains unchanged by rescaling with an arbitrary constant b, then the object will qualify as being non-trivially scale invariant:

$$c(br) \sim b^{-\alpha} c(r), \quad 0 < \alpha < d \qquad (2)$$

where α is a non-integer scaling constant between zero and the Euclidean dimension, d, in which the object is embedded. The only function that satisfies this condition is a power law $c(r) \sim r^{-\alpha}$, indicating an algebraic decay in the local density.[18] For an isotropic object, as most random fractals are, one can see that integrating $c(r)$ over a sphere of radius R gives the number of particles $N(R)$ contained within the sphere:

$$N(R) \sim \int_0^R c(\vec{r}) d^d r \sim R^{d-\alpha}. \qquad (3)$$

Comparing this to the earlier power law formula, $N(R) \sim R^D$, where D is the fractal dimension of the object, one arrives at $D = d - \alpha$. This formula may be used to determine the fractal dimension from the density correlations within a random fractal.

Applying the pair correlation function to real data such as the position of the growth hillock cores involves choosing a hillock vertex and counting the number of other vertices that fall within a circular shell of radius $r + \delta r$, where δr is typically ten percent of r. This calculation is repeated with each of the vertices as the origin, and for varying sizes of r. Using circular shells in this manner is equivalent to determining the derivative of $N(r)$, which essentially gives $c(r)$. The results were normalized according to the number of vertices and the areas of the circles for each scan. Vertices close to the outer edges were discarded as counting centers in order to avoid area truncation effects. Thus the fractal behavior of any real object is only valid within a limited scaling range determined by the object's size. Fig. 5 displays the method just described using the (010) outlines and hillock coordinates for all layers of the z-stack analyzed. A useful feature of this method for calculating the fractal dimension is that it improves the counting statistics by averaging distances

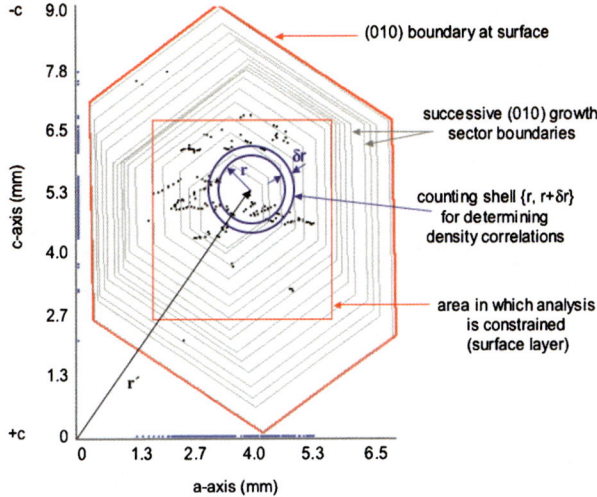

Fig. 5 Pair correlation function analysis. Hillock locations are denoted by black dots. Blue dots on the coordinate scales are the projections of the hillock locations onto the a-axis and c-axis of the crystal. Hexagonal outlines display the areas of the (010) face as the crystal grew between optical z slices where the innermost border is near the bottom of the (010) sector and the outermost (red) border is the fully-grown surface. The red box encloses the central 30–40% of the (010) section of interest (here shown for the surface layer). Any hillocks lying outside this box were discarded in the analysis so as to avoid area truncation effects. The blue ring gives an example of how pair correlations are counted for various r; all points falling within the ring are counted for that r from each center.

between many points within a single cluster; there are $n(n-1)/2$ sample points rather than just n, where n is the number of vertices.

The fractal dimension is obtained by plotting $c(r)$, the normalized number of points that fall within the shell of radii r and $r + \delta r$, *versus* the shell radii used, r, on a log–log scale (Fig. 6). The slope of this line within the appropriate length range gives the scaling parameter α, from which the fractal dimension is determined as $D = d - \alpha$, where d is the embedding Euclidean dimension (in this case $d = 2$).

After mapping the sequence of hillocks for all layers of our z-series data, the D for the hillock patterns on each optical section was determined. The resulting plot of the fractal dimensions for all the layers in our crystal showed that $D \sim 1.4 \pm 0.2$, even when 'dead hillocks' were removed from the sequence and new ones added (Fig. 7a).

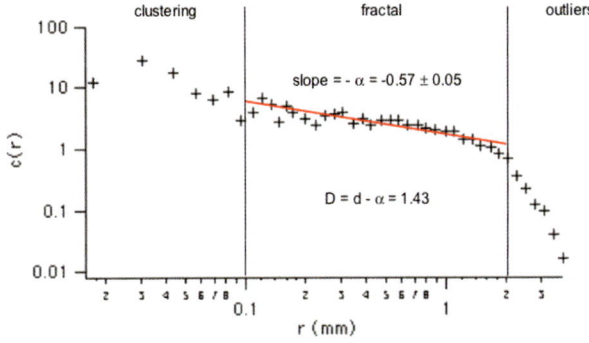

Fig. 6 Pair correlation plot for one optical z section of hillock data. Three distinct regions are identified within the range of 0.1 mm and 2 mm pair separations that display the characteristic power-law for a random fractal. The region at large r tails off due to boundary effects, while the region at smaller length scales displays evidence of hillock clustering.

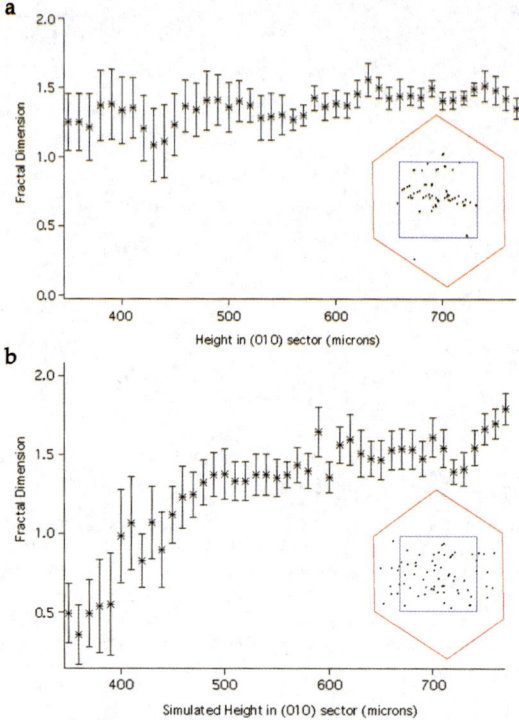

Fig. 7 Fractal dimensions for all optical sections. The fractal dimension and its uncertainty, determined from fitting the slope of the pair correlation plots, is shown as a function of vertical growth of the (010) sector for both (a) the hillock data and (b) the Monte Carlo data. The fractal dimension is fairly consistent (1.4 ± 0.2) throughout the plot for the hillock data (a), whereas it gradually increases towards 2 (the Euclidean dimension) for the Monte Carlo data (b). The insets in (a) and (b) are examples of the hillock locations on a single layer of the actual crystal and from the Monte Carlo simulation, respectively.

Furthermore, a Monte Carlo simulation was created to mimic the growth of the crystal with random events (or 'hillock' locations), limited by the boundaries of the growing (010) face. The simulation took the numbers of hillock "births" and "deaths" from the actual crystal optical sections. The positions of new hillocks and the selection of those to be removed were both randomized. Fig. 7b shows the resulting plot of the fractal dimensions for all corresponding layers of the Monte Carlo simulation. In this case the randomly generated data continue to fill more space as the simulated growth progresses, gradually increasing towards a fractal dimension of 2, which is expected since this is the same as the Euclidean dimension that encloses the pattern. The comparison of these data indicates that there is a form of stability in the information content of the developing pattern for real hillocks, even if the individual locations are not predetermined and the number of growth active hillocks varies from layer to layer.

These results show there is indeed fractal behavior within a limited scaling range in the pattern of growth-active hillocks that emerge on the (010) surface of KAP. The results also reveal clustering that occurs at smaller length scales (pair separations below 0.1 μm). Furthermore, by projecting the hillock locations onto the a-axis (non-polar) and c-axis (polar), a large number of hillock cores were observed to run in 'channels' along [100]. There were correlations, albeit up to 13 times less frequent, along [001] as well. These correlations are absent in the Monte Carlo data. The relevance of this observation will become evident in the Discussion.

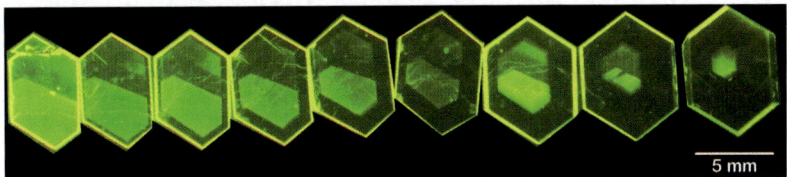

Fig. 8 Successive slices (∼200 μm) of KAP dyed with **1** from top surface (left) to bottom (right) of a (010) growth sector. Cross sections reveal the development of the luminescent dye inclusion. In this case, growth was dominated by a single hillock.

2C Cleavage and growth

Fig. 8 shows a crystal grown with **1** that was cut into successive 200 μm slices by placing the edge of a sharp razor parallel to the cleavage plane as did Borc and Sangwal.[19] In this way, the development of the luminescent (010) growth sector and its hillocks was revealed in cross sections. The (010) cleavage would be characterized by the descriptive crystallographer as 'perfect'.

Next we tried to establish if the defect pattern is inheritable, that is, passed on to daughter fragments. Fig. 9 shows a cartoon of crystal reproduction in order to recapitulate the idea that generation of screw dislocations in a crystal, followed by cleavage and growth, can lead to the propagation of information. Here, we explicitly indicate that the 'bits' are the 2D spatial positions of hillocks.

In KAP, screw dislocations passing through cleavage planes will expose corresponding high-energy cores following cleavage. As KAP has a diad axis parallel to c, complementary cores may be related by symmetry. It was expected that the defects that exist on the exposed surfaces of cleaved seeds would propagate during further growth and be readied for observation by luminescence microscopy after a hillock recognizing dye was added to the growth solution.[20]

We cleaved a pure KAP crystal into 13 150 μm slices. These slices were then used as seeds to grow daughter crystals from a KAP solution doped with 4×10^{-5} M **2** and supersaturation 0.8% at 26 °C. The resulting daughter crystals were imaged using fluorescence microscopy. The photo inset in Fig. 9 shows two of the daughters grown from sequential slices. Indeed, seven labeled hillocks, identified at the lower right corners of the triangles in Fig. 9, were shared by each of the daughter crystals. That said, we demonstrated the proposition of CS that the information encoded as

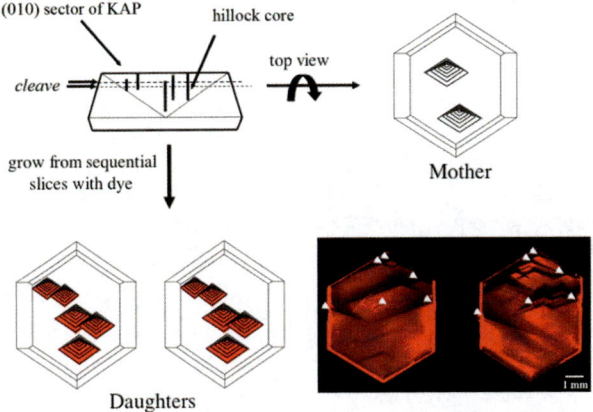

Fig. 9 KAP has perfect (010) cleavage perpendicular to the screw dislocation cores. Cleavage and subsequent growth generates daughter crystals that in principal retain their mother's genes (cartoon). Inherited dislocations are indicated in the photomicrographs by the lower right corners of the white triangles in the patterns of luminescence from dopant **2**.

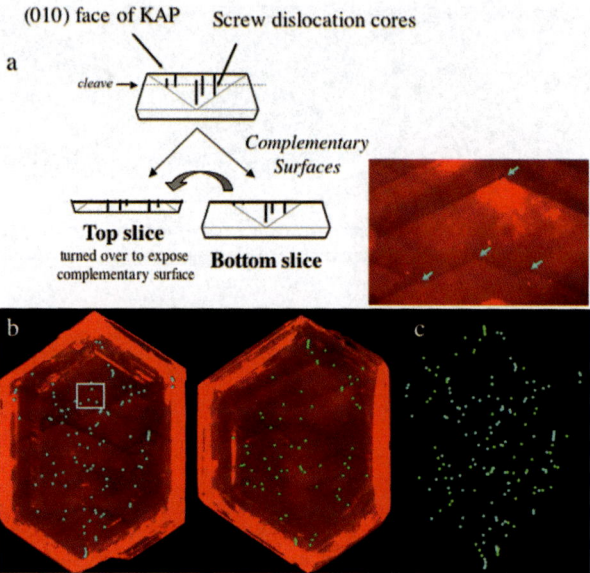

Fig. 10 (a) Cartoon of complementary surfaces from 'perfect' cleavage. Complementary surfaces expose the same screw dislocation cores on each surface and should result in exact replication of growth active hillocks during subsequent growth of daughters. (b) Daughters grown with **2** from complementary surfaces. Blue and green dots mark the hillock cores. (c) Superposition of green and blue dots showing there is little if any correspondence between daughters. Inset above (c) shows a magnified section of the crystal on the left indicated by the box. Crystals = 6 mm × 9 mm.

the spatial disposition of screw dislocations can be transferred from one crystal to another. However, a greater number of new hillocks, 'mutations', were also observed that were apparently not present or active in the parent or in the twin daughters. For this work to be a compelling stimulus to other scientists, the number of preserved hillocks in the daughter crystals must exceed the number of new hillocks that are created after cleavage. In other words, for KAP crystals to resemble genes, there must be more inheritance than mutation in successive generations. The great number of mutations led to experiments aimed at the elucidation of the fundamental mechanisms that give rise to the birth of new growth active hillocks.

Real crystals often grow through the emergence, competition, and poisoning of many hillocks. In order to ensure that the seeds used for subsequent growth share the same 'genes' (punches or screw dislocations) we grew daughter crystals from *complementary* surfaces produced from a single cleave, rather than sequential surfaces (Fig. 10) so as to produce enantiomorphous twins. Daughters were grown for four hours from a KAP solution (supersaturation 1.0% at 23 °C) doped with 10^{-5} M **2**. Complementary surfaces resulted in even more mutation than sequential surfaces. This is true even when extra care is taken not to expose the fresh cleaved surfaces to atmosphere, by cleaving the mother in solution and thereby avoiding the collection of particulates that might nucleate new dislocations. To investigate this further we undertook a study of the surfaces of cleaved KAP.

KAP crystal layers separate as easily as mica. However, the razor places stresses on the separating halves. After cutting the crystals, we examined the surfaces by DIC and atomic force microscopy[21] (AFM). The images of cleaved surfaces showed ridges and terraces of unit cell height or greater, indicating that the cleavage process is inherently imperfect at the nanometer scale (Fig. 11). AFM showed similar surface features to those observed previously:[19] disorganized breakage, terraced macrosteps, and V-shaped elementary steps oriented along low-index crystallographic directions.

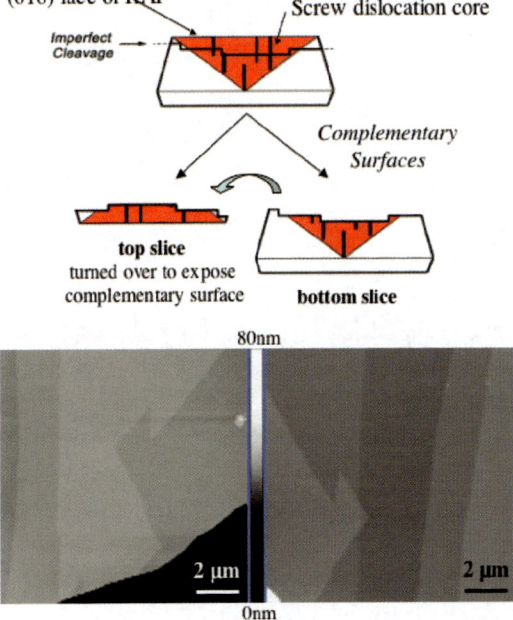

Fig. 11 Complementary surfaces from imperfect cleavage. Surfaces exposed by cleavage will have, in principle, complementary topographies. Hills will have abutted valleys and *vice versa*. This is shown, imperfectly, by complementary AFM[21] images at bottom.

Complementary halves will leave a screw dislocation emergent on a hill on one surface and in a valley on the other. Thus, mass transport will undoubtedly be affected by differences in fluid dynamics associated with these topographical features. Furthermore, hillocks in valleys are more easily subject to overgrowth from macrosteps, whereas dislocations emerging on higher terraces have a greater chance of surviving to continue growth. After re-growth, these unavoidable macrosteps become a prolific source of new dislocations and false positive inheritance or mutation. After just two minutes of re-growth from a cleaved surface in a 1% supersaturated pure KAP solution, DIC micrographs show that new hillocks form along newly created macroedges (Fig. 12), principally on the upper terraces. Daughter growth from imperfectly cleaved, complementary surfaces is doomed to fail. Cleavage by its very nature serves to generate new, and non-reciprocal hillocks.

We further attempted to grow daughters from unadulterated seeds so as to minimize the generation of new dislocations. To accomplish this, seeds were freshly harvested from a growth solution at 30 °C and immediately transferred to a new solution at the same temperature without cleavage and without exposing the surfaces to air. In order to clearly discern which hillocks were transferred from the seed and which ones were mutations, we first labeled the seed with red fluorescing **3**. Growth was continued for 1 h from a solution (σ = 0.8%, 30 °C) containing green fluorescing **1**, as in Fig. 13. Results still show more mutation than inheritance.

3. Discussion

The lag time between experiments lacking a theoretical foundation or theories without supporting experiments has been dramatically diminished in modern times given the exponential rise of the scientific enterprise. The hypotheses of CS are unusual because they have resisted experiment for decades. The principal researchers and commentators in the origin of life community are clearly enamored with GT,

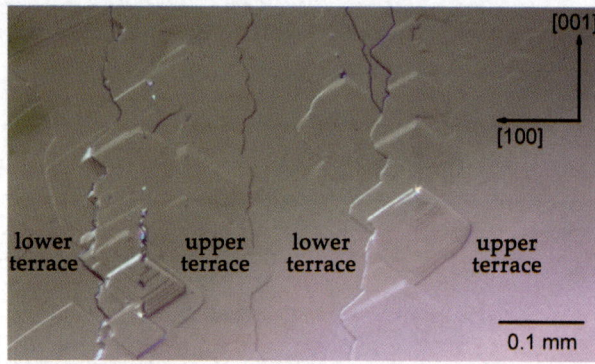

Fig. 12 DIC image of KAP re-growth after cleavage. Newly formed hillocks are mainly on the upper terraces running along cracks and macrosteps induced by the cleave. The jagged macrosteps seen to the left of the columns of hillocks is the growth front that propagated from the newly formed macro-edges.

and seem poised to embrace it if only they could be provided with a reason for so doing.

Herein, we have presented an experimental test of CS's hypothesis of defect-laden crystals as analogs of prebiotic genes. Our goal was not to solve the problem of the origin of life but was much more modest: to rescue the proposals of CS from the 'limbo of uninvestigated hypotheses'.[5] Along the way, we came to realize that transferring information encoded in the disposition of screw dislocations as imagined by CS is complicated by several factors. Foremost, the process of cleavage (crystalline mitosis) gives rise to many new growth active hillocks or 'mutations'. Macrosteps appear to be abundant sources of new hillocks, particularly on the elevated edges of the steps. The cartoon in Fig. 1a derived from GT is an oversimplification. Even crystals with 'perfect' cleavage produce imperfect, complementary halves. The growth dynamics of complementary surfaces will be intrinsically different from identical surfaces that, in any case, are idealizations.

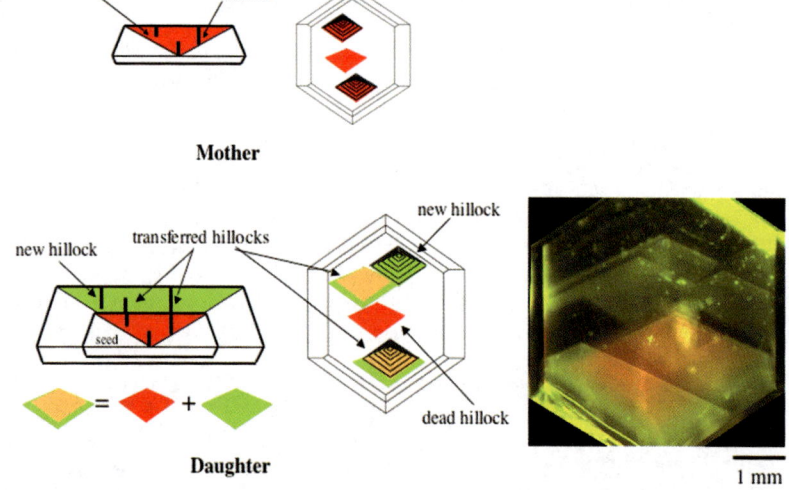

Fig. 13 Inheritance without cleavage. Hillocks in the parent crystal were labeled with a red fluorescent dye (**3**). Subsequent growth from a solution containing a green fluorescent dye (**1**) allowed for the distinction between transferred hillocks (information preserved), buried hillocks (information lost), and new hillocks (mutations or information added) based on colour.

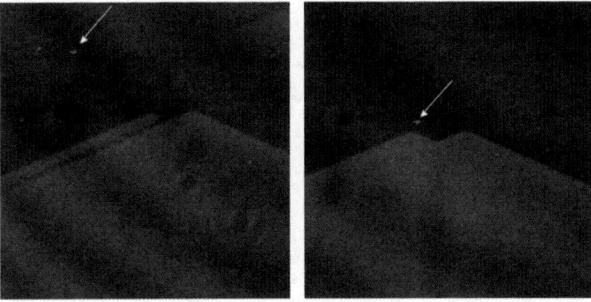

Fig. 14 Confocal image of crystal grown with 1 µm red fluorescent polystyrene beads. Arrow points to an embedded microbead. There is no correspondence between beads and hillocks. Image sizes = 230 µm × 230 µm.

These observations forced us to confront the fundamental mechanisms that give rise to screw dislocations. A common presumption is that dislocations in freely grown crystals arise from internal stresses generated by heterogeneous impurity particles.[22] However, such observations fall far short of molecular mechanisms of dislocation generation.

Despite the greatest of care taken to not expose the fresh seeds to atmosphere, and even in the absence of cleavage, new hillocks still abound. In order to test directly whether these newly formed hillocks result from the inclusion of foreign particles, we grew KAP in the presence of water soluble red-fluorescing CdSe quantum dots (30–50 nm)[23] and red-fluorescing rhodamine-coupled polystyrene microspheres[24] (200 nm and 1 µm). Growth was continued by slow evaporation in a dry, 30 ± 0.02 °C chamber with 2×10^{-5} M **1** to label the hillocks and 10^{-9} M or less of the luminescent particulate. Were the particles to generate hillocks during overgrowth then we would see bright red luminescing spots at the hillock cores (luminescent vertices). Beads (Fig. 14) and quantum dots were indeed embedded within the crystal but there was no correspondence between particulate inclusion locations and dyed hillocks, even for the 1 µm beads. (However, fluid inclusions in KAP have been observed by Halfpenny *et al.* to act as sources for the appearance of new dislocations.[25])

While the data in Fig. 14 do little to support the notion that particulate impurities nucleate screw dislocations, it is clear that sometimes, when growth hillocks do arise, they do so in pairs with precisely the same location along the *c*-axis. Such pairs are consistent with the X-ray topographic analysis of Halfpenny *et al.* who made detailed analyses of the dislocations in KAP.[25] In the {010} sectors, they observed mixed screw/edge dislocations (Burger's vectors = $\langle 110 \rangle$) nucleated in V-shaped pairs with about 20° between arms that were equally inclined with respect to [010] and that lie in the (001) plane. Such dislocations are consistent with the *a*-axis correlations evident in the fractal analysis but are inconsistent with the immobility and sharpness of the hillock vertices. Halfpenny *et al.* do indeed presuppose that pure screw dislocations with Burger's vectors parallel to [010] would be difficult to observe in (010) plates by X-ray topography. Needless to say, there remains much that can be learned about dislocation nucleation and structure in KAP which is indeed essential in using it to evaluate GT.

The punched card model of CS is an idealization that does not well represent the hillock dynamics in KAP. There may be another substance, better suited to the idealization. However, the screw dislocation mechanism is only one of many proposed crystal genetic mechanisms laid out in GT. We hope that the demonstrations herein will give more confidence than caution to others so that they might expand upon our work in more biologically relevant crystal systems while trying to tackle other proposals in GT. In doing so, we may finally break past the fatalism that pervades any discussion of bringing GT within the realm of experimental science.

References

1. A. G. Cairns-Smith, *Genetic Takeover and the Mineral Origins of Life*, Cambridge University Press, Cambridge, 1982; These ideas were first articulated in the following: A. G. Cairns-Smith, *J. Theor. Biol.*, 1966, **10**, 53–58; A. G. Cairns-Smith, in *Towards a Theoretical Biology. I. Prolegomena*, ed. C. H. Waddington, Edinburgh University Press, Edinburgh, 1968; A. G. Cairns-Smith, *The Life Puzzle: On Crystals and Organisms and the Possibility of a Crystal as an Ancester*, Oliver and Boyd, Edinburgh, 1971; A. G. Cairns-Smith, *Proc. R. Soc. London, Ser. B*, 1975, **189**, 249–274; A. G. Cairns-Smith, *Origins Life*, 1975, **6**, 2657.
2. A. G. Cairns-Smith, in *Frontiers of Life*, ed. D. Batlimore, R. Dulbecco, F. Jacob and R. Levi-Montalcini, Academic Press, San Diego, 2001, vol. 1, p. 169.
3. A. G. Cairns-Smith, *Seven Clues to the Origin of Life*, Cambridge University Press, Cambridge, 1985.
4. See for example: D. W. Deamer and G. Fleischaker, *Origins of Life: The Central Concepts*, Jones and Bartlett Publishers, London, 1994; I. Fry, *The Emergence of Life on Earth: A Historical and Scientific Overview*, Rutgers University Press, Piscataway, 2000; P. Davies, *The 5th Miracle: The Search for the Origin and Meaning of Life*, Touchstone, New York, 1999; R. M. Hazen, *Genesis*, National Academic Press, Washington, DC, 2005.
5. L. E. Orgel, *Trends Biochem. Sci.*, 1998, **23**, 491–495.
6. F. Dyson, *Origins of Life*, Cambridge University Press, Cambridge, 1999.
7. B. Kahr and R. W. Gurney, *Chem. Rev.*, 2001, **104**, 893–951.
8. T. A. Eremina, N. G. Furmanova, L. F. Malakhova, T. M. Okhrimenko and V. A. Kuznetsov, *Crystallogr. Rep.*, 1993, **38**, 554.
9. L. A. M. J. Jetten, B. van der Hoek and W. J. P. van Enckevort, *J. Cryst. Growth*, 1983, **62**, 603.
10. Y. Okaya, *Acta Crystallogr.*, 1965, **19**, 879.
11. W. J. P. van Enckevort and L. A. M. J. Jetten, *J. Cryst. Growth*, 1982, **60**, 275.
12. A. G. Shtukenberg and Yu. I. Punin, in *Optically Anomalous Crystals*, ed. B. Kahr, Springer, Berlin, 2007.
13. T. Bullard, M. Kurimoto, S. Avagyan, S. H. Jang and B. Kahr, *ACA Trans.*, 2004, 39. Other examples of the use of confocal fluorescence microscopy to study impurity distributions in crystals include: Y. Iimura, I. Yoshizaki, H. Nakamura, S. Yoda and H. Komatsu, *Cryst. Growth Des.*, 2005, **5**, 301; Y. Iimura, I. Yoshizaki, S. Yoda and H. Komatsu, *Cryst. Growth Des.*, 2005, **5**, 295. Intrasectoral zoning in minerals by cathodoluminescence was pioneered by Reeder *et al.* See: J. Rakovan and R. J. Reeder, *Geochim. Cosmochim. Acta*, 1996, **60**, 4435 and references therein.
14. S. W. Paddock, *Mol. Biotech.*, 2000, **16**, 127.
15. G. Ester, R. Price and P. J. Halfpenny, *J. Cryst. Growth*, 1997, **182**, 95.
16. Confocal image made with a Zeiss LSM 510 NLO microscope at the University of Washington Nanotechnology User Facility, a member of National Nanotechnology Infrastructure Network supported by NSF. A 10x/0.5D FLUAR objective at 0.7 zoom and with 488 nm excitation light from an Argon laser were used.
17. This crystal was imaged using a Leica TCS SP/NT on a DMIRBE incident-light inverted microscope through a 5× objective, pinhole width of 1 Airy unit, and with 568 nm excitation light from a 25 mW krypton ion laser at the University of Washington W. M. KECK Center for Advanced Studies in Neural Signaling. http://depts.washington.edu/keck/.
18. T. Vicsek, *Fractal Growth Phenomena*, World Scientific Publishing, New Jersey, 2nd edn, 1992.
19. J. Borc and K. Sangwal, *Surf. Sci.*, 2004, **555**, 1.
20. L. N. Rashkovich, E. V. Petrova, O. A. Shustin and T. G. Chernevich, *Phys. Solid State*, 2003, **45**, 400.
21. Cleaved KAP crystals were examined with a Veeco Dimension 3100 SPM with Nanoscope IV a controller from the University of Washington Nanotechnology User Facility (see note 16). The AFM data was collected under ambient atmosphere in tapping mode.
22. D. Hull and D. J. Bacon, *Introduction to Dislocations*, Butterworth Heinemann, Oxford, 4th edn, 2001.
23. Quantum Dot Corporation water soluble fluoresecent Qtracker non-targeted quantum dots were used. CdSe core with ZnS shell, polyethylene glycol derivatized and polyacrylic acid coated. Excitation light used was 543 nm, emission at 655 nm.

24 Invitrogen Microprobes water soluble FluoSpheres were used. Carboxylate modified polystyrene beads with rhodamine dyes chemically attached. Excitation light used was 543 nm, emission at 605 nm.
25 G. R. Ester, R. Price and P. J. Halfpenny, *J. Phys. D: Appl. Phys.*, 1999, **32**, A128; G. R. Ester and P. J. Halfpenny, *Philos. Mag. A*, 1999, **79**, 593.

Precipitation of α L-glutamic acid: determination of growth kinetics

Jochen Schöll,[a] Christian Lindenberg,[a] Lars Vicum,[a] Jörg Brozio[b] and Marco Mazzotti*[a]

Received 7th November 2006, Accepted 30th January 2007
First published as an Advance Article on the web 24th April 2007
DOI: 10.1039/b616285a

Growth kinetics of α L-glutamic acid was determined based on seeded batch desupersaturation experiments. The growth rate correlation applied in this study accurately describes the growth process in a temperature range of 25–45 °C and in a supersaturation range of 1–3. The newly developed approach for the growth rate characterization has the advantage of a high robustness especially with respect to the influence of competing particle formation mechanisms as nucleation or agglomeration. The efficient technique employs *in situ* process analytical technologies, *e.g.* attenuated total reflection Fourier transform infrared spectroscopy (ATR-FTIR) and focused beam reflectance measurement (FBRM), different *ex situ* analytical tools and population balance modeling combined with a non-linear least squares optimization algorithm to determine the growth kinetics. The growth mechanism was identified to be integration controlled and of birth and spread (B + S) type. The quality of the determined growth rate correlation was assessed by comparison with the experimental data and with literature data.

1 Introduction

Precipitation from solution involves several fundamental mechanisms, namely nucleation, growth, agglomeration and breakage, that determine the particle size distribution, shape and polymorphic form of the precipitated product. The ability to measure the kinetics of these mechanisms is of crucial importance for process design and development. Various methods for determining growth rate kinetics for crystallization processes can be found in the literature. Besides single crystal methods, where the growth mechanism and kinetics of different crystal faces are usually determined by optical or atomic force microscopy (AFM) under different flow and supersaturation conditions,[1] seeded batch multiparticle experiments are employed for process design purposes.[2]

Recently, the use of *in situ* process analytical technologies (PAT) has become more and more frequent for the characterization of crystallization and precipitation processes: growth kinetics of monosodium glutamate were determined combining several PATs,[3] a method for the direct measurement of crystal growth kinetics of sodium carbonate using the focused beam reflectance measurement (FBRM) was proposed,[4] and the combination of an *in situ* concentration measurements and

---

[a] *Institute of Process Engineering, ETH Zurich, 8092, Zurich, Switzerland.*
  E-mail: mazzotti@ipe.mavt.ethz.ch.; Fax: +41 44 6321141; Tel: +41 44 63222456
[b] *Chemical & Analytical Development, Novartis Pharma AG, 4002, Basel, Switzerland*

FBRM with population balance modeling was employed to determine the growth kinetics of paracetamol.[5]

This work focuses on a PAT based characterization of the growth kinetics of α L-glutamic acid precipitated from monosodium glutamate solutions upon addition of hydrochloric acid. This system was already investigated in a previous work with respect to the nucleation kinetics.[6]

We consider seeded growth experiments under different conditions and estimate the parameters of the growth rate correlation by combining the offline measurement of the particle size distribution of seeds and the time resolved measurement of the concentration in solution through ATR-FTIR with population balance modeling and an optimization routine. The chord length distribution of particles in suspension is monitored during the experiments using FBRM, to make sure that no new particles are formed and thus to confirm that the desupersaturation process is dominated by particle growth and not by nucleation. In order to asses the role of agglomeration, simulation results for the final particle size distributions are compared with experimental data and scanning electron micrographs (SEM) of the product particles are analyzed.

2. Materials and methods

Monosodium glutamate monohydrate ($\geq$ 98%, Sigma-Aldrich, Buchs, Switzerland), fuming hydrochloric acid solution (37–38%, L. T. Baker, Deventer, The Netherlands) and deionized water were used in all experiments. L-Glutamic acid has two monotropically related polymorphs, the metastable α and the stable β form.[7,8] In this work, the polymorphic purity of the solid fraction was characterized by X-ray powder diffraction, Raman spectroscopy and scanning electron microscopy, as reported elsewhere.[6,9] Contrary to a previous work,[9] the solvent mediated polymorphic transformation from the metastable α to the stable β form was not observed here due to the relatively low temperature (less than 45 °C) and rather short duration (less than one hour) of the experiments.

2.1 Batch crystallizer set-up

A jacketed 500 mL borosilicate glass reactor with an inner diameter of 100 mm from LTS (Basel, Switzerland) was used for all experiments. The 4-blade glass stirrer from LTS (Basel, Switzerland) with 45° inclined blades had a diameter of 50 mm, was positioned 10 mm above the reactor bottom, and was operated at 250 rpm to ensure a homogeneous dispersion of the crystals in the reactor. The temperature in the crystallizer was controlled using a Pt 100 together with a CC230 thermostat from Huber (Offenburg, Germany). The position of the immersion probes (ATR-FTIR and FBRM, see below) was chosen in the zone of high fluid velocities, *i.e.* close to the impeller tips, to minimize clogging of the probe windows, thus optimizing the quality of the measured data.[6]

2.2 Measurement of concentration

Combined attenuated total reflection Fourier transform infrared (ATR-FTIR) spectroscopy and focused beam reflectance measurement (FBRM) have been recently applied successfully to monitor the liquid and solid phase during crystallization processes.[6,10] ATR-FTIR allows for the acquisition of liquid phase IR spectra in the presence of solid material due to the low penetration depth of the IR beam into the liquid phase.[11] An ATR-FTIR ReactIR 4000 system from Mettler-Toledo (Schwerzenbach, Switzerland), equipped with a 11.75 "DiComp" immersion probe and a diamond ATR crystal, was used for all experiments.

In this work, ATR-FTIR was used to monitor the concentration of L-glutamic acid using the two bands in the IR spectra at 1224 and 1408 cm^{-1}, which correspond

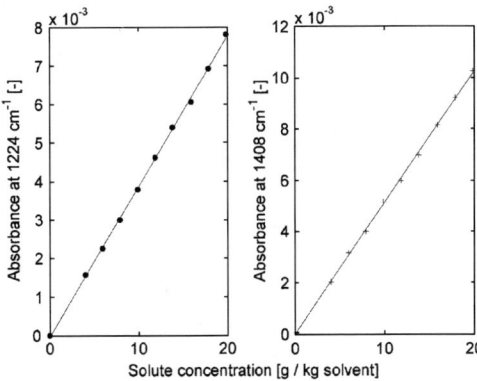

Fig. 1 Calibration data at 25 °C for the absorbance values at 1224 and 1408 cm^{-1} that correspond to the protonated and dissociated carboxylate group of the L-glutamic acid molecule.

to the protonated and dissociated forms of the carboxylic groups of the molecule, respectively.[12] Independent calibrations were performed for these two bands at 25, 35 and 45 °C using a set of solutions with known L-glutamic acid concentrations (always at its isoelectric point) and applying the law of Beer–Lambert.[6] Fig. 1 shows both calibrations at 25 °C, which achieve values of the regression coefficient R^2 of 0.9997. The differences in concentrations determined based on these two calibrations were on the order of less than 4%, which was considered to be acceptable. The supersaturation was calculated in this work by averaging the concentration values calculated with the two calibrations and dividing the result by the solubility of the metastable α form:

$$S = \frac{c_{1224} + c_{1408}}{2c^*(T)}; \quad (1)$$

here c_{1224} and c_{1408} denote the concentration values calculated with the IR bands at 1224 and 1408 cm^{-1}, respectively, and c^* is the solubility, which is a function of temperature. The solubilities of α L-glutamic acid, $c^*(T)$, are 73.9 mol m^{-3} (10.9 g per kg solvent) at 25 °C, 104.0 mol m^{-3} (15.4 g per kg solvent) at 35 °C and 146.0 mol m^{-3} (21.7 g per kg solvent) at 45 °C; they were measured using ATR-FTIR spectroscopy as reported and discussed earlier.[9]

With reference to eqn (1) a remark is worth making. As discussed in section 3.1, all experiments reported in this work have been carried out at the pH level corresponding to the isoelectric point of L-glutamic acid. During growth the pH value does not change. In aqueous solution, L-glutamic acid is present as the neutral form, also called free acid, together with its protonated, dissociated and twice dissociated species. At the isoelectric point the concentration of the free acid is higher than that of the other species.[6] The concentration of the free acid, [Glu], is linearly related to the nominal L-glutamic acid concentration, c, i.e. the total concentration of all species of L-glutamic acid in solution, by the following equation:

$$[\text{Glu}]\left(1 + \frac{[\text{H}^+]}{K_\alpha} + \frac{K_\gamma}{[\text{H}^+]} + \frac{K_b K_\gamma}{[\text{H}^+]^2}\right) = c, \quad (2)$$

where K_α, K_γ and K_b are the equilibrium constants of the dissociation and protonation reactions (the values at 25 °C are $K_\alpha = 6.46 \times 10^{-3}$, $K_\gamma = 5.62 \times 10^{-5}$ and $K_b = 2.14 \times 10^{-10}$).[6,13] At constant pH value and assuming an ideal solution the proportionality constant is independent of the total concentration: at the isoelectric point and 25 °C about 84% of the L-glutamic acid in solution is present as free acid according to eqn (2). In eqn (1) and in the calibration illustrated in Fig. 1 the nominal concentration c is used. However, thanks to the linear relationship given by eqn (2) and the fact that the pH value is constant during growth, the supersaturation ratio in eqn (1) is the same whether it is calculated with the nominal concentration c or with the free L-glutamic acid concentration [Glu].

2.3 Measurement of particle size: online and offline

The present study employs a Multisizer 3 from Beckman Coulter (Nyon, Switzerland), which applies the electrical sensing zone or Coulter method for the offline characterization of the particles grown in solution. The measurement principle of electrical sensing zone particle analyzers have been discussed in detail elsewhere.[14] The present study employs the Multisizer 3 since this device both counts the measured particles and measures very accurately their volume with an unchallenged resolution; this information is then used to calculate the particle size of volume equivalent spheres.[15] An aqueous solution saturated with respect to α L-glutamic acid was used as an electrolyte for the sample analysis. Each PSD presented in this work was measured several times and represents at least 30 000 counted particles.

FBRM is an established technique to record the chord length distributions (CLDs) of a suspended particle population even at high solid concentrations.[16,17] The FBRM was used to monitor the CLD of the solid phase and to assure that no significant nucleation occurred over the course of the experiments, as described in section 3.1. All measurements in this work were carried out using a laboratory scale FBRM 600L from Mettler-Toledo (Schwerzenbach, Switzerland).

3. Desupersaturation experiments

Desupersaturation experiments have been carried out by various authors to determine overall growth kinetics.[18,19] In this work a similar but improved approach is proposed, that is based on the accurate measurement of the PSD of the seed particles and of the solute concentration time profile, which are combined with population balance modeling and an optimization routine to determine the growth kinetics. The experimental technique as well as the limitations of this method are described and discussed in detail in section 3.1. An important prerequisite of the experimental technique is the production, preparation, and characterization of seed crystals, as highlighted in section 3.2. Finally, the measurement results of the desupersaturation experiments are presented in section 3.3.

3.1 Experimental procedure

All desupersaturation experiments were conducted in the following way: first, a supersaturated solution was created by mixing equimolar solutions of monosodium glutamate and hydrochloric acid in the temperature controlled reactor. It is worth noting that the mixture is at the isoelectric point of L-glutamic acid under these conditions, *i.e.* pH = 3.22 at 25 °C. Second, the *in situ* monitoring with ATR-FTIR and FBRM was started and afterwards a certain amount of dry seeds was introduced into the reactor. Fig. 2 shows measured solute concentration profiles, in short desupersaturation curves, for two repeated runs of experiments with different initial conditions. It can be observed that the reproducibility is satisfactory. Nevertheless, the experiments being used for the parameter estimation were performed twice. The reproducibility of the PSD at the end of the desupersaturation experiments was similar but is not explicitly shown here, since the PSDs were not used for parameter estimation. This aspect will be discussed in detail in section 4.3.

The method requires that experimental conditions are chosen in such a way that particle growth is dominating the desupersaturation process and other competing mechanisms, *e.g.* nucleation, agglomeration and breakage, are of negligible influence. To confirm that no nucleation occurred the CLD was monitored during the

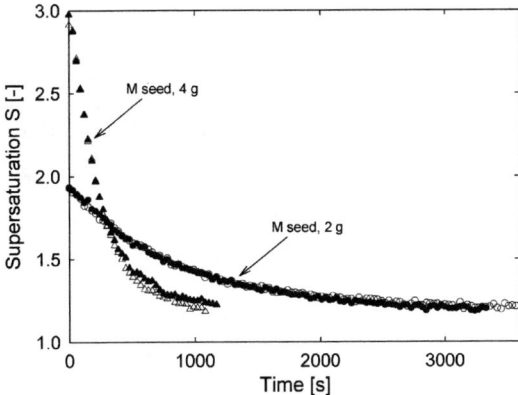

Fig. 2 Repeatability of two sets of desupersaturation experiments presented in this work. ●, ○, ▲, and △ represent run 5, 6, 11, and 12 of the set of experiments given in Table 1, respectively. The different initial conditions are highlighted in the graph.

process using the FBRM and the SEM microphotographs of the final particles were analyzed. A typical time evolution of the CLD of the particle population during a growth experiment is shown in Fig. 3. It can be observed that the number of particle counts in the small size range is virtually constant during the experiment, thus indicating that no nucleation occurred during the process. Similar behavior was observed in all experiments carried out in this work. Breakage could be avoided by using a glass impeller with rounded stirrer blade tips. The occurrence of agglomeration could not be prevented, but it will be shown in section 4.3 that agglomeration does not affect the results in this study. Furthermore, it is assumed that the speciation of L-glutamic acid proceeds instantaneously upon mixing of monosodium glutamate and hydrochloric acid solutions. The mixing time in this reactor is in the range of a few seconds, as monitored using ATR-FTIR.[6] Thus, mixing and reaction are much faster than growth and do not influence the course of the growth process.

3.2 Preparation and characterization of seed crystals

The main objective of seed preparation was to obtain strain free, undamaged, and non agglomerated crystals of the metastable α form of L-glutamic acid in three different size fractions. Consequently, the seeds were produced by precipitation at

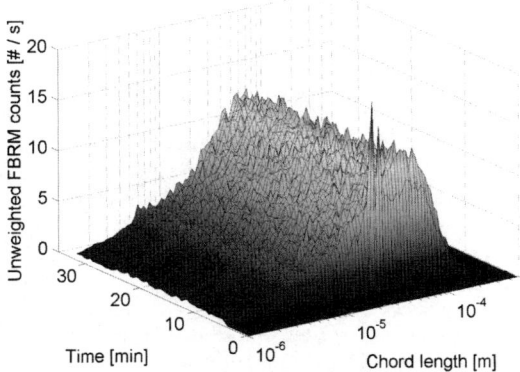

Fig. 3 FBRM data of run 9. It can be observed that no significant nucleation occurred over the course of the experiment since the counts of small chords remain at a constant low level.

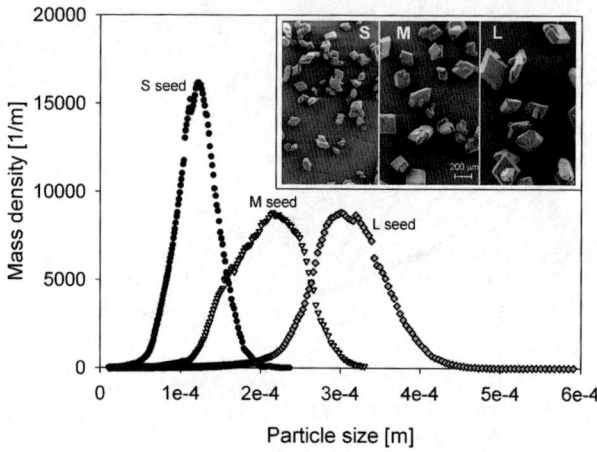

Fig. 4 Particle size distributions of the three seed fractions S, M, and L. The inset shows the corresponding scanning electron micrographs.

low initial supersaturation, and the different size fractions were obtained by wet-sieving the grown crystals. Monosodium glutamate and hydrochloric acid solutions of 0.5 mol L^{-1} concentration were prepared and purified by filtration with a 0.22 μm filter. Both solutions were mixed in a 2 L stirred batch reactor, where particle formation occurred. The precipitated crystals were wet-sieved using four sieves with nominal mesh sizes of 64, 125, 250, and 355 μm, respectively. Three seed fractions, labeled S, M, and L, were obtained by collecting crystals triple sieved in the size ranges of 64–125 μm (S), 125–250 μm (M), and 250–355 μm (L), respectively. Crystals coarser than 355 μm or finer than 64 μm were discarded. Finally, the crystals of all three fractions were washed, filtered, and dried. The α polymorphic form of the seeds was confirmed using X-ray powder diffraction, SEM, and Raman spectroscopy as described elsewhere.[6] The PSDs of the three seed fractions were characterized using a Coulter Multisizer 3 (see Fig. 4). Additionally, scanning electron micrographs were taken of the three seed fractions, which are shown as the inset in Fig. 4; it can be observed that all fractions consist mostly of single crystals and only a few agglomerates.

3.3 Experimental results

A series of experiments was conducted at 25, 35 and 45 °C using different seed fractions, seed masses, and initial supersaturation values. The operating conditions of all experiments are listed in Table 1. The mass of the solution was 400 g in each experiment.

The influence of changing either seed fraction, seed mass or initial supersaturation on the measured desupersaturation curves is illustrated for 25 °C in Fig. 5–7, respectively. In these figures the experimental data are shown together with simulation results which will be discussed in section 4. Experimental data are represented by symbols, whereas simulation results are plotted as lines. It can be observed that an increase of the total surface area of the seed crystals, either by reducing seed size as shown in Fig. 5 or by increasing seed mass as shown in Fig. 6, leads as expected to a faster decrease in supersaturation. Yet, the experimental desupersaturation curves exhibit only small differences for a change in seed mass, whereas an increase in crystal surface by seeding with smaller crystals results in a faster decrease of supersaturation. This can be related to the difference in available seed surface, which is more significant in the experiments with different seed fractions. Moreover, it can be seen that during run 21, *i.e.*, the experiment with an initial supersaturation

Table 1 Experimental conditions of the desupersaturation experiments and corresponding mean residual values R_m (eqn (8)) of the two optimized growth correlations given in section 4.2. The fractions S, M, and L, correspond to small, medium and large seeds and the corresponding experimental PSDs and SEM microphotographs are shown in Fig. 4, respectively

Run	S_0	Temperature/°C	Seed fraction	Seed mass/g	R_m (B + S)	R_m (BCF)
1	1.94	25	S	2	4.0×10^{-2}	3.3×10^{-2}
2	1.94	25	S	2	4.6×10^{-2}	4.3×10^{-2}
3	1.94	25	M	2	2.4×10^{-2}	4.7×10^{-2}
4	1.94	25	M	2	1.8×10^{-2}	4.5×10^{-2}
5	1.94	25	L	2	2.1×10^{-2}	3.6×10^{-2}
6	1.94	25	L	2	1.9×10^{-2}	4.0×10^{-2}
7	1.94	25	M	3	1.1×10^{-2}	1.9×10^{-2}
8	1.94	25	M	3	3.7×10^{-2}	3.6×10^{-2}
9	1.94	25	M	4	3.2×10^{-2}	4.2×10^{-2}
10	1.94	25	M	4	2.6×10^{-2}	2.4×10^{-2}
11	2.95	25	M	4	3.0×10^{-2}	3.9×10^{-2}
12	2.95	25	M	4	2.4×10^{-2}	3.4×10^{-2}
13	1.96	35	M	2	2.5×10^{-2}	3.0×10^{-2}
14	1.96	35	M	2	3.3×10^{-2}	3.9×10^{-2}
15	2.98	35	M	4	6.0×10^{-2}	6.6×10^{-2}
16	2.98	35	M	4	3.7×10^{-2}	4.6×10^{-2}
17	1.62	45	M	2	1.3×10^{-2}	2.9×10^{-2}
18	1.62	45	M	2	1.4×10^{-2}	2.9×10^{-2}
19	2.46	45	M	4	3.5×10^{-2}	5.7×10^{-2}
20	2.46	45	M	4	5.6×10^{-2}	7.3×10^{-2}
21	3.86	25	M	4		

of $S_0 = 3.86$ illustrated in Fig. 7, significant clogging of the probe window occurred which induced corruption of the *in situ* liquid phase data.

The desupersaturation profiles for 35 and 45 °C are shown in Fig. 8 and 9, respectively. The comparison of the desupersaturation profiles for different temperatures as given by Fig. 7–9 shows a faster depletion of the supersaturation at elevated temperatures and hence a higher growth rate, as expected.

At the end of each experiment the mass of the grown crystals was determined after filtration, washing and drying. The yield of all experiments was within 92 and 97% of the calculated masses. The PSD of the grown particles was measured after drying

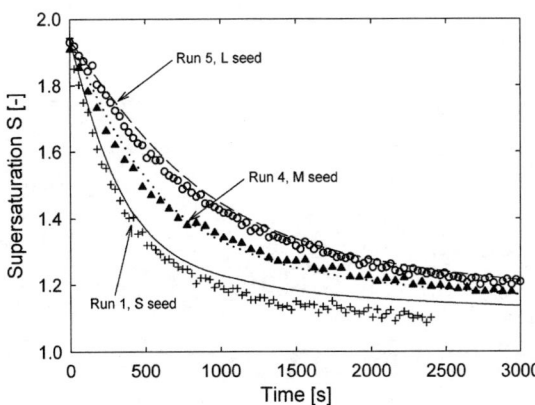

Fig. 5 Effect of seed size for two experiments at 25 °C. Symbols: experimental data; lines: simulation results.

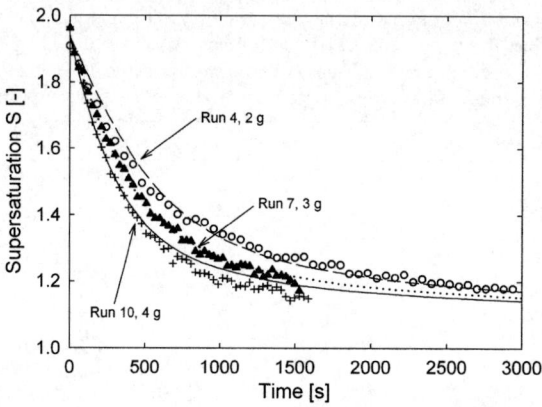

Fig. 6 Effect of seed mass for two experiments at 25 °C. Symbols: experimental data; lines: simulation results.

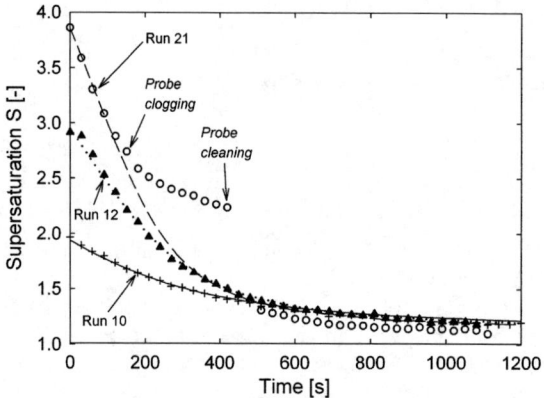

Fig. 7 Effect of initial supersaturation for two experiments at 25 °C. Symbols: experimental data; lines: simulation results.

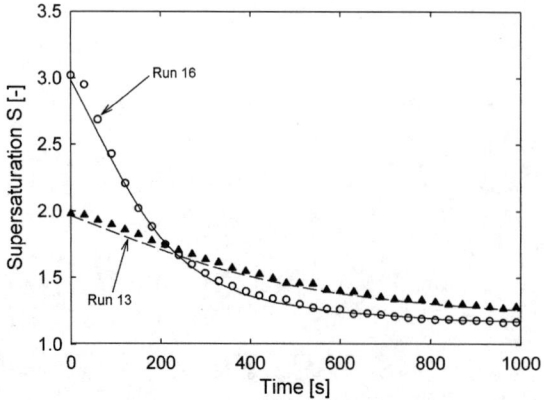

Fig. 8 Effect of initial supersaturation for two experiments at 35 °C. Symbols: experimental data; lines: simulation results.

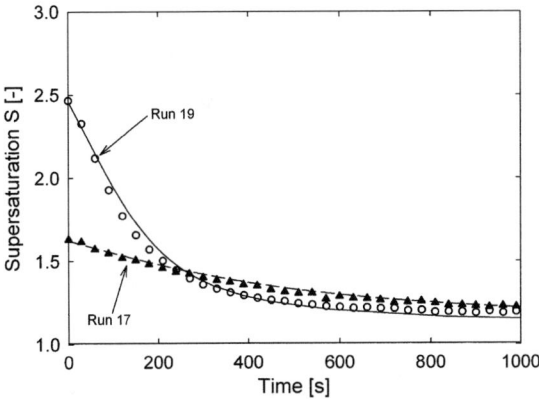

Fig. 9 Effect of initial supersaturation for two experiments at 45 °C. Symbols: experimental data; lines: simulation results.

using the Multisizer 3. The measured PSDs corresponding to the experiments shown in Fig. 5 are shown in Fig. 10. These PSDs will be used to verify the assumptions made and thus to validate the growth rate parameters estimated and discussed in the next section.

4. Growth kinetics of α L-glutamic acid

In the following the experimental results presented in the previous section are analyzed using a population balance equation (PBE) model to determine a growth rate correlation for α L-glutamic acid. Section 4.1 presents the model equations and the parameter estimation procedure. The chosen growth mechanism and the estimated parameters are presented in section 4.2. Then, in section 4.3 the quality of the model is discussed in terms of description capability of the experiments. Finally, the results reported in this study are discussed and compared with available data in the literature in section 4.4.

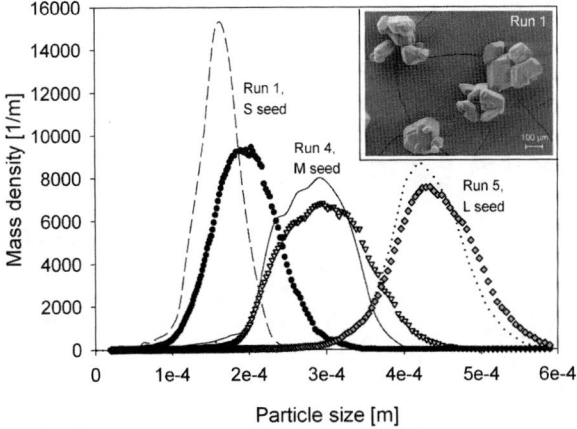

Fig. 10 Final particle size distributions of three experiments at 25 °C with different seed fractions S, M, and L. Symbols: experimental data; lines: simulation results. The inset shows a SEM picture of agglomerates in the final particles of run 1 consisting of several single particles.

4.1 PBE model and parameter estimation procedure

The model used in this work consists of a population balance equation and a material balance equation. The population balance equation for the particles can be written as follows:

$$\frac{\partial n}{\partial t} + G \frac{\partial n}{\partial L} = 0, \quad (3)$$

where L is a characteristic dimension of the crystal, t is the time, and $n(L, t)$ is the particle size distribution, i.e. $n(L, t)dL$ is the number concentration of crystals of size between L and $L + dL$.[20] It is worth noting that L is the diameter of a volume equivalent sphere. The crystal growth rate G is assumed to be size independent. This assumption could be validated by carefully designed experiments in this work, as discussed in section 4.3.

The material balance equation for the solute can be written as follows:

$$\frac{dc}{dt} = -\frac{3}{M} k_v \rho_c G \int_0^\infty n L^2 \, dL, \quad (4)$$

where c is the molar concentration of the solute, M is the molar mass of the solute, ρ_c and k_v are the solid density and the volume shape factor of the α crystals, respectively. In this work, we have used the values of $\rho_c = 1540$ kg m^{-3} for the solid density and $k_v = \pi/6$ for the volume shape factor. The following initial and boundary conditions apply for the two equations above:

$$c(0) = c_0 \quad (5)$$

$$n(L, 0) = n_0(L) \quad (6)$$

$$n(0, t) = 0, \quad (7)$$

with c_0 and $n_0(L)$ being the initial solute concentration and PSD, respectively. The growth rate G depends on supersaturation, i.e. on the solute concentration c, as discussed below. It is worth noting that neither nucleation nor agglomeration nor breakage are accounted for in the model, since experiments are run under conditions where these phenomena are absent or negligible.

The model equations above were solved using the discretization method proposed by Kumar and Ramkrishna.[21] This technique combines the discretization of the PBE with the method of characteristics and allows for the control of grid resolution and computational efficiency, while overcoming problems of numerical diffusion and instability.

The optimization procedure used in this work for the parameter estimation is based on a weighted non-linear least squares algorithm, which is described by

$$\min \sum_{m=1}^{N_e} R_m^2 = \min \sum_{m=1}^{N_e} \left[\frac{1}{t_m^{end}} \int_0^{t_m^{end}} [S_m^{exp}(t) - S_m^{mod}(t)]^2 \, dt \right], \quad (8)$$

where N_e is the number of experiments, R_m the mean residual, S_m^{exp} and S_m^{mod} are the experimental and calculated supersaturation values, respectively, and t_m^{end} is the duration, all referred to the m-th experiment. The optimization problem was solved using the *lsqnonlin* algorithm of the MATLAB optimization toolbox.

4.2 Growth mechanism and growth rate parameters

Different crystal growth mechanisms have been proposed in the literature to characterize the growth kinetics of α L-glutamic acid during cooling crystallization. Tai et al. compared an integration-controlled BCF screw dislocation mechanism with the empirical two-step model and found a better description of the experiments using the BCF correlation.[22] Kitamura compared two integration-controlled mechanisms, the BCF and the surface nucleation based birth and spread (B + S) mechanism, and found that the B + S correlation only yielded meaningful results.[1] Since the growth mechanism is not unambiguously reported in the literature, the optimization algorithm was used together with the desupersaturation data to estimate the parameters for both surface integration-controlled growth mechanisms, B + S and BCF. It is worth noting that run 21 could not be used for the parameter estimation due to the partly corrupted desupersaturation data. The growth parameters were estimated simultaneously for the three temperatures by using run 1 to run 20. The equations used to describe the B + S and BCF mechanism can be cast as follows:[23]

$$G_{B+S} = A\,T\exp\left(\frac{-B}{T}\right)(S-1)^{2/3}(\ln S)^{1/6}\exp\left(\frac{-C}{T^2\ln S}\right) \quad (9)$$

$$G_{BCF} = D\,T\exp\left(\frac{-E}{T}\right)(S-1)(\ln S)\tanh\left(\frac{F}{T\ln S}\right), \quad (10)$$

where A, B, C, D, E and F are parameters (three for each model) that are estimated from experiments. The optimization of the growth rate parameters yielded the following values for the B + S mechanism:

$$A = 3.63 \times 10^{-4}\,\text{m}\,\text{s}^{-1}\,\text{K}^{-1} \quad (11)$$

$$B = 3.72 \times 10^3\,\text{K} \quad (12)$$

$$C = 5.42 \times 10^4\,\text{K}^2 \quad (13)$$

and for the BCF correlation:

$$D = 1.57 \times 10^{-3}\,\text{m}\,\text{s}^{-1}\,\text{K}^{-1} \quad (14)$$

$$E = 4.34 \times 10^3\,\text{K} \quad (15)$$

$$E = 4.15 \times 10^2\,\text{K} \quad (16)$$

Changing the initial values of the parameters in the optimization procedure over several orders of magnitude always produced the same final results, thus indicating that the values in eqn (11)–(16) correspond to a global optimum.

In contrast to findings reported in the literature[1] the parameter estimations for both growth mechanisms yield physically meaningful results. The overall mean residuals describing the deviation between measured and simulated desupersaturation curves for the entire set of experiments, as calculated using eqn (8), are 6.0×10^{-1} and 8.1×10^{-1} for the B + S and BCF mechanisms, respectively. This indicates that the B + S growth mechanism describes the observed growth behavior of the experimental set more accurately, which is in agreement with AFM studies, where screw dislocation growth sites typical for the BCF mechanism could not be observed, whereas patterns characteristic for the B + S mechanism could be identified.[24] Consequently, the kinetics described by the B + S mechanism will be used in the following sections.

We have also checked whether the growth kinetics was influenced by diffusional limitations. To this aim, the mass transfer coefficient k_d was predicted using the Sherwood correlation:[23]

$$k_d = \frac{D}{L}\left(2 + 0.8\left(\frac{\bar{\varepsilon}L^4}{\nu^3}\right)^{1/5}Sc^{1/3}\right), \quad (17)$$

where D is the diffusivity, L the crystal size, $\bar{\varepsilon}$ the average power input, ν the kinematic viscosity and Sc the Schmidt number, i.e. $Sc = \nu/D$. With typical values for water of $D = 2 \times 10^{-9}$ m^2 s^{-1} and $\nu = 1 \times 10^{-6}$ m^2 s^{-1}, average particle size of $L = 2 \times 10^{-4}$ m, and characteristic average power input for our crystallizer of $\bar{\varepsilon} = 5.9 \times 10^{-2}$ W kg^{-1}, the mass transfer coefficient k_d from eqn (16) is 2.2×10^{-4} m s^{-1}. Diffusion controlled crystal growth rates can be calculated using the following equation:[23]

$$G = k_d \frac{k_a M}{3 k_v \rho_c}(c - c^*), \qquad (18)$$

where ρ_c is the crystal density, k_a and k_v denote the surface and volume shape factors, respectively. In this work we used a value of $k_a = \pi$ for the surface shape factor. The experimental growth rates were found to be at least one order of magnitude smaller than the diffusion limited growth rates computed with the mass transfer coefficient predicted by the Sherwood correlation. Therefore, it was concluded that under the conditions studied in this work the growth of α L-glutamic acid is controlled by the surface integration of new molecules in the crystal lattice.

4.3 Accuracy of the growth rate model

The accuracy of the model predictions and the validity of the initial assumptions can be evaluated by comparison of the computed model results with the experimental data, e.g., the desupersaturation curve and the final PSD. It is worth noting that the desupersaturation curve is time-resolved experimental information yielding several measurement information over the course of an experiment while the final PSD is an integral one. However, the desupersaturation data does not describe which particle formation mechanisms are responsible for the solute consumption over time. Consequently, besides accurate information about the crystal surface at the beginning of the experiment, one needs additional information about the mechanisms involved. This information can be obtained by comparison of simulated and experimental final PSD for each experiment. The final PSD contains valuable information about whether and to what extent undesired phenomena like nucleation or agglomeration occurred during the experiment.

First, let us consider the experimental and simulation results obtained at 25 °C by using the different seed fractions, shown in Fig. 4, but otherwise identical operating conditions. The desupersaturation curves and final PSDs corresponding to the different seed fractions are shown in Fig. 5 and 10, respectively. It can be observed in Fig. 5 that the computed desupersaturation curves are in good agreement with the experimental data, which is also reflected by the low values of the mean residuals given in Table 1. If we compare the final PSDs for the three experiments with different seed fractions shown in Fig. 10, it can be observed that all PSDs computed for run 1, 4, and 5 agree reasonably well with the experimental results. However, simulated PSDs are shifted towards smaller particle sizes compared to the experimental ones, and the deviation is larger for smaller seeds. Such behavior can be attributed to agglomeration of particles during growth, i.e. a phenomenon that is not accounted for by the model used in this work. The existence of agglomeration is supported by SEM micrographs of grown crystals at the end of run 1 as shown in the inset of Fig. 10.

A model describing the agglomeration of L-glutamic acid is presented in ref. 25. This model accounts for a size-dependent agglomeration kernel and for the shear rate distribution in the stirred tank. It is used here to show the influence of agglomeration on the final PSD and on the desupersaturation curve. In Fig. 11a the experimental and computed final PSDs of run 4 are shown. The experimental PSD (symbols) is compared with three simulated PSDs, namely the PSD accounting for sole growth and computed with the model presented in this work (solid line), the PSD calculated accounting for agglomeration as reported in ref. 25 (dashed line) and

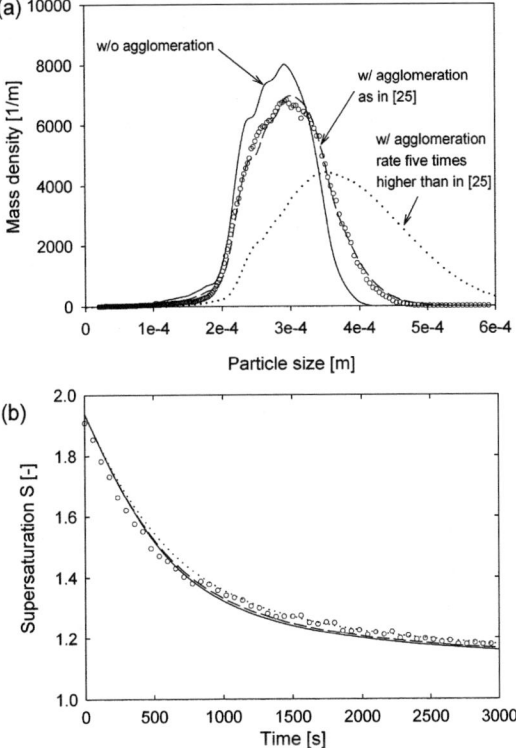

Fig. 11 (a) Effect of agglomeration on final particle size distributions (PSDs) of run 4. Symbols: experimental data; lines: simulation results. (b) Effect of agglomeration on desupersaturation curves of run 4. Circles: experimental desupersaturation curve; continuous line: computed desupersaturation curve accounting for sole growth; dashed line: computed desupersaturation curve for the agglomeration rate given in ref. 25; dotted line: computed desupersaturation curve for a five times larger agglomeration rate.

the PSD calculated assuming that the agglomeration rate is five times larger than that determined in ref. 25 (dotted line). It can be seen that the difference in the final PSDs between model and experiment is mainly caused by agglomeration. For the agglomeration rate given in ref. 25 the inclusion of agglomeration in the model improves the prediction of the final PSD quite significantly. When simulating the process with a five times higher agglomeration rate the final PSD is broadened and shifted towards larger particle sizes. Let us now consider the effect of agglomeration on the desupersaturation profile, which is shown in Fig. 11b. It can be seen that even at high agglomeration rates the desupersaturation curve is not significantly affected by agglomeration. Therefore, it is concluded that in this case the growth rate can be estimated from solute concentration data with significantly less error than from solid phase data.

When comparing the model prediction quality for the solid and liquid phase properties, *i.e.*, the supersaturation in Fig. 5 and the PSDs in Fig. 10, it can be clearly observed that supersaturation data can be predicted with higher accuracy than the PSDs. Consequently, the use of desupersaturation data as reported here represents a robust method for determining growth kinetics parameters. However, it is necessary to check the final PSDs to verify whether and to what extent the basic conditions of minimizing concurrent phenomena were respected during the experiments. Finally, in combination with the experiments using different seed fractions it can be concluded that size dependent crystal growth, a phenomena observed for other systems,[26] cannot be observed in the size range of the three seed fractions used.

The final PSD of run 1 is most affected by agglomeration compared to all other experiments. Therefore, the final PSDs of these experiments are not displayed and it is assumed that the influence of agglomeration on growth rate estimation can be neglected too. Furthermore, no indication of nucleation was given by the final PSDs since there was neither an increased number of particles in the small size range of the Multisizer 3 nor a bimodal PSD, as in the case of significant nucleation.

The experimental and modeling results obtained using different seed masses of fraction M of either 2, 3, or 4 g, are illustrated in Fig. 6. Similar to the desupersaturation curves with different seed fractions, the computed profiles for different seed masses are also in good agreement with the experimental values, although the differences between the three runs are rather small.

The computed and experimental desupersaturation curves obtained at different initial supersaturations are shown in Fig. 7. In the case of run 21, whose data were not used for the parameter optimization procedure, the corresponding model results can be considered as predictive. Similar to the experiments with varying crystal surface presented above, the model results for different initial supersaturation are in good agreement with the experimental desupersaturation curves. It is worth noting that this holds true also for run 21 for the data points where the experimental values are not corrupted by probe clogging. An interesting feature of the system can be observed by closely comparing the course of the different experiments shown in Fig. 7. It can be observed that the measured supersaturation values of run 21 drop below the values of both other experiments. This crossover is also predicted by the simulation results. The computed desupersaturation curves also exhibit a crossover between runs 10 and 12, which is too small to be verified experimentally. The reason for the crossing over of the desupersaturation curves is related to the significant change of available crystal surface over the course of the experiment. The initial crystal surface is the same in all three cases since the same seed fraction and seed mass were used. Yet, due to the higher crystal growth rates at higher supersaturation the available crystal surface area increases faster in runs 12 and 21 when compared to run 10. This results in a higher mass deposition rate and leads ultimately to the crossover behavior observed experimentally.

Unlike cooling crystallization, where the driving force of the process is induced by a change in temperature the supersaturation was created in this work by a chemical reaction. Thus, the growth kinetics of α L-glutamic acid were studied at constant temperature. The temperature dependence of the growth rate was determined by conducting the experiments at different temperatures, as can be seen in Table 1. Experimental and simulated desupersaturation profiles are shown in Fig. 8 and 9 for 35 and 45 °C, respectively. For both temperatures the experiments were carried out at two different initial supersaturations. It can be observed that also at elevated temperatures experimental and simulation results are in good agreement.

4.4 Comparison with literature data

In this section we aim at comparing the overall growth kinetics determined in this work with the overall growth rate of α L-glutamic acid measured at 15 °C[27] and with the growth rate of individual faces of α L-glutamic acid measured at 25 °C.[1]

To the best of our knowledge, the overall growth kinetics of α L-glutamic acid has been determined from batch experiments at 15 °C[27] and 45 °C only.[9] The growth rate at 15 °C is plotted together with that calculated using eqn (9) for the temperatures 15, 25, 35 and 45 °C as a function of supersaturation in Fig. 12. Although the application of eqn (9) to 15 °C is an extrapolation, the predicted growth rate is in excellent agreement with the empirical growth rate expression proposed by Tai and Shei.[27] The growth rate of α L-glutamic acid was determined at 45 °C based on unseeded batch experiments by Schöll et al.[9] The comparison of this growth rate with that being presented here shows some discrepancy. We believe this is due to the fact that in the previous work nucleation

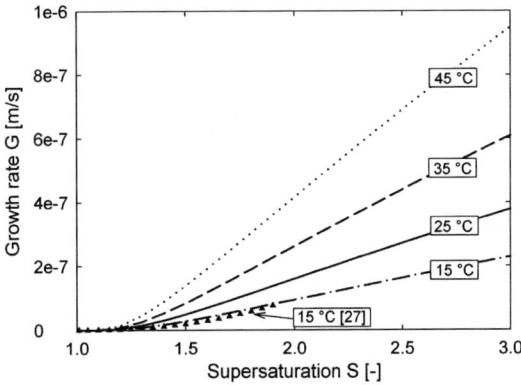

Fig. 12 Comparison of the growth kinetics of α L-glutamic acid determined in this work for different temperatures with kinetics found in the literature.[27]

and growth rates had been determined together.[9] Based on the new results we should conclude that nucleation and growth rate are highly correlated. Hence, the accuracy of each individual rate was lower than that of the overall model, which in that paper was used to describe the formation of metastable α crystals and their transformation to stable β crystals.[9] Therefore, the overall growth kinetics should be measured independently in order to improve the accuracy and prediction quality of a model, as it is done here.

To allow for the direct comparison of our correlation for the overall growth rate given by eqn (9) with measured growth rates of the single faces of α L-glutamic acid published in literature for 25 °C,[1] these face growth rates have to be used to calculate a corresponding overall growth rate of a volume equivalent sphere. This conversion is presented in the following. The volumetric change of a crystal by growth can be written as:

$$\frac{dV}{dt} = \sum_{i=1}^{N_{faces}} v_i A_i = \sum_{i=1}^{N_{faces}} s_i k_a L^2 v_i, \qquad (19)$$

where v_i is the growth rate, A_i the surface area and s_i the surface fraction of i-th face, L is a characteristic dimension of the particle and k_a the corresponding surface shape factor. The number of faces N_{faces} with distinctly different growth rates is 3 for α L-glutamic acid and these have the Miller indices (001), (111) and (011).[1] The volumetric change of a crystal can also be described by

$$\frac{dV}{dt} = 3k_v L^2 \frac{dL}{dt} = 3k_v L^2 G_L, \qquad (20)$$

where k_v is the volume shape factor of the particle and G_L is the growth rate of the characteristic dimension L. By combining eqn (19) and (20) the growth rate G_L can be expressed in terms of the growth rate of individual faces as well as of their surface fraction:

$$G_L = \frac{k_a}{3k_v} \sum_{i=1}^{3} s_i v_i. \qquad (21)$$

To allow for the comparison with the growth rate given by eqn (9), the growth rate of a volume equivalent sphere has to be determined from eqn (21) using

$$G = G_L \left(\frac{6k_v}{\pi}\right)^{1/3}, \qquad (22)$$

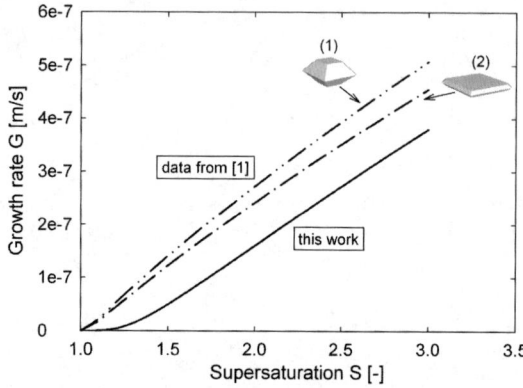

Fig. 13 Comparison of the growth kinetics of α L-glutamic acid determined in this work for 25 °C with kinetics calculated with eqn (21) and (22) from face growth rates from literature.[1] Shape 1: $s_1 = A_{(001)}/A_{crystal} = 0.14$; $s_2 = A_{(111)}/A_{crystal} = 0.80$; $s_3 = A_{(011)}/A_{crystal} = 0.06$. Shape 2: $s_1 = 0.67$; $s_2 = 0.31$; $s_3 = 0.02$.

with G being the growth rate of the volume equivalent sphere. Based on the growth rate correlations for the different faces of α L-glutamic acid published in the literature[1] and data for the surface fractions s_i of the corresponding faces determined from a geometric particle model the growth rate G can be determined using eqn (21) and (22). The calculated growth rates for two specific crystal shapes are shown in Fig. 13 together with the growth rate given by eqn (7) at 25 °C. All crystal geometries observed in the seeded batch experiments are somewhere within the range of the thick and compact crystals represented by shape (1) and the thin platelets represented by shape (2) in Fig. 13. Thus, the corresponding growth rate curves can be understood as the lower and upper bounds for the growth rates that different crystals would experience during the desupersaturation experiments. The overall growth rate determined in this work and given by eqn (9) is quite close to those obtained using growth rates of individual faces, although there is a deviation. Considering the quite different nature of the two measurement approaches, with one being a bulk crystallization involving a large number of particles and the other being essentially a single particle characterization method, the agreement can still be considered satisfactory especially for larger supersaturations.

5. Conclusions

A method for the fast and robust measurement of growth rate kinetics is presented in this work. The method is based on two *in situ* process analytical technologies, namely ATR-FTIR spectroscopy and FBRM to monitor the liquid and the solid phase during seeded batch desupersaturation experiments. Additionally, a Coulter particle analyzer employing the electric zone sensing method is used to characterize the initial PSD of the different seed fractions. The experimental requirements for an independent growth kinetics determination, *i.e.*, the absence of other mechanisms such as nucleation, agglomeration and breakage, are checked using the Coulter particle analyzer and scanning electron microscopy. It could be shown that the presented characterization method was also robust in cases of agglomeration. For the first time, the overall growth kinetics of α L-glutamic acid were determined independently at 25, 35 and 45 °C when the supersaturation is created by pH-shift. The growth mechanism considered was birth and spread in agreement with other authors. The comparison of the determined growth kinetics to correlations proposed in literature at different temperatures has shown satisfactory agreement.

Notation

A	Particle surface area [m^2]
c	Concentration [mol m^{-3}]
c^*	Solubility [mol m^{-3}]
D	Diffusivity [m^2 s^{-1}]
G	Growth rate [m s^{-1}]
k_a	Surface shape factor [-]
k_d	Mass transfer coefficient [m s^{-1}]
k_v	Volume shape factor [-]
K	Equilibrium constant [-]
L	Crystal size [m]
M	Molar mass [kg mol^{-1}]
n	Number density [# m^{-4}]
R_m	Mean residual [-]
s	Surface fraction [-]
$S = c/c^*$	Supersaturation
Sc	Schmidt number [-]
t	Time (s)
T	Temperature [K]
V	Particle volume [m^3]
$\bar{\varepsilon}$	Average power input [W kg^{-1}]
ν	Kinematic viscosity [m^2 s^{-1}]
ρ_c	Crystal density [kg m^{-3}]
v	Growth velocity of a face [m s^{-1}]

References

1 M. Kitamura, *J. Cryst. Growth*, 2000, **209**, 138–145.
2 J. Garside and A. Mersmann, *Measurement of Crystal Growth and Nucleation Rates*, Institution of Chemical Engineers, Rugby, UK, 2nd edn, 2002.
3 H. Grön, P. Mougin, A. Thomas, G. White, D. Wilkinson, R. B. Hammond, X. J. Lai and K. J. Roberts, *Ind. Eng. Chem. Res.*, 2003, **42**(20), 4888–4898.
4 A. A. Shaikh, A. D. Salman, S. Mcnamara, G. Littlewood, F. Ramsay and M. J. Hounslow, *Ind. Eng. Chem. Res.*, 2005, **44**(26), 9921–9930.
5 J. Worlitschek and M. Mazzotti, *Cryst. Growth Des.*, 2004, **4**(5), 891–903.
6 J. Schöll, L. Vicum, M. Müller and M. Mazzotti, *Chem. Eng. Technol.*, 2006, **29**(2), 257–264.
7 J. D. Bernal, *Z. Kristallogr.*, 1931, **78**, 363–369.
8 S. Hirokawa, *Acta Crystallogr.*, 1955, **8**(10), 637–641.
9 J. Schöll, D. Bonalumi, L. Vicum, M. Müller and M. Mazzotti, *Cryst. Growth Des.*, 2006, **6**(4), 881–891.
10 B. O'Sullivan and B. Glennon, *Org. Process Res. Dev.*, 2005, **9**(6), 884–889.
11 D. D. Dunuwila, L. B. Carroll and K. A. Berglund, *J. Cryst. Growth*, 1994, **137**(3–4), 561–568.
12 F. J. Bergin, L. Rintoul and H. F. Shurvell, *Can. J. Spectrosc.*, 1990, **35**(2), 39–44.
13 D. R. Lide, *CRC Handbook of Chemistry and Physics*, CRC Press LLC, Boca Raton, FL, 85th edn, 2004.
14 R. Schuhmann and R. H. Muller, *Pharm. Ind.*, 1998, **60**(2), 157–163.
15 R. L. Xu and O. A. Di Guida, *Powder Technol.*, 2003, **132**(2–3), 145–153.
16 A. Ruf, J. Worlitschek and M. Mazzotti, *Part. Part. Syst. Char.*, 2000, **17**(4), 167–179.
17 J. Worlitschek and M. Mazzotti, *Part. Part. Syst. Char.*, 2003, **20**(1), 12–17.
18 Y. F. Qiu and A. C. Rasmuson, *Chem. Eng. Sci.*, 1991, **46**(7), 1659–1667.
19 H. Glade, A. M. Ilyaskarov and J. Ulrich, *Chem. Eng. Technol.*, 2004, **27**(7), 736–740.
20 A. Randolph and M. A. Larson, *Theory of Particulate Processes*, Academic Press, San Diego, CA, 2nd edn, 1988.
21 S. Kumar and D. Ramkrishna, *Chem. Eng. Sci.*, 1997, **52**(24), 4659–4679.

22 C. Y. Tai, C. S. Cheng and Y. C. Huang, *J. Cryst. Growth*, 1992, **123**(1–2), 236–246.
23 *Crystallization Technology Handbook*, ed., A. Mersmann, Marcel Dekker Inc., New York, 2nd edn, 2001, vol. 2.
24 M. Kitamura and K. Onuma, *J. Colloid Interface Sci.*, 2000, **224**(2), 311–316.
25 C. Lindenberg, J. Schöll, L. Vicum, J. Brozio and M. Mazzotti, L-Glutamic acid precipitation: agglomeration effects, *Cryst. Growth Des.*, 2007, submitted.
26 J. Garside and S. J. Jancic, *Chem. Eng. Sci.*, 1978, **33**(12), 1623–1630.
27 C. Y. Tai and W. L. Shei, *Chem. Eng. Commun.*, 1993, **120**, 139–152.

PAPER | www.rsc.org/faraday_d | Faraday Discussions

Precursor structures in the crystallization/precipitation processes of CaCO₃ and control of particle formation by polyelectrolytes

J. Rieger,*[a] T. Frechen,[a] G. Cox,[a] W. Heckmann,[a] C. Schmidt[b] and J. Thieme[b]

Received 30th January 2007, Accepted 7th February 2007
First published as an Advance Article on the web 10th May 2007
DOI: 10.1039/b701450c

The formation of $CaCO_3$ is usually discussed within the classical picture of crystallization, *i.e.* assuming that the formation of $CaCO_3$ crystals proceeds *via* nucleation and growth. This may be true for the case of low supersaturation. In this work it is shown that the formation process is far more complex at high supersaturation, *i.e.* during precipitation. New insight into the mechanisms of precipitation is obtained by analyzing structure formation with a time resolution down to the millisecond range from the initiation of the reaction. The techniques used are scanning electron microscopy, electron diffraction, X-ray microscopy and cryo-transmission electron microscopy combined with a special quenching technique. It is seen that upon mixing $CaCl_2$ and Na_2CO_3 solutions (0.01 M) first an emulsion-like structure forms. This structure decomposes to $CaCO_3$-nanoparticles. These nanoparticles aggregate to form vaterite spheres of some micrometers in diameter. The spheres transform *via* dissolution and recrystallization to calcite rhombohedra. Once a suitable amount of additive, in our case polycarboxylic acid, is present during the precipitation the nanoparticles are stabilized against compact aggregation; instead they form flocs. This stabilization is either of a temporary nature if the amount of polymer is insufficient to cover the surface of the nanoparticles formed or more long lived if there is enough polymeric material present. By means of Ca-activity measurements it can be shown that the polymers are partially incorporated into the forming crystals.

1. Introduction

Precipitation/crystallization reactions play an important role in nature, in the household and in many industrial processes. Crystallization in nature may be summarized under the label biomineralization. This is meant to describe processes where inorganic reactions occur in the presence of organic matter, *e.g.*, proteins, resulting in ordered structures with interesting properties.[1–3] Examples are the formation of seashells, bone, teeth, *etc*. In the household unwanted precipitation annoys on the one hand in the form of undesired scaling in water pipes and on heating devices in kettles and washing machines and on the other hand as

---

[a] *BASF Aktiengesellschaft, Polymer Physics, 67056 Ludwigshafen, Germany. E-mail: jens.rieger@basf.com; Fax: +49 621 60 66 73731; Tel: +49 621 60 73731*
[b] *University of Göttingen, Institute of X-Ray Physics, Friedrich-Hund-Platz 1, 37077 Göttingen, Germany*

incrustation on fabric during the washing process.[4,5] The incrustation (scaling) in sea water desalination plants is caused by the deposition (precipitation and crystal growth) of inorganic salts like $CaCO_3$, $CaSO_4$, $Mg(OH)_2$, *etc.* which become insoluble at elevated temperatures. Scaling can be delayed by the introduction of polymers—as is evidenced by the successful use of polycarboxylates in sea water desalination plants.[6] Finally, precipitation is used in industry to produce fillers, pigments, and catalysts. Industrial interest stems from the fact that it might be possible to precipitate, *e.g.*, $CaCO_3$ in the presence of suitable additives in order to obtain $CaCO_3$ filler particles of well-defined crystal habit and modification or to develop new materials by using suitable additives.[7]

Despite the importance of precipitation and an upsurge of activities in the last decade it is astonishing to note how little is still known about the time evolution of precipitation reactions. Textbooks usually describe precipitation reactions according to the following scheme:[8] from the supersaturated solution nuclei are formed which subsequently are assumed to grow according to the classical picture of crystal growth until crystals of mesoscopic size (in the range of micrometers) have been formed. This linear approach has greatly influenced the way in which precipitation reactions have been investigated experimentally and theoretically: it is often assumed that it is possible to obtain sufficient information about the precipitation reaction by looking at the crystal structure, size and shape of the final products. This approach has been popular because of the ease with which the resulting crystals are accessible to investigation by scanning electron microscopy (SEM), wide angle X-ray scattering (WAXS) or light microscopy. A thorough discussion of the classical approach of nucleation and growth—and its shortcomings—is found in recent reviews.[9,10]

There have been some attempts to clarify the complexity of structure formation during precipitation of $CaCO_3$. It was realized early by Ogino *et al.* that before the actual formation of the thermodynamically stable calcite crystals, aggregation of smaller precursor particles takes place.[11] The present authors observed these precursor nanoparticles by initial time-resolved studies of the formation of $CaCO_3$ during precipitation.[4] The subsequent formation of vaterite spheres and calcite rhombohedra was also described. The growth of the mentioned nanoparticles was measured by Bolze *et al.* using *in situ* small angle X-ray scattering experiments.[12] Analysis of the scattering intensities revealed that the nanoparticles were monodisperse, amorphous and strongly hydrated. Using transmission X-ray microscopy the present authors observed *in situ* the process of complete dissolution of these intermediate amorphous particles.[13] It was postulated that the ions released recrystallized in the form of calcite. For the case of boehmite AlO(OH) and the organic pigment chinacridone it was shown by means of cryo-transmission electron microscopy (cryo-TEM) that upon precipitation initially an unstructured hydrated, amorphous phase forms which restructures, densifies and recrystallizes to build the final crystalline particles.[14] Faatz *et al.* put forward a liquid–liquid (spinodal) phase separation mechanism in order to explain the occurrence of similar nanospheres.[15] On the other hand, Navrotsky argued that the initial formation of amorphous nanoparticles could be explained by the classical nucleation and growth approach;[16] assuming that the surface energy becomes smaller as the phase becomes more metastable (*i.e.* amorphous as compared to crystalline) a lower barrier to nucleation is expected favouring the formation of the metastable phase on kinetic grounds.

When discussing the influence of organic (surfactants and polymers) or inorganic (cations) additives the classical approach of considering the adsorption of these molecules to crystal growth surfaces still prevails, where the additives are assumed to affect growth rate and crystal habit. But again there is increasing evidence that more subtle effects are active. Naka *et al.* observed that vaterite, which is a less stable polymorph of $CaCO_3$, is stabilized by the addition of polyacrylic acid either in a spherulitic or nanoparticulate form, depending on the moment when the polymer is interfering with the crystallization.[17] Wang *et al.* proposed a mechanism based on the assumption of mesocrystals in order to explain peculiar shapes of $CaCO_3$ crystals crystallized in the presence of polystyrene sulfonate.[18] The polymer affected the process by complexing ions, by nucleating, by stabilizing intermediate amorphous particles and by controlling the assembly of so-called mesocrystals. Similarly, Zhang et al. reported the formation of an amorphous liquid-like $CaCO_3$-polypeptide precursors, the crystallization and stabilization of polypeptide-capped nanoparticles and their final aggregation to micrometer-sized spheres.[19] The idea of a liquid-like amorphous precursor goes back to L. Gower who was the first to postulate a polymer-induced liquid precursor phase when initiating the crystallization of $CaCO_3$ in the presence of a polypeptide (polyaspartic acid).[20] This concept was recently exploited to prepare nacre-like thin films.[21] Furthermore, an idea was put forward to rationalize how polyacrylic acid prolongs the lifetime of amorphous $CaCO_3$ by sequestering locally Ca^{++}-ions.[22] At this stage it is interesting to note that meanwhile several structurally different phases of amorphous $CaCO_3$ have been identified in biogenic phases.[23]

Thus evidence for the occurrence of precursors, recrystallization and aggregation processes in the course of $CaCO_3$-precipitation has been given. And for a number of different systems there is strong evidence that precursor formation[14] and aggregation[10,24,25] is an important process in particle growth. But, still the view of particle formation due to nucleation and growth dominates the discussions in the field of precipitation reactions.

In this contribution we show that the appearance of amorphous $CaCO_3$ spheres is preceded by even another precursor stage where a spinodal like phase separation between a denser and a less dense phase occurs. Only by restructuring of this short-lived structure do the $CaCO_3$ spheres form. These spheres aggregate to micrometer sized vaterite spheres, which in turn recrystallize to calcite. Furthermore evidence is given that polycarboxylate molecules present during the precipitation stabilize the $CaCO_3$ nanoparticles. This stabilization is either of a temporary nature if the amount of polymer is insufficient to cover the surfaces of the nanoparticles formed or more long lived if there is enough polymeric material present. The term precipitation is used in line with the description given by Mullin denoting fast crystallization "initiated at high supersaturation, resulting in fast nucleation and the consequent creation of large numbers of very small primary crystals".[8] Supersaturation values are not discussed in the following since the solubilities of the precursor structures of $CaCO_3$ are not known and it is the supersaturation with respect to these structures that is responsible for the resulting particle formation processes.

2. Experimental

The precipitation of $CaCO_3$ was started by mixing solutions of $CaCl_2$ and Na_2CO_3. The experiments shown here were performed with 0.01 M solutions. The salts were obtained from Merck and are analytical grade. The water was deionized and was not degassed prior to the experiments. As an additive we used a copolymer made from 70 wt% acrylic acid and 30 wt% maleic acid (M_w = 70 000 g Mol^{-1}) that was added to the Na_2CO_3 solution prior to the precipitation. This polymer is the same as the one used in earlier studies[4,13] and was chosen because of its relevance as a scale inhibitor. The pH of the system was adjusted by adding NaOH such that the precipitation occurred at pH = 11. The mixing was performed by pumping both solutions through a mixing cell at a rate of 4 kg h^{-1} using pumps showing no oscillations in the pumping rate. Different mixing cells were used, as, e.g., T-mixers and radial mixers of different dimensions. We observed no influence of the type of mixing cell on the course of the precipitation. Thus, it may be safe to assume that the mixing is performed on a sufficiently short time scale resulting in a sufficient homogeneity of the mixed solutions. In some cases, the mixing was performed by pouring one solution into the other one in a beaker under stirring. The course of the reaction was monitored by Ca^{++}-sensitive electrodes (ZABS, Marburg, Germany) in order to ensure, besides other aspects, reproducibility.

For the analysis of the state of the precipitation at short times a shock-freezing cryo-transmission electron microscopic technique was used.[14] For each sampling a TEM grid was mounted on a hammer-type device ("guillotine"), constructed by us, which was driven by a pneumatic cylinder under photoelectric barrier control and was slowed down after entering the cryogen. The TEM grid was moved through the free jet at a speed of about 15 m s^{-1}, during which it acquired a suspension specimen which became shock-frozen within about 3 ms. For safety reasons, supercooled nitrogen (−210 °C) was used instead of liquid ethane as the cryogen for this method. The copper grids with the shock-frozen specimens were transferred to a Zeiss EM 910 transmission electron microscope in a cooled holder (CT3500, Oxford Instruments). In most cases, the liquid film frozen onto the grid was too thick for transmission. Part of the ice matrix was therefore sublimed off under observation in the electron microscope by increasing the specimen temperature to approximately −100 °C, until structures could be identified, and then lowering it again to about −160 °C. The TEM imaging was carried out at an accelerating voltage of 100 kV. Using the information about the rates of flow mixing and quenching we estimate that we can freeze the precipitation reaction in its state at the earliest 100 ms after mixing. Information about the state of the reaction at later stages can be obtained by increasing the distance between the mixing device and the sampling position or by working without the free jet in the following way. In some cases the precipitating stream was filled into a 100 ml beaker. Samples were taken at deliberate time intervals by wetting a TEM copper grid which was then quenched in liquid nitrogen and processed as described above. In addition the crystal structure was characterized by electron beam diffraction in the TEM (Hitachi 8100). Or, material was collected on a suitable substrate, as, *e.g.*, metal or cotton, dried in air, sputtered with a thin gold layer and transferred to a scanning electron microscope (SEM, Zeiss DSM 960 and Hitachi S4000FE).

X-Ray microscopy, as used for these studies in the form of transmission X-ray microscopy (TXM), is an imaging method similar to light microscopy or transmission electron microscopy (TEM). Due to the much shorter wavelength X-ray microscopy provides an approximately tenfold higher resolution than light microscopy. The ability to study samples under ambient pressure with virtually no sample preparation allows aqueous systems to be studied. In addition, time-resolved microscopic studies can be carried out. For the studies described here a transmission X-ray microscope was used, which was developed and is operated by the Institute for X-ray Physics at the University of Göttingen.[26] It utilizes synchrotron radiation of the electron storage ring BESSY II in Berlin. In the wavelength range between the K-absorption edge of oxygen at $\lambda = 2.34$ nm and carbon at $\lambda = 4.38$ nm radiation is weakly absorbed by water compared to the absorption within other organic or inorganic substances. This gives rise to a natural contrast in images of aqueous colloidal systems, so no further preparation steps such as drying or staining are necessary. For the experiments described here the resolution is in the order of 40 nm. The samples are prepared by placing one drop of the suspension to be studied on a 0.15 μm thin support (polyimide foil). The drop is covered with a second foil of the same type and then adjusted to a thickness of about 10 μm by applying a slight downward pressure. From commencement of preparation to the first exposure takes about 3 min. The object section visible at any one moment is 15 μm in diameter. The image data are read from a CCD camera and stored digitally for further processing. Further experimental details are given in ref. 26.

Attempts have been made to characterize the crystal modification of the precipitate at the early stages of the reaction using wide angle X-ray scattering (WAXS). These efforts partly failed because it turned out that it is difficult to stop the transformation of the material without inducing artefacts. These artefacts may occur because filtering the precipitate through 0.2 μm-filters leads to erroneous results

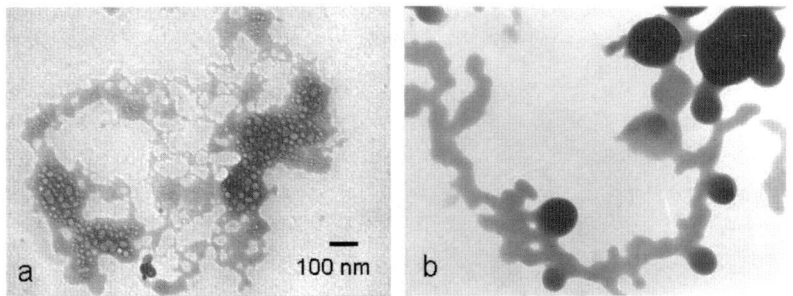

Fig. 1 Cryo-TEM micrographs of the state of the $CaCO_3$-precipitation about (a) 100 ms and (b) 2 min after mixing 0.01 M solutions of $CaCl_2$ and Na_2CO_3.

since $CaCO_3$-nanoparticles were drawn through the filter thus leaving only a non-representative fraction of the sample on the filter; filters with a smaller pore size were clogged.

3. Results and discussion

3.1 Precipitation of $CaCO_3$ without additive

Fig. 1 shows cryo-TEM micrographs of the precipitate in its state at about 100 ms and 2 min after mixing, respectively. Two types of structures can be discerned. The precipitate in Fig. 1a resembles an emulsion, though it must be stressed that no organic molecules were involved in the experiments discussed in this section. It is evident that no nuclei are to be seen, contrary to the predictions of classical nucleation and growth models. There are three tentative explanations for the occurrence of such a structure. Either the phase separation follows the spinodal route; the structure in Fig. 1a looks like being induced by spontaneous phase separation. This situation occurs according to the thermodynamics of phase transitions[27] when the homogeneous supersaturation is achieved fast enough to push the system into the spinodal region of the phase diagram without allowing for nucleation when passing through the binodal region. This route was also proposed by Faatz et al.,[15] though they did not observe the initial state of phase separation. The second explanation assumes that the state of homogeneous supersaturation is not reached but that the system starts to react at the interface of the two liquid educts once they intermingle while being mixed. The homogeneous appearance of some parts of the phase separated structure makes this explanation less plausible though some evidence for this approach was given in ref. 14. Thirdly, it must be asked whether this structure is an artefact. In ref. 14 evidence was given for another inorganic system (boehmite) that this explanation is less probable and that the method of cryo-TEM yields consistent results. The micrograph in Fig. 1b shows a later stage of the phase-separated structure. Here the structure resembles more closely that of aggregates of particles; but still a less dense and less structured phase connecting the particles is seen. We assume that the structure in Fig. 1b forms from structures like the one given in Fig. 1a by densification *via* expulsion of water. Fig. 2 is a cryo-TEM micrograph taken about 1 min after mixing exhibiting well-defined spherical particles that are supposed to form once the systems passed the stages depicted in Fig. 1. The result of an electron diffraction experiment on the spheres given in the inset indicates that the spheres are indeed amorphous. The seeming discrepancy with respect to the sampling times given, *i.e.* 2 min for Fig. 1b and 1 min for Fig. 2 are assumed to be due to uncertainties in the sample preparation procedure, as *e.g.* differences in (local) cooling rates when preparing the sample for cryo-TEM. The amorphous structures examined so far by TEM-techniques were also prepared for SEM. A sample holder was used to collect the precipitated material, then drawn

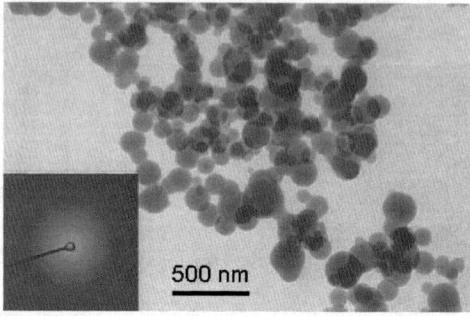

Fig. 2 Cryo-TEM micrograph of the state of the $CaCO_3$-precipitation 1 min after mixing 0.01 M solutions of $CaCl_2$ and Na_2CO_3. The inset shows the selected area electron diffraction pattern of one of the spheres.

from the reacting system after appropriate times and prepared for SEM as indicated above. Here the material was not prepared by inducing the precipitation in a mixing chamber but by pouring the educts together in a beaker while gently stirring. The result of such an experiment is shown in Fig. 3. Spherical particles looking akin to the objects in Fig. 1b and 2 are recognized. These particles seem to spread on the substrate, they are partially fused and in certain areas a film has been formed. Furthermore, comparably large globular objects are seen. They might be formed by the fusion of smaller spheres. The observed softness of the particles and structures is in seeming contradiction to data from Faatz *et al.* who measured the Young's modulus of amorphous hydrated $CaCO_3$ by Brillouin scattering and found a value of 37 Gpa,[28] which is about one order of magnitude larger than the modulus of common glassy polymers that do not flow. These data taken together with our observations is another hint at the existence of different types of amorphous $CaCO_3$. These structures might differ, *e.g.*, with respect to their water content.

Examination of the precipitation after some minutes yields results like the one shown in Fig. 4 where the familiar spheres and rhombohedra of the vaterite and calcite form of $CaCO_3$ are observed. Fig. 5 proves the identification of the respective phases *via* electron diffraction. A remarkable feature in Fig. 5 is the occurrence of nanoparticles around the edges of the vaterite spheres. This point deserves attention since it has been discussed for quite some time controversially whether the vaterite spheres are built from $CaCO_3$-particles as identified above or whether the spheres are built *via* spherulitic growth. Recently, Andreassen argued in favour of the latter assumption.[29] But it must be noted that Andreassen performed his precipitation runs at much higher concentrations leading to a gelatinous precursor which served as a

Fig. 3 SEM micrograph of the state of the $CaCO_3$-precipitation 1 min after gently mixing 0.01 M solutions of $CaCl_2$ and Na_2CO_3.

Fig. 4 SEM micrograph of CaCO$_3$-particles formed 5 min after the start of the precipitation without any additives.

matrix in which the vaterite spheres grew. Here we argue that initially amorphous CaCO$_3$-nanoparticles form which then agglomerate to form vaterite spheres. So far it is not clear whether the particles crystallize in the dispersed state or whether solid-state crystallization takes place upon aggregation. Further evidence for the aggregation hypothesis is given in Fig. 6 where the SEM-micrograph of the precipitation product is shown that was obtained by stopping the reaction after 60 min in acetone. Clearly, the aggregate structure of the vaterite spheres and of fragments thereof is seen. Given the evidence for both growth modes it might be assumed that both spherulitic growth and aggregation may occur depending on the reaction conditions such as, *e.g.*, supersaturation. The vaterite spheres in turn dissolve providing the ions for the crystallization of calcite from solution. As is seen in Fig. 6 the calcite crystals grow in the immediate vicinity of the vaterite spheres.

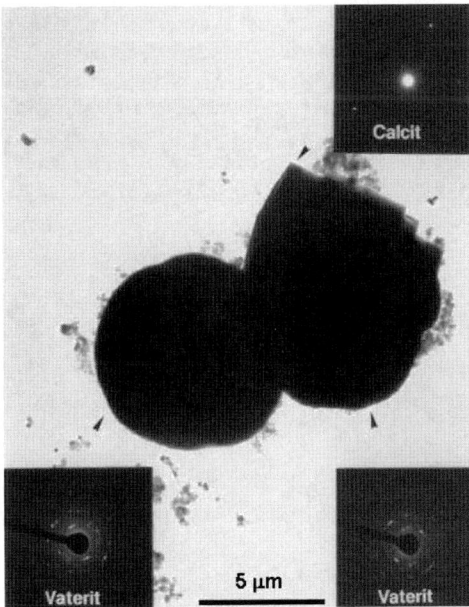

Fig. 5 TEM micrograph of CaCO$_3$-particles formed 10 min after the start of the precipitation. Insets: electron diffraction patterns at selected areas marked by arrows in the TEM-micrograph.

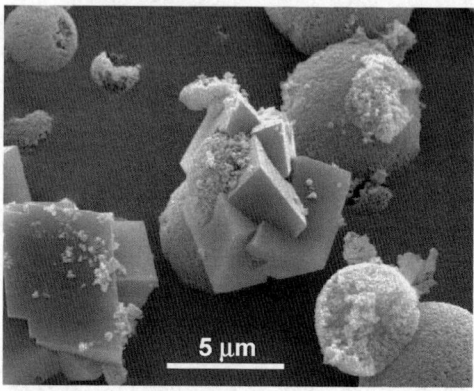

Fig. 6 SEM micrograph of CaCO$_3$-particles formed 60 min after start of the precipitation (0.05 M solutions of CaCl$_2$ and Na$_2$CO$_3$ were used). The sample was prepared by stopping the reaction in acetone.

3.2 Precipitation of CaCO$_3$ in the presence of polycarboxylate

Adding polycarboxylate to the Na$_2$CO$_3$ solution prior to mixing initially leads to a phase-separated structure (Fig. 7) that is similar to the one observed in the case without polymer (Fig. 1). In the polymer case, the boundaries of the phase-separated regions seem to be slightly better defined and the connectivity seems to be higher. From this phase-separated structure again nanoparticles form which are stabilized against compact aggregation, as is deduced from Fig. 8 where a cryo-TEM micrograph of the state of the precipitation after 5 min is shown; the reaction was performed in the presence of 600 ppm polycarboxylate. The primary particles are covered by a polymer coating which interconnects the particles to build a floc structure.[13] This floc structure extends to a size of up to 100 μm as was measured by Fraunhofer diffraction (data not shown here). Accordingly, the precipitating system appears turbid to the naked eye. The amount of additive present during the precipitation plays a decisive role for the further evolution of the precipitate. If the amount of polymer is insufficient to cover the surfaces of the nanoparticles efficiently, dissolution of the primary particles occurs resulting in recrystallization of CaCO$_3$ to crystals of mesoscopic size. This effect is illustrated in Fig. 9 where a time sequence taken with TXM during a time interval of 4 min with exposure times of a few seconds is reproduced. First, the CaCO$_3$ nanoparticles embedded in a network

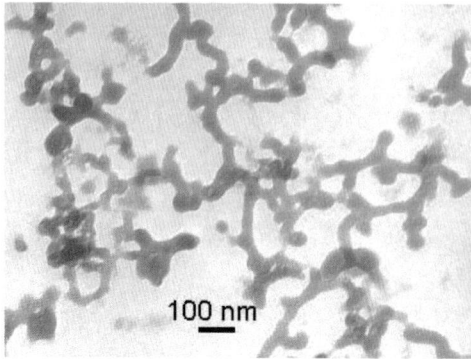

Fig. 7 Cryo-TEM micrograph of the state of the CaCO$_3$-precipitation about 100 ms after mixing 0.01 M solutions of CaCl$_2$ and Na$_2$CO$_3$ where the Na$_2$CO$_3$-solution contains 100 ppm of the polycarboxylate.

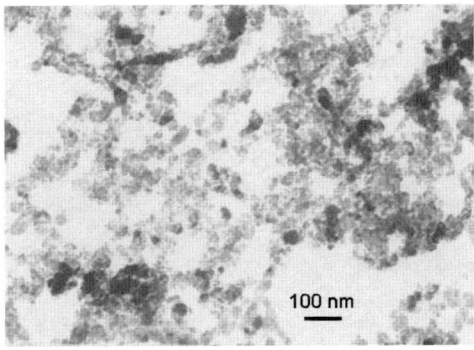

Fig. 8 Cryo-TEM micrograph of the state of the $CaCO_3$-precipitation about 5 min after mixing 0.01 M solutions of $CaCl_2$ and Na_2CO_3 where the Na_2CO_3-solution contains 1200 ppm of the polycarboxylate.

of polymers are seen to be in agreement with the TEM-image in Fig. 8. Such TXM-data have been described in detail previously.[13] The second micrograph in Fig. 9 demonstrates how the particles start to dissolve and that objects with dimensions of about 1 μm form. Given the regular shape of these objects, which becomes even clearer in the third picture, it is assumed that the Ca- and CO_3-ions from the dissolved nanospheres recrystallize to $CaCO_3$. Presently, we cannot state which polymorph forms. Yet, it is safe to assume that it is not calcite because these crystals dissolve again as is evident from the fourth and fifth micrographs. Since there must be a sink for the ions released we conclude that crystals that are more stable than the ones seen in pictures 2–4 of Fig. 9 are forming during the process outside the area observed with the TXM. The polymers are no longer seen on the micrographs although they must still be present. Since the polycarboxylates are able to dissolve $CaCO_3$ due to their higher binding strength they will be fully complexed with Ca^{++}-ions and precipitate.[30] That they are not recognized as a separate phase in the micrographs in Fig. 9 might be due to the adhesion of single molecules to the substrate thus preventing aggregation of the complexed polymers. The influence of radiation damage can be ruled out since the phenomena described so far were verified by checking other previously not illuminated sample spots. The timescale of the processes observed by TXM are usually somewhat retarded due to the almost two-dimensional character of the system with a thickness of approximately 10 μm.

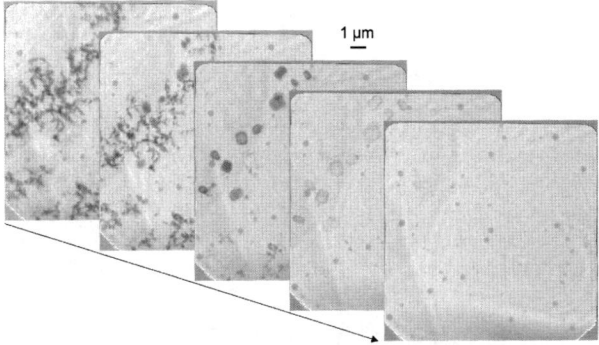

Fig. 9 Direct *in situ* imaging of the state of the $CaCO_3$-precipitation as aqueous suspensions by means of TXM. The first micrograph was taken about 10 min after mixing 0.01 M solutions of $CaCl_2$ and Na_2CO_3 where the Na_2CO_3-solution contains 400 ppm of the polycarboxylate. The five micrographs cover a time interval of 4 min.

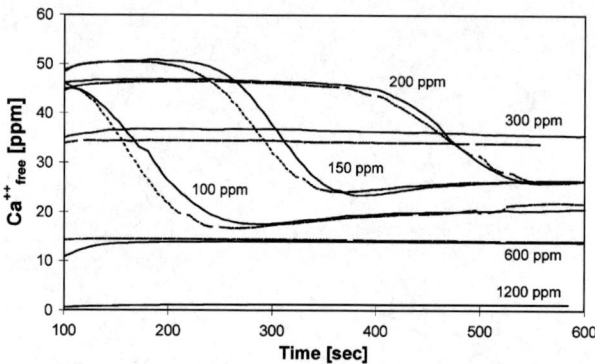

Fig. 10 Time evolution of the activity of free Ca^{++}-ions after mixing 0.01 M solutions of $CaCl_2$ and Na_2CO_3 where the Na_2CO_3-solution contained the indicated amounts of polycarboxylate. The dashed lines are reproductions. The data for a polymer concentration of 50 ppm are not shown, they superimpose with the data for 100 ppm.

More information about the role of the polymer during the formation of the solid phase can be gained by means of Ca^{++}-activity measurements performed with ion-selective electrodes. The calibration of the Ca^{++}-sensitive electrodes was performed such that in a first order approach the activity can be given in terms of the concentration of free Ca^{++}-ions, i.e. Ca^{++}-ions that are neither bound to polymers nor involved in $CaCO_3$ particles. Fig. 10 shows the evolution of the Ca^{++}-activity as a function of time for different polymer concentrations where the polymer is again the acrylic acid/maleic acid polycarboxylate used above. For polymer concentrations c_P lower than 300 ppm a step is observed whereas for concentrations equal to or higher than 300 ppm, a steady behaviour occurs. The initial drop of the Ca^{++}-activity from 0.005 M/200 ppm to the respective values seen is not shown because the response time of the electrodes does not allow for a reliable measurement in this window. It is noteworthy that the activity reaches a constant level after this initial drop before decreasing again for lower polymer concentrations. In one case (c_P = 200 ppm) the amount of polymer in the serum was determined by first separating the precipitated phase by centrifugation and then determining the polymer concentration in the residual liquid phase by means of a TOC (total organic content) determination. A value of c_P = 9 ppm was obtained, i.e. 5% of the polymer are not bound to the precipitated $CaCO_3$ particles but are dispersed in the water phase. This amount of polymer can bind approximately 2 ppm Ca^{++}-ions (data not shown), which is negligible compared to the amount discussed in the following. Recalling the evidence given above it is clear that the steps in Fig. 10 indicate the dissolution of the $CaCO_3$ nanoparticles that were temporarily stabilized by adsorbed polymers. Since these stabilized nanoparticles are in a thermodynamically metastable state the Ca^{++}-activity remains constant until dissolution. The absence of a step in the Ca^{++}-activity for polymer concentrations equal to or higher than 300 ppm indicates that the $CaCO_3$-nanoparticles were stabilized by a sufficient amount of polymer. The data in Fig. 10 furthermore show that at polymer concentrations below c_P the time interval to dissolution of the temporarily stabilized particles increases with polymer concentration indicating that a higher polymer concentration allows for a more efficient surface coverage and thus for a better temporary stabilization of the $CaCO_3$-particles.

Further insight is gained when plotting the values of the Ca^{++}-activity plateaux given in Fig. 10 as a function of the polymer concentration, cf. Fig. 11. From this figure the following points can be deduced: first, a critical amount of polymer c_P is necessary to stabilize the $CaCO_3$-nanoparticles against dissolution. This implies that the polymer forms a shell around the particles. Second, $CaCO_3$ recrystallized in the

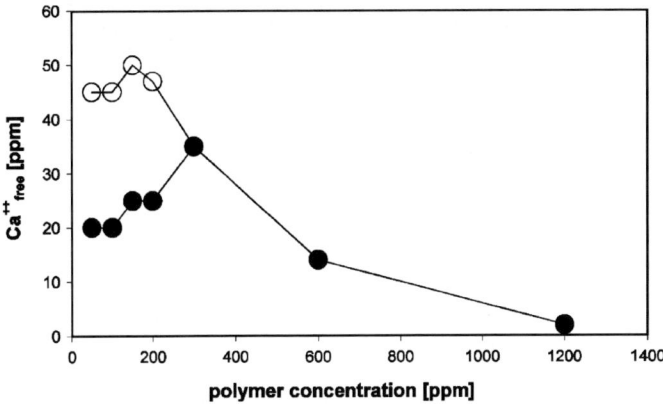

Fig. 11 Activity of free Ca^{++} ions after mixing 0.01 M solutions of $CaCl_2$ and Na_2CO_3 where the Na_2CO_3-solution contained the indicated amounts of polycarboxylate. The split data on the left side are values before and after the step observed in Fig. 10.

presence of polycarboxylate still exhibits a higher solubility than pure calcite, above and below the critical polymer concentration. Interestingly, the values for Ca^{++}-activity measured after the step for polymer concentrations below the critical threshold increase with increasing polymer concentration up to the value where the step vanishes. Third, when comparing the slope of the data in Fig. 11 with the slope of the curve corresponding to the Ca^{++}-binding capacity of the polycarboxylate used (0.2 ppm Ca^{++} per 1 ppm polymer) it becomes evident that a larger amount of polymer present in the precipitation does not only lead to complexation of free Ca^{++}-ions but that a significant fraction of polymer must be incorporated into the $CaCO_3$-particles. When increasing the amount of polymer present during the precipitation the data in Fig. 11 indicate that 2/3 of the polymers are consumed by the forming particles and only 1/3 of the additional polymers are complexing Ca^{++}-ions. The increase in solubility with increasing polymer concentration seen below the threshold value of c_P supports the hypothesis that polymer is consumed by the $CaCO_3$ particles which is in this case the dominating effect.

At the present stage of understanding it would of course be interesting to know more about the molecular details of how the polymers interfere with the phase separation processes, with particle formation and with stabilization of the particles described above. Preliminary results from time-resolved FTIR-studies and molecular modelling examining the interaction of polycarboxylates with water and Ca^{++}-molecules indicate that even on this level of simplification where the CO_3^--ions are omitted as a reaction partner, an unexpected complexity of phenomena with regards to the interaction between the cations and the functional carboxyl groups of the polycarboxylates is unravelled.[31,32] It became evident that the local chemistry of the polymers used plays a far greater role than expected. It is obviously not sufficient to consider the polycarboxylate molecules as simple polyelectrolytes with a certain charge density.

4. Conclusions and outlook

In this contribution we showed that the appearance of amorphous $CaCO_3$ nanoparticles is preceded by even another precursor stage where a spinodal like phase separation between a denser and a less dense phase occurs. Only by restructuring of this short-lived structure do the $CaCO_3$ spheres form. These spheres aggregate to micrometer sized vaterite spheres, which in turn recrystallize to calcite. Furthermore evidence is given that polycarboxylate molecules present during the precipitation stabilize the $CaCO_3$ nanoparticles. This stabilization is either of a temporary nature if the amount of polymer is insufficient to cover the surface of the nanoparticles formed efficiently or more long lived if there is enough polymeric material present.

Despite the insight that has been gained into the nano- and mesoscopic details of the morphological time evolution during precipitation many questions still remain unanswered: Does the model of spinodal decomposition apply to the early stages of precipitation? How is the phase separation leading to hydrated $CaCO_3$ affected by the mixing conditions? What is the exact structure of the gelatinous phase that forms upon mixing concentrated solutions of $CaCl_2$ and Na_2CO_3? How many different forms of amorphous $CaCO_3$ occur—even in the absence of any additive? Is there a continuum of forms parameterized, *e.g.*, by the water content? Are there indeed two paths for the formation of vaterite, namely spherulitic growth and aggregation? If the latter is the case when does the crystal structure of vaterite form, in the dispersed $CaCO_3$-particles or during aggregation? What is the nucleating/inhibiting mechanism of additives? How do they interfere with the particle formation (nucleation) on a molecular level? What role does the local chemistry of the added polymer play, *e.g.* stereochemistry, as opposed to physical parameters such as charge density?

Finally, we note that it is of great importance to understand the initial stages of precipitation reactions because it is at this stage where additives such as incrustation inhibitors or additives for influencing the further course of the precipitation are active. This point has been mostly disregarded in the past for two reasons; firstly, the scientific approach in the domain of precipitation is still dominated by pictures and models taken from the field of crystallization thus neglecting special transitional states. Secondly, it is only now possible to investigate the whole time–space field of the precipitation. As is evident above it is necessary to cover the time domain from milliseconds to hours from the start of the precipitation. In parallel, the resulting structure must be characterized starting from the crystal structure up to the micrometer level. Of course, as concerns experimental methods, the spatial aspect is less difficult to cover than the time resolution. It must be emphasized that it is by no means sufficient to perform precipitations and to discuss the final products when the aim is to understand the temporal evolution of the reaction.

Acknowledgements

We thank Ms Heidrun Debus for her tedious efforts to handle all the technical difficulties with the innumerable precipitations we performed over the years.

References

1 *Biomineralization I*, ed. K. Naka, *Top. Curr. Chem.*, 2007, vol. 270.
2 *Biomineralization I*, ed. K. Naka, *Top. Curr. Chem.*, 2007, vol. 271.
3 L. Addadi, D. Joester, F. Nudelman and S. Weiner, *Chem.–Eur. J.*, 2006, **12**, 980.
4 J. Rieger, E. Hädicke, I. U. Rau and D. Boeckh, *Tenside Surfactants Deterg.*, 1997, **34**, 430.
5 J. Detering, W. Bertleff, M. Essig and A. Kistenmacher, *Tenside Surfactants Deterg.*, 1999, **36**, 399.
6 H. Glade, C. Wildebrand, S. Will, M. Essig, J. Rieger, K.-H. Büchner and G. Brodt, *Int. Desalination Assoc. World. Congress, Proc. Singapore 2005, IDA*, 2005.
7 H. Cölfen, *Curr. Opin. Colloid Interface Sci.*, 2003, **8**, 23.
8 J. W. Mullin, *Crystallization*, Butterworth-Heinemann, Oxford, 3rd edn, 1992.
9 D. Horn and J. Rieger, *Angew. Chem., Int. Ed.*, 2001, **40**, 4330.
10 H. Cölfen and S. Mann, *Angew. Chem., Int. Ed.*, 2003, **42**, 2350.
11 T. Ogino, T. Suzuki and K. Sawada, *Geochim. Cosmochim. Acta*, 1987, **51**, 2757.
12 J. Bolze, B. Peng, N. Dingenouts, P. Panine, T. Narayanan and M. Ballauff, *Langmuir*, 2002, **18**, 56.
13 J. Rieger, J. Thieme and C. Schmidt, *Langmuir*, 2000, **16**, 8300.
14 H. Haberkorn, D. Franke, Th. Frechen, W. Goesele and J. Rieger, *J. Colloid Interface Sci.*, 2003, **112**, 259.
15 M. Faatz, F. Gröhn and G. Wegner, *Adv. Mater.*, 2004, **16**, 996.

16 A. Navrotsky, *Proc. Natl. Acad. Sci. USA*, 2004, **101**, 12096.
17 K. Naka, S.-C. Huang and Y. Chujo, *Langmuir*, 2006, **22**, 7760.
18 T. Wang, M. Antonietti and H. Cölfen, *Chem.–Eur. J.*, 2006, **12**, 5722.
19 Z. Zhang, D. Gao, H. Zhao, C. Xie, G. Guan, D. Wang and S.-H. Yuh, *J. Phys. Chem. B*, 2006, **110**, 8613.
20 L. Gower and D. J. Odom, *J. Cryst. Growth*, 2000, **210**, 719.
21 D. Volkmer, M. Harms, L. Gower and A. Ziegler, *Angew. Chem., Int. Ed.*, 2005, **44**, 639.
22 E. DiMasi, S.-Y. Kwak, F. F. Amos, M. J. Olszta, D. Lush and L. B. Gower, *Phys. Rev. Lett.*, 2006, **97**, 045503.
23 Y. Levi-Kalisman, S. Raz, S. Weiner, L. Ad dadi and I. Sagi, *Adv. Funct. Mater.*, 2002, **12**, 43.
24 T. Sugimoto, in *Fine Particle Science and Technology*, ed. E. Pelizzetti, Kluwer, Dordrecht, 1996, p. 61.
25 J. Park, V. Privman and E. Matijevic, *J. Phys. Chem. B*, 2001, **105**, 11630.
26 P. Guttmann, B. Niemann, S. Rehbein, C. Knöchel, D. Rudolph and G. Schmahl, *J. Phys. IV*, 2003, **104**, 85.
27 P. G. DeBenedetti, *Metastable Liquids*, Princeton University Press, Princeton, 1996.
28 M. Faatz, W. Cheng and G. Wegner, *Langmuir*, 2005, **21**, 6666.
29 J.-P. Andreassen, *J. Cryst. Growth*, 2005, **274**, 256.
30 C. Geffroy, J. Persello, A. Foissy, F. Tournilhac and B. Cabane, *Rev. Inst. Fr. Pet.*, 1997, **52**, 183.
31 F. Fantinel, J. Rieger, F. Molnar and P. Hübler, *Langmuir*, 2004, **20**, 2539.
32 F. Molnar and J. Rieger, *Langmuir*, 2005, **21**, 786.

Does supercooled liquid Si have a density maximum?

Masahito Watanabe,[a] Masayoshi Adachi,[a] Tetsuya Morishita,[b] Kensuke Higuchi,[a] Hidekazu Kobatake[c] and Hiroyuki Fukuyama[c]

Received 9th November 2006, Accepted 7th February 2007
First published as an Advance Article on the web 10th May 2007
DOI: 10.1039/b616394g

We have performed precise measurements of the density of supercooled liquid silicon (l-Si) in the temperature range of 1530–1800 K using an electromagnetic levitation (EML) technique with static magnetic fields. We also performed first-principles molecular dynamics simulation (FPMD) of supercooled l-Si. The observed density of the supercooled l-Si and the FPMD results show good agreement in the temperature range of 1530–1800 K. The structure of the supercooled l-Si also showed good agreement between the experimental measurements and FPMD simulations. Based on these results, we discuss nucleation in supercooled l-Si and also the existence of a density maximum in the supercooled l-Si, which is well known in water at 4 °C.

1. Introduction

The structure and properties of liquid silicon (l-Si) have been of interest for a long time from the viewpoint of applications in high-quality large-diameter crystal growth as well as basic materials science. From the viewpoint of nucleation, it is very important to clarify the structure of l-Si. However, the structure of supercooled l-Si is still unclear. Recently, the possible existence of two distinct liquid forms, low-density liquid (LDL) and high-density liquid (HDL), in the supercooled regime has been a topic of intense research. Much work has been aimed towards exploring the transition between HDL (low tetrahedrality) and LDL (high tetrahedrality) in supercooled l-Si as well as in other liquids having tetrahedral coordination.[1,2] The density and structure of supercooled l-Si have been measured using various levitation techniques, including electromagnetic levitation (EML),[3–5] electrostatic levitation (ESL),[6–9] and conical-nozzle gas levitation (CNL) techniques.[10,11] However, experimental results have not shown evidence of the HDL–LDL transition in supercooled l-Si. If the HDL–LDL transition exists in supercooled l-Si, there should be some change in the structure with the transition. However, different temperature dependence behavior of the l-Si structure has been reported in the supercooled region, and thus, clear evidence of the HDL–LDL transition has not yet been obtained. On the other hand, from the viewpoint of nucleation in supercooled liquid,

[a] *Department of Physics, Gakushuin University, 1-5-1 Mejiro, Tokyo 171-8588, Japan. E-mail: masahito.watanabe@gakushuin.ac.jp; Fax: +81 3 5992 1029; Tel: +813 2986 0221*
[b] *Research Institute for Computational Sciences (RICS), National Institute of Advanced Industrial Science and Technology (AIST), 1-1-1 Umezono, Tsukuba, Ibaraki 305-8568, Japan*
[c] *Institute of Multidisciplinary Research for Advanced Materials (IMRAM), Tohoku University, 2-1-1 Katahira, Aoba-ku, Sendai 980-8577, Japan*

knowledge of the supercooled l-Si structure is very important for understanding the effect of the liquid structure on nucleation phenomena. Therefore, we need to investigate the density and structure of supercooled l-Si. For this purpose, we use the EML technique to maintain the supercooled state with high supercooling. From the above viewpoint for the investigation of supercooled l-Si, the density of the supercooled l-Si needs to be precisely measured. We have therefore developed a new measurement method using electromagnetically levitated liquid with reduced surface oscillations using a static magnetic field. In the normal EML technique, since the levitated liquid droplets have surface oscillation with large amplitudes, the density of the levitated liquid droplet shows large error values in the temperature dependence data due to the asymmetry in the surface oscillation of the levitated liquid droplets. Since our new measurement technique reduces surface oscillation effects, we can obtain the precise temperature dependence of the l-Si density. We also performed first-principles molecular dynamics (FPMD) simulations of supercooled l-Si to investigate the precise structure of supercooled l-Si by comparison with previous high-energy X-ray diffraction experiments combined with the EML technique.[4,5] From the measurement of density and structure, and the FPMD simulations, we discuss nucleation in supercooled l-Si and also discuss the existence of a density maximum of l-Si which is well known in water at 4 °C.

2. Experimental

2.1 Density measurement of supercooled l-Si in static magnetic fields

Density measurements were carried out using an image processing method,[4,12] wherein the volume of the sample is measured by its shape and the density is then calculated as mass over volume. The mass of the sample was measured before and after levitation in order to confirm the volume change by evaporation of the samples. Finally, the l-Si density was calculated as the mass divided by the volume. However, using the conventional method of measuring the density of an electromagnetically levitated liquid droplet, the measured value has a large error in the temperature dependence of the density due to asymmetric surface oscillations. The radius r of an electromagnetically levitated liquid droplet with surface oscillation at time t is described by the following equation,

$$r(t) = R + \varepsilon Y_l^m(\theta, \varphi) = R + \varepsilon P_l^m(\theta) \cos(m\varphi) \cos(\omega_{l,m} t), \qquad (1)$$

where R is the radius of a true sphere, ε is the amplitude of oscillation, Y_l^m is the spherical harmonic function, P_l^m is the Legendre polynomial, $\omega_{l,m}$ is the frequency of surface oscillation, and l and m are labels that indicate types of oscillations. The deviation in the levitated liquid droplet shape by surface oscillation leads to a change in the apparent volume of the droplet. We found, from analysis of the time-dependent droplet shape change, that the periodic change in the apparent volume is caused by the $(l, m) = (2,2)$ mode.[13] Thus, for precise measurement of the density, we must reduce the effects of surface oscillation. For this purpose, we have developed a new EML facility to measure the density using the surface oscillation reduction method by using static magnetic fields.[12,14] If we apply a static magnetic field to the electromagnetically levitated liquid droplet, the surface oscillation can be reduced by the Lorentz force, and we can easily measure the volume of the electromagnetically levitated liquid sample from its shape without deformation generated by surface oscillations. EML experiments were performed using an RF power supply with 250 kHz and 10 kW, using the facility shown in Fig. 1. The solid Si samples were set on the carbon pre-heater in the vacuum chamber. The chamber was evacuated to 10^{-4} Pa and then backfilled with Ar gas with 6N purity. The levitated sample shape was observed by a high-speed camera (Read Tech Motion Scope, PCI 1000S) with a 500 Hz frame rate from the side position of the sample. The levitated sample picture was recorded with a resolution of 512 × 512 pixels for one image. Two polarization filters

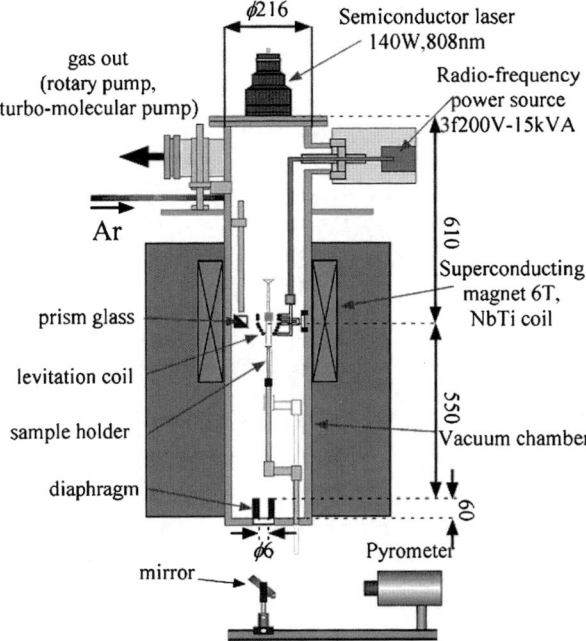

Fig. 1 Experimental set-up of an electromagnetic levitation system for the measurement of the density of supercooled l-Si in static magnetic fields generated by a superconducting magnet.

were used to keep the brightness of the sample constant at any temperature. The temperature of the levitated sample was measured by a two-color pyrometer (CHINO, IR-CAQ) with wavelengths of 850 and 1000 nm, and using the value of 1.0 for the emissivity ratio.[15] Fig. 2 shows the apparent volume of the electromagnetically levitated molten Si in various static magnetic fields. From these results, we find that in a 0.25 T static magnetic field, the volume of the molten Si droplet changed with time, similar to the result seen for a 0 T magnetic field. However, in a 1.5 T static magnetic field, the l-Si volume did not changed. This indicates that the surface oscillation of electromagnetically levitated l-Si droplets can be reduced by a static magnetic field stronger than 1.5 T.

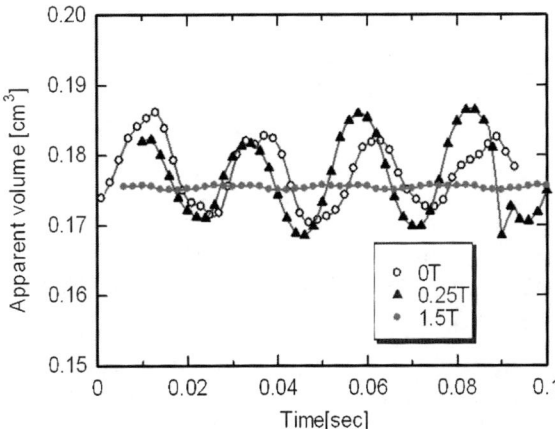

Fig. 2 Time dependence of the apparent volume of electromagnetically-levitated molten Si at 1680 K under various static magnetic fields.

2.2 First-principles molecular dynamics simulation of supercooled l-Si

A series of isothermal–isobaric first-principle molecular dynamics (FPMD) simulations, each for 23–130 ps, was carried out with a 64-atom supercell of l-Si over a temperature range of 950–1700 K. Details of the simulation method are described elsewhere.[16,17] The electronic state calculation was performed within the local density approximation of the density functional theory (DFT). The electronic wave functions were expanded in a plane wave basis with an energy cutoff of 21.5 Ry, and the norm-conserving pseudopotential was used to describe the electron–ion interaction.[18,19] In this work, we aimed to elucidate the temperature dependence of the liquid properties under normal pressure. However, previous experimental values of the density at T_m range from 2.52–2.59 g cm^{-3}. We thus examined the density dependence of the liquid properties by running FPMD simulations at two pressures, p_0 and p_1; the former pressure p_0 yields a value of the density of 2.59 g cm^{-3}, while the latter p_1 yields a value of the density of 2.52 g cm^{-3} at 1700 K. The pressure values of p_0 and p_1 strongly depend on the plane wave cutoff. For instance, an energy cutoff of 21.5 Ry gives p_0 = 7.2 GPa (p_1 = 5.5 GPa), while an energy cutoff of ~12 Ry gives p_0 = 0 GPa (p_1 = -2.1 GPa), although both these energy cutoffs essentially give the same structural information. It is known that "shifted" pressures are often necessary to yield experimental densities in DFT calculations.[20–22] We found that although the stability of the supercooled state strongly depends on the external pressure below 1100 K, both pressures yield essentially the same results at temperatures ≥ 1100 K. Previous simulations did not pay attention to the pressure effect by the change of energy cutoff on the l-Si structure. Our attention in this research will therefore be focused on l-Si under pressure p_0 using the cutoff energy of 21.5 Ry.

3. Results and discussion

Fig. 3 shows the structure factor $S(Q)$ obtained by FPMD simulations. In the figure, the experimental results of $S(Q)$ at 1350 and 1800 K by high-energy X-ray diffraction combined with the EML technique[3] are also shown. It is found that the FPMD results agree with the experimental results. The previous FPMD results do not quite agree with our experimental results, unlike our present FPMD results. Thus, we confirmed that our FPMD method can precisely predict the properties of supercooled l-Si. Fig. 4 shows the coordination number of l-Si in the temperature range from 1100–1800 K. The coordination number does not changed in the temperature range from 1800–1200 K. This result agrees with our experimental results[4] and those of another research group.[6] Therefore, from the present results, we can clearly conclude that the nearest coordination number of l-Si is not drastically changed in the temperature range from 1800–1200 K. However, below 1100 K, which represents the very highly supercooled region, the coordination number of l-Si decreases with decreasing temperature. Moreover, the first nearest neighbor distance is not drastically changed with a change in temperature. This means that the nearest neighbor short-range structure of l-Si is arranged with tetrahedral coordination. In our analytical model of the supercooled l-Si structure, we used the model structure combined with the β-tin structure and diamond structure based on tetrahedral coordination to fit the experimental $S(Q)$.[5] Our model structure for analysis is almost the same structure obtained by FPMD simulations based on tetrahedral coordination. However, it is well known that the density of l-Si is larger than that of crystalline Si, and the density increases with decreasing temperature, as shown in Fig. 4. This density change is attributed to the continuous development of tetrahedral order. However, the nearest short-range structure shows no significant change above ~1200 K. This can be explained by the temperature dependence of the average distances of the first eight neighbors. From FPMD simulations, we clearly observe that the distances of the first four or five neighboring atoms gradually

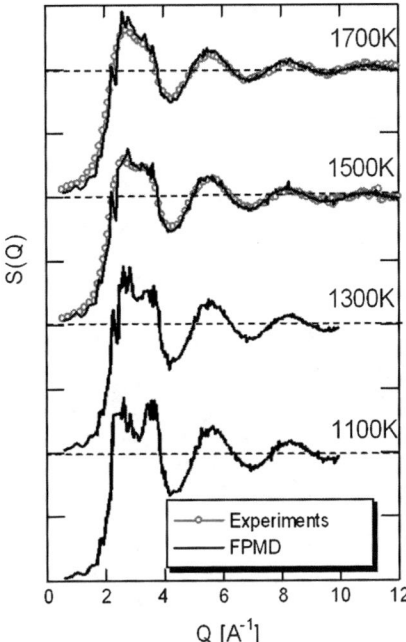

Fig. 3 Structure factor $S(Q)$ obtained by high-energy X-ray diffraction with the EML technique[4] at 1700 and 1500 K, and by FPMD simulations at 1700, 1500, 1300, and 1100 K.

decrease with temperature, while those of the 6th–8th neighboring atoms remain unchanged except at 1100 K, which should increase to form a highly tetrahedral network.

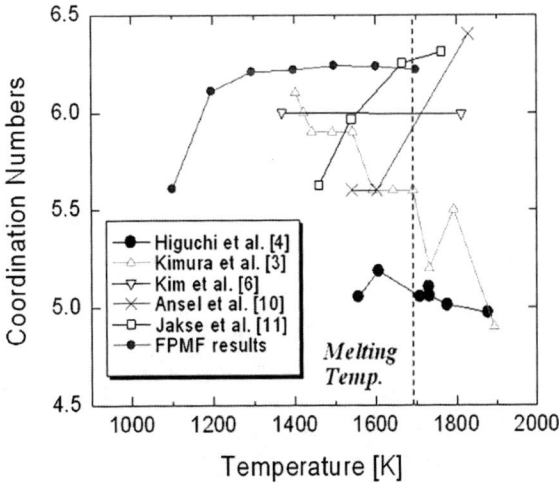

Fig. 4 First nearest coordination number of l-Si obtained by FPMD simulations. In the figure, previously reported values are also shown. The results of Kim et al.[6] and our experimental results (Higuchi et al.[4]) show no significant change in coordination number in the temperature range from 1500–1900 K. The coordination values of our experimental data are lower than the other results. This is because the first minimum position in the radial distribution function $g(r)$ was not very clear; therefore the coordination values had a large error. The values plotted in the figure were the smallest values at each temperature.

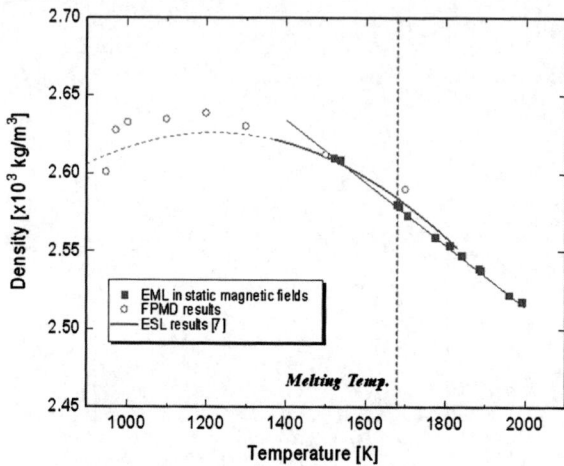

Fig. 5 Temperature dependence of l-Si density in the temperature range from 950–2000 K. FPMD simulation results, our experimental data measured by EML in static magnetic fields and previous results using the ELS technique[8] are shown.

Fig. 5 shows the temperature dependence of the density of l-Si obtained by using EML with a static magnetic field and also by FPMD simulations. In the present experiments, we succeeded in measuring the l-Si density in the temperature range from 1530–1800 K. Here, the deviation in density with temperature change was very small compared to previous results.[3] From these results, we were able to obtain the temperature dependence of the l-Si density, as described by the following equation,

$$\rho(T) = 2.578 - 2.0 \times 10^{-4}(T - T_m). \qquad (2)$$

In the temperature range from 1530–1700 K, the experimental and FPMD results showed good agreement. Thus, we clearly confirmed that the present FPMD technique can correctly simulate the l-Si structure and density in supercooled regions. Our results showed a linear relationship between the density and temperature, however, the FPMD results strayed from the linear relationship below 1300 K. The present density measurements had a temperature limit of 1530 K, therefore, the temperature dependence of the density described by eqn (1) cannot be fitted below 1400 K. In the figure, the previous results were obtained by using an electrostatic levitation technique.[7] The ESL results of the temperature dependence of the l-Si density in the temperature range from 1370–1870 K are described by:

$$\rho(T) = 2.583 - 1.851 \times 10^{-4}(T - T_m) - 1.984 \times 10^{-4}(T - T_m)^2 \qquad (3)$$

Our EML data agree with the ESL results in the temperature range from 1530–1900 K. If the ESL density results are extrapolated to 1000 K, it is found that the ESL data agree with the FPMD data. Although the temperature dependence of l-Si density has been the topic of much discussion, we can now conclude from the present results that the density of l-Si has a maximum value at around 1200 K. If we can obtain high supercooling at 1200 K by the levitation technique, we can obtain the maximum point of the density of l-Si. However, it is very difficult to achieve the supercooled state of l-Si at 1200 K using the conventional levitation technique. Our present EML technique using static magnetic fields has the potential to achieve high supercooling more easily than the conventional levitation technique, therefore, we are trying to achieve high supercooling of l-Si close to 1200 K. However, the temperature dependence of supercooled l-Si density shown in Fig. 5 is not sufficient, on average, for the establishment of the tetrahedral order beyond the first four neighbors, *i.e.*, the LDL–HDL transition would not occur. Finally, we discuss the nucleation phenomena in supercooled l-Si. From the structural results, we find that l-Si in the temperature range from 1200–1800 K has short-range order based on tetrahedral coordination. The tetrahedral coordination of Si atoms grows with decreasing temperature. This means that the tetrahedrally coordinated atoms would be the center of homogeneous nucleation for crystal growth. Thus, the nucleus of crystal growth would have the same structure of crystalline Si. If the supercooling (temperature below the melting temperature) increases, the number of tetrahedrally coordinated atoms in the liquid state increases, and the tetrahedrally coordinated network of atoms grows as well. The increasing number of tetrahedrally coordinated atoms and the growing tetrahedral network would increase the number of nucleation sites, and thus, the nucleation rate would also be increased. Therefore, in the highly supercooled liquid, the tetrahedrally coordinated atoms would make it difficult to form bulk amorphous Si from the supercooled liquid state. However, if we achieve a highly supercooled liquid state below 1200 K, the liquid state has a high density, and therefore will be able to form bulk amorphous Si from this liquid state. In this paper, we have described and discussed nucleation in supercooled l-Si phenomenologically. However, based on this research, we will be able to revise the classical nucleation theory by connecting the microscopic viewpoint based on the atomic structure and dynamics to the macroscopic viewpoint based on thermodynamics. In the next step, we will try to obtain the interfacial energy between the supercooled liquid and the crystalline state with dependence on the l-Si structure in order to obtain the nucleation rate from classical nucleation theory.

4. Conclusions

We performed density measurements of l-Si in the supercooled liquid state using the EML method in a static magnetic field. We also performed FPMD simulations. The structure of l-Si obtained from the FPMD results agreed well with the experimental data of high-energy X-ray diffraction and also agreed with the density of l-Si obtained by the present experiments and previously reported data. Based on these results, we discussed the supercooled l-Si structure and also the existence of a density maximum. Our conclusions show that l-Si has a density maximum around 1200 K. It is necessary to carry out more simulations and experiments to find the precise temperature of the density maximum of l-Si in the future. From this density change of l-Si, we also discussed the nucleation phenomena in supercooled l-Si.

References

1 P. H. Poole, T. Grande, C. A. Angell and P. F. Mcmillan, *Science*, 1997, **275**, 322.
2 O. Mishima and H. E. Stanley, *Nature*, 1998, **396**, 329.
3 H. Kimura, M. Watanabe, K. Izumi, T. Hibiya, D. Holland-Moritz, T. Schenk, K. R. Bauchspie, S. Schneider, I. Egry, K. Funakoshi and M. Hanfland, *Appl. Phys. Lett.*, 2001, **78**, 604.
4 K. Higuchi, K. Kimura, A. Mizuno, M. Watanabe, Y. Katayama and K. Kuribayashi, *Meas. Sci. Technol.*, 2005, **16**, 381.
5 M. Watanabe, K. Higuchi, A. Mizuno, K. Nagashio, K. Kuribayashi and Y. Katayama, *J. Cryst. Growth*, 2006, **294**, 16.
6 T. H. Kim, G. W. Lee, B. Sieve, A. K. Gangopadhyay, R. W. Hyers, T. J. Rathz, J. R. Rogers, D. S. Robinson, K. F. Kelton and A. I. Goldman, *Phys. Rev. Lett.*, 2005, **95**, 085501.
7 Z. Zhou, A. Mukherjee and W. K. Rhim, *J. Cryst. Growth*, 2003, **257**, 350.
8 W. K. Rhim, S. K. Chung, A. J. Rulison and R. E. Spjut, *Int. J. Thermophys.*, 1997, **18**, 459.
9 K. Osaka, S. K. Chung, W. K. Rhim and J. C. Holzer, *Appl. Phys. Lett.*, 1997, **70**, 423.
10 S. Ansell, S. Krishnan, J. J. Felten and D. L. Price, *J. Phys.: Condens. Matter*, 1998, **10**, L73.

11 N. Jakse, L. Hennet, D. L. Price, S. Krishnan, T. Key, E. Artacho, B. Glorieux, A. Pasturel and M.-L. Saboungi, *Appl. Phys. Lett.*, 2003, **83**, 4734.
12 M. Adachi, K. Higuchi, A. Mizuno, M. Watanabe, H. Kobatake and H. Fukuyama, *Int. J. Thermophys.*, accepted.
13 D. L. Cummings and D. A. Blackburn, *J. Fluid Mech.*, 1991, **224**, 395.
14 H. Kobatake, H. Fukuyama, I. Minato, T. Nakamura, T. Tsukada and S. Awaji, *Proceedings of 16th Symposium on Thermophysical Properties*, Boulder, 2006.
15 H. Kawamura, H. Fukuyama, M. Watanabe and T. Hibiya, *Meas. Sci. Technol.*, 2005, **16**, 386.
16 T. Morishita and S. Nosé, *Prog. Theor. Phys. Suppl.*, 2000, **138**, 251.
17 T. Morishita, *Phys. Rev. Lett.*, 2006, **97**, 165502.
18 D. R. Hamann, M. Schlüter and C. Chiang, *Phys. Rev. Lett.*, 1979, **43**, 1494.
19 I. Štich, R. Car and M. Parrinello, *Phys. Rev. B*, 1991, **44**, 4262.
20 A. R. Oganov, J. P. Brodholt and G. D. Price, *Nature*, 2001, **411**, 934.
21 T. Morishita, *Phys. Rev. B*, 2002, **66**, 054204.
22 L. M. Ghiringhelli and E. V. Meijer, *J. Chem. Phys.*, 2005, **122**, 184510.

PAPER

Simulating ice nucleation, one molecule at a time, with the 'DFT microscope'†

Angelos Michaelides[abc]

Received 15th November 2006, Accepted 1st February 2007
First published as an Advance Article on the web 11th April 2007
DOI: 10.1039/b616689j

Few physical processes are as ubiquitous as the nucleation of water into ice. However, ice nucleation and, in particular, heterogeneously catalyzed nucleation remains poorly understood at the atomic level. Here, we report an initial series of density functional theory (DFT) calculations aimed at putting our understanding of ice nucleation and water clustering at metallic surfaces on a firmer footing. Taking a prototype hydrophobic metal surface, Cu(111), for which scanning tunneling microscopy measurements of water clustering have recently been performed, possible structures of adsorbed clusters comprised of 2–6 H_2O molecules have been computed. How the water clusters in this size regime differ from those in the gas phase is discussed, as is the nature of their interaction with the substrate.

1. Introduction

H_2O–metal interactions are important in a wide variety of phenomena in materials science, catalysis, corrosion, electrochemistry, and so on. H_2O–metal interfaces also provide fertile ground for understanding basic physical processes such as ice nucleation and water clustering. In this respect a wide variety of surface science style studies have been performed, focussing on understanding the structures and properties of thin ice films and water overlayers that form on well-defined single-crystal metal surfaces under ultra-high vacuum (UHV) conditions. By now such overlayer systems have been interrogated with almost every available experimental surface science probe as well as a variety of theoretical approaches.[1–3]

Many of the recent studies in this area have focussed on determining the nature and structures of the first wetting layer of H_2O–ice that may form. Due to the competing influences of H_2O–metal bonding and H_2O–H_2O epitaxial mismatch between the overlayer and the substrate a wide variety of extended overlayer structures have been proposed, such as 2D ice-like overlayers[4–9] quasi-2D overlayers,[10] 1D chains,[11] and mixed H_2O–OH overlayers.[12–19] Several of these overlayer systems have been examined with a range of complementary experimental and theoretical techniques such as low-energy electron diffraction (LEED), vibrational spectroscopies, photoemission, and density functional theory (DFT) leading to a clear atomic-scale characterisation of their structures.

Much less is known, however, about the structures of small H_2O clusters that form prior to the creation of the extended overlayers; dimers, trimers, tetramers, *etc.* Although on a few metal surfaces some of the clusters have been reported in infrared

---

[a] *London Centre for Nanotechnology, University College London, London, UK WC1E 6BT*
[b] *Department of Chemistry, University College London, London, UK WC1E 6BT*
[c] *Fritz-Haber-Institut der Max-Planck-Gesellschaft, Faradayweg 4-6, 14195, Berlin, Germany*

† The HTML version of this article has been enhanced with colour images.

reflection absorption (IRAS) spectroscopy[20–22] and scanning tunnelling microscopy (STM)[3,23–25] studies their internal structure or registry with the substrate remains, in every case, unclear. Moreover, it is not known how the structures and relative energies of the adsorbed clusters differ from the equivalent clusters in the gas phase. It is in this regard that first principles electronic structure theories can make valuable contributions. And, indeed, adsorbed H_2O monomers and small H_2O clusters have recently been computed with DFT on a number of close-packed metal surfaces providing predictions as to what structures may form as well as insight and understanding as to why particular structures are favoured over others.[26–34]

The study reported here is in a similar spirit to the previous DFT studies cited above. However, here we examine, for the first time, adsorbed clusters right up to the H_2O hexamer; dimers, trimers, tetramers, pentamers, and hexamers have all been computed. In addition we examine how the structures and stability of these small H_2O clusters differ from those in the gas phase, thus shedding light on the 'catalytic' role of the Cu(111) substrate in H_2O cluster formation. We have chosen Cu(111) for this work mainly because recent low temperature STM experiments of H_2O and D_2O clustering have been reported on this surface.[35] In addition, Cu(111) is of interest because it is a substrate upon which H_2O does not normally dissociate and is an hexagaonl surface which can, in principle, support an adsorbed H_2O–ice bilayer with a lattice mismatch of $<2\%$.[1,36]

The plan for the remainder of this paper is the following. Details of our first principles total energy calculations are outlined below. Following this structures and energetics for the adsorbed H_2O clusters are presented (section 3). In section 4 we briefly discuss our results, paying particular attention to the electronic structures of adsorbed H_2O dimers and hexamers. In section 5 we close by drawing some conclusions.

2. Approach and computational details

The majority of the calculations reported here have been performed within the DFT framework as implemented in the CASTEP code.[37] The electron–ion interactions were treated with Vanderbilt ultrasoft pseudopotentials,[38] which were expanded within a plane-wave basis set up to a cut-off energy of 400 eV. Electron exchange and correlation effects were described by the Perdew Burke Ernzerhof (PBE)[39] generalized gradient approximation (GGA).

The Cu(111) surface was modelled by a periodic array of Cu slabs, separated by a vacuum region in excess of 12 Å. The majority of the adsorption calculations were performed in $p(4 \times 4)$ or $p(5 \times 5)$ unit cells. Such large cells were required in order to minimize the interaction between water clusters in adjacent unit cells. Because of the relatively large unit cells employed and the desire to explore a wide variety of structures for each cluster, thin 3 layer Cu slabs were used throughout. The significance of this apparent compromise in accuracy was, however, carefully checked with test calculations for adsorbed water clusters on slabs of up to 9 layers thickness. The main results of these tests are shown in Table 1, where it can be seen that the absolute adsorption energies obtained on the thicker slabs deviated by <10 meV per H_2O and bond distances deviated by <0.1 Å. Monkhorst–Pack k-point meshes with the equivalent of at least $8 \times 8 \times 1$ sampling within the surface Brillouin zone of a $p(1 \times 1)$ unit cell were used throughout.[40]

In all optimizations the bottom two Cu layers were fixed at their *ab initio* bulk truncated positions (3.647 Å (expt 3.615 Å[41])) and the remaining atoms were allowed to fully relax. For each cluster a large number of trial structures were optimized; as many as 30 trial structures for the pentamers and hexamers. In addition, the occasional simulated annealing *ab initio* molecular dynamics simulation was performed. However, we caution that configurational space becomes so large for systems with >2–3 adsorbed and interacting H_2O molecules that we can in no way guarantee that the structures identified are the global minimum energy

Table 1 Selected results of the layer and *k*-point sampling convergence tests for H_2O monomer and H_2O cluster (dimer and hexamer) adsorption on Cu(111)

Adsorbate	Layers	*k* mesh[a]	E_{ads} (meV/H_2O)	Cu–O[b] (Å)	O–O[b] (Å)
Monomer [$p(2 \times 2)$ cell]	3	8 × 8 × 1	151	2.345	—
	3	12 × 12 × 1	146	2.370	—
	3	24 × 24 × 1	145	2.357	—
	9	8 × 8 × 1	157	2.347	—
Dimer [$p(3 \times 3)$ cell]	3	12 × 12 × 1	321	2.202/3.032	2.740
	3	18 × 18 × 1	327	2.199/3.026	2.740
	9	12 × 12 × 1	332	2.152/2.934	2.716
Hexamer [$p(4 \times 4)$ cell]	3	8 × 8 × 1	440	2.772	2.697
	3	16 × 16 × 1	435	2.771	2.697
	9	8 × 8 × 1	444	2.802	2.696

[a] $p(1 \times 1)$ equivalent sampling. [b] Average values are given for the hexamer.

structures. At best the structures identified serve merely as plausible candidates for the lowest energy adsorption structures of H_2O clusters that may form on Cu(111).

Adsorption energies (E_{ads}) per H_2O molecule are calculated from,

$$E_{ads} = nE_{H_2O} + E_{Cu} - E_{nH_2O/Cu}, \quad (1)$$

where E_{H_2O}, E_{Cu} and $E_{nH_2O/Cu}$ are the total energies of an *isolated* H_2O molecule, the clean Cu(111) surface, and the nH_2O/Cu(111) adsorption system, respectively. In this definition positive adsorption energies correspond to an exothermic adsorption process. The reference energy of the isolated gas phase H_2O molecule is calculated by placing it in a 20 Å^3 cell. Calculations of gas phase H_2O clusters have also been performed within the same 20 Å^3 cell, for which their gas phase binding energy (E_{bind}) is defined as,

$$E_{bind} = nE_{H_2O} - E_{nH_2O}, \quad (2)$$

where E_{nH_2O} is the total energy of the nH_2O gas phase cluster.

3. Results

We now consider the structures and energies of the low energy water clusters predicted by DFT, which represents the main body of results of this study.

3.1 H_2O monomer adsorption

The most stable structure for the H_2O monomer on Cu(111) is displayed in Fig. 1(a) and (b). From Fig. 1(a) it can be seen that H_2O adsorbs preferentially at an atop site and from the side view in Fig. 1(b) it can be seen that the molecular plane lies almost parallel to the surface. This structure has been reported before as the most stable one according to DFT for water adsorbed on Cu(111) and several other close-packed transition metal surfaces.[26] We show it again here simply because it is important to know how a single water adsorbs on Cu(111) in order to understand the structures of the adsorbed clusters that we come to next. The Cu–O bond length is 2.35 Å and the adsorption energy is 0.15 eV per H_2O. The internal structure of the H_2O molecule deforms little upon adsorption: the HOH angle is 1° less that its computed gas phase value of 104.6° and the OH bond lengths are essentially unchanged from their computed gas phase value of 0.98 Å. Indeed also in the H_2O clusters we do not observe any significant changes of the internal structure of the H_2O molecules and thus will not discuss this issue further.

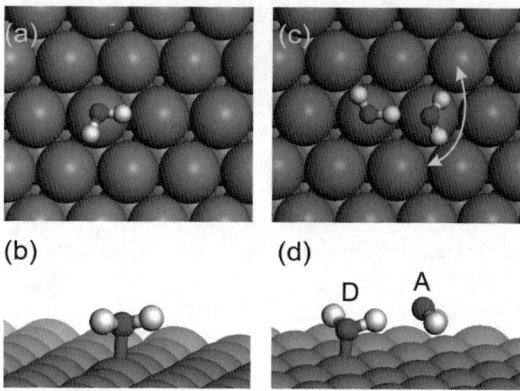

Fig. 1 Top and side views of the lowest energy adsorbed H_2O monomer and H_2O dimer structures identified. In (d) the H_2O molecule which donates (accepts) the H bond is labelled D (A), and in (c) the arrow is to illustrate that rotation of the acceptor H_2O in a plane about the surface normal is facile.

3.2 H_2O dimer adsorption

The most stable adsorbed H_2O dimer structure identified is displayed in Fig. 1(c) and (d). The adsorption energy of this dimer (relative to two isolated H_2O molecules as defined in eqn (1)) is 0.32 eV per H_2O. In this structure both H_2O molecules adsorb above atop sites with the H_2O molecule that donates the H bond (H bond donor, D) noticeably closer to the surface than the H_2O molecule that accepts the H bond (H bond acceptor, A). Specifically the shortest Cu–O bond lengths for the donating and accepting H_2O molecules are 2.20 and 3.00 Å, respectively. The O–O distance in this configuration is 2.74 Å. As was found previously on Pt(111)[28] we see here on Cu(111) that there is essentially free rotation of the high-lying acceptor H_2O about a plane normal to the surface, *i.e.*, the H bond acceptor is not constrained to remain above the precise atop site. For example, rotating the dimer so that the acceptor is located over a bridge site (with the donor still at the atop site) costs a negligible 2 meV.

The lowest energy dimer structure reported here is similar to the structures previously reported from DFT for dimers on Pt(111)[29] and Pd(111).[28] It was also reported in the previous studies that the H_2O donor interacts more strongly with the substrate than the H_2O acceptor. We provide a general explanation for these observations below.

3.3 H_2O trimer adsorption

Two low energy trimer structures of essentially identical adsorption energy have been identified. These are labelled (a) and (b) in Fig. 2, and are displayed along with a third adsorbed trimer to be discussed in a moment. Also shown in Fig. 2 are the relative energies and structures of all three of these trimers in the gas phase. The two low energy adsorption structures are similar to each other in as much as they are bent structures with only two H bonds connecting the three H_2O molecules. They differ, however, in that structure (b) has one H_2O which donates two H bonds whereas structure (a) does not. The adsorption energy for both trimers is 0.37 eV per H_2O.

It is interesting to see that the two bent adsorbed trimers identified here are not the lowest energy isomers for H_2O trimers in the gas phase. Instead this is a cyclic structure with three H bonds between the three H_2O molecules; each H_2O donates and accepts a single H bond, as can be seen from the upper part of Fig. 2(c).

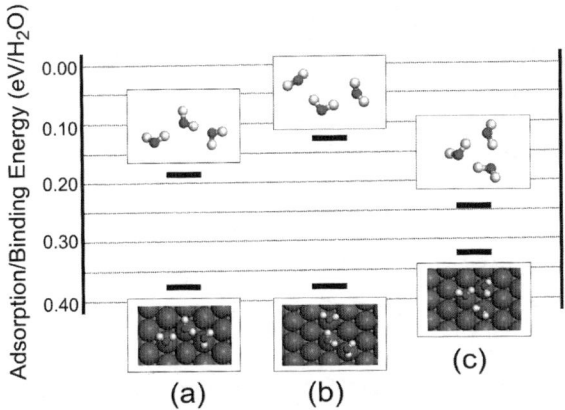

Fig. 2 Relative energies and structures of three gas phase and adsorbed H_2O trimers. Adsorption energies and binding energies are relative to isolated gas phase H_2O molecules as defined in eqn (1) and (2), respectively.

However, from the lower part of Fig. 2(c), it can be seen that when adsorbed the cyclic structure is ~0.05 eV per H_2O less stable than structures (a) and (b). Thus it is clear from structures (a), (b), and (c) that the surface has a significant effect on the relative energies of these isomers, favouring those with only two H bonds. Essentially then the current calculations indicate that one H bond in the trimer breaks upon adsorption.

3.4 H_2O tetramer adsorption

The lowest energy structure identified for the H_2O tetramer is displayed in Fig. 3(a) and (b). This structure resembles one of the low energy trimers (trimer b) but with a fourth H_2O added as a H bond donor to the central H_2O. The binding energy of this cluster is 0.41 eV per H_2O. Again the adsorbed structure differs significantly from the lowest energy gas phase isomer which is, like the trimer, a cyclic structure and is comprised of four H bonds. The O–O distances in the adsorbed structure range from 2.66 to 2.75 Å and the Cu–O distances from 2.20 to 3.10 Å.

3.5 H_2O pentamer adsorption

The most stable H_2O pentamer identified is displayed in Fig. 3(c) and (d). In this structure each H_2O acts as a single H bond donor and a single H bond acceptor,

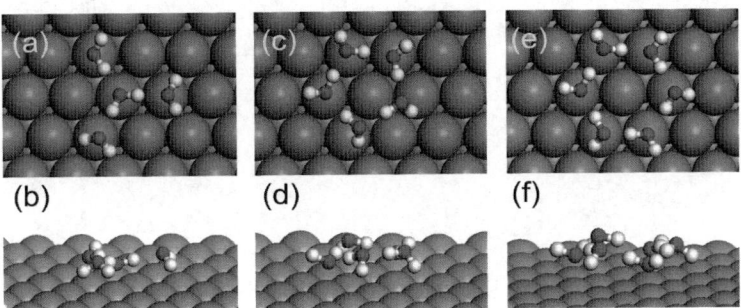

Fig. 3 Top and side views of the lowest energy structures identified for an adsorbed H_2O tetramer ((a)–(b)), pentamer ((c)–(d)), and hexamer ((e)–(f)) on Cu(111).

rather similar to the low energy structure of the gas phase pentamer. Three of the H_2O molecules are located over atop sites, whereas the two others are closest to threefold sites. There is a considerable buckling in the heights of the H_2O molecules above the surface; a 0.8 Å difference between the lowest and highest H_2O molecules. The O–O distances in the adsorbed structure range from 2.60 to 2.80 Å and its adsorption energy is 0.44 eV per H_2O.

3.6 H_2O hexamer adsorption

The most stable hexamer identified is the one shown in Fig. 3(e) and (f). It is a cyclic hexamer with a binding energy of 0.44 eV per H_2O. The symmetry and registry of this cyclic hexamer with the substrate appears to be consistent with recent STM experiments in which it was observed.[35] From the top view of the hexamer in Fig. 3(e) it is clear that all six H_2O molecules are located approximately above substrate atop sites and act as single H bond donors and single H bond acceptors. From the side view in Fig. 3(f) it can further be seen that this structure is significantly buckled with the H_2O molecules residing at two distinct heights above the surface: the vertical displacement between adjacent H_2O molecules is ~ 0.76 Å. Further, and unlike in the gas phase, the six nearest neighbour O–O distances are not equal in the adsorbed hexamer. Instead they alternate between two characteristic values: 2.76 and 2.63 Å. This symmetry-breaking dimerization is reminiscent of the alternating single and double C–C bonds in the Kekulé model of benzene.

Previous DFT reports of an isolated cyclic hexamer on Ru(0001)[30] and a cyclic hexamer as part of a quasi-2D water overlayer on Pd(111)[10] did not predict the buckled structure identified here. Instead a planar cyclic hexamer with all molecules at the same height was found. Here, on Cu(111), our PBE calculations indicate that the planar hexamer is ~ 0.12 eV per H_2O less stable than the buckled one. Since it is not inconceivable that this difference is a result of our chosen computational set-up or exchange–correlation functional, we performed a series of tests in order to assess the reliability of this result. Specifically we compared the energy difference between the low energy buckled hexamer identified here and a hypothetical planar cyclic hexamer on: (i) a Cu(111) slab with the RPBE[42] functional; (ii) a close-packed Cu_{10} cluster with the PBE functional, the hybrid DFT PBE0[43] functional, and with Møller–Plesset perturbation theory to second order (MP2).[44] The results of these test calculations are listed in Table 2, with the conclusion of all of them being essentially the same: there is a considerable energetic preference (>0.1 eV per H_2O) for buckling on this surface.

Table 2 Selected results of the test calculations of the energy difference, ΔE (meV per H_2O), between the buckled and planar cyclic H_2O hexamers on Cu(111). A positive ΔE indicates that the buckled hexamer is more stable than the planar one, which is always the case

Approach	ΔE (meV/H_2O)
Cu(111) PBE	122[a]
Cu(111) RPBE	178[a]
Cu_{10} cluster PBE	170[b], 175[c]
Cu_{10} cluster PBE0	170[b], 173[c]
Cu_{10} cluster MP2	194[b], 186[d]

[a] Pseudopotential plus plane-wave approach with CASTEP as described in Section II. [b] All-electron Gaussian03 calculation with a 6-311+G(2df,pd) basis set. [c] All-electron Gaussian03 calculation with a 6-311++G(3df,3pd) basis set. [d] All-electron Gaussian03 calculation with a 6-311++G(2df,pd) basis set.

4. Discussion

Let's now briefly investigate some details of the results presented above. First we consider the relative stability of the clusters identified and then several aspects of their electronic structures.

4.1 Relative energies of adsorbed and gas phase clusters

To begin, we compare the stability of the H_2O clusters when adsorbed to when they are in their equilibrium gas phase configurations. We show this in Fig. 4, where the adsorption (binding) energies of adsorbed (gas phase) H_2O clusters ranging from 2–9 H_2O molecules are plotted. The structures of the larger 7–9 molecule gas phase clusters are based on the Hartree–Fock structures reported by Maheshwary et al.[45] Here we have simply re-optimized these gas phase structures within the current DFT-PBE set-up. The structures of the 7–9 molecule adsorbed clusters are those recently identified in a combined STM and DFT study of H_2O on Cu(111).[35]

From Fig. 4 it can be seen that in the entire regime examined the adsorbed clusters are more stable than the gas phase ones. It is clear, however, that as the clusters grow in size the energy difference between the adsorbed and gas phase clusters decreases. The increased stability of the adsorbed clusters is obviously because of their binding with the substrate. However, it is interesting to note that the adsorbed clusters remain more stable than their gas phase counterparts even for those clusters which have fewer H bonds when adsorbed. For example the H_2O trimer and tetramer have one less H bond when adsorbed as compared to in the gas phase. And for the larger clusters with 7–9 H_2O molecules several more H bonds are broken.

Also clear from Fig. 4 is that the adsorption energy of the adsorbed clusters gradually decreases until around the pentamer and then seems to level off for the larger clusters. The adsorption energies for the clusters with 5, 6, 7, 8, and 9 molecules are 0.44, 0.44, 0.45, and 0.44 eV per H_2O, respectively. The limiting value of the adsorption energy is at ~ 0.44 eV per H_2O similar to the adsorption energy reported for a hypothetical 2D ice-like bilayer on Cu(111).[36]

4.2 Electronic structures

Next we briefly consider the nature of the binding between some of the clusters and the Cu(111) substrate. In particular we focus on trying to understand the buckled structure of the H_2O dimer and H_2O hexamer.

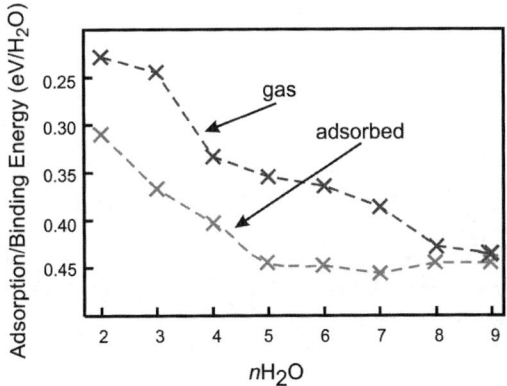

Fig. 4 Plot of the adsorption energies and binding energies for adsorbed nH_2O (n = 2–9) clusters on Cu(111) and for gas phase nH_2O (n = 2–9) clusters. The structures of the 7–9 H_2O molecule gas phase and adsorbed clusters are taken from ref. 45 and ref. 35, respectively. The dashed lines are merely guides to the eye.

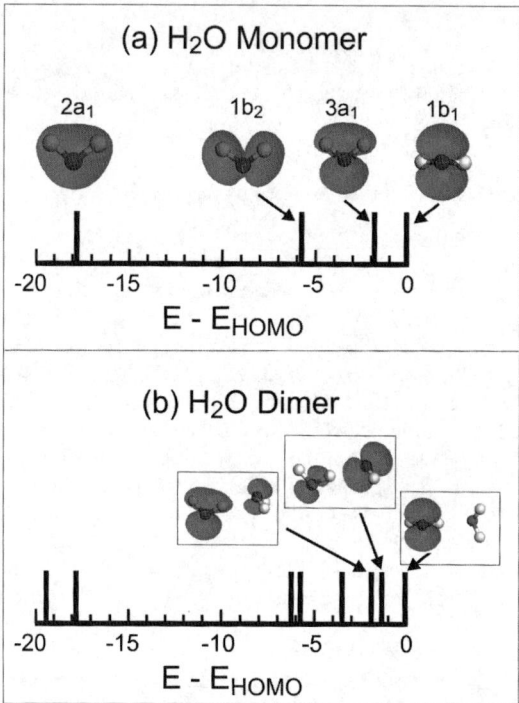

Fig. 5 DFT computed energy level diagrams and selected Kohn–Sham orbitals for the highest occupied orbitals of an isolated (gas phase) H_2O monomer (a) and an isolated (gas phase) H_2O dimer (b). The energy of the highest occupied molecular orbital (E_{HOMO}) is set to zero eV in each system.

A consideration of how the H_2O monomer interacts with the Cu(111) substrate provides a basis for understanding the clusters and so we recall the four highest energy occupied molecular orbitals of an isolated gas phase H_2O (Fig. 5). In order of increasing energy these are labelled, according to C_{2v} symmetry, $2a_1$, $1b_2$, $3a_1$ and $1b_1$. The highest occupied molecular orbital (HOMO) of H_2O is the $1b_1$ orbital. Previous studies have shown that the interaction of the H_2O monomer with close-packed transition metal surfaces is mediated mainly through the $1b_1$ orbital.[17,33] Indeed this is what we find here again for Cu(111). We demonstrate this in Fig. 6(a) with the electron density difference plot for a H_2O monomer on Cu(111). The electron density difference ($\Delta\rho$) is defined here as

$$\Delta\rho = \rho_{nH_2O/Cu} + \rho_{Cu} - \rho_{H_2O}, \qquad (3)$$

where $\rho_{nH_2O/Cu}$, ρ_{Cu}, and ρ_{nH_2O} are the electron densities of the particular nH_2O/Cu(111) adsorption system under consideration, the isolated Cu(111) surface, and the isolated H_2O molecule(s) each in the exact structure they adopt in the adsorption system. Electron density difference plots such as this capture the rearrangement of the electron density upon making the adsorption bond and in this case demonstrate the key role played by the $1b_1$ orbital.

Moving to the H_2O dimer we show in Fig. 6(b) a similar electron density difference plot. From this is it is clear that the H bond donor interacts more strongly with the substrate than the acceptor does, which is what we would expect based on the structure alone. Further, it can be seen that the donor of the dimer interacts in a similar manner as the H_2O monomer does, *i.e.*, through the H_2O $1b_1$ orbital. Since the nature of the H bond in the adsorbed and gas phase dimers is similar (*cf.* Fig. 6(c)

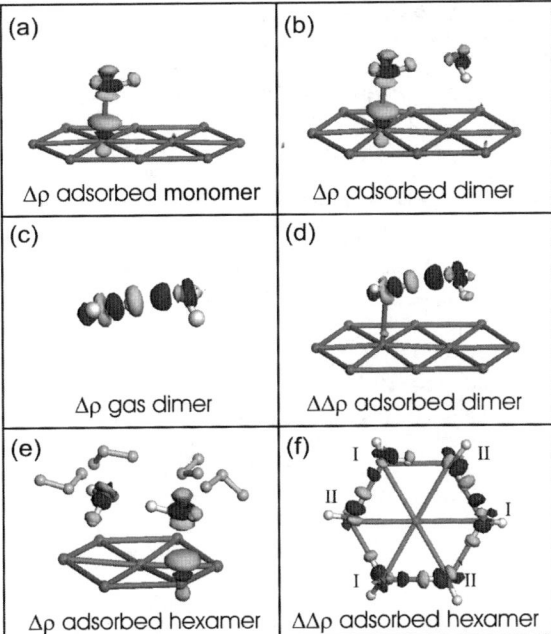

Fig. 6 Isosurfaces of constant electron density difference ($\Delta\rho$) and electron density "rearrangement" ($\Delta\Delta\rho$) as defined in eqn (3) and (4), respectively. $\Delta\rho$ is displayed for H$_2$O monomer adsorption (a), H$_2$O dimer adsorption (b), a gas phase H$_2$O dimer (c), and H$_2$O hexamer adsorption (e). $\Delta\Delta\rho$ is displayed for the adsorbed H$_2$O dimer (d) and the adsorbed H$_2$O hexamer (f). Dark isosurfaces correspond to regions of electron accumulation and light isosurfaces to regions of electron depletion. In (a), (b), (c), and (e) the units are 2×10^{-2} e Å^{-3}, and in (d) and (f) the units are 5×10^{-2} e Å^{-3}. For clarity in (e) $\Delta\rho$ is only displayed around the two front-most H$_2$O molecules and Cu atoms.

and (d)) it remains reasonable to seek insight into the nature of the adsorption bond by considering the electronic structure of the adsorbate in the gas phase, as we did for the H$_2$O monomer. Thus in Fig. 5(b) we display the energy level diagram for the high energy occupied orbitals of the gas phase H$_2$O dimer. Fig. 5(b) provides an immediate explanation for why the donor interacts more strongly with the substrate than the acceptor does and why this interaction is, like the monomer, mediated through the 1b$_1$ orbital. Specifically, it can be seen that this is because the HOMO of the gas phase dimer is a 1b$_1$-like orbital located on the H bond donor, and that this 1b$_1$-like orbital remains a non-bonding orbital, not being involved in the H bond. On the other hand the 1b$_1$-like orbital on the acceptor is not free being shifted to a lower energy through the formation of the H bond. Thus the 1b$_1$-like orbital of the H bond acceptor of the dimer is, in a chemical sense, "saturated" through its participation in the H bond and less inclined to bond with the substrate than the H bond donor.

A similar reasoning explains the symmetry broken buckled structure of the adsorbed hexamer. Indeed the structure of the hexamer is best understood by considering it as being comprised of three weakly interacting dimers. Within the (equilibrium) buckled adsorption structure the two types of water molecules interact differently with the substrate, as revealed by the electron density difference plot in Fig. 6(e). Again the low-lying H$_2$O molecules in the adsorbed hexamer interact with the surface through their 1b$_1$ molecular orbitals whereas the high-lying H$_2$O molecules do not. Since, as we have seen above, the 1b$_1$ orbital of H$_2$O is also implicated when H$_2$O acts as a H bond acceptor, we can anticipate that the low-lying H$_2$O molecules are rendered poor H bond acceptors through their bonding with the substrate. Indeed this is precisely what we see in the structure of the adsorbed hexamer with the longer (2.76 Å) H bonds being formed when the low-lying H_2O molecules act mainly as H bond acceptors, whereas the shorter (2.63 Å) H bonds form when the high-lying H_2O molecules act mainly as H bond acceptors. The two types of H bond in the adsorbed hexamer is clear not only from the atomic structure, but also in the electronic structure. To illustrate this we examine the electron density "rearrangement" ($\Delta\Delta\rho$) within the adsorbed hexamer. This is defined here as

$$\Delta\Delta\rho = \rho_{6H_2O/Cu} + \rho_{Cu} - \rho_{3H_2O - I/Cu} - \rho_{3H_2O - II/Cu}, \qquad (4)$$

where $\rho_{6H_2O/Cu}$ and ρ_{Cu} are the electron densities of the hexamer/Cu(111) adsorption system and the clean Cu(111) slab, respectively. $\rho_{3H_2O-I/Cu}$ and $\rho_{3H_2O-II//Cu}$ are the electron densities of two subsets of the 6 adsorbed H_2O molecules, as labelled in Fig. 6(f). The two subsets of H_2O molecules are selected so that the quantity $\Delta\Delta\rho$ essentially reveals the H_2O–H_2O interaction in the adsorbed hexamer. Clearly from Fig. 6(f) it can be seen that the stronger H bonds form when the high-lying waters act mainly as the acceptors of the H bond, whereas the weaker H bonds form when the low-lying waters act mainly as acceptors. Essentially what the buckled structure of the hexamer tells us is that there is a competition between the ability of a H_2O molecule to form a 'strong' bond with the surface and its ability to act as a 'strong' H bond acceptor.

5. Conclusions

In conclusion the initial results of a DFT study of H_2O clustering on Cu(111) have been reported. Clusters with 2–6 H_2O molecules have been examined and for each cluster low energy adsorption structures reported. In addition, several novel concepts have been illustrated such as the fact that high energy gas phase isomers with fewer H bonds can be stabilized upon adsorption and that the nature of the competition between water–water bonding and water–metal bonding is more subtle than previously realised with a clear distinction having been identified between H_2O molecules that act either as H bond donors or as H bond acceptors.

Acknowledgements

This work was conducted as part of a EURYI scheme award. See www.esf.org/euryi. Matthias Scheffler is thanked for valuable discussions and for carefully reading this manuscript.

References

1 P. A. Thiel and T. E. Madey, *Surf. Sci. Rep.*, 1987, **7**, 211, and references therein.
2 M. A. Henderson, *Surf. Sci. Rep.*, 2002, **46**, 1, and references therein.
3 A. Verdaguer, G. M. Sacha, H. Bluhm and M. Salmeron, *Chem. Rev.*, 2006, **106**, 1478, and references therein.
4 H. Ogasawara, B. Brena, D. Nordlund, M. Nyberg, A. Pelmenschikov, L. G. M. Pettersson and A. Nilsson, *Phys. Rev. Lett.*, 2002, **89**, 276102.
5 S. Meng, L. F. Xu, E. G. Wang and S. W. Gao, *Phys. Rev. Lett.*, 2002, **89**, 176104.
6 C. Clay and A. Hodgson, *Curr. Opin. Solid State Mater. Sci.*, 2005, **9**, 11.
7 T. Schiros, S. Haq, H. Ogasawara, O. Takahashi, H. Ostrom, K. Andersson, L. G. M. Pettersson, A. Hodgson and A. Nilsson, *Chem. Phys. Lett.*, 2006, **429**, 415.
8 G. Zimbitas, S. Haq and A. Hodgson, *J. Chem. Phys.*, 2005, **123**, 174701.
9 C. Clay, S. Haq and A. Hodgson, *Chem. Phys. Lett.*, 2004, **388**, 39.
10 J. Cerdá, A. Michaelides, M. L. Bocquet, P. J. Feibelman, T. Mitsui, M. Rose, E. Fomin and M. Salmeron, *Phys. Rev. Lett.*, 2004, **93**, 116101.
11 T. Yamada, S. Tamamori, H. Okuyama and T. Aruga, *Phys. Rev. Lett.*, 2006, **96**, 036105.
12 A. Michaelides and P. Hu, *J. Am. Chem. Soc.*, 2001, **123**, 4235.
13 A. Michaelides and P. Hu, *J. Chem. Phys.*, 2001, **114**, 513.
14 G. Held, C. Clay, S. D. Barrett, S. Haq and A. Hodgson, *J. Chem. Phys.*, 2005, **123**, 064711.
15 P. J. Feibelman, *Science*, 2002, **295**, 99.

16 G. S. Karlberg, F. E. Olsson, M. Persson and G. Wahnstrom, *J. Chem. Phys.*, 2003, **119**, 4865.
17 A. Michaelides, A. Alavi and D. A. King, *J. Am. Chem. Soc.*, 2003, **125**, 2746.
18 C. Clay, S. Haq and A. Hodgson, *Phys. Rev. Lett.*, 2004, **92**, 046102.
19 G. A. Kimmel, N. G. Petrik, Z. Dohnatek and B. D. Kay, *Phys. Rev. Lett.*, 2005, **95**, 166102.
20 M. Nakamura and M. Ito, *Chem. Phys. Lett.*, 2004, **384**, 256.
21 M. Nakamura and M. Ito, *Chem. Phys. Lett.*, 2005, **404**, 346.
22 S. Yamamoto, A. Beniya, K. Mukai, Y. Yamashita and J. Yoshinobu, *J. Phys. Chem. B*, 2005, **109**, 5816.
23 K. Morgenstern and J. Nieminen, *Phys. Rev. Lett.*, 2002, **88**, 066102.
24 T. Mitsui, M. K. Rose, E. Fomin, D. F. Ogletree and M. Salmeron, *Science*, 2002, **297**, 1850.
25 K. Morgenstern and K.-H. Rieder, *J. Chem. Phys.*, 2002, **116**, 5746.
26 A. Michaelides, V. A. Ranea, P. L. de Andres and D. A. King, *Phys. Rev. Lett.*, 2003, **90**, 216102.
27 A. Michaelides, *Appl. Phys. A*, 2006, **85**, 415.
28 V. A. Ranea, A. Michaelides, R. Ramirez, P. L. de Andres, J. A. Verges and D. A. King, *Phys. Rev. Lett.*, 2004, **92**, 136104.
29 S. Meng, E. G. Wang and S. Gao, *Phys. Rev. B*, 2004, **69**, 195404.
30 S. Haq, C. Clay, G. R. Darling and A. Hodgson, *Phys. Rev. B*, 2006, **73**, 115414.
31 P. Vassilev, R. A. van Santen and M. T. M. Koper, *J. Chem. Phys.*, 2005, **122**, 054701.
32 V. A. Ranea, A. Michaelides, R. Ramirez, J. A. Verges, P. L. de Andres and D. A. King, *Phys. Rev. B*, 2004, **69**, 205411.
33 A. Michaelides, V. A. Ranea, P. L. de Andres and D. A. King, *Phys. Rev. B*, 2004, **69**, 075409.
34 D. Sebastiani and L. Delle Site, *J. Chem. Theory Comput.*, 2005, **1**, 78.
35 A. Michaelides and K. Morgenstern, (submitted).
36 A. Michaelides, A. Alavi and D. A. King, *Phys. Rev. B*, 2004, **69**, 113404.
37 M. D. Segall, P. J. D. Lindan, M. J. Probert, C. J. Pickard, P. J. Hasnip, S. J. Clark and M. C. Payne, *J. Phys.: Condens. Matter*, 2002, **14**, 2717.
38 D. Vanderbilt, *Phys. Rev. B*, 1990, **41**, R7892.
39 (*a*) J. P. Perdew, K. Burke and M. Ernzerhof, *Phys. Rev. Lett.*, 1996, **77**, 3865; (*b*) J. P. Perdew, K. Burke and M. Ernzerhof, *Phys. Rev. Lett.*, 1997, **78**, 1396.
40 *Handbook of Chemistry and Physics*, ed. D. R. Lide, CRC, London, 80th edn, 1999.
41 H. J. Monkhorst and J. D. Pack, *Phys. Rev. B*, 1976, **13**, 5188.
42 B. Hammer, L. N. Hansen and J. K. Nørskov, *Phys. Rev. B*, 1999, **59**, 7413.
43 C. Adamo and V. Barone, *J. Chem. Phys.*, 1999, **110**, 6158.
44 The calculations on the Cu clusters were performed with the Gaussian03 code, M. J. Frisch, G. W. Trucks, H. B. Schlegel, G. E. Scuseria, M. A. Robb, J. R. Cheeseman, J. A. Montgomery, Jr., T. Vreven, K. N. Kudin, J. C. Burant, J. M. Millam, S. S. Iyengar, J. Tomasi, V. Barone, B. Mennucci, M. Cossi, G. Scalmani, N. Rega, G. A. Petersson, H. Nakatsuji, M. Hada, M. Ehara, K. Toyota, R. Fukuda, J. Hasegawa, M. Ishida, T. Nakajima, Y. Honda, O. Kitao, H. Nakai, M. Klene, X. Li, J. E. Knox, H. P. Hratchian, J. B. Cross, V. Bakken, C. Adamo, J. Jaramillo, R. Gomperts, R. E. Stratmann, O. Yazyev, A. J. Austin, R. Cammi, C. Pomelli, J. Ochterski, P. Y. Ayala, K. Morokuma, G. A. Voth, P. Salvador, J. J. Dannenberg, V. G. Zakrzewski, S. Dapprich, A. D. Daniels, M. C. Strain, O. Farkas, D. K. Malick, A. D. Rabuck, K. Raghavachari, J. B. Foresman, J. V. Ortiz, Q. Cui, A. G. Baboul, S. Clifford, J. Cioslowski, B. B. Stefanov, G. Liu, A. Liashenko, P. Piskorz, I. Komaromi, R. L. Martin, D. J. Fox, T. Keith, M. A. Al-Laham, C. Y. Peng, A. Nanayakkara, M. Challacombe, P. M. W. Gill, B. G. Johnson, W. Chen, M. W. Wong, C. Gonzalez and J. A. Pople, *GAUSSIAN 03 (Revision C.02)*, Gaussian, Inc., Wallingford, CT, 2004.
45 S. Maheshwary, N. Patel, N. Sathyamurthy, A. D. Kulkarni and S. R. Gadre, *J. Phys. Chem. A*, 2001, **105**, 10525.

Nucleation in alkali metal chloride solution observed at the cluster level

Akihiro Wakisaka*

Received 2nd November 2006, Accepted 2nd February 2007
First published as an Advance Article on the web 12th April 2007
DOI: 10.1039/b615977j

Nucleation processes of alkali metal chlorides (MCl) in methanol were studied by specially designed electrospray ionization mass spectrometry at the cluster level. From solutions of LiCl, NaCl, KCl and RbCl, the $M^+(MCl)_n$ clusters with n = 4, 13 and 22 were observed prominently as magic number clusters. These clusters correspond to a square-planar $3 \times 3 \times 1$, cubic $3 \times 3 \times 3$, and cuboid $3 \times 3 \times 5$ structure, respectively. On the other hand, from the solution of CsCl, the $Cs^+(CsCl)_n$ clusters were almost monotonically decreased with an increase in the cluster size n, without showing any magic number species. This has a good correlation with the difference in the crystalline structures. When nucleation occurs in the condensed phase, the difference in the crystalline structure will be realised at the cluster level. Moreover, it was demonstrated that the mass distributions of $K^+(KCl)_n$ and $Cl^-(KCl)_{n'}$ clusters look very similar. When these inter-cluster interactions take place efficiently, the crystal growth will be accelerated rapidly.

1. Introduction

Structures of ionic crystal of alklai metal chlorides (MCl) can be classified into two groups, that is, the 6-coordinate rock salt structure (LiCl, NaCl, KCl, RbCl) and the 8-coordinate caesium chloride structure (CsCl), as shown in Fig. 1.[1] The switch from the rock salt structure to the caesium chloride structure is related to the radius ratio between the cation and the anion. The difference in the crystalline structure between the rock salt structure and the caesium chloride structure can be investigated precisely through X-ray diffraction experiments. However, there is little information about the difference in the nucleation process. Therefore, we would like to focus on the nucleation processes of alkali metal halides on the basis of the cluster structures observed through specially designed electrospray ionization mass spectrometry (ESI-MS).

The alkali halide clusters ($M^+(MX)_n$ and $X^-(MX)_n$) have been investigated extensively by ESI-MS in order to discern the mechanism of ESI.[2–4] Mechanisms for the ESI process have been proposed previously: the ion evaporation model of Iribarne et al.[5] and the charge residue model of Dole et al.[6] Furthermore, Wang and Cole proposed the extended charge residue model.[4] Recently, March et al.[2] and Kebarle et al.[3] reported that $M^+(MX)_n$ and $X^-(MX)_n$ clusters at n = 4, 13 and 22 show unusual high intensities (magic numbers). Kebarle et al. suggested that even though the ESI mechanism in the gas phase is not known in sufficient detail, it is

National Institute of Advanced Industrial Science and Technology, Onogawa 16-1, Tsukuba, Ibaraki, 305-8569, Japan. E-mail: akihiro-wakisaka@aist.go.jp; Fax: 81 29 861 8252; Tel: 81 29 861 8088

Rock salt structure (6-coordination): LiCl, NaCl, KCl, RbCl

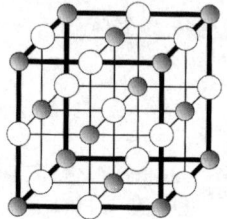

CsCl structure (8-coordination): CsCl

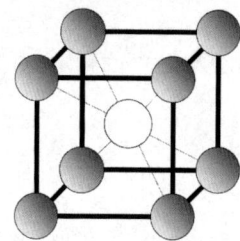

Fig. 1 Structures of ionic crystals of alkali metal chlorides.

likely that the observed distribution of clusters will reflect at least qualitatively the relative concentrations of different microcrystallites in the solution.[3]

Besides ESI, there are many reports on the magic number alkali metal halide clusters generated through fast atom bombardment (FAB) to alkali halide crystals,[7] cooling of alkali halide vapour[8] and laser vaporization of alkali halide rods.[9] In order to determine the structures of the alkali metal halide clusters, Diefenbach and Martin[10] carried out model calculations of the total energy of alkali halide clusters. The most stable configurations of clusters $M^+(MX)_n$, $n = 1$–18 were calculated. For the magic number $n = 4$, a square-planar-like $3 \times 3 \times 1$ configuration was calculated, but it was not completely planar. For the magic number $n = 13$, a cubic $3 \times 3 \times 3$ configuration was calculated. Through the experimental approaches, especially mass spectrometry, it is difficult to determine the structures of the clusters. However, they could demonstrate that the magic-numbered alkali metal halide clusters ($M^+(MX)_n$, $n = 4, 13, 22$) were relatively stable. March et al.[2] and Kebarle et al.[3] demonstrated that, with ESI-MS, the distribution of $M^+(MX)_n$ clusters was influenced by the drift field potential of the ESI-MS. At relatively high drift field potential, the $M^+(MX)_n$ clusters were accelerated, collided energetically with background gas molecules, and underwent collisional dissociation. This led to the formation of magic number clusters with high signal intensities. Moreover, Ens et al.[7] demonstrated that, with FAB, $Cs^+(CsI)_{13}$ clusters had a long lifetime. All these experimental observations showed that the magic number $M^+(MX)_n$ clusters were particularly stable. Since these clusters are formed through electrostatic forces, it is reasonable that symmetrical structures will be particularly stable.

The difference in the crystal structures of the rock salt and caesium chloride structures was not reflected in these previous observations. We speculated that this might be due to the fact that these alkali halide clusters were generated in the gas phase. If the clustering processes are controlled by the interaction in the condensed phase, the resulting cluster structure will reflect the difference in the crystal structures as their nucleation. Here we would like to report mass spectrometric analyses of alkali metal chlorides in methanol by means of ESI-MS with a five-stage (three-skimmer system) or a seven-stage (five-skimmer system) differentially pumped vacuum system. In the observed mass spectra, the magic-numbered clusters of alkali metal chlorides were found to be related to their bulk crystalline structures. We have already demonstrated that the structures of the alkali halide clusters in methanol are determined by the balance of the ion–counterion and the ion–solvent (or substrate) interactions in solutions by means of ESI-MS.[11] It was observed that the solvation of the ions increased with a decrease in the ion–counterion interactions, and decreased with an increase in the ion–counterion interactions. Such findings, which we call 'complementary relationship',[11,12] indicate that the observed clusters reflect the interactions inherent in the solutions. By use of this advantage of our ESI-MS, the nucleation processes of alkali metal chlorides will be discussed here.

2. Experimental

Mass spectrometric analyses of positively and negatively charged clusters generated from alkali metal chloride–methanol solutions were carried out by means of a specially designed mass spectrometer. The experimental setup is shown schematically in Fig. 2, and details has been reported previously.[11] It is composed of a home-made electrospray interface made of a fused silica capillary tube (i.d. 0.1 mm) and a platinum electrode, a quadrupole mass filter (Extrel C50), and a specially designed differentially pumped vacuum system. During the measurements, a methanolic solution of an alkali metal chloride was injected into the high electric field between the nozzle and the first skimmer through the fused silica capillary tube at a flow rate of 0.005–0.01 cm^3 min^{-1}. Based on the electrospray principle, positively and negatively charged liquid droplets including excess cations or anions were generated according to the polarity of the electric field. The resulting multi-charged liquid droplets entered into the second, third, and fourth chambers travelling under the influence of the potential and pressure gradients. The multi-charged liquid droplets were then fragmented into clusters *via* adiabatic expansion and electrostatic repulsion during their flight. The charged clusters, which reached the quadrupole mass filter, were mass-analyzed without further external ionization.

For obtaining the electrospray, electric voltage was supplied to the nozzle and the skimmers. It should be noted that the clustering process is remarkably dependent on the electric field given by these skimmer potentials, as described in the following section. The electric potentials used are shown in Table 1. In order to investigate the effect of electric field on the clustering, the five-skimmer system is used. Other measurements are carried out by use of the three-skimmer system shown in Fig. 2. To maintain an appropriate pressure for the electrospray, nitrogen gas was flowing during the measurement. The pressure of each room is shown in Table 2.

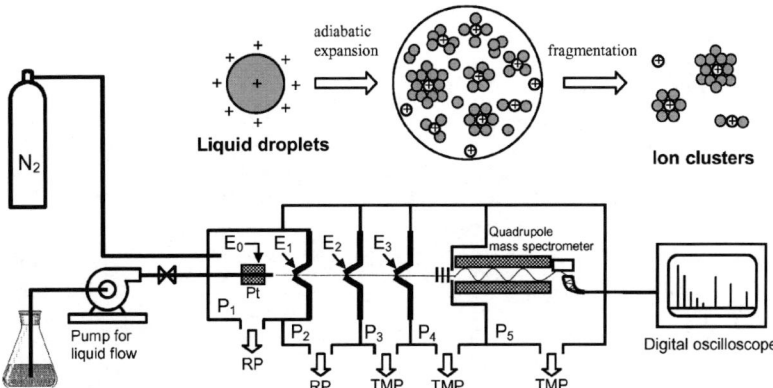

Fig. 2 Schematic illustration of the specially designed electrospray ionization mass spectrometry for clusters isolated from solutions equipped with three skimmers for the measurements shown in Fig. 3 and 4. E: electric potential, P: pressure, Pt: platinum electrode, RP: rotary pump, TMP: turbo molecular pump.

Table 1 Electric potentials supplied to the nozzle and skimmers used in the measurements, E_o: nozzle potential, E_{1-5}: skimmer potential. Fig. 3 and 4: three skimmers, Fig. 6 and 7: five skimmers

	Fig. 3	Fig. 4 (negative)	Fig. 6	Fig. 7
E_0	+3890	−3550	+3990	+3950
E_1	+346	−300	+315	+285
E_2	+335	−277	+317	+298
E_3	+227	−191	+207	+264
E_4			+186	+232
E_5			+163	+167

Chemicals of the highest purity were used without further purification: LiCl (anhydrous, special grade, Wako), NaCl (99.5%, Wako), KCl (99.5%, Wako), RbCl (95%, Wako), CsCl(99.0%, Wako), methanol (99.8%, Wako).

3. Results and discussion

3.1 Mass spectrometric analyses of alkali metal chloride solutions

In order to observe the difference in the nucleation of alkali metal chlorides, LiCl, NaCl, KCl, RbCl and CsCl, their solutions in methanol solvents were analyzed by ESI-MS. The mass spectra observed through the positive-ion mode are shown in Fig. 3. In each spectrum, a series of salt clusters $(M^+(MCl)_n, n = 1, 2, 3, \ldots)$ and a series of solvated ions $(M^+(CH_3OH)_s, s = 0, 1, 2, 3, 4)$ are clearly observed.

For the $M^+(MCl)_n$ clusters, there is obvious difference in the mass distributions between CsCl and the others. For LiCl, NaCl, KCl and RbCl, the magic number $M^+(MCl)_n$ with $n = 4$, 13 and 22 are observed as prominent peaks. The magic number clusters with $n = 4$, 13 and 22 will correspond to a square-planar $3 \times 3 \times 1$ structure, a cubic $3 \times 3 \times 3$ structure and a cuboid $3 \times 3 \times 5$ structure, respectively, if the highly symmetrical structures make stable ones. In particular, the cubic structure is the same as a unit cell of the rock salt crystal as shown in Fig. 1. For LiCl and NaCl, $n = 10$ and 16 also show the magic number property. These clusters will have terraced structures. On the other hand, CsCl does not form such magic number clusters. The peak intensities of $Cs^+(CsCl)_n$ clusters are decreased almost monotonically with an increase in the cluster size. The difference in the mass spectra between CsCl and the others shows good correlation with the difference in the crystalline structures as shown in Fig. 1. Accordingly, it is suggested that the formation of clusters in the solutions is related to the nucleation process connecting with the crystalline structure.

Doubly and triply charged salt clusters are also observed in Fig. 3. It has been reported that the formation of these multiply charged clusters is dramatically influenced by the drift field potential of the ESI-MS.[2,3] It is known that the multiply

Table 2 Pressures in the vacuum chambers of differentially pumped system during the measurements. Fig. 3 and 4: five stages with three skimmers, Fig. 6 and 7: seven stage with five skimmers

	Fig. 3	Fig. 4 (negative)	Fig. 6	Fig. 7
P_1	691.6	749.4	673.0	673.0
P_2	15.79	17.38	12.62	12.62
P_3	0.012	0.015	2.33	2.33
P_4	2×10^{-5}	2×10^{-5}	0.437	0.437
P_5	4×10^{-7}	4×10^{-7}	9.5×10^{-4}	9.5×10^{-4}
P_6			6×10^{-6}	6×10^{-6}
P_7			3×10^{-7}	3×10^{-7}

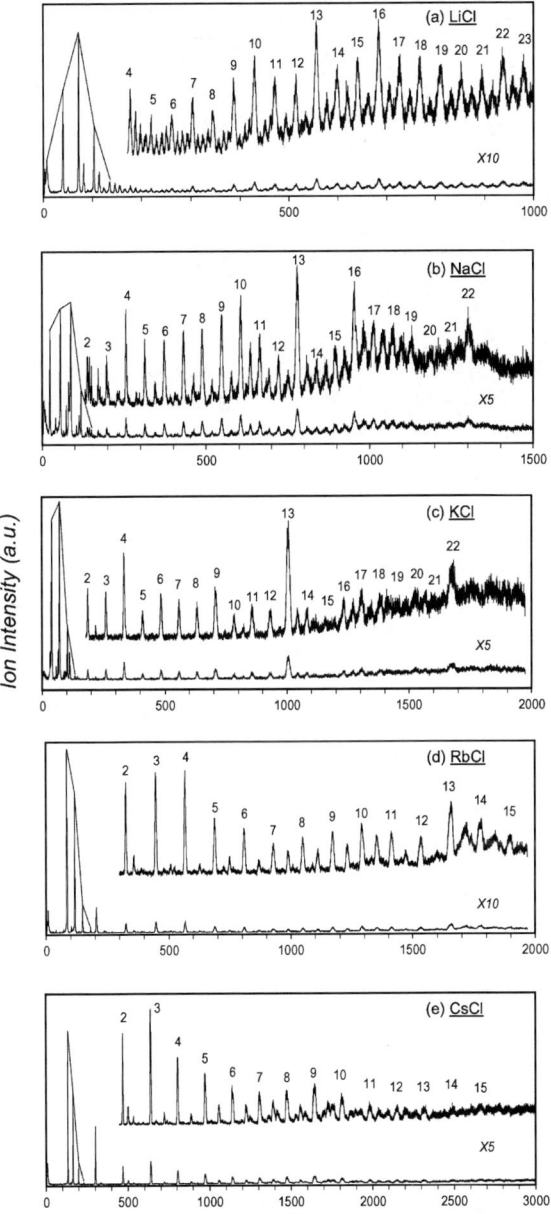

Fig. 3 Mass spectra of clusters generated from alkali metal chloride–methanol solutions: (a) LiCl (0.02 mol dm^{-3}), (b) NaCl (0.005 mol dm^{-3}), (c) KCl (0.005 mol dm^{13}), (d) RbCl (0.0032 mol dm^{-3}), and (e) CsCl (0.0034 mol dm^{-3}). The numbers shown on the peaks represent n of $M^+(MCl)_n$. A series of solvated ions $M^+(CH_3OH)_s$ $s = 0, 1, 2, 3, 4$ are connected by the lines. The electric potentials and pressures used are shown in Tables 1 and 2, respectively.

charged clusters are decomposed through the collisional interaction with background gas at a high electric field. The effect of the drift field potential on the resulting clusters is discussed in the following section.

As for the solvated ions, it is demonstrated that the alkali metal ion–methanol molecule interactions are controlled by the charge densities of the alkali metal ion. For LiCl solution, the intensity of the naked Li$^+$ peak is very small, and Li$^+$

solvated by methanol molecules ($Li^+(CH_3OH)_s$) is observed prominently. With an increase in ionic radius, that is, with a decrease in the charge density, the naked ion is observed prominently rather than that solvated by methanol. In the case of the CsCl–methanol solution (Fig. 3e), the salt clusters are easy to generate even at low concentration owing to the weak solvation to Cs^+.

3.2 Growth of clusters

Positively and negatively charged clusters. The salt clusters observed in Fig. 3 were measured by the positive-ion mode; therefore, each observed cluster had one or a few excess cations. In order to see the symmetry of cluster structures between positive and negative ion clusters, we compared mass spectra measured using the positive-ion mode and the negative-ion mode. Since the magic numbers of positive ion clusters with n = 4, 13 and 22 for KCl were observed most clearly in Fig. 3, the KCl–methanol solution was measured by both the positive- and negative-ion modes. Fig. 4 shows these mass spectra. In the negative-ion mode, a series of salt clusters ($Cl^-(KCl)_{n'}$) and a series of solvated ions ($Cl^-(CH_3OH)_{s'}$, s = 0, 1, 2) are observed. The magic-numbered $Cl^-(KCl)_{n'}$ with n' = 4, 13 and 22 are confirmed easily in Fig. 4. Furthermore, the mass distribution of other salt clusters also looks very similar between the $K^+(KCl)_n$ and the $Cl^-(KCl)_{n'}$ clusters. This finding indicates that the positively and negatively charged KCl clusters will coexist with their almost symmetrical distribution in methanol. It will be suggested that the inter-cluster

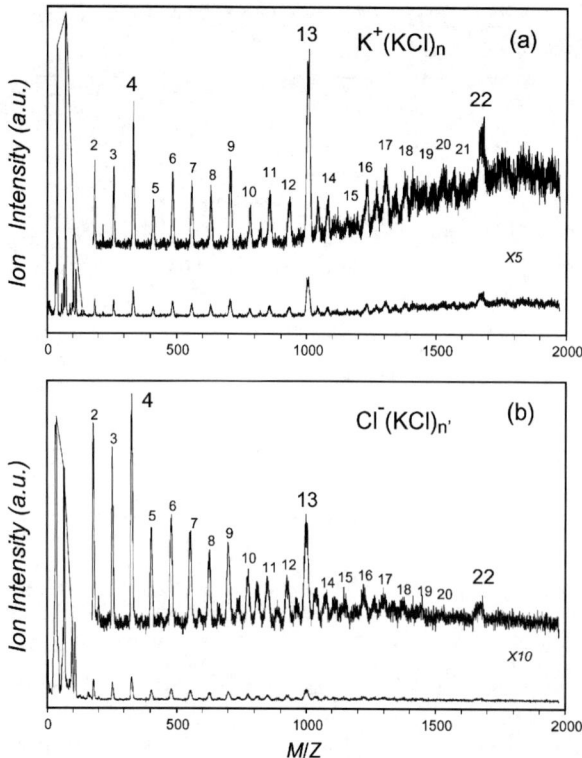

Fig. 4 Mass spectra of clusters generated from a KCl (0.005 mol dm^{-3})–methanol solution measured by the positive-ion mode (a) and the negative ion mode (b). The numbers shown on the peaks represent n and n' of $K^+(KCl)_n$ and $Cl^-(KCl)_{n'}$. A series of solvated ions $K^+(CH_3OH)_s$, s = 0, 1, 2, 3, and $Cl^-(CH_3OH)_{s'}$, s' = 0, 1, 2 are connected by the lines. The electric potentials and pressures used are shown in Tables 1 and 2, respectively.

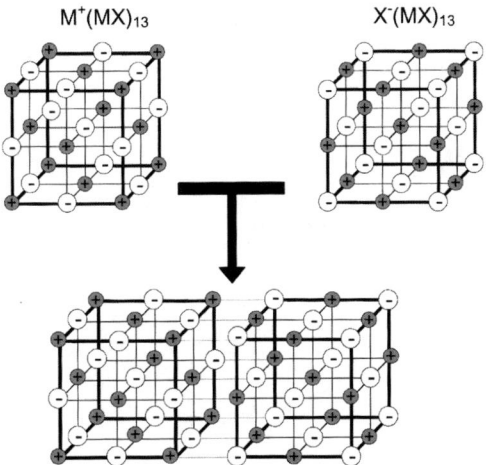

Fig. 5 A schematic illustration of the growth of nucleation through the interaction between the positively and negatively charged clusters. This is one of the possible interactions between $M^+(MX)_{13}$ and $X^-(MX)_{13}$ clusters.

interaction between the positively and negatively charged clusters can accelerate the growth of nucleation from solutions, as shown schematically in Fig. 5.

Effect of concentration on the salt cluster formation. Since the nucleation should be dependent on the concentration, the concentration effect on the KCl salt cluster formation is examined. Fig. 6 shows mass spectra of $K^+(KCl)_n$ clusters measured for the KCl (5×10^{-3}, 2×10^{-3}, 5×10^{-4} mol dm^{-3})–methanol solutions. This concentration effect was studied by use of the five-skimmer system, which can control the electric field more precisely as listed in Table 1, because the result will also be discussed from the viewpoint of the effect of the electric field. The mass distribution of $K^+(KCl)_n$ clusters at 5×10^{-3} mol dm^{-3} reproduces that in Fig. 4 well, and the magic numbers can also be observed. With a decrease in KCl concentration, the magic numbers become ambiguous. It will be reasonable that the magic number cluster, especially $n = 13$ corresponding to the $3 \times 3 \times 3$ cubic unit cell structure, will form significantly in almost saturated or supersaturated solutions. Even though the solutions (5×10^{-3} mol dm^{-3}) are not saturated with KCl, the extensive evaporation of the droplets during the drift in the vacuum chamber of the ESI-MS will lead to the saturated and/or supersaturated condition. This evaporation was remarkably dependent on the drift field potential given by the electric potentials of the skimmers. One can expect that the collisional interaction with background gas in the drift field will induce the evaporation of the liquid droplets to concentrate the salt.

Effect of the drift field potential. In our ESI-MS, the drift field potential between skimmer 2 and skimmer 3, E_{2-3} had significant influence on the evaporation of the liquid droplets generated by the electrospray. As shown in Fig. 3, 4 and 6, when E_{2-3} was over 80 V as listed in Table 1, the salt clusters were generated markedly through the evaporation of methanol. However, when E_{2-3} was lower than 35 V, the resulting clusters were drastically changed. Fig. 7 shows clusters generated from the same KCl–methanol solutions as used for Fig. 6a under the lower drift field potential (Table 1). In Fig. 7, the naked and solvated ions ($K^+(CH_3OH)_s$, $s = 0, 1, 2$) are observed predominantly instead of the salt clusters. This means that the liquid droplets just before the fragmentation to the clusters will not be saturated due to the decrease in the evaporation under the lower drift field potential. In other words, the

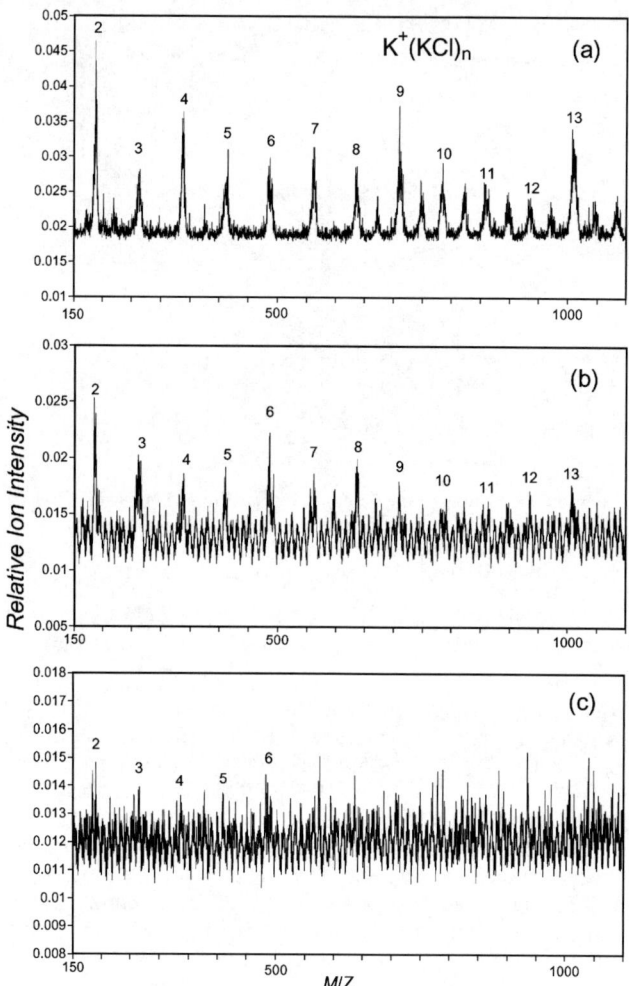

Fig. 6 The effect of concentration on the formation of $K^+(KCl)_n$ clusters in methanol: (a) 5×10^{-3} mol dm^{-3}, (b) 2×10^{-3} mol dm^{-3}, (c) 5×10^{-4} mol dm^{-3}. The electric potentials and pressures used are shown in Tables 1 and 2, respectively.

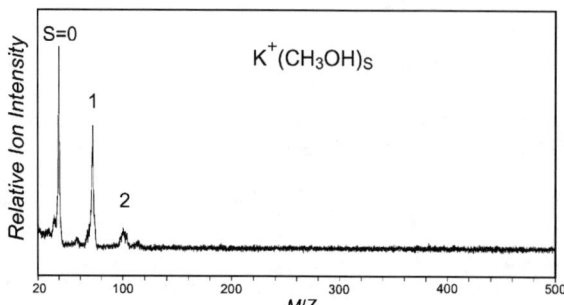

Fig. 7 A mass spectrum of clusters generated from a KCl (5×10^{-3} mol dm^{-3})–methanol solution under the lower drift field potential. The electric potentials and pressures used are shown in Tables 1 and 2, respectively. The sample solution is the same as that used in Fig. 6a.

salt clusters are generated efficiently only when the salt concentrations in the liquid droplets attain saturation or supersaturation through the collisional interaction.

It should be noted that the effect of the drift field potential has been extensively studied. March et al.[2] and Kebarle et al.[3] have already reported that the mass distributions of alkali metal chloride solutions were markedly influenced by the magnitude of the drift field potential for the liquid droplets. In their experiment, multi-charged clusters were generated at the low drift field potential, which represents the initial distribution of ionic clusters without having energetic collisional interaction; whereas stable clusters such as magic number species were observed at the high drift field potential, which were formed through the collisional dissociation of the liquid droplets.

There is something in common between these previous reports and our results on the effect of the drift field potential. In both experiments the collisional interaction of the liquid droplets with background gas molecules is restrained at the low drift field potentials. In our ESI-MS, the electrospray nozzle is situated coaxial to the five skimmers to reduce collisional interaction more effectively, and the five-skimmer potentials can generate a smooth electric field. Due to this special design, the solvated K^+, $K^+(CH_3OH)_s$, was observed predominantly at the lower drift field potential, as shown in Fig. 7. This reflects the ion–molecule interaction in the liquid droplets before the saturated condition. At the higher drift field potential, however, the collisional interaction of the liquid droplets with the background gas molecules cause the evaporation of methanol from the liquid droplets. This leads to the formation of salt clusters via saturated or supersaturated conditions.

3.3 Nucleation in the condensed phase

It is well known that the structures of ionic crystal of alkali metal chlorides can be classified into two groups, that is, the 6-coordinate rock salt structure and the 8-coordinate CsCl structure. However, in previous reports, the mass spectra of clusters of all alkali halides showed the same magic numbers, and the difference in the crystalline structure was not reflected. On the basis of those results, Diefenbach and Martin[8] and El-Sayed et al.[7] suggested that the nature of binding differences which led to distinct structures in the condensed phase was insignificant in the cluster size regime. This is contrary to our results shown here.

We think that this difference will depend on the environment where the nucleation takes place. In the gas phase, the nature of binding differences leading to crystal structures may not be significant in cluster formation.[7–9] On the contrary, when the nucleation takes place in the condensed phase, the difference in the crystalline structures is also reflected in the cluster structures. Our mass spectrometric analyses give information reflecting interactions in solutions. This has already been demonstrated by the findings of 'complementary relationships' among the interactions working at the same time in solutions.[11,12] Accordingly, we could observe the difference in the nucleation at the cluster level. It should be noted that the difference in the crystalline structure is realised at the cluster level when the clusters form in the condensed phase.

4. Conclusions

We have observed the nucleation of alkali metal chlorides in methanol at the cluster level by means of specially designed mass spectrometry. The $M^+(MCl)_n$ salt clusters were found to be concentration and electric field dependent. The salt cluster with $n = 13$ corresponding to the $3 \times 3 \times 3$ cubic unit cell for a rock salt crystal was observed as a representative magic number cluster from the LiCl, NaCl, KCl and RbCl solutions, but no magic number cluster was observed from the CsCl solution. This is in good agreement with the difference in their crystalline structures. The concentration and electric field effects indicate that the salt clusters will form in the liquid droplets at the saturated or supersaturated conditions generated in the drift field of the ESI-MS. The difference in the crystalline structures is realised at the nucleation process when the nucleation starts in the condensed phase. Moreover, it was demonstrated that the mass distributions of $K^+(KCl)_n$ and $Cl^-(KCl)_{n'}$ clusters look very similar. The crystal growth will be accelerated when these inter-cluster interactions take place frequently.

Acknowledgements

The author thanks M. Nakagawa for the data processing.

References

1 C. Kittel, *Introduction to Solid State Physics*, Wiley, New York, 1986.
2 C. Hao, R. E. March, T. R. Croley, J. C. Smith and S. P. Rafferty, *J. Mass Spectrom.*, 2001, **36**, 79.
3 A. T. Blades, M. Peschke, U. H. Verkerk and P. Kebarle, *J. Am. Chem. Soc.*, 2004, **126**, 11995.
4 (*a*) G. Wang and R. B. Cole, *Anal. Chim. Acta*, 2000, **406**, 53; (*b*) G. Wang and R. B. Cole, *Anal. Chem.*, 1998, **70**, 873.
5 (*a*) J. V. Iribarne and B. A. Thomson, *J. Chem. Phys.*, 1976, **64**, 2287; (*b*) B. A. Thomson and J. V. Iribarne, *J. Chem. Phys.*, 1979, **71**, 4451; (*c*) C. K. Meng and J. B. Fenn, *Org. Mass Spectrom.*, 1991, **26**, 542; (*d*) J. B. Fenn, *J. Am. Soc. Mass Spectrom.*, 1993, **4**, 524.
6 M. Dole, L. L. Mack, R. L. Hines, R. C. Mobley, L. D. Ferguson and M. B. Alice, *J. Chem. Phys.*, 1968, **49**, 2240.
7 (*a*) T. S. Ahmadi and M. A. El-Sayed, *J. Phys. Chem. A*, 1997, **101**, 690; (*b*) T. M. Barlak, J. E. Campana, J. R. Wyatt and R. J. Cotton, *J. Phys. Chem.*, 1983, **87**, 3441; (*c*) W. Ens, R. Beavis and K. G. Standing, *Phys. Rev. Lett.*, 1983, **50**, 27.
8 (*a*) J. Diefenbach and T. P. Martin, *J. Chem. Phys.*, 1985, **83**, 2238; (*b*) R. Pflaum, K. Sattler and E. Recknagel, *Chem. Phys. Lett.*, 1987, **138**, 8.
9 (*a*) P. Dugourd, R. R. Hudgins and M. F. Jarrold, *Chem. Phys. Lett.*, 1997, **267**, 186; (*b*) R. R. Hudgins, P. Dugourd, J. M. Tenenbaum and M. F. Jarrold, *Phys. Rev. Lett.*, 1997, **78**, 4213.
10 J. Diefenbach and J. P. Martin, *J. Chem. Phys.*, 1985, **83**, 4585.
11 (*a*) S. Mochiduki and A. Wakisaka, *J. Phys. Chem. A*, 2002, **106**, 5095; (*b*) T. Megyes, T. Radnai and A. Wakisaka, *J. Phys. Chem. A*, 2002, **106**, 8059.
12 (*a*) A. Wakisaka and T. Ohki, *Faraday Discuss.*, 2005, **129**, 231; (*b*) A. Wakisaka and K. Matsuura, *J. Mol. Liq.*, 2006, **129**, 25.

General Discussion

Professor Davey opened the discussion of Professor Kahr's paper: In the choice of material with which to test this hypothesis it might be important to consider the effects of mechanical properties in the generation of new dislocations. This was explored in the 1980s by Sherwood and his collaborators.[1]

1 J. N. Sherwood and T. Shripathi, *Faraday Discuss.*, 1993, **95**, 173–182.

Professor Kahr responded: We should test a full range of conceivable replicating systems. Only by investigation will we be able to sort the wheat from the chaff. The dislocation/punched-card model, in hindsight, seems like chaff, but we couldn't know this at the outset. Likewise, it was inconceivable to presuppose the mechanism of action of DNA without first developing its structure. It seems daunting to guess at the characteristics of primitive genetic systems without first playing-out some of these ideas in the laboratory. Polytypism plays a much larger role in the theory of Genetic Takeover than do dislocations. However, we began with dislocations merely because we felt that we had chemistry in hand to rigorously test this idea.

Dr Ristic said: I find this work extremely interesting. The authors managed to cast a considerable amount of light on the Cairns-Smith's 'crystals-as-genes' hypothesis. Listening to the presenter I realised that in the recent past of our research work, my colleagues and I had been somehow involved in one aspect of this intriguing problem, but with a different motivation. Our aim was to find experimental conditions at which the existing screw or mixed dislocations in a crystal seed would continue to propagate through the interface between the original seed and its re-grown part and continue to serve as active sources of growth steps. The re-growing face was observed by *in situ* laser interferometry which would enable us to see the growth hillocks at the same sites as those at the original face. In addition, X-ray topography would confirm that these hillocks originated from the same dislocations, which were propagating from the original seed to the new re-growing part of the crystal. These experiments were performed on different materials such as a brittle sodium chlorate and potash alum, and ductile sodium nitrate. In all cases the transfer of the arrangement of the growth units (ions, molecules) within a dislocation 'line' from a seed to its re-growing part would always take place under very low supersaturation. On the contrary, at moderate and high levels of supersaturation, a markedly strained interface between seed and the re-grown part would develop. In this case the original seed dislocations would end up in the strained region and cease to exist. From this region new screw or mixed type dislocations would form and act as the generators of the growth steps. Finally, it might be interesting to mention that by stretching a plate of a ductile sodium nitrate during its growth, it was possible to induce mechanical dislocations in the seed and transfer them to the re-growing part in which they would act as new sources of growth steps.

Professor Kahr answered: Is there a better choice of material? There most certainly may be. Professor Ristic already suggested several possibilities such as sodium nitrate. Moreover, I am not disputing that macroscopic inclusions can be a source of dislocations. However, we grew KAP in the presence of luminescent nanoparticles and microspheres, while also adding luminscent dyes that light-up the hillocks. If the particles cause new dislocations, then we would have expected to see bright spots at the vertices of the luminescent chevrons marking the hillocks. While the particles were easily overgrown by KAP, they were never found at the apices of the vicinal faces of the hillocks. Particles do not create dislocations in this case. When examining cleaved surfaces that have been re-grown, we observed that new hillocks formed largely on the upper terraces. Presumably material is more easily carried to the upper surfaces. But, fundamentally, what causes hillocks? What are the molecular mechanisms?

Dr Cairns-Smith commented: It is good to see this serious piece of experimental exploration on the question of how complicated information might be transferred from pre-existing structures to newly forming ones through crystal growth. The main motive for seeking efficient systems of this sort is that they might provide novel genetic materials, and hence novel physicochemical systems able to evolve through natural selection.

In 1914 Troland had suggested that some sort of crystal growth process might be part of how the genes of today's organisms pass on hereditary information to offspring,[1] and Schrödinger in the 1940s used an analogy for what he imagined our genes to be like.[2] They would turn out to be crystals of a sort but not merely repetitious like wall paper, rather they would be interestingly complicated like a tapestry. They should be "aperiodic crystals". DNA indeed turned out to be aperiodic, although more like a text than a tapestry and using some 2 million Daltons of enzymatic machinery for this "text" to be replicated by means that bear little resemblance to ordinary crystal growth processes. Particularly in connection with the origin of life on the earth we should be looking for altogether more primitive genetic materials, which is to say ones that operate without the aid of highly evolved machinery. Professor Kahr explores an experimentally sophisticated but conceptually simple version of how a 2D pattern (a "tapestry" or perhaps we should say a pile of tapestries) might replicate a particular aperiodicity. He identifies two problems which the design of any potential crystal gene must cope with. There must be occasional mutations for genetic information to evolve, but the main problem is to keep mutation rates *down* so that evolved information is preserved through repeated copying. The second point which is highlighted in this study is that crystal cleavage, an essential part of a complete replication cycle, is particularly prone to introducing new dislocations in such 2D systems.

I rather like the idea of a crystal gene based on a disordered mixed layer or polytypic material.[3] The "information" would be in the form of a particular aperiodic stacking sequence. It would be 1D, like DNA in this respect, but unlike the elaborate procedures of DNA replication we imagine this crystal gene reproduces by simple units adding only to its edges in conformity with the pre-existing stacking pattern. The result would be an extensive, perhaps branching, flat stack of uniform thickness and with the same stacking pattern displayed everywhere on its edges. Such a primitive gene would be reproducing vegetatively like a clover plant covering a field, although bits of the stack might break off from time to time to seed similar colonies elsewhere. The damage likely to be caused by such breaking off would be less serious here than for a 2D model since here the cleaved surfaces do not have to be able to re-grow. Here the process of breaking off part of the structure could leave most of the edge surfaces untouched.

1 L. Troland, *Monist*, 1914, **24**, 92–133.
2 E. Schrödinger, *What is Life?*, Cambridge University Press, 1944.
3 A. G. Cairns-Smith, *Elements*, 2005, **1**, 157–161.

Professor Kahr responded: As mentioned above, in the theories of Cairns-Smith, polytypism in layered materials plays a more prominent role than the rather off-hand proposition of dislocation propagation that we focused on. It is certainly true that our study is not a representative exploration of the mineral origins of life. We seized on the crystal-as-punched card idea only because we felt that we had the chemistries in hand to properly evaluate it in just one crystal. The stacking model surely deserves equal if not more attention. At the very least, I hope that we have shown that it is possible to design experiments that can put these ideas on a firm experimental foundation. But one study is surely not sufficient to cover the range of ideas over which Dr Cairns-Smith's imagination has travelled.

Given advances in scanning probe microscopies among other technologies, we can now test the ideas of Cairns-Smith that were not imaginable 20 or 30 years ago. Some ideas are clearly before their time. Now is the time to bring this hypothesis under the umbrella of experiment.

Dr Cairns-Smith said: Mansfield and Bailey[1] found that the unit layers of vermiform kaolinite had a kind of crazy-paving domain structure, a domain corresponding to one of three possible positions for the octahedral vacancies in the kaolinite structure, and Williams and Garey[2] found mosaic patterns in replicas of etched kaolinite surfaces. So kaolinite might provide a model for a 2D genetic material—if the same pattern is written in successive layers stacked on top of each other. Perhaps some of the newer techniques used by Professor Kahr, such as AFM, might be of help here.

1 C. F. Mansfield and S. W. Bailey, *Am. Mineral.*, 1972, **57**, 411–425.
2 D. G. Williams and C. L. Garey, *Clays Clay Miner.*, 1974, **22**, 117–125.

Professor Breu asked: Storing information and replicating it is only the first step in genetic take over. Additionally, the data need to be read out, meaning the information stored needs to be transcribed into specific chemistry. Do you have an idea how different patterns of dislocations could be transcribed into selectively different chemistry/catalytic activity?

Professor Kahr responded: We have observed that riboflavin and flavin adenine dinucleotide (FAD) can be adsorbed and overgrown by KAP. One can imagine crystals concentrating such small biomolecules and catalyzing their condensation. FAD can lead to poly-A in this way. We are not suggesting that this is how it happens. But Dr Cairns-Smith talks at length about the adsorption, concentration, and catalytic condensation of small organics by clays. We must however recognize that there may be innumerable genetic systems that precede oligonucleotides. We must resist the temptation to rush to DNA.

Dr Cairns-Smith commented: Watson and Crick were quickly able to identify DNA as a genetic material, from its structure, without having any knowledge of how the information in it might work. I think we should take a similar attitude to the origin of life in our attempts to identify truly primitive genetic materials. Of course what their messages might mean, how they might benefit the systems holding them, will be the next question, but very much the next question. The first thing to look for, the *sine qua non*, is the potential to hold and replicate information efficiently.

Professor Roberts commented: The issue of screw dislocations, particularly those generated through growth, is a surprisingly complex process being the result of the conflicting 'demands' of the real (lattice direction defects) and reciprocal (lattice growth planes) crystal lattices. The former reflect closest atom–atom (or molecules) distances between equivalent Bravais lattice sites. The latter for facetted crystals is driven by the need for slow-growing and hence close-packed crystal faces. Hence, it is often the case that crystal habit faces have a form for which the shortened lattice vector defining the screw dislocation Burgers vector does not lie along the growth normal and hence pure screw dislocation are most unlikely. There are many examples for this:

Material	Dominant crystal habit plane	Lowest area Burgers vector
Si, C (diamond)	{111}	$\frac{1}{2}\langle 110 \rangle$
Potash alum.	{111}	$\langle 100 \rangle$
KDP	{100}	$\frac{1}{2}\langle 111 \rangle$
Ammonium sulfate	{111}	$\langle 001 \rangle$
Benzophenone	{110}	$\langle 001 \rangle$

The directions of growth dislocation has been described in the seminal work of Klapper.[1] Broadly speaking, these can be defined with respect to their line (*l*), Burgers vector (*b*) and growth direction (*n*) thus

If **b** is parallel to **n** then **l** is parallel to **n**
If **b** is perpendicular to **n** then **l** is parallel to **n**
If **b** is not perpendicular or parallel to **n** then **l** lies between **b** and **n**

The latter case gives rise to dislocations that we often see in solution grown crystals where mixed component dislocations travel at an angle to the normal to the growth face having only a partial screw component. A related outcome is that growth spirals from such dislocation are somewhat eccentric in nature.

A related generic point is that for symmetry below cubic the growth normal is often non-parallel to the associated lattice direction. The exception to this is in the cube directions for orthogonal lattices (tetragonal and orthorhombic lattices) for **b** = $\langle h00\rangle$, $\langle 0k0\rangle$ or $\langle 00l\rangle$.

In the extreme case of a triclinic lattice then **b** is never parallel to **n** and hence pure screw dislocations can be expected. In this case **b** can be perpendicular to **n** and hence pure edge dislocations are possible. We should not close out a discussion on growth dislocation without reflecting that quite a lot of experimental evidence now exists to support the view that edge dislocations promote the growth process.

Thus, in considering the growth dislocations in crystals as templates *via* fractures in promoting genetic material, a focus on dislocations in general and not just specifically screw dislocations, might have some value.

1 H. Klapper, in *The Characterisation of Crystal Growth Defects by X-Ray Methods*, ed. B. K. Tanner and D. K. Bowen, Plenum Press, New York, 1980.

Dr Schön asked: What is the original motivation behind looking at the fractal dimension of the hillock formation/distribution? What concerns me is the fact that your distribution in Fig. 6 exhibits the fractal behavior only over one order of magnitude, which is rather short on first sight—thus it would be helpful to have a growth model that leads to fractal behavior in a "natural" fashion. Do you expect *e.g.* some diffusion limited aggregation process to be involved? Perhaps such a process could follow from the issue mentioned earlier that the axis of the screw-dislocation and the axis of the surface growth are mis-aligned? Or is it connected to your observation that there is periodical splitting of the screw-dislocation into two new dislocations?

Professor Kahr replied: We struggled with defining the kind of information a primitive crystal gene might carry. The fractal dimension seemed like a number or "score" that could be used to compare crystals grown under different conditions. However, we are dealing with random fractals that have a stochastic character, not geometric fractals that are deterministic. We are certain open to better strategies for characterizing the dispositions of all of the growth active hillocks that appear in space (xyz) and time. Since we don't know what causes the formation of new screw dislocations in KAP it is hard to presuppose what mechanisms might result in the observed correlations of their positions. There are some substantial differences between our images of hillocks and the complementary information that is available from X-ray topography. The notion of V-shaped dislocations comes from Halfpenny[1] and could explain correlations of hillock positions along the *a*-axis. However, movement of cores in the *ac* plane would blur our luminescence micrographs. This is not observed.

1 G. R. Ester and P. J. Halfpenny, *Philos. Mag. A*, 1999, **79**(3), 593–608.

Professor Roberts commented: One factor to be considered is that mixed character dislocations, as they do not have dislocation line vectors parallel to the growth normal, can tend to intersect with growth sector boundaries and 'refract' into another sector thus removing a step-creating screw component and lowering the growth rate and changing the crystal morphology. Hence, the dislocation substructure can effect variation in the growth process causing an oscillating crystal morphology with time.

Professor Jones said: I should like to add that some 25 or so years ago transmission electron microscopy was successfully used to identify defects in "nanocrystals" (although that phrase was not used at the time) and indeed was also successful in looking at stress induced phase changes and microtwinning. I believe that this is an area that we should now return to with regard to organic nanocrystals and perhaps a better understanding of the consequences of mechanical treatment of organic and similar materials.[1]

1 See W. Jones and J. M. Thomas, Applications of Electron Microscopy to Organic Solid State Chemistry, *Prog. Solid State Chem.*, 1979, **12**, 101–124.

Dr Hare communicated: The line $\log c(r) = -\alpha \log r$ does seem to provide a close fit in the central region, but is there a continuous function that will give an even closer fit over a fuller range of r values? For example, I notice that if before plotting $e^{-\alpha \ln r}$ is first multiplied by a factor $1 + \phi$ where $\phi = 1 - \exp(kr^{\alpha})$, then there is the prospect of a fit across the whole range (here, k appears to be approximately 0.4; though it could conceivably be as high as α). Could this have any physical significance?

From the outset the authors stress that their test subject, KAP, does not resemble a pre-biotic mineral. Perhaps the jury is still out on clay. Whatever further test results arise (in an "amenable" non-clay system), though, and the verdict on these, could the properties of today's clay resemble those of the pre-biotic mineral? If, for the moment, Cairns-Smith's original question remains unanswered, then could the gardener, digging through clay with a spade, even today be preparing a scaffold for the post-human era? Or do we think that, with all the false positives and mutations that there could be, little or nothing will remain of the human legacy once all our brick buildings have collapsed and weathered away? Or, setting aside clay again, might the chemist be able to choose a material better suited to transmitting (our human) information into a future age than either yesterday's clay or today's discarded hard disks?

Professor Kahr communicated in reply: The data to me seem to indicate two distinct behaviors at the small and mid-length scales. If there is a continuous function it would be a convolution of two separate functions, one that is more fractal in nature at the middle length scale and another that would be physically appropriate for the clustered region. As for the longer length scale, the outlier region, the data is distorted by limited area. For example, in an infinite crystal this outlier region should disappear and the fractal region should extend further.

Is it appropriate here to invent an arbitrary function (or so it seems to me) to fit the data? Wouldn't it be better to identify functions with a corresponding physical significance and then evaluate how well those functions match the data?

We tried to be conservative in speculating in our manuscript about past "life". We will certainly be conservative by not speculating here about future life.

Dr Hammond opened the discussion of Professor Mazzotti's paper: Could you comment on the way in which the parameters in the detailed, surface-specific growth model that has been employed to look at the growth of all the surfaces of the crystals simultaneously, should be interpreted given that, presumably, the growth mechanisms on the different crystal faces could be significantly different. Would a simpler kinetic model not fit the available experimental data equally well and be more valid for interpreting data of this kind?

Professor Mazzotti replied: The model given by eqn (19)–(22) was used to convert growth velocities of single faces published in the literature[1] to overall growth rates of the crystal that can be compared with our correlation for the overall growth rate. The growth mechanism of the single faces was identified by atomic force microscopy in another paper by the same author as being for all faces "birth and spread".[2] Therefore, we have decided to use the same model for the overall growth rate that we have estimated, and we have compared it only with the growth rate expression of the

BCF model. Another simple model could in principle also work, as asked by Dr Hammond, but we strongly doubt that this would bring more insight.

1. M. Kitamura, *J. Cryst. Growth*, 2000, **209**, 138.
2. M. Kitamura and K. Onuma, *J. Colloid Interface Sci.*, 2000, **224**, 311.

Professor Unwin said: I would like to follow up on the issue of mass transport. My main point is that the neglect of mass transport relies on using a Sherwood correlation (eqn (17)) to estimate mass transport coefficients. How accurate are such correlations? Moreover, under what conditions do they apply? For example, what is the concentration of particles (in terms of fractional volume, fractional mass or other) and size range of particles in solution for which the correlation can be used? A more minor point concerns the value of $D = 2 \times 10^{-9}$ m^2 s^{-1} for α-glutamic acid quoted in the paper and used in the correlation. Where does this come from? It seems rather high based on values for similar small acids.[1,2] The value of D for such acids can, of course, be measured accurately with dynamic electrochemistry methods.[3]

1 W. J. Albery, A. R. Greenwood and R. F. Kibble, *Trans. Faraday Soc.*, 1966, **63**, 360.
2 B. E. Bidstrup and C. J. Geankoplis, *J. Chem. Eng. Data*, 1963, **8**, 170.
3 See for example R. D. Martin and P. R. Unwin, *J. Electroanal. Chem.*, 1995, **397**, 325 and references therein.

Professor Mazzotti answered: Correlations of mass transfer coefficients are usually given as correlations of dimensionless numbers and are based on a large number of experimental data. In general they can be applied to many different physical systems, represented by Re and Sc numbers, as long as the characteristic dimensions, *e.g.* the characteristic length, are chosen correctly. An overview about the different correlation can be found in standard textbooks.[1,2] The accuracy for solid–liquid interfaces is typically in the range of ± 10%.[2] According to Mersmann the most severe drawback in the prediction of mass transfer coefficients is the accurate knowledge of the diffusivities.[3] In our study we have calculated the rate of diffusion controlled growth based on eqn (17) and (18) for a broad range of particle sizes and supersaturations. For a value of $D = 2 \times 10^{-9}$ m^2 s^{-1} used in this paper and also for smaller diffusivities (taken from Albery[4] and Bidstup[5]) the experimental growth rates were found to be at least one order of magnitude smaller than the diffusion limited growth rates.

1 T. K. Sherwood, R. L. Pigford and C. R. Wilke, *Mass Transfer*, McGraw-Hill, New York, 1975.
2 E. L. Cussler, *Diffusion*, Cambridge University Press, Cambridge, 2nd edn, 1997
3 A. Mersmann, *Crystallization Technology Handbook*, Marcel Dekker Inc., New York, 2nd edn, 2001.
4 W. J. Albery, A. R. Greenwood and R. F. Kibble, *Trans. Faraday Soc.*, 1966, **63**, 360.
5 B. E. Bidstrup and C. J. Geankoplis, *J. Chem. Eng. Data*, 1963, **8**, 170.

Dr Vonk addressed Professor Mazzotti and Professor Unwin: Industrial experience with similar components like α-gluacid shows that mass transfer is usually much faster than surface integration.

Professor Mazzotti answered: Whether or not the growth of a substance is surface integration or mass transfer controlled depends very much on the substance and on the operating conditions. However, we have also found that for other small organic compounds the growth process is indeed integration controlled.

Professor Hyne asked: The data presented in Fig. 12 should permit an Arrhenius type plot of crystal growth rate *vs.* temperature to be constructed and the equivalent of enthalpies and entropies of activation for the growth process to be extracted (also by eqn (9)). Do such additional "pseudothermodynamic" parameters provide useful information about the crystallization process or is it comprised of too many discreet steps to be associated with any one?

Professor Mazzotti replied: The growth rate expression that we have used describes a two step process: formation of nuclei on the surface and spreading of the layer at the kink sites of these nuclei. Hence, the growth rate expression given by eqn (9) comprises of a nucleation and a surface diffusion term. The parameter C in eqn (9) is related to surface nucleation and depends on the surface tension between the crystal and the solution. The parameter B in eqn (9) is related to surface diffusion and depends on the activation energy between two neighboring equilibrium points.[1] This is how the parameters in the growth rate equation should be interpreted. In general, an Arrhenius type plot is drawn to estimate the parameters in an Arrhenius type equation. In this case the growth rate expression is more complex than that, and the parameters are estimated as described in the paper hence there is no need for an Arrhenius plot.

1 A. Mersmann, *Crystallization Technology Handbook*, Marcel Dekker Inc., New York, 2nd edn, 2001.

Professor Addadi opened the discussion of Dr Rieger's paper:

(1) In the amorphous calcium carbonate phases observed in biological environment, the stable amorphous calcium carbonate (ACC) phases are hydrated with composition $CaCO_3 \cdot H_2O$, while the transient ACC phases are anhydrous. What do you think is the role of water in your system?

(2) The Asprich proteins also stabilize ACC. They are composed of 50–70% carboxylate-containing amino acids, which makes them more acidic than polyacrylate, but they are much more efficient than polyacrylate in ACC induction and stabilization. Do you have any explanation for this effect?

(3) You suggest that the stabilization is due to coverage of the particles by the polymer. Do you know that as a fact or is this a suggestion? What would prevent nucleation inside the particle bulk if the polymer is coating the particle only? How do you envisage the stabilization mechanism?

Dr Rieger replied:

(1) Since we do not have sufficient information about the evolution of the water content in the amorphous precursor structures I refrain from speculating about any molecular processes. The simple phenomenological explanation relies on referring to Ostwald's rule of stages stating that "an unstable system does not necessarily transforms directly into the most stable state, but into one which most closely resembles its own",[1] where it is sometimes implied that this resemblance refers to the local structure of the original and the forming phase.

(2) In some instances it has been reported that aspartic acid-rich polymers exhibit nucleating and dispersing efficiencies which differ from those of polyacrylic acid. To my knowledge there is yet no convincing (molecular) explanation for this finding.

(3) Presently we have only indirect evidence for the hypothesis that the polymer is forming a shell around the $CaCO_3$ particle: we observe that a critical amount of polymer is necessary in order to stabilize the metastable precursor particles. When working in the under-critical concentration range we can tune the duration of the metastable state by varying the amount of polymer added. It may be assumed that part of the polymer is located within the precursor particles, thus preventing nucleation. Again, we do not have any direct proof.

1 J. W. Mullin, *Crystallization*, Butterworth-Heinemann, Oxford, 1997, p. 201.

Professor Breu asked: What is the role of ACC? Is it a precursor that is transformed *via* a solid–solid reaction into crystalline forms? Or is it just a buffer system that fixes Ca^{2+} and CO_3^{2-} concentrations to a level that speeds up nucleation and/or growth of one or the other polymorph?

Dr Rieger answered: Both cases seem to occur: in an earlier study we observed by means of X-ray microscopy *in situ* how the amorphous calcium carbonate precursor particles dissolve completely while material recrystallizes at a different spot (the "buffer" mechanism),[1] whereas in other cases micron-sized vaterite spheres seem to be assembled from precursor particles which must have undergone a solid-state transformation, Fig. 5 of our paper). The degree of supersaturation and the mode of mixing seem to be the decisive parameters to switch between the two modes in the case of calcium carbonate.

1 J. Rieger, J. Thieme and C. Schmidt, *Langmuir*, 2000, **16**, 8300.

Professor Unwin said: Dr Rieger's paper contains significant new data which may clearly have major implications for the way in which we describe crystal growth from solution. The experiments in which emulsion structures are observed at short times are deliberately at high concentration to induce significant precipitation. Have any experiments been carried out at lower concentrations of Ca^{2+} and CO_3^{2-}, similar to those that are typically encountered in studies of seeded growth or growth at single crystals? It would be interesting to know the nominal threshold saturation ratio needed to produce the emulsion structures found in this study.

Dr Rieger replied: We did not vary the educt concentrations because we focused our interest on the role of additives on the formation of calcium carbonate under conditions which resemble applications using very hard water. Concerning the proposed supersaturation threshold between the formation of precursor structures and classical nucleation and growth we recommend a recently published article by Meldrum *et al.*[1] where it is postulated that the transition between ion-by-ion growth as compared to aggregation of nanoscopic building blocks is continuous.

1 A. N. Kulak, P. Iddon, Y. Li, S. P. Armes, H. Cölfen, O. Paris, R. M. Wilson and F. C. Meldrum, *J. Am. Chem. Soc.*, 2007, **129**, 3729.

Professor Davey asked:
(1) In the case of the amorphous calcium carbonate system, does the dispersion behave in a similar way to other colloidal dispersions? Can it be stabilized by simple surfactants?
(2) Do the dispersed particles have to aggregate before vaterite can form?
(3) Do other systems show this behaviour?

Dr Rieger replied:
(1) Once the spherical particles are formed from the inorganic emulsion structure they can be considered as colloidal particles. Surfactants might be suitable to stabilize these particles against aggregation but I assume that they do not provide a dense enough shell to stabilize the precursor particles against dissolution.
(2) We still were not able to clarify this point, the problem being that it is very difficult to sample enough of the precursor nanoparticles for XRD right before they aggregate to form the micron-sized vaterite spheres.
(3) There are a number of systems where colloidal particles are formed from nano-sized precursor particles; this field was pioneered by Prof. Matijevic and Prof. Sugimoto. Some references are given in our review on p. 4337.

1 D. Horn and J. Rieger, *Angew. Chem., Int. Ed.*, 2001, **40**, 4330.

Professor Kahr addressed Professor Davey and Dr Rieger: Likewise, Peter Vekilov has observed disordered aggregates of a variety of proteins that form prior to crystallization.[1]

1 O. Gliko, W. Pan, P. Satsonis, N. Neumaier, O. Galkin, S. Weinkauf and P. G. Vekilov, *J. Phys. Chem. B*, 2007, **111**(12), 3106–3114.

Professor Addadi addressed Professor Davey: In answer to your question: Are there other known cases of hydrated amorphous precursor phases before crystallization? Yes, there are: Calcium phosphate is a classical case, following a cascade of phase transformations from amorphous calcium phosphate to octacalcium phosphate, which is hydrated and crystalline, to apatite. Other known cases are for iron oxides and iron oxides hydroxides (ferrihydrite).

Dr Rieger said: I would like to point out that we observed similar precursor phases as discussed here for the case of calcium carbonate when investigating the precipitation of quinacridone (an organic red pigment) and boehmite (aluminium oxide hydroxide), though we could not obtain any quantitative information on the water content of these initial phase-separated structures.[1]

1 H. Haberkorn, D. Franke, Th. Frechen, W. Goesele and J. Rieger, *J. Colloid Interface Sci.*, 2003, **259**, 112.

Professor Roberts addressed Professor Addadi: Do we know why we get this sequence of hydrated complexes and amorphous precursors which seem to be prevalent for a number of mineral and organic systems? What is the current state of our fundamental understanding and are these effects generic in your view?

Professor Addadi replied: We do believe that the existence of amorphous precursor phases in the pathway to crystallization of many biogenic minerals may be much more widespread than what was believed until now. The existence of various forms of amorphous calcium carbonate (ACC) precursors to calcite has been proven in several sea urchin skeletal parts (larval spicules, adult spines and teeth) as well as in sponge spicules.[1] Mathias Epple's work[2] and Ingrid Weiss' work[3] have demonstrated ACC precursors to the formation of aragonite in larval mollusc shells. Other biogenic amorphous precursor phases include, among others, amorphous calcium phosphate as a precursor to apatite and ferrihydrite as a precursor to magnetite in chiton teeth.[4]

1 I. Sethmann, R. Hinrichs, G. Worheide and A. Putnis, *J. Inorg. Biochem.*, 2006, **100**, 88.
2 J. C. Marxen, W. Becker, D. Finke, B. Hasse and M. Epple, *J. Mollus. Stud.*, 2003, **69**, 113.
3 I. M. Weiss, N. Tuross, L. Addadi and S. Weiner, *J. Exp. Zool.*, 2002, **293**, 478.
4 H. A. Lowenstein and S. Weiner, *On Biomineralization*, Oxford University Press, New York, 1989.

Professor Addadi then addressed Dr Rieger and Professor Roberts: If I remember correctly, the first suggestion of a liquid amorphous calcium carbonate phase was from Aksay and Groves back in 1998.[1] They followed formation of calcite under monolayers of carboxylated porphyrins. They observed that an amorphous film is formed first, which subsequently crystallized.

1 G. Xu, N. Yao, I. A. Aksay and J. T. Groves, *J. Am. Chem. Soc.*, 1998, **120**, 11977.

Dr Rieger replied: In their 1998 paper Aksay and Groves indeed describe the "phase transformation from an initially deposited amorphous phase to crystalline calcite during the film formation". Their approach differs from ours insofar as they initiate the solid state formation at "a porphyrin template/subphase interface" in water—in all cases in the presence of polyacrylic acid. In our experiments we investigated the precipitation in the bulk system without any interfering interfaces; furthermore we also observed the formation of amorphous calcium carbonate in the absence of any additive.

Professor Roberts responded: This is most interesting but how can one be sure that the phase is truly amorphous and not just nanocrystalline. For example, a crystalline cluster of say 50 molecules of $CaCO_3$ would not be easy to identify *via* X-ray diffraction methods. In principle X-ray absorption spectroscopy (see *e.g.* ref. 1) can be used but for $CaCO_3$ this is not trivial given the weakly scattering second coordination shell Ca–Ca interactions, the absence of which would be needed to clearly quantify the amorphous state.

1 K. J. Roberts, *Mol. Cryst. Liq. Cryst.*, 1994, **248**, 207–242

Professor Harding addressed Dr Rieger: In your paper you mention some MD simulations performed on polycarbonates interacting with calcium ions in solution. Could you please expand on what these simulations found about the binding of polycarbonates?

Dr Rieger responded: We studied the binding of Ca ions to polycarboxylates both by means of MD simulations as well as by attenuated total reflectance Fourier transform infrared dialysis spectroscopy, *c.f.* ref. 31 and 32 in our paper. This complementary approach gave consistent results: an ionic chelating complex evolves either to a unidentate complex, which can form a pseudobridge by coordinating a water molecule, or a pseudobridge with a sodium ion. The process leads to a chelating bidentate product, the yield of this step may be not complete. Within this general scheme an influence of the degree of polymerization has been observed. Concerning the MD simulations it must be stressed that it is mandatory to include the water molecules and the local chemistry of the oligomer/polymer in the simulation in order to obtain meaningful results. It is by no means sufficient to rely on considerations of charge density of the polymer, ionic strength, *etc.*

Professor Davey commented: In the context of the initial phase separation of inorganic materials it is important not to confuse this with liquid–liquid separation seen in protein solutions. The latter is due to a liquid–liquid miscibility gap that lies hidden beneath the solubility curve.

Dr Rieger answered: Professor Davey is of course right and we thank him for this clarification. At the same time we would like to stress that still virtually nothing is known about the process of initial phase separation in inorganic systems during precipitation. The complexity arises from the influence of several timescales: meso- and micromixing, diffusion, the reacting ions, phase formation, phase restructuring, and expulsion of water from the initially formed phase. As a further complication it must be recognized that the local viscosity of the system changes during solid state formation. The latter point renders any quantitative consideration very difficult since it affects all the above points.

Professor Heyes said: It has been known for about 15 years that for quenched fluid systems, if the attractive interaction between the particles is short-ranged compared to their diameter, the crystallisation phase separation occurs first through an amorphous phase, which is consistent with your experiments. The fluid–fluid separation which occurs first is metastable with respect to and within the solid-vapour coexistence. This behaviour could be "tuned" by changing the ionic strength of the solution. Have you investigated the effects of ionic strength change on the morphology of the initial $CaCO_3$ precipitate?

Dr Rieger replied: At the present state of knowledge I would not compare the phase behaviour of dispersions made from colloidal particles/proteins with ions dissolved in water. Nevertheless, it would be interesting to know more about the influence of ionic strength on the precipitation. But due to the ion specificity of the calcium carbonate precipitation this endeavour would open another large parameter field (apart from concentration of educts, temperature, mode of mixing, type and concentration of additive, *etc.*) which we did not attempt to tackle.

I agree that it might be interesting to change the interaction between the precipitating ion by varying the ionic strength in order to compare with phase separation mechanisms known from protein crystallization. The problem is that, once we change one parameter of the system the precipitation behaviour of the whole system is affected. To make things worse, sometimes step-like changes occur upon continuous variations of parameters like ionic strength, temperature, etc. This differentiates the precipitation of ions (Ca^{++}, CO_3^{--}) from the crystallization of proteins. The latter system can be discussed in terms of colloidal hard sphere systems with short-range interaction whereas the former evades such an approach.

Professor Heyes then asked: Have you measured the viscosity of the polyelectrolyte $CaCO_3$ nanoparticle aggregates? The intrinsic viscosity, for example, would give you some idea of the rigidity of the aggregates formed. You could use for example the Mark–Houwink equation (see, *e.g.*, ref. 1). The intrinsic viscosity characterises the dilute solution behaviour, which could be particularly appropriate here.

1   K. Kamide and T. Dobashi, *Physical Chemistry of Polymer Solutions*, Elsevier, Amsterdam, 2000.

Dr Rieger answered: Due to the difficult handling of the precipitating systems and the rather short timescales involved we did not attempt to measure the evolution of the viscosity though these data might indeed yield valuable information if available.

Professor Hodnett asked: Did you measure the solubility of calcium in equilibrium with the amorphous calcium carbonate? How does this value compare with the starting concentration (0.01 M) and the equilibrium concentration for calcite?

Dr Rieger replied: The data can be derived in a first order approximation from Fig. 11 in our paper. If we extrapolate the activity of free calcium ions to a polymer concentration of zero we arrive at a value of roughly 70 ppm, which corresponds to 1.75 mM, compared to 5mM l^{-1} Ca ions at the start of the precipitation (0.01 M l^{-1} educt solutions are mixed in equal volumes).

Dr Lewtas asked: The stability of very small particles in the presence of polymers depends upon the absolute size and the relative size compared to the polymer. How does the proposed mechanism vary with changes in polymer MW and also polymer MWD?

Do you see a change from dispersion to bridging/flocculation? If so, at what MW? Is there a relationship? MWD should also have an effect— do you have any data?

Is there a transition from stabilization/dispersion to nucleation with the rasing of polymer MW?

Dr Rieger responded: To a certain extent the known rules of colloid science apply here, *i.e.* higher molar mass polymers tend to flocculate, among other effects. The situation is complicated by the fact that polyacrylic acid strongly interacts with both the calcium ions in solution and the calcium carbonate in the form of dispersed solid particles. Neither in our case where we precipitate calcium carbonate in the presence of polyacrylic acid nor in the classical field of dispersing hardly soluble salts, such as calcium carbonate, is there a complete understanding of how the different effects affect each other.

Professor Mazzotti asked: With reference to Fig. 9, are these images of the same spot in the precipitator? If yes, how can you rule out the effect of concentration gradients?

With regard to the effect of supersaturation, which is not considered in the manuscript, can one assume supersaturation levels that are consistent with the phenomenological picture that has been presented?

Dr Rieger answered: The five micrographs shown in Fig. 9 are indeed taken at the same spot of the sample; related experiments were performed in order to ensure that the result is not impaired by radiation damage of the sample in the X-ray microscope. It is safe to assume that there are concentration gradients in the sample, but at the same time we assume that variations of the gradients due to the different geometries (bulk *vs* 10 µm thick layer) will not affect the outcome of the experiments qualitatively. In some cases we observed that the timescale of morphological transformations was stretched in the latter case. With the educt concentrations used our system falls into the class of precipitation–crystallization reactions where it is known that classical crystallization theory only applies to a limited extent. A typical example for the complication encountered is the occurrence of several intermediate steps preceding the final crystallization.

Dr Hare communicated: One unanswered question was why growth was observed at right angles to the expected direction. In the absence of other suggestions—and purely "off the wall"—might one line of enquiry be to ask whether there is *any* growth in the expected direction? If there were, but it were slow, then could it be that to achieve the stablest outcome, the bulk of the crystal needs "to go around" the obstacle of the slow-growing crystal already growing, *i.e.*, must the bulk of it begin by setting off in the perpendicular direction? Might this also be consistent with a "two paths" mechanism (in which the first has a low followed by a high activation-energy barrier, while the second has one of medium height)? Could it really be, perhaps, that the *inanimate* ensemble is adopting an animal-like strategy?

Professor Catlow opened the discussion of Professor Watanabe's paper: First let me congratulate you on the quality of the data you have obtained in these experiments. Could you discuss in more detail the coordination numbers obtained from simulation and experiment? Fig. 4 indicates an apparent discrepancy, while the S (Q) data indicates good agreement between theory and experiment. Could you also give more details as to how the coordination numbers change with temperature?

Professor Watanabe replied: Our conclusion from experiments and simulations is that there is no change in the temperature range from 1200–1800 K. It is too difficult to discuss more details regarding coordination number change, because the coordination number obtained by both experiments and simulation is an averaged value from moving atoms. Thus the number contains large errors. In future, we must develop the technique to obtain precise atom coordinates from one shot pictures of moving atoms.

Professor Roberts said: It is interesting to note the observation of 6-fold coordination in liquid Si related to 4-fold tetrahedral short-range correlating and an average of 2 for the second shell. The first 3 shells in the radial distribution function from the bulk crystal Si structure should be 4, 12 and 6. Hence, this implies the fraction of second shell co-ordination (2/12) which may be indicative of diffusional motion between the Si tetrahedra. Its important to note that crystal size plays a role and that when small (<1 µm) the radial distribution function will be reduced by surface effects, *i.e.* allowing for the effects of undercoordinated surface atoms, and that the calculated coordination numbers need to be corrected for this to allow for this effect.

Professor Watanabe answered: We did not consider the effect of surface on the radial distribution function. It is very difficult to consider the surface effect on the radial distribution function by conventional experimental and analysis techniques.

Professor Catlow asked: Could you give more details of the 6 coordinated structures? What is the detailed geometry of the coordination shell?

Professor Watanabe replied: From the one shot picture of MD simulations, we identify tetrahedral coordination of atoms. We think that the 6 coordinated structure is not important to short-range structure of liquid Si. Our ideas about the short-range structure of l-Si are that tetrahedral coordination is important to the short-range structure of l-Si. Since the tetrahedral coordinated atoms are moving in the liquid, the average nearest neighbor coordination number was 6. Thus, we think that the rigid 6 coordinated structure would not exist in liquid Si.

Dr Schön said: You model the liquid state using *ab initio* MD for a 64-atom supercell. While this is already quite a challenge, I am concerned about the typical effects of a rather small simulation cell (we have encountered such effects ourselves, too) such as a considerably elevated melting temperature or an overly strong prevalence of solid-like local coordinations. This is of particular concern, since you are discussing the issue of changes in the local coordination number and the jump in the density upon melting. Did you check for or notice such effects? You might perhaps be seeing some transition states from a four-fold to a six-fold coordinated state that exhibit local ordering with 4 + 1- or 4 + 2-coordinations. Studies of the energy landscape of *e.g.* alkali halides have shown that there are many local minima with comparable energies that correspond to crystalline configurations with 4-, 5- or 6-fold coordinations.[1] Similar analogous configurations might appear during the simulations at high temperatures in the Si-system.

Quite generally, what criterion do you use to recognize that you are dealing with a liquid-like state? Do you observe a rapid increase of the diffusion constant near the melting point?

1 J. C. Schön and M. Jansen, *Comput. Mater. Sci.*, 1995, **4**, 43.

Professor Watanabe replied: In our simulation, we start from crystalline state to liquid state with temperature increasing. In this sequence, we clearly observed an energy jump at melting temperature, thus we can identify the transition from the crystalline to liquid state. In the cooling cycle from high temperature liquid state, we did not observe any drastic change at the melting temperature. Thus, we can achieve a supercooling state from the high temperature liquid state. In these conditions, a rapid increase in the diffusion constant was not observed at the melting temperature.

Dr Murray asked: Could the magnetic field influence the freezing process?

Professor Watanabe replied: No.

Professor Roberts commented: The key issue relates to how magnetic levitation might affect nucleation. If the material is diamagnetic then the effects should be quite small or the field should just effect dipolar rotation (Larmor effect) and providing the field is small, which it is, then such effects should be small. The main utility of the levitation technique reflects the effect that the sample size is small and the "container" is removed thus providing little opportunity for heterogeneous nucleation and hence providing an opportunity to study a purely homogeneous process.

Dr Hughes addressed Professor Roberts and Professor Watanabe: Examples exist in the literature where the effect of a magnetic field on crystallization has been to orient the crystals but not to affect the polymorphism. The particular example I have in mind is a study on α-glycine by M. Sueda *et al.*.[1]

1 M. Sueda, A. Katsuki, Y. Fujiwara and Y. Tanimoto, *Sci. Tech. Adv. Mater.*, **7**, 2006, 380–384.

Professor Watanabe replied: I did not know of Sueda's paper. However, the effect of magnetic field on the Si crystallization is different from the glycine. In the case of Si, magnetic fields suppress melt flow by the Lorentz force. Therefore, the Si crystal growth behavior is unchanged from that in the case with no magnetic field.

Professor Roberts responded: Yes, this is a good point and indeed several commercial devices have been produced, *e.g.* for anti-water scaling applications, in which claims are made concerning the effect of magnetic fields in suppressing scale-forming crystallisation processes. The science basis for such action remains, though, quite weak and there is much still to be done in order to understands such "effects". To my knowledge, there have been no substantial work concerning the effect of magnetic field on polymorph formation and its control.

Dr Ristic asked: A static magnetic field of 6 T generated by a superconductive magnet appears to be very strong compared to the maximum strength of the magnetic field that can be achieved by a conventional electromagnet (≈ 1 T). Have you observed any influence on the crystallisation of silicon at 6 T?

Professor Watanabe responded: Magnetic fields of 6 T cannot be generated by a normal electromagnet. Magnetic fields can suppress the Si melt flow by the Lorentz force. We have not observed any difference in the crystal growth behavior.

Professor Heyes asked:
(1) Classical MD simulations with a small number of particles (*e.g.* 64) readily show crystallisation. The periodic boundary conditions promote crystallisation. Your simulations do not show this (*e.g.* no density change at the experimental melting temperature). Is this difference a consequence of the first principles method or some other reason?
(2) SiO_2 and H_2O manifest density maxima above the normal melting temperature. The tentative density maximum for Si suggests a density maximum below the normal melting temperature. Do you have any ideas why there should be this qualitative difference?

Professor Watanabe answered:
(1) Many classical MD simulations showed supercooling liquid state (see, *e.g.*, ref. 1), the same as our results. Therefore, the supercooled state does not depend on simulation technique.
(2) For the case of l-Si, the reason for the temperature of maximum density existing below the melting temperature would be the balance of thermal energy and binding energy. The balance of thermal energy and binding energy may decide the maximum density temperature. This is just my impression, in future we must clarify the reason for the difference by qualitative analysis.

1 C. A. Angell, S. Borick and M. Grabow, *J. Non-Cryst. Solids*, 1996, **205–207**, 463.

Dr Hammond said: It would, I think, be interesting to extract information about the motion of silicon atoms in the molten state from the FPMD simulations that you have performed. Have you calculated time correlation functions to describe this motion?

Professor Watanabe replied: Yes, we can obtain a time correlation function from the MD results. Please see the previous paper evaluating diffusivity of liquid Si under high pressure conditions.[1]

1 T. Morishita, *Phys. Rev. E*, 2005, **72**, 021201.

Professor Catlow asked: Are you able to calculate the Si diffusion coefficient in your MD simulations? If so, what values do you obtain?

Professor Watanabe answered: Yes, we can obtain a diffusion constant from our MD results. Our diffusion coefficient is of the same order as normal liquid metals (10^{-4} cm^2 s^{-1}).

Dr Wakisaka opened the discussion of Dr Michaelides' paper: Are there any differences in the following cases: (i) water clusters in the gas phase have a soft-landing on the Cu surface; (ii) monomeric H_2O adsorb onto the Cu surface?

Dr Michaelides replied: As far as I am aware the experiments which could provide an answer to this question have not been performed.

Dr Murray asked: Real atmospheric particles have defects and faults. Have you looked at the clustering of H_2O molecules on defects on surfaces?

Dr Michaelides answered: No, not yet.

Professor Unwin asked: My question is related to the previous one. The STM image presented in the overview of this paper appeared to show extensive decoration of steps with entities that did not look like water clusters, as well as the water clusters of interest which were distributed at various locations on the terraces. Are the steps preferential binding sites for water on such surfaces and have you tried to model the adsorption of water at steps?

Dr Michaelides responded: We have not looked into this issue in any great detail. It's an interesting question and one we would like to address. Knowledge of molecular adsorption (including water adsorption) on other metal surfaces would lead one to expect that binding at the step sites will be greater than on the terraces and that whatever the barrier for water dissociation on the terraces is, it is likely to be lower than on the steps.[1–3] Here we have simply focused on attempting to understand the structures of the clusters on the terraces. In the future we would like to turn our attention to the steps, although it's worth pointing out that to examine clusters such as the hexamer adsorbed at a step site will be computationally challenging because a very large simulation cell will be required.

1 M. Morgenstern, T. Michely, and G. Comsa, *Phys. Rev. Lett.*. 1996, **77**, 703.
2 A. Michaelides, A. Alavi, and D.A. King, *Phys. Rev. B*, 2004, **69**, 205411.
3 A. Michaelides, Z.-P. Liu, C. J. Zhang, A. Alavi, D. A. King and P. Hu, *J. Am. Chem. Soc.*, 2003, **125**, 3704.

Professor Bensch asked: Do you obtain different results if the nucleation of H_2O occurs on different orientations of the Cu surface?

Dr Michaelides answered: The structure of the substrate does indeed play an important role. On Cu(110), for example, one dimensional chains, which differ greatly from the structures observed on Cu(111), have been observed in similar low-temperature STM experiments[1] to those performed by K. Morgenstern and discussed in my paper.[2] Further, by altering the surface structure one will also alter its reactivity and make it more likely that the water molecules will dissociate. Some discussion along these lines can be found in ref. 3. Likewise the chemical nature of the substrate[4,5] and the presence of co-adsorbates play important roles in the structures which form.[6]

1 T. Yamada, S. Tamamori, H. Okuyama, and T. Aruga, *Phys. Rev. Lett.*, 2006, **96**, 036105.
2 A. Michaelides and K. Morgenstern, *Nat. Mater.*, 2007, DOI: 10.1038/nmat1940.
3 A. Michaelides, *Appl. Phys. A*, 2006, **85**, 415.
4 J. Cerda, A. Michaelides, M.-L. Bocquet, P. J. Feibelman, T. Mitsui, M. Rose, E. Fomine and M. Salmeron, *Phys. Rev. Lett.*, 2004, **93**, 116101.
5 A. Michaelides, A. Alavi, and D. A. King, *J. Am. Chem. Soc.*, 2003, **125**, 2746.
6 A. Michaelides and P. Hu, *J. Chem. Phys.*, 2001, **114**, 513.

Dr Schön said: I want to follow up on the earlier discussion on the accuracy of the calculations, both in the gas phase and on the surface. I would agree that DFT is not sufficient for catching the subtle effects of the gas phase energies of the water oligomers. Probably post-Hartree–Fock methods such as coupled cluster calculations are necessary. Have you looked at such calculations?

Concerning the oligomers on the surface, such accuracy is probably difficult to reach (perhaps MP2 level is possible). Have you checked how your results vary with the use of different kinds of pseudo-potentials or functionals, in order to get some feeling regarding the robustness of your results? Regarding Fig. 4, why is there an apparent convergence of the two quantities plotted at about $n = 9$? Is a convergence expected to follow from the definition of these quantities? (The shape of the $n = 9$ cluster in the gas phase is surely quite different from the one on the surface?)

Dr Michaelides replied: The gas phase energies of the water oligomers and how well different DFT functionals do at describing their relative energies is still an open question. In particular for the gas phase water hexamer in which there are at least 4 isomers (cage, book, prism, cyclic) all within a few meV per H_2O of each other. We are currently looking into this issue and using MP2 at the complete basis set limit to generate our reference data; a paper by Santra, Michaelides, and Scheffler is currently being prepared on this topic. For the adsorbed hexamers our DFT PBE calculations find a preference for the cyclic hexamer. Since this is also what is observed in experiment it gives us some confidence that DFT is accurate enough to reliably treat these systems. Also, in a few previous publications we found "good" agreement between DFT GGA calculations and experiment: specifically, by good I mean that DFT GGA calculations correctly predict the correct binding site for water monomer adsorption on several metal surfaces[1] and the correct intermediate in the water formation reaction on Pt.[2,3] However, for the specific issue of the buckling predicted by DFT for the adsorbed hexamer on Cu(111) we checked our DFT results against MP2. The MP2 calculations were all-electron and performed on small Cu clusters. Some of these results are reported in Table 2, where it can be seen that the all-electron MP2 calculations and the pseudopotential DFT calculations agree that there is a strong preference for the H_2O hexamer to buckle when adsorbed. Also in this table are results with two other DFT functionals ("RPBE" and the hybrid "PBE0" functional) and in a forthcoming paper results with B3LYP as well as additional tests on the numerical accuracy of our results will be reported.[4] I think the "convergence" of the gas phase and adsorbed phase cluster energies is just a coincidence.

1 A. Michaelides, V. A. Ranea, P. L. de Andres and D. A. King, *Phys. Rev. Lett.*, 2003, **90**, 216102.
2 A. Michaelides and P. Hu, *J. Chem. Phys.*, 2001, **114**, 513.
3 A. Michaelides and P. Hu, *J. Am. Chem. Soc.*, 2001, **128**, 4235.
4 A. Michaelides and K. Morgenstern, *Nat. Mater.*, 2007, DOI: 10.1038/nmat1940.

Professor Catlow asked: Have you been able to calculate activation energies for water migration on the copper surface?

Dr Michaelides answered: Yes and no. A reasonable estimate of the diffusion barrier for a molecule across a surface can be obtained by comparing the binding energy of the molecule at the most stable and next-most stable adsorption sites. On Cu(111) and many other metal surfaces we have made such estimates.[1] If you look in ref. 1, for example, you will find such an estimate for the diffusion barrier on Cu(111) (and several other metals) giving a value of 0.05 eV. However, when we looked at this issue in detail for water diffusion on Al(100)[2] in which we explicitly mapped out the potential energy surface between the most stable and next-most stable adsorption sites we found that the actual diffusion barriers were a little bit larger (10-100 meV) than the binding energy difference and that the diffusion barriers depended quite strongly on the orientation of the water molecule. For Cu(111) we have not performed such a complete scan of the potential energy surface as we did for Al(100). Also, I note that when water clusters diffuse much more complex and interesting mechanisms are predicted.[3,4]

1. A. Michaelides, V. A. Ranea, P. L. de Andres and D. A. King, *Phys. Rev. Lett.*, 2003, **90**, 216102.
2. A. Michaelides, V. A. Ranea, P. L. de Andres and D. A. King, *Phys. Rev. B*, 2004, **69**, 075409.
3. V. A. Ranea, A. Michaelides, R. Ramirez, J. A. Verges, P. L. de Andres and D. A. King, *Phys. Rev. Lett.*, 2004, **92**, 136104.
4. A. Michaelides, *Appl. Phys. A*, 2006, **85**, 415.

Professor Kahr asked: Your trimer is chiral as are some of the other clusters. Recently there have been extensive investigations into enantioselective processes of high index metal surfaces such as the Cu (643). This surface is naturally chiral.[1] Do you think that you can calculate enantiomer discriminating energies of chiral clusters on these surfaces?

1. A. J. Gellman, J. D. Horvath, *J. Am. Chem. Soc.*, 2002, **124**, 2384.

Dr Michaelides replied: This is a very interesting issue and something we plan to look into. It is certainly possible to compute water cluster, in particular chiral hexamer, adsorption on these intrinsically chiral surfaces.

Professor Roberts asked: It is interesting to note the rather beautiful water hexameric structures on Cu surfaces. Is there any evidence for these in other structures provided for bulk 3D X-ray/neutron crystallographic structures? It would be interesting to search, *e.g.*, the CCDC to see if such structures have been found, *e.g.* in the crystallography of organic hydrates.

Dr Michaelides replied: Water hexamers are not uncommon; regular hexagonal and cubic ice are built out of them. As for the presence of water hexamers and the finer details of their structures in organic hydrates and other bulk hydrates I cannot comment. I agree, however, that it would be interesting and worthwhile to check the CCDC for evidence of water hexamers in other environments.

Professor Vlieg asked: Can you perform calculations with similar accuracy on surfaces of salt?

Dr Michaelides replied: Yes. See, for example, ref. 1 or 2.

1. B. Li, A. Michaelides, and M. Scheffler, *Phys. Rev. Lett.*, 2006, **97**, 046802.
2. Y. Yang, S. Meng, and E.G. Wang, *Phys. Rev. B*, 2006, **74**, 245409.

Dr Hare communicated: Given the characterisation of the adsorbed hexamer as a weak "triple-dimer" structure, would it be worthwhile to study the tetramer specifically as a double dimer (neglecting both trimer and pentamer), not necessarily seeking low-energy configurations but with a view to exploring possible paths from single to triple dimer?

Given the adsorption-energy minimum at or near $n = 6$ (Fig. 4), do the hexamer's electron-density isosurfaces (Fig. 6) suggest a symmetry for the crystal nucleus? Once a hexamer has formed, could this survive throughout bulk growth into the ice crystals we see with hexagonal shape, such as in a snowflake? If we are seeking a model for the crystal growth of ice, do we need look at values beyond $n = 6$, or, might it be possible to study aggregation of the hexamer, allowing it to buckle and unbuckle? To understand the melting of ice, do we need to establish an activation-energy surface for its nucleation and growth over a range of values from small to extremely large $6n$?

Dr Michaelides communicated in reply: Addressing each question in turn: (i) This is an interesting suggestion but not something we have looked at. My feeling, however, is that the most likely growth mechanisms of these clusters is by the sequential addition of individual water molecules and not necessarily through the grouping of dimers. Probably the main reason I favour a mechanism of sequential water addition is the nice set of experiments by Salmeron and co-workers[1] in which they saw adsorbed clusters (on Pd(111)) grow one molecule at a time with STM.

(ii) Yes, the hexamer structures do survive in the bulk; cyclic water hexamers are often called the "smallest piece of ice" since they are the building blocks of ice Ih. In a closely related study to this one on Pd(111) we saw that the local hexameric structure and hydrogen bonding pattern within an individual adsorbed hexamer directly impacted upon the mesoscopic (2D) ice structures which formed on this surface.[2,3] (iii) I fear that these systems are so complex and that the relative energies between different water clusters with different numbers of molecules and even clusters with the same number of molecules but with different structures are so small (meV per water) that a clear picture of aggregation will require knowledge of many more clusters than just the cyclic hexamer. (iv) I don't know.

1 T. Mitsui, M. K. Rose, E. Fomin, D. F. Ogletree, and M. Salmeron, *Science*, 2002, **297**, 1850.
2 J. Cerda, A. Michaelides, M.-L. Bocquet, P. J. Feibelman, T. Mitsui, M. Rose, E. Fomin, and M. Salmeron, *Phys. Rev. Lett.*, 2004, **93**, 116101.
3 A. Michaelides, *Appl. Phys. A*, 2006, **85**, 415.

Dr Hughes opened the discussion of Dr Wakisaka's paper: The method discussed in the paper can only observe charged clusters with odd numbers of ions and cannot see neutral clusters with even numbers of ions. Nonetheless, a model for crystallization is proposed which ignores neutral clusters. Hence, I wanted to know if the energies of the neutral clusters had been calculated, along with those of the charged ones, in order to determine how common such species might be.

Dr Wakisaka responded: By means of mass spectrometry, the charged clusters with odd numbers of ions can be observed, but the neutral clusters with even numbers of ions cannot be observed. Therefore, we cannot compare the stability of clusters with charged and neutral clusters experimentally. Diefenbach and Martin calculated the stability of these clusters in the gas phase.[1] The calculation suggests that the neutral clusters are more stable than the charged clusters. However, in the liquid phase, the charged clusters may become more stable due to the solvation. Fig. 5 of our paper (model of crystallization) shows just an example. We do not exclude the neutral clusters from the crystallization processes.

1 J. Diefenbach and T. P. Martin, *J. Chem. Phys.*, 1985, **83**, 2238.

Professor Vlieg said: For larger NaCl crystals, two morphologies are known to occur: {100} which is charge neutral, and {111}, which is polar and should be charged. However, inside a liquid, the {111} morphology turns out to be stabilized through an electrochemical double layer. We have evidence from surface X-ray diffraction for this.[1] Thus: how does this compare to your results? And would you not expect neutral species to be at least as stable?

In an STM experiment combined with DFT calculations[2] it was found that a three layer NaCl crystal, Na–Cl–Na, is in fact charge neutral because the Na atoms were found to have only charge 1/2, making the system stable through a different charge transfer than normal.

1 N. Radenović, D. Kaminski, W. J. P. van Enckevort, W. S. Graswinckel, I. Shah, M. in 't Veld, R. Algra and E. Vlieg, *J. Chem. Phys.*, 2006, **124**, 164706.
2 W. Hebenstreit, M. Schmid, J. Redinger, R. Podloucky and P. Varga, *Phys. Rev. Lett.*, 2000, **85**, 5376.

Dr Wakisaka replied: In our mass spectrometry, liquid droplets charged by the excess cations or anions are generated by the electrospray method. The charged clusters are formed through the fragmentation of these charged liquid droplets. Since the electrospray is a kind of electrochemical process, our results should have some correlation with your study. As for the neutral clusters, we do not exclude them from the crystallization processes.

Dr Michaelides addressed Professor Vlieg: With regard to your point about the double layer at salt surfaces, I note that very recently dynamic scanning force microscopy and Kelvin probe force microscopy revealed that a double layer exists even on the (001) surface of alkali halide surfaces such as NaCl(001) due to the presence of divalent impurity ions in the bulk which causes the surface to carry a net negative charge.[1]

1 C. Barth and C. R. Henry, *Phys. Rev. Lett.*, 2007, **98**, 136804.

Professor Kahr opened a general discussion, addressing Dr Michaelides: Richard Saykally has for many years studied small water clusters in the gas phase by rotational-vibrational spectroscopy.[1] How do your clusters on metal surfaces compare with what has been observed in the gas phase? It appears that, for example, your trimer and Saykally's trimer are diastereomers.

1 N. Pugliano and R. J. Saykally, *Science*, 1992, **257**, 1937.

Dr Michaelides responded: Referring to Fig. 2 in the paper, with our current DFT set-up, we are able to predict the correct ground-state for the gas phase trimer. This is the (chiral) cyclic trimer observed by Saykally to which you refer.[1] Upon adsorption on Cu(111) we predict that two other isomers [(a) and (b) in Fig. 2] become more stable than the cyclic structure. The analysis of the electronic structures in section 4.2 provides some explanation for why this is the case. I think the low energy adsorbed trimers predicted here are probably best described as constitutional isomers of the cyclic trimer since it is the connectivity of the water molecules that differs between the structures.

In general, our calculations predict that upon adsorption the relative energetic ordering of different isomers of a given nH$_2$O molecule cluster are liable to change from the gas phase and that the internal O–O bond distances of a H$_2$O cluster can be different when in the gas phase and when adsorbed. Notably, for the H$_2$O dimer we find that, both here on Cu(111) and on Pd(111),[2] the O–O distance decreases by about 0.2 Å upon adsorption.

1 N. Pugliano and R. J. Saykally, *Science*, 1992, **257**, 1937; F. N. Keutsch, J. D. Cruzan and R.J. Saykally, *Chem. Rev.*, 2003, **103**, 2533.
2 V. A. Ranea, A. Michaelides, R. Ramirez, J. A. Verges, P. L. de Andres and D. A. King, *Phys. Rev. Lett.*, 2004, **92**, 136104.

Dr Schön addressed Dr Wakisaka: The magic numbers in your MS measurements indicate that there are very likely clusters in the gas phase which correspond to blocks from the rock-salt structure. Our *ab initio* calculations on bulk crystalline modifications of alkali halides have shown that there is a very strong competition between, in particular, the wurtzite and the rock-salt modification in essentially all systems.[1] Thus it would be interesting to know whether other classes of clusters are present besides rock-salt based ones. Would some of the other (non-magic) peaks correspond to blocks taken from the wurtzite structure?

The 3 × 3 × 3 clusters with + and − charge are suggested to match *via* one of their 3 × 3 faces. Since these clusters are most likely quite distorted in the gas phase, might their union look more like a dumb-bell, a fully merged approximately "spherical" block, or do you expect them to form a 3 × 3 × 6 block as your picture suggests?

How do these clusters come into existence in the first place during the electrospray process? Do you first form individual atoms in the gas phase which then react, or does the process produce larger clusters from the solution right away? What is the influence of the solvent on the process?

1 Z. Cancarevic, PhD Thesis, Universität Stuttgart, 2006.

Dr Wakisaka replied: In our experiment, clusters are formed through fragmentation of liquid droplets generated by the electrospray. When the alkali halides in the liquid droplets are concentrated *via* evaporation of the solvent molecules from the liquid droplets during the flight in the vacuum chamber, the salt clusters are formed efficiently. The concentration and electric field effects suggest this mechanism for the cluster formation. The clusters formed in the liquid phase are isolated in the vacuum chamber. Therefore, the cluster formation was dependent on the solvent. The salt clusters were more efficiently formed from methanol than from water. Fig. 5 suggests a possible mechanism for cluster growth in the liquid phase.

Dr Hare communicated: Given a suitable choice of salt solution, would it also be feasible to study, for a given concentration, the effect of a chemical reagent, possibly as a function of time? Even as long ago as *Faraday Discussions* 61, on Precipitation, there was interest in cluster formation. With aluminium salt solutions in mind, and models of octahedra (in which the Al^{3+} cation is surrounded by 6 ligands), I remarked to P. de Bruyn that the ultimate crystal form in bulk *could perhaps* be determined as early as the event of the 3rd cation attaching itself to a dimer in solution (whether in a straight line, or at a 60° angle). I recall Peter de Bruyn agreeing that "this would be a beautiful experiment", but I never could devise one. Now I am wondering if Akihiro (Dr Wakisaka) has come very close to doing so? Suppose that next he were to choose $AlCl_3$, instead of KCl; is the mass spectrum sensitive enough to identify, and distinguish between, clusters that might contain the hypothetical cyclic complex $Al_6(OH)_{12}(OH_2)_{12}^{6+}$ or linear series like $Al_n(OH)_{2(n-1)}(OH_2)_{2(n+2)}^{(n+2)+}$? What then might be the effect of introducing the smallest aliquots of KOH into the solution (and could the $K^+(KCl)_x$ contribution now be subtracted out in some way) ? Over time, could we anticipate small n values superseded by large ones, for example?

A bigger question, I suppose, might be whether the salt solution (or indeed, melt) is really nothing more than "a crystal waiting to happen" (*i.e.*, comprises a predetermined set of nuclei of given geometry), or whether the reagent—or impurity, or other field or perturbation—must first radically alter the structure of the solution (or melt) to induce its own characteristic nucleus.

Dr Wakisaka communicated in reply: This suggestion and comment are very important. I am going to observe clusters from an aqueous $AlCl_3$ solution. We are thinking that there is a relationship between the cluster structure in the solution and the crystallization. At the cluster level, ions and molecules cannot be mixed homogeneously in the solution. Since the clusters in the liquid phase are determined by the balance of interactions in the solution, these structures are sometimes different from these in the gas phase. The balance of interactions in the condensed phase controls the clustering, nucleation and crystallization. The important point is to know what kind of balance of interactions is working in the solution.

PAPER

The effect of oxygen-containing reagents on the crystal morphology and orientation in tungsten oxide thin films deposited *via* atmospheric pressure chemical vapour deposition (APCVD) on glass substrates†

Geoffrey Hyett, Christopher S. Blackman and Ivan P. Parkin*

Received 1st November 2006, Accepted 19th January 2007
First published as an Advance Article on the web 11th April 2007
DOI: 10.1039/b615877c

Thin films of monoclinic WO_3 and WO_{3-x} have been synthesized by atmospheric pressure chemical vapour deposition from WCl_6 and three oxygen containing precursors; water, ethanol and ethanoic anhydride. A wide variation in the colour, crystal morphology and preferred orientation of the films was observed, depending on the chosen oxygen source. In particular contrast were the films formed from WCl_6 and ethanol, which were blue and had needle-like crystallites, and those formed from WCl_6 and water, which were yellow and had hexagonal shaped crystallites. Studies were also undertaken to form films from WCl_6, ethanol and water simultaneously, in which the ratio of ethanol to water was varied, and this led to films in which the crystal morphology and orientation could be controlled.

Introduction

Thin films of WO_3 are of particular interest because of their electrochromic and photochromic properties and have been the focus of a number of previous publications.[1–3] The properties of such films were first investigated systematically by Deb et al.,[4] who found that the colour of WO_3 films could be changed *via* the application of an electric current (electrochromism), or by exposure to UV light (photochromism). In both cases the change in colouration is from a yellow WO_3 film to a blue WO_{3-x} film, caused by the injection of electrons to form reduced tungsten species.

WO_3 films have been successfully synthesized, amongst other methods, by vacuum thermal deposition synthesis[5] and chemical vapour deposition.[6] Our interest is in the CVD synthesis, which has been carried out using a range of precursors,[7] but of particular interest is the synthesis using WCl_6 as this is inexpensive, safe and applicable to industrial scale up.

The ability to control the morphology and orientation of crystallites on planar surfaces would be an important development in thin-film technology, because it has been previously found that these characteristics have significant effects on the functional properties of thin films. It has been demonstrated that TiO_2 films

Christopher Ingold Laboratory, University College London, 20 Gordon Street, London, United Kingdom WC1H 0AJ. E-mail: i.p.parkin@ucl.ac.uk

† The HTML version of this article has been enhanced with colour images.

deposited by sol–gel from Degussa P25 are much more photocatalytic than films deposited by CVD, because the larger crystallites of the Degussa films produce a rougher surface than the nano-crystalline films formed by CVD.[8] Gavrilyuk et al. have determined that crystal orientation can affect the surface reactivity of WO_3 with the (001) plane having a favourable free energy for chemisorption and the diffusion of protons—important processes for surface reactions,[9] and the ability to use the materials as electrochromic windows.

Presented here are investigations into WO_3 thin films grown from WCl_6 and three oxygen sources: ethanol, water and ethanoic anhydride. It is demonstrated that the use of these different oxygen sources produced different crystallite morphologies determined by SEM, and different crystallographic orientations, as determined by preferred orientation observed in the X-ray diffraction patterns. It will then be shown that a level of control can be exerted over these properties, morphology and orientation, by using a mixture of ethanol and water as oxygen sources—producing films with tuneable properties between the extremes found when either water or ethanol were used as single oxygen sources.

Experimental

Two different atmospheric pressure CVD reactors were used to produce the films discussed in this work. The first apparatus used contained a 170 mm long reactor with a single mixing chamber and two reactant bubblers. This rig was used for synthesizing films from WCl_6 (99.9%, Strem) and one oxygen source—either water (double distilled, in-house), ethanol (99.7%, BDH) or ethanoic anhydride (98%, Aldrich). The second, and larger, apparatus had two mixing chambers each with two reactant bubblers, allowing a total of four reactants to be introduced into the reactor. This apparatus was used to produce films of tungsten oxide using WCl_6 with both ethanol and water simultaneously as oxygen sources. Single mixing chamber reactors like the 170 mm apparatus used have been described in detail previously;[10] the larger 320 mm dual mixing chamber rig will be described in detail below.

The films of tungsten oxide were deposited on the larger reactor onto glass slides of dimensions 89 mm × 225 mm × 3.2 mm (width, length, thickness). The slides were coated on the top surface with a barrier layer of SiO_2 to prevent ions in the glass from migrating into the synthesized film.

The larger apparatus was a cold walled atmospheric pressure chemical vapour deposition (APCVD) reactor composed of a 170 mm diameter, 320 mm length quartz cylinder, containing a semi-circular section carbon block and top plate to channel the vapour streams over the glass slides. The reactor was heated by three Whatman heater cartridges, inserted into the carbon block.

The carrier gas was supplied to the rig from a single 99.99% N_2 cylinder (BOC) which was then split into six lines each with an individual needle valve flow controller. All six N_2 gas lines passed through several metres of coiled tubing inside a cylindrical furnace set at 200 °C, to preheat the gas. Four of these lines supplied the precursor bubblers, while the other two supplied each mixing chamber with a plain flow.

In each mixing chamber the downstream precursor containing flows from two bubblers were combined with a plain flow of N_2, which then entered the inlet manifold. The mixing chambers were arbitrarily labelled A and B, with precursor bubblers 1 and 2 feeding into mixing chamber A, and bubblers 3 and 4 into mixing chamber B. All the piping on the rig was heated to 200 °C, with heater tape, and the mixing chambers were heated to 250 °C to prevent any of the vaporized precursors from condensing. The gas streams from each of the separate mixing chambers remained segregated as they passed through the inlet manifold. This meant that vapours from different mixing chambers could not combine until they were in the reactor itself, and also that asymmetry could be introduced into the system by

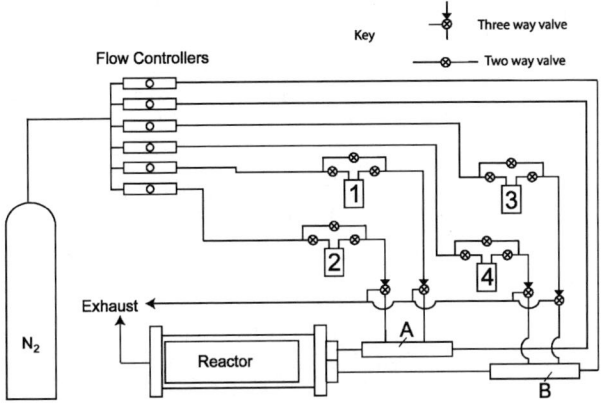

Fig. 1 Schematic diagram of the four bubbler atmospheric pressure reactor, showing bubblers 1–4 and mixing chambers A and B.

passing a precursor from only one side of the reactor. A schematic diagram of the 320 mm rig and gas flow is shown in Fig. 1.

For the work done using the larger apparatus to investigate mixed oxygen source reactions three different bubbler configurations were set up, taking advantage of the two separate vapour entry points to the reactor. Five films were deposited using differing flow rates with each configuration. In the first configuration WCl_6 was placed in bubbler one, ethanol in bubbler two and water in bubbler three, so that the WCl_6 and ethanol vapour flows could mix before entering the reactor while water entered solely from the other side of the reactor. The second configuration switched the water and ethanol so that the WCl_6 and water would be mixed before entry into the reactor. In the final configuration the WCl_6 was placed in bubbler one, and the ethanol and water in bubblers three and four, such that neither oxygen source could premix with the metal chloride until they were in the reactor. These bubbler configurations are represented schematically in Fig. 2.

In all of the experiments the WCl_6 bubbler was heated to 250 °C and the vapour transported using a nitrogen gas flow of 4 L min^{-1}. This was equivalent to a molar flow of 0.0083 mol min^{-1}, determined using the mass–flow equation.[11] This rate of WCl_6 flow was kept constant and used in all of the configurations. The ethanol and water bubblers were heated to 40 and 60 °C, respectively, such that their vapour pressures were approximately equal at 140 mm Hg. The N_2 gas flow rates through the oxygen and ethanol bubblers were then set at either 0, 0.4 or 1 L min^{-1} giving molar flow equivalents of 0, 0.0037 and 0.0092 mol min^{-1}. These different oxygen source flow rates were combined in pairs to give five different combinations, as shown in Table 1, with varying ratios of ethanol to water flowing into the reactor. Each of these five combinations was used with each of the three bubbler configurations (Fig. 2) to generate a total of 15 films.

The plain line flow rates in these depositions were varied from 1 to 6 dm^3 min^{-1}, such that the combined N_2 gas flow through each mixing chamber was always 6 dm^3

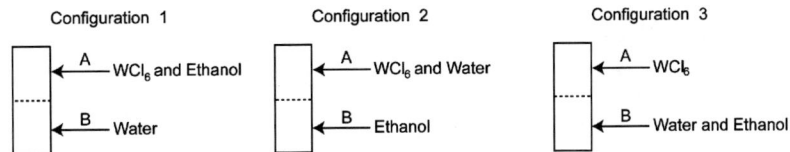

Fig. 2 Schematic diagram of the different bubbler configurations used on the larger 320 mm rig, showing which reagents enter from the two separate entry points to the reactor from mixing chambers A and B.

Table 1 Flow rates through the water and ethanol bubblers for each of the five different reaction conditions used in each of the bubbler configurations for the depositions on the large 320 mm reactor. The figure in brackets is the molar percentage of the oxygen source made up by that reactant

Conditions set	1	2	3	4	5
Ethanol molar flow/mol min^{-1}	0 (0%)	0.0037 (29%)	0.0092 (50%)	0.0092 (71%)	0.0092 (100%)
Water molar flow/mol min^{-1}	0.0092 (100%)	0.0092 (71%)	0.0092 (50%)	0.0037 (29%)	0 (0%)

min^{-1}. Depositions were conducted using a substrate temperature of 600 °C and a deposition time of 180 s.

The smaller apparatus used was a 170 mm long cold walled APCVD reactor, had only one mixing chamber fed by two reactant bubblers and a plain N$_2$ line, and had a single undivided baffle inlet. Depositions conducted using this apparatus were onto 45 mm × 150 mm × 3.2 mm sheets of glass, also coated with a SiO$_2$ barrier layer.

For the initial reactions carried out on the smaller apparatus the WCl$_6$ bubbler was heated to 290 °C with a 0.4 L min^{-1} N$_2$ gas flow through the bubbler, equivalent to 0.0075 mol min^{-1}. This was kept constant for the reactions with ethanol, water and ethanoic anhydride, as was the plain line flow of 10 dm^3 min^{-1}. The bubbler temperatures, flow rates and molar flow rates for the oxygen source in each of the three reactions are detailed in Table 2. In each of the reactions the substrate was heated to 625 °C with a deposition time of 1 min.

Analytical methods

After deposition the films were investigated using X-ray diffraction conducted using a tight X-ray beam focus. This meant that discrete areas of only 3–5 mm^2 of the film were illuminated by the beam and it was possible to record diffraction patterns of specific, isolated areas, or spots, of the film. For each slide 168 of these 'spots' were analysed, in a 21 × 7 grid of spots 10 mm apart in both x and y directions, effectively allowing any variation in film composition to be measured across the substrate.

The photocatalytic activity of the films was measured by spin coating a layer of stearic acid onto the film from a 0.02 M solution in methanol. The quantity of stearic acid deposited was determined by recording an IR spectrum of the film over the range 3000–2750 cm^{-1}, in which the CH stretches for stearic acid intensity appear and where an integrated absorbance area of 1 cm^{-1} is equivalent to $\sim 3.17 \times 10^{15}$ molecules cm^{-2}.[12] The film was then irradiated with 254 nm UV light for several hours, and the IR spectrum re-recorded periodically to determine if photo-activated destruction of the stearic acid was occurring, and to monitor its progress.

The hydrophobicity/hydrophilicity of the films was tested by placing a 1 µL drop of water onto the surface of the film, then measuring the horizontal spread of the droplet. This could then be converted into a contact angle.

Table 2 Reaction conditions for the depositions conducted on the smaller 170 mm apparatus

Reactant	Bubbler temp/°C	N$_2$ flow rate/dm^3 min^{-1}	Molar flow of reactant/mol min^{-1}	Molar flow of WCl$_6$/mol min^{-1}
Ethanol	70	0.4	0.075	0.0075
Water	90	0.4	0.375	0.0075
Ethanoic anhydride	100	0.4	0.2	0.0075

Results and discussion

Investigations using the 170 mm reactor

The reactions of WCl_6 with water, ethanol or ethanoic anhydride were investigated by APCVD. All three reactions successfully produced films which were adherent and passed the 'scotch tape test'; a piece of sticky tape was placed on the film then removed—the film did not lift off with the tape. The films formed using water and ethanoic anhydride with WCl_6 were yellow in colour, while the film formed with ethanol and WCl_6 was blue in colour. Additionally the film formed from water was noticeably thinner—as judged by eye and measured by side-on SEM, which will be discussed below.

Sections of the films were analysed using X-ray diffraction, and the resulting patterns can be seen in Fig. 3. The patterns of the yellow films formed with ethanol and WCl_6 and with ethanoic anhydride and WCl_6 could be indexed as monoclinic WO_3, a distorted ReO_3 structure[13] in the $P12_1/n1$ space group. The blue film formed from WCl_6 and ethanol, however, had a diffraction pattern in which only two peaks could be identified between $10 < 2\theta < 66°$. The peaks could also be indexed in the $P12_1/n1$ space group with the same structure as the yellow films, but with a strong preferred orientation in the (010) plane, such that only the 020 and 040 peaks were observed—Fig. 3(b). As the film is blue it must contain oxygen non-stoichiometry, WO_{3-x}. Non-stoichiometric forms of WO_{3-x} have been found with this structure as far as $x = 0.13$ and have also been reported with at least four different structure types[14–16] with an x value up to 0.3; all are based on the monoclinic distortion of the ReO_3 structure but with the addition of columns of pentagonal bipyramids, with a notional stoichiometry of $WO_{2.67}$[17] to accommodate the missing oxygen atoms. The match between the two peaks that can be observed in the diffraction pattern and the (010) oriented $P12_1/n1$ model is, however, sufficiently good to strongly suggest that this is in fact the structure of the WO_{3-x} present in the film, thereby limiting the value of x to less than 0.13. Unfortunately as only two peaks appear in the diffraction pattern of the WCl_6–ethanol film it is not possible to definitively determine the structure by diffraction. As such, for clarity, we will assume that the

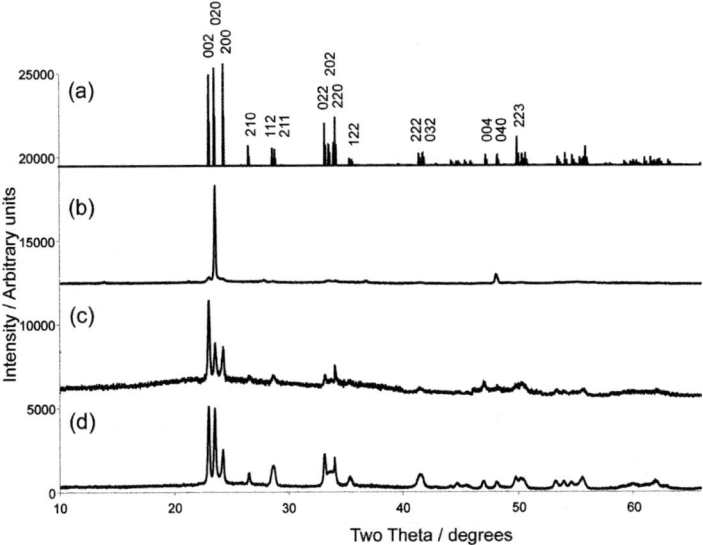

Fig. 3 X-Ray diffraction patterns of films synthesized on the smaller 170 mm apparatus. (a) Model pattern of monoclinic WO_3. Patterns from films synthesized with WCl_6 and (b) ethanol, (c) water and (d) ethanoic anhydride.

Fig. 4 SEM images of the films synthesized using the 100 mm reactor from WCl$_6$ and (a) ethanol, (b) ethanoic anhydride, (c) water.

film has the $P12_1/n1$ structure identical to the stoichiometric films of WO$_3$, but with oxygen vacancy and a large degree of preferred orientation.

SEM imaging was also conducted on sections of the three films and these are reproduced here in Fig. 4. The film formed with ethanol and WCl$_6$, shown in Fig. 4(a), was composed of needle-like crystallites greater than 1 micron in length but only 150 nm in width. The film formed from WCl$_6$ and water had crystallites that appeared more isotropic in shape, with hexagonal angles between many of their edges, approximately 500 nm diameter. The film synthesized using ethanoic anhydride and WCl$_6$ had plate-like crystallites, also isotropic in shape; these were approximately 1 micron in diameter. Cross-sectional SEM was also conducted and found that the ethanol film was 800 nm thick, the ethanoic anhydride film a similar 1000 nm, but the water film was thinner at 225 nm. All films were deposited in 1 min.

In summary, the work conducted on the smaller reactor has found that the choice of oxygen source used in synthesizing tungsten oxide from WCl$_6$ has a profound effect on the crystal growth and orientation properties of the subsequent film. The three oxygen sources highlighted have differences in the shape of their crystallites, in the orientation of those crystallites, and in the thickness of the tungsten oxide film deposited, however despite these differences in the crystal morphology all three films are composed of monoclinic WO$_3$ or WO$_{3-x}$. Motivated by these differences, work was attempted on the larger 320 mm rig where multiple precursor streams enabled three reactants to be used simultaneously—WCl$_6$, ethanol and water—to synthesize films with properties intermediate between the films made from water and from ethanol and thus attempt to exert some control over the crystal growth phenomena.

Synthesis in the 320 mm reactor, configuration 1

In the first configuration bubblers 1 and 2 were loaded with WCl$_6$ and ethanol, and combined with the plain line flow in mixing chamber A (see Fig. 1), allowing these two reactants to mix before entering the reactor and separate from the water vapour which, loaded into bubbler 3, entered the reactor through mixing chamber B (see Fig. 2). All five different oxygen source concentration conditions (see Table 1), which varied the ethanol to water ratio, led to the formation of WO$_3$ or WO$_{3-x}$ films with at least partial coverage of the glass slides. Photographs of these five films can be seen in Fig. 5.

In this configuration in the two extremes of conditions, where either only water or only ethanol were used as the sole oxygen source, the results of preferred orientation and crystallite morphology analysis agreed with those, already discussed, from the smaller rig. In the case where ethanol was the only oxygen source, being introduced from the left-hand side of the reactor through the same mixing chamber as WCl$_6$, a blue coloured film of WO$_{3-x}$ formed over the majority of the substrate. Of the 168 spots used in the diffraction mapping, one hundred and eleven had film coverage, giving of total substrate coverage of 66% over a section of the substrate located

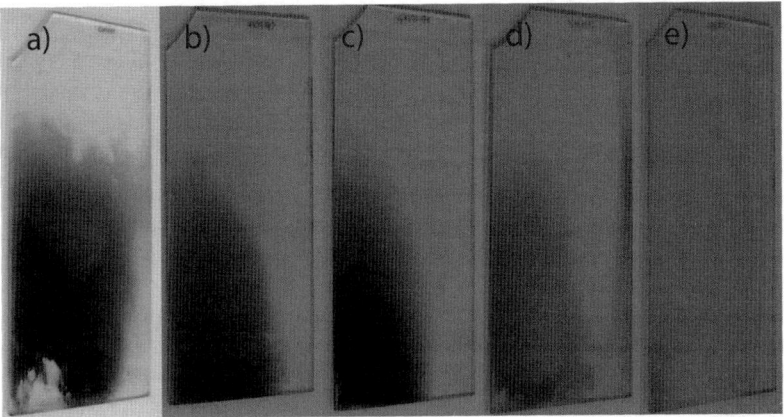

Fig. 5 Photographs of the five films produced with bubbler configuration 1 with the following oxygen sources (a) 100% ethanol; (b) 71% ethanol, 29% water; (c) 50% ethanol 50% water; (d) 29% ethanol, 71% water; (e) 100% water.

nearest to the WCl_6 inlet. This can be seen in the far left-hand photograph in Fig. 5(a). Such partial coverage is due to a depletion of one or all of the reactant precursors as the gas flow progresses down the reactor. At the opposite extreme of conditions, the WCl_6 and water vapour formed a yellow film covering 165 of the diffraction spots, an almost total coverage of 98% (Fig. 5(e)). Two explanations are possible for the seemingly faster rate of precursor depletion with the ethanol film, and hence smaller substrate coverage. Firstly, the WCl_6 and ethanol were introduced through the same mixing chamber, so there may have been some deposition in the baffle manifold, consuming some of the reactants; in the water–WCl_6 film in this configuration the two reactants entered from different mixing chambers, and so could not begin to react and deplete until over the substrate within the reactor. Secondly, it has been determined (using side-on SEM) that the film formed from ethanol and WCl_6 on the 170 mm apparatus was thicker than that formed from WCl_6 and water. This will also deplete the reactants more quickly, giving rise to the difference in observed substrate coverage.

X-Ray diffraction found that the blue film was composed of monoclinic WO_{3-x} with a large degree of preferred orientation, such that only the 020 and 040 peaks were observed, (see Fig. 6(d))—the same effect observed in the WCl_6 and ethanol film made on the 170 mm apparatus. Although the figure shows only one pattern from the film, this is representative and all of the patterns showed the same degree of orientation. As the film is coloured with a strong blue tint it is composed of non-stoichiometric WO_{3-x}, rather than the fully oxidised WO_3, as the W^{6+} ion is pale yellow in colour, while reduced W^{5+} ions in the film can act as colour centres, providing electron donor sites 1.37 eV below the conduction band.[18] The diffraction patterns of the yellow film synthesized with water were also indexed as monoclinic WO_3, but no preferred orientation was observed in the patterns—the same result as found for the film synthesized on the smaller apparatus. An example diffraction pattern taken from one of the spots is given in Fig. 6(a), representative of all of the patterns recorded on the yellow film. The same pattern of relative peak intensities was found uniformly for the whole of the covered substrate, with the only changes observable being in the overall intensity of the pattern—caused by variation in the thickness. Hence the phase composition or preferred orientation of the film does not depend on the position on the substrate—this is in contrast to previous work on TiO_2 films that also used this X-ray mapping method to investigate changes across a substrate, in which both the phase composition and orientation of films were found to vary with position.[19]

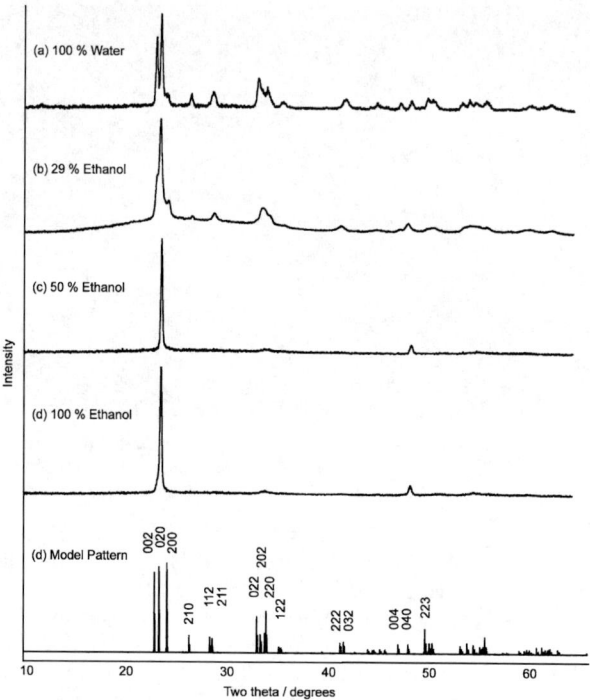

Fig. 6 X-Ray diffraction patterns of representative spots from films formed using configuration 1, showing the range of preferred orientation that can be found. (a) Film made with 100% water; (b) film made with 71% water, 29% ethanol; (c) film made with 50 : 50 water : ethanol; (d) film made from 100% ethanol oxygen source. (e) Model pattern of powder WO$_3$ without preferred orientation and indexing up to 50° 2θ.

A further analysis was carried out on the X-ray diffraction patterns of the film formed from ethanol and WCl$_6$ to determine the extent of the preferred orientation observed in the films. The diffraction patterns were analyzed using Rietveld refinement, and the preferred orientation modelled using the March–Dollase approximations.[20] The March–Dollase r factor is a measure of the degree of preferred orientation, with $r = 1$ being the case when no preferred orientation is observed; deviations above or below this value indicates the presence of preferred orientation. This analysis found that the one hundred and eleven spots within the blue section of the film showed the same preferred orientation, in the (010) plane, and that the March–Dollase factor, r, for spots on this film averaged 0.6 with a standard deviation of 0.1; a strong degree of preferred orientation, producing a diffraction pattern like that observed in Fig. 6(d). There was no change in the extent of the preferred orientation with position on the substrate. A similar analysis was conducted on the yellow film produced from water; all the patterns could be modelled well and produced an average r factor of only 0.9(1), numerically confirming that almost no preferred orientation was found in the film—as would be found in a powder sample of WO$_3$.

SEM imaging carried out on the sample synthesized using only ethanol can be seen in Fig. 7(a) and using only water in Fig. 7(i). These show that the yellow film synthesized with water is composed of tightly packed crystallites, indicative of island growth, and with no particular elongation. The side-on SEM images in Fig. 7(j) indicate that the film thickness is 660 nm. As the crystallites seem to be isotropic in shape, they are consistent with the lack of preferred orientation observed in the diffraction patterns and also match the results found in the WCl$_6$–water film

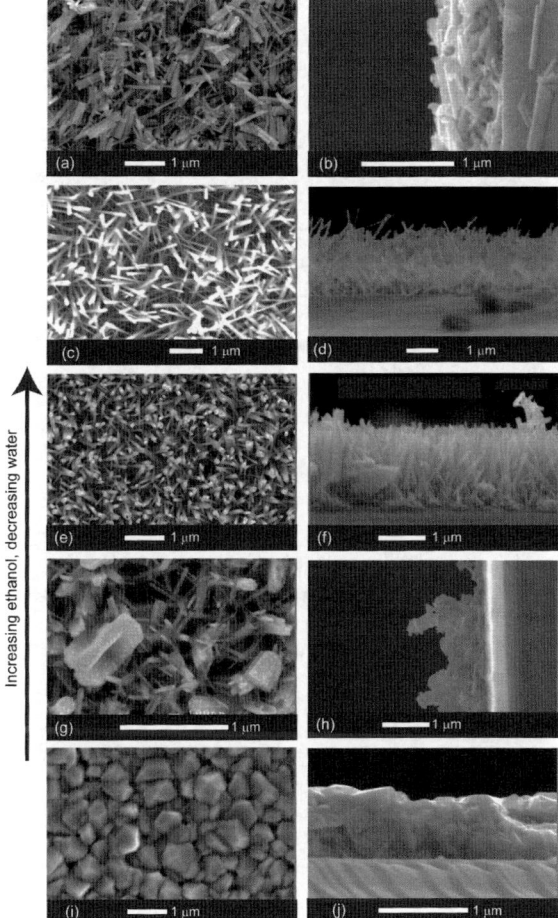

Fig. 7 Scanning electron microscope images of the tungsten oxide films produced in configuration 1 from WCl_6 and: 100% ethanol (a) top down, (b) side-on; 71% ethanol, 29% water (c) top down and (d) side-on; 50% ethanol, 50% water (e) top down and (f) side-on; 29% ethanol, 71% water (g) top down and (h) side-on; and 100% water (i) top down and (j) side-on.

synthesized on the 170 mm reactor. The film grown on the larger reactor, however, is much thicker and this may be the result of experimental difference between the two setups. In the smaller reactor ethanol and WCl_6 were combined in the mixing chamber, whereas in the larger apparatus, in this configuration, the two reactants entered through different mixing chambers.

In contrast, the SEM images of the blue film synthesized with ethanol (Fig. 7(a)) show that it is composed of crystals with a needle-like morphology—as was also seen in the ethanol film synthesized using the smaller rig. The needles have a diameter in the range 60–230 nm, and a length between 750–1000 nm. Side-on SEM measurements in Fig. 7(b) give the film a thickness of approximately 700 nm. Both top down and side-on SEM images suggest that the needles are predominantly orientated so that their long axis is parallel to the substrate.

The results of the analysis of the films formed from ethanol only and water only as oxygen sources are consistent with the work from the smaller reactor, which also found that films formed from ethanol showed strong preferred orientation in the (010) plane, while those formed from water had randomly orientated crystallites.

In the intermediate films between these two extremes, where *both* ethanol and water were present in different amounts as oxygen sources a different type of film coverage was observed. In all three intermediate cases a section of the substrate was covered with a blue film, but with much less coverage than in either of the single oxygen source films. This can be observed in Fig. 5(b)–(d).

The Rietveld refinement of the X-ray diffraction patterns found that the films formed from WCl_6 and water–ethanol mixtures showed strong preferred orientation in the (010) plane, like the film formed exclusively from ethanol and WCl_6, but with less substrate coverage, ranging only from 30–36%.

In the film where the oxygen source was 71% ethanol and 29% water the film covered 36% of the substrate and the modelled diffraction patterns had an average r factor of 0.52(7). For the film formed with equal amounts of each oxygen source, 50% water and 50% ethanol, the film coverage dropped to 30%, but a large degree of preferred orientation was still observed with an average r factor from the refinements of 0.5(1). In both the 71% and 50% ethanol films there was little variation around this mean value of about $r = 0.5$, and so the high degree of preferred orientation was found in all the spots of the films.

Both of these mixed oxygen source films, then, share the same characteristics as the ethanol only film, with the increasing water content apparently having no effect on the film, except for the reduced substrate coverage—although the SEM imaging discussed below does reveals further differences. It should also be noted that the preferred orientation is uniform across the substrates and does not vary with position across the plate.

In the film formed from 71% water and 29% ethanol a blue film is also formed, with 33% substrate coverage. Differences between this film and the higher ethanol content films became apparent when the Rietveld modeling was conducted, where the average r factor for the film was found to be 0.79(4). This is a lesser degree of preferred orientation than in the 50%, 71% and 100% ethanol films, but more than when water is the sole oxygen source. It should be noted that this is *not* an intermediate average caused by a range of preferred orientations across the spots of the film, some with no preferred orientation ($r = 1$) and some with low preferred orientation ($r = 0.5$). Instead, the statistical spread of values is low, with most of the spots actually having this intermediate level of preferred orientation of about $r = 0.79$, between a maximum and minimum value $0.87 > r > 0.71$. An example diffraction pattern of the 71% water, 29% ethanol film is shown in Fig. 6(b). So this film, at the level of resolution available on the diffractometer, does have crystallite orientation intermediate between the films formed only with ethanol and those formed only with water.

SEM images were recorded on the intermediate films to investigate their crystallite morphology. The top down and side-on images for the 71% ethanol, 29% water film are shown in Fig. 7(c) and (d), respectively. These show the same needle-like morphology as the WCl_6–ethanol only film, but instead of laying parallel to the substrate, the crystal needles appear to grow diagonally upwards in a mesh, (this can be seen best in the side-on SEM), resembling a carpet pile. The film thickness was 2 μm. Significantly, *only* needle-like crystallites could be observed in the SEM image, with none of the more isotropic particles found in the water–WCl_6 synthesized film.

The 50% water, 50% ethanol film SEM images are shown in Fig. 7(e) top down, and side-on in Fig. 7(f). Again the film is made of needle shaped crystallites, like the films formed using only ethanol, despite the oxygen source being 50% water. The needles do appear to be standing end on, however, growing up from the surface of the glass in the 'carpet' like morphology, but with even more vertical orientation than in the 71% ethanol film. Again, only needle-like crystallites could be observed. The side-on SEM image suggests that the film is 1.7 μm thick.

The side-on SEM images show that both the 71% and 50% ethanol films are twice as thick as the film formed from WCl_6 and ethanol only, and this helps to explain the reduction in substrate coverage—deposition of thicker films will deplete the precursors at a greater rate. The mixed water–ethanol oxygen source also promotes vertical growth of the needles, compared with the horizontal alignment seen in the ethanol only film, and the extent of this vertical alignment scales with the amount of water present. Thus it has been shown that the crystal growth can be controlled by altering the reaction conditions, although the exact mechanistic nature of water in causing this change in the crystal growth pattern is, however, still uncertain.

The final film in the intermediate set, made with 29% ethanol and 71% water had SEM images that showed the needle shaped crystallites found in the other WO_{3-x} films but also larger, more isotropically shaped particles (several 100 nm in diameter) amongst the smaller needles. This observation explains the intermediate level of preferred orientation seen in the X-ray diffraction patterns, in the previous films the needle-like crystallites correlated with a high level of preferred orientation, while the larger isotropic crystallites are likely to have little or none. As these two different types of crystallite are mixed at the sub-microscopic level, the diffractometer probing spots with 10^{-3} m resolution will 'see' an average of the orientation of these two types of crystallite. The side-on SEM images give a film thickness of 350 nm, but also show that the surface is extremely uneven, and in places is over 1000 nm thick. This is caused by localized build up of the larger isotropic particles, on top of a base layer of the needles-like crystallites. Given the results observed in the other films it seems likely that WCl_6 reacts with ethanol to form the base layer of needle-like crystals, with the larger isotropic particles formed by gas-phase reaction between WCl_6 and water, which then 'snow' onto the needle layer.

The results from the intermediate films, based on their colour, preferred orientation and crystallite shape suggest that the films are predominantly formed from the reaction of ethanol and WCl_6 (until the ethanol level is reduced to only 29% of the oxygen source available), no doubt because the vapour of these two reagents are combined in the mixing chamber before water vapour mixes with them. This suggests that the first step of the reaction is in the vapour phase.

The photocatalytic properties of all of the films were analysed by testing a spot of each of the films using stearic acid degradation testing, with 254 nm UV light. This found that all of the films were photoactive, and the rates of destruction could be modeled using zeroth order kinetics. The calculated rate constants from this data for stearic acid destruction are shown in Table 3. These values may be compared with the known photocatalyst, TiO_2, which has been shown previously to have a rate of 4×10^{12} to 8×10^{12} molecules min^{-1} cm^{-2} (figures adjusted for UV light intensity).[21,22] These TiO_2 films were also synthesized by APCVD and were of a similar thickness of 600 nm. The rate constants thus show that all of the films have a high rate of photocatalysis, but with the mixed oxygen source films having slightly lower rates than the films made with a single oxidant. The water contact angle tests found that all of the films were also hydrophilic with contact angles of between 5–21°. These values are again comparable with those found in TiO_2 films of 13–25°, and are important because hydrophilicity aids the cleaning mechanism in self-cleaning coatings.

The results of the tests conducted on the films formed using the 320 mm reactor in configuration 1 have reproduced the results found on the 170 mm rig in films formed from ethanol and water as sole oxygen sources, and additionally shown some interesting results when mixtures were used. With the mixed ethanol–water films the first noticeable effect was a reduction in substrate coverage, dropping to 30–36%

Table 3 Results of photocatalytic ability tests conducted on the films formed in configuration 1, showing the rate of destruction of applied stearic acid films

Ethanol concentration (%)	100	71	50	29	0
Water concentration (%)	0	29	50	71	100
Film colour	Blue	Blue	Blue	Blue	Yellow
Rate of destruction/molecules min^{-1} $cm^{-2} \times 10^{12}$	11(2)	9(1)	7.4(5)	8.9(3)	9.8(6)

coverage from over 66% when only a single type of oxygen source was used, but with an increase in film thickness. SEM imaging also revealed that the introduction of water into the WCl_6–ethanol system produced an unexpected vertical alignment of the needle crystallites, and this became more pronounced as the water content increased.

Configuration 2

In the second configuration the water and ethanol bubbler positions were reversed such that water and WCl_6 were allowed to mix in the left-hand mixing chamber, and their combined vapours entered the reactor on the left-hand side. Ethanol vapour entered the system alone, save for the plain gas flow, from the right-hand side. This can be seen schematically in Fig. 2. Five combinations of differing flows rates for the oxygen precursors were used, the same as those used in configuration 1.

The results of these reactions present something of a contrast to those investigated in configuration 1. In configuration 2 none of the substrates were found to be covered by blue film in any of the combinations of oxygen source except when water was excluded and only ethanol was used. With the first addition of water into the oxygen source mix at 29% water, no blue film was formed at all, and the film that did form was thin, powdery and poorly adherent, unlike the yellow film formed in configuration 1 from water. In the remaining films (50%, 71% and 100% water as oxygen source) no tungsten oxide film at all could be observed either visually or by X-ray diffraction mapping.

The film produced in this configuration using only ethanol generated a blue film covering 83 spots. The X-ray diffraction patterns of these spots showed the same preferred orientation in the (010) plane as that observed in the blue films formed in configuration 1. The Rietveld modelling of the diffraction patterns found an average March–Dollase r factor of 0.5(1)—essentially showing the same properties as the previous films synthesized using ethanol and WCl_6.

These experiments show in comparison with those conducted using configuration 1 that the way in which precursors are combined prior to entering the reactor can have dramatic effects on the subsequent film formation. The reaction of WCl_6 and water successfully produced a thick (>600 nm) adherent film covering over 98% of the substrate in configuration 1, when the reagents could not mix until they entered the reactor; while in configuration 2 where WCl_6 and H_2O premixed in the mixing chamber the reaction did not produce an adherent film. Similarly in the 170 mm reactor, when water and WCl_6 were premixed the film formed was much thinner than when ethanol was used. When allowed to premix like this the WCl_6 and H_2O undergo a gas phase reaction, leading to a formation of WO_3 powder particles, which then do not adhere to the substrate but are carried through the exhaust. This would consume the WCl_6 without forming a film, also explaining the absence of any blue film either, in any of the mixed oxygen source films attempted.

Configuration 3

In the final configuration the metal source was placed in bubbler 1, and the oxygen sources in bubblers 3 and 4. This meant that the metal and oxygen sources were always kept separate until they entered the reactor. The ethanol only and water only films in this configuration were identical to the end members of configuration 1 and 2, and so will not be discussed here. Instead, the mixed oxygen source films will be considered. The film coverage, colour and average preferred orientation are reviewed in Table 4. These show the poor levels of substrate coverage that have been observed in the other mixed oxygen source films, but also the same correlation between oxygen source, colour and preferred orientation. As WCl_6 is not premixed with either oxygen source in this configuration we do not see, however, the domination of the mixed films by reaction with one of the precursors that was seen in the other

Table 4 Film coverage, colour and preferred orientation of the films synthesized using configuration 3 on the larger reactor. Preferred orientation was determined by March–Dollase modeling in Rietveld refinement. Photocatalytic rate determined by stearic acid testing

Ethanol concentration (%)	71	50	29
Water concentration (%)	29	50	71
Substrate coverage	47 spots (28%)	29 spots (17%)	38 spots (23%)
Average r factor	0.5(1)	1.0(1)	1.0(1)
Colour	Blue	Yellow	Yellow
Photocatalytic rate/$\times 10^{12}$ molecules min^{-1} cm^2	21.2(2)	8.6(8)	10.0(5)

Fig. 8 SEM images of the films synthesized using configuration 3 with WCl$_6$ and (a) 29% ethanol, 71% water; (b) 50% ethanol, 50% water; (c) 71% ethanol, 29% water.

configurations (*e.g.* the configuration 1 mixed oxygen source films all being blue). Instead when ethanol was the majority oxygen source a blue film was formed with high preferred orientation, but when water was the majority oxygen source, a yellow film was formed with no preferred orientation. When both oxygen sources were present in equal amounts, the film formed predominantly had the characteristics of the water only films. This shows that, all else being equal, the reaction between water and WCl$_6$ was more favourable.

SEM imaging was also carried out on the films and these are shown in Fig. 8. These show that the 71% water and 50% water film are composed predominantly of particles which are broadly isotropic, although some needles are observable in both images. The film synthesized with ethanol as the majority oxygen source is composed of the expected needle-like crystals, but with a reasonable number of larger isotropic crystallites also present. The crystallite compositions reflect what would be expected from the respective oxygen sources used to synthesize them, but it should be noted that both needle and isotropic crystallites are present in all three images to some extent—something that was not the case when one of the oxygen sources was allowed to premix with WCl$_6$, and hence dominate the type of crystal morphology observed. This further supports the idea that a gas phase interaction is an important step in the mechanism of tungsten oxide film formation, irrespective of the oxygen source.

Photocatalytic testing was also carried out on the films, and the results of this are given in Table 4. These show that these films are also photocatalytic, with the rate for the 71% ethanol, 29% water film showing particularly high activity. The water contact angle testing also found all the films to be very hydrophilic with contact angles of 3–5°.

Conclusions

In this paper we have considered some of the crystal growth phenomena and physical properties of WO$_{3-x}$ and WO$_3$ thin films grown from WCl$_6$ and varying mixtures of ethanol and water as oxygen sources. We have confirmed that formation with WCl_6 and either water or ethanol gave different types of tungsten oxide film with distinctions in colour, preferred orientation and crystallite morphology. Those formed with ethanol gave blue films, with needle-like crystallites and high levels of preferred orientation in the diffraction patterns. The films formed with water were yellow, had isotropic crystallites and no preferred orientation in their diffraction patterns. The mechanistic reason for this difference, however, still remains unclear.

The various studies undertaken of mixed oxygen source systems have found that a gas-phase step is important in the formation of films from both water and ethanol with WCl_6, as allowing the WCl_6 vapour to premix with one of the oxygen sources leads to the films formed being dominated by the characteristics of that oxygen source—*i.e.* films formed when WCl_6 and ethanol were premixed were almost exclusively blue in colour with needle-like crystallite morphology. This gas phase reaction is also more rapid and complete with water, such that gas phase nucleation of particles occurs if water and WCl_6 are premixed, preventing the formation of thick or adherent films in these cases.

When WCl_6 and ethanol were premixed, although the gas phase reaction step meant that the films formed were blue and composed of the needles, characteristic of the ethanol reaction, introduction of water did have a noticeable effect. In the presence of water the needles began to change their orientation to the substrate from horizontal to vertical. This effect was also found to scale with the amount of water; increasing the water present increased the extent of horizontal orientation of the crystallites.

The physical property tests conducted on the films found that all of the tungsten oxide films were photocatalytic and hydrophilic, both important properties for producing self-cleaning films. The rate of photocatalysis was also high, even when compared with TiO_2 thin films, the current industrial choice for self-cleaning coatings. This shows that the films have physical properties that could potentially be exploited.

In conclusion, it has been demonstrated that the morphology of tungsten oxide films formed using WCl_6 is dependant on the choice of oxygen source, and that it is possible to control to a certain extent the crystal growth phenomena in these films by altering the available oxygen sources.

Acknowledgements

The EPSRC is thanked for financial support. Pilkington glass are thanked for supplying glass substrates. IPP is a Wolfson Royal Society merit award holder.

References

1   A. I. Gavrilyuk, Photochromism in WO_3 thin films, *Electrochim. Acta*, 1999, **44**(18), 3027–3037.
2   J. V. Gabrusenoks, P. D. Cikmach, A. R. Lusis, J. J. Kleperis and G. M. Ramans, Electrochromic color-centers in amorphous tungsten trioxide thin-films, *Solid State Ionics*, 1984, **14**(1), 25–30.
3   S. H. Lee, H. M. Cheong, C. E. Tracy, A. Mascarenhas, D. K. Benson and S. K. Deb, Raman spectroscopic studies of electrochromic a-WO_3, *Electrochim. Acta*, 1999, **44**(18), 3111–3115.
4   S. K. Deb, Optical and photoelectric properties and color centers in thin-films of tungsten oxide, *Philos. Mag.*, 1973, **27**(4), 801–822.
5   S. Santucci, L. Lozzi, M. Passacantando, S. Di Nardo, A. R. Phani, C. Cantalini and M. Pelino, Study of the surface morphology and gas sensing properties of WO_3 thin films deposited by vacuum thermal evaporation, *J. Vac. Sci. Technol., A*, 1999, **17**(2), 644–649.
6   R. U. Kirss and L. Meda, Chemical vapor deposition of tungsten oxide, *Appl. Organomet. Chem.*, 1998, **12**(3), 155–160.
7   C. Blackman, C. J. Carmalt, I. P. Parkin, S. O'Neill, L. Apostolico, K. C. Molloy and S. Rushworth, Titanium phosphide coatings from the atmospheric pressure chemical vapor deposition of $TiCl_4$ and RPH_2 (R = t-Bu, Ph, Cy-Hex), *Chem. Mater.*, 2002, **14**(7), 3167–3173.

8 A. Mills, S. K. Lee, A. Lepre, I. P. Parkin and S. A. O'Neill, Spectral and photocatalytic characteristics of TiO_2 CVD films on quartz, *Photochem. Photobiol. Sci.*, 2002, **1**(11), 865–868.
9 A. I. Gavrilyuk, B. P. Zakharchenya and F. A. Chudnovskii, Photochromism in WO_3 films, *Sov. Technol. Phys. Lett.*, 1980, **6**(10), 512–513.
10 L. S. Price, I. P. Parkin, A. M. E. Hardy, R. J. H. Clark, T. G. Hibbert and K. C. Molloy, Atmospheric pressure chemical vapor deposition of tin sulfides (SnS, Sn_2S_3, and SnS_2) on glass, *Chem. Mater.*, 1999, **11**(7), 1792–1799.
11 R. J. Betsch, Parametric analysis of control parameters in MOCVD, *J. Cryst. Growth*, 1986, **77**(1–3), 210–218.
12 Y. Paz, Z. Luo, L. Rabenberg and A. Heller, Photooxidative self-cleaning transparent titanium-dioxide films on glass, *J. Mater. Res.*, 1995, **10**(11), 2842–2848.
13 B. O. Loopstra and H. M. Rietveld, Further refinement of structure of WO_3, *Acta Crystallogr., Sect. B*, 1969, **B25**, 1420–1421.
14 E. Gebert and Rj. Ackerman, Substoichiometry of tungsten trioxide—crystal systems of $WO_{3.00}$, $WO_{2.98}$ and $WO_{2.96}$, *Inorg. Chem.*, 1966, **5**(1), 136–142.
15 M. Sundberg, Crystal and defect structures of $W_{25}O_{73}$, a member of homologous series W_nO_{3n-2}, *Acta Crystallogr., Sect. B*, 1976, **32**, 2144–2149.
16 A. Magneli, Structures of the ReO_3-type with recurrent dislocations of atoms - homologous series of molybdenum and tungsten oxides, *Acta Crystallogr.*, 1953, **6**(6), 495–500.
17 M. M. Dobson and R. J. D. Tilley, A new pseudo-binary tungsten-oxide, $W_{17}O_{47}$, *Acta Crystallogr., Sect. B*, 1988, **44**, 474–480.
18 G. Hyett, M. Green and I. P. Parkin, X-Ray diffraction area mapping of preferred orientation and phase change in TiO_2 thin films deposited by chemical vapor deposition, *J. Am. Chem. Soc.*, 2006, **128**(37), 12147–12155.
19 W. A. Dollase, Correction of intensities for preferred orientation in powder diffractometry—application of the March model, *J. Appl. Crystallogr.*, 1986, **19**, 267–272.
20 A. Mills, A. Lepre, N. Elliott, S. Bhopal, I. P. Parkin and S. A. O'Neill, Characterisation of the photocatalyst Pilkington Activ (TM): a reference film photocatalyst?, *J. Photochem. Photobiol., A*, 2003, **160**(3), 213–224.
21 A. Mills, N. Elliott, I. P. Parkin, S. A. O'Neill and R. J. Clark, Novel TiO_2 CVD films for semiconductor photocatalysis, *J. Photochem. Photobiol., A*, 2002, **151**(1–3), 171–179.

Stabilization of metastable phases in spatially restricted fields: the case of the Fe_2O_3 polymorphs

Martí Gich,†*[a] Anna Roig,[a] Elena Taboada,[a] Elies Molins,[a] Caroline Bonafos[b] and Etienne Snoeck[b]

Received 6th November 2006, Accepted 5th February 2007
First published as an Advance Article on the web 8th May 2007
DOI: 10.1039/b616097b

In this work we show how the confinement of particles in a silica matrix with pores, acting as nano-vessels, plays an important role in the formation of pure ε-Fe_2O_3 nanoparticles and their thermal stability. In particular, a HRTEM study of a series of Fe_2O_3–SiO_2 xerogels annealed at different temperatures reveals that, at low temperatures, γ-Fe_2O_3 nanoparticles are formed and only transform to ε-Fe_2O_3 after subsequent annealing at higher temperatures. These data are complemented by measurements of the SiO_2 matrix porosity as well as with calorimetric and structural analysis of the nanoparticles after matrix removal. The gathered data indicate that the SiO_2 matrix acts as a barrier, hindering thermal diffusion and particle growth even at the high temperatures where the formation of ε-Fe_2O_3 appears to be favoured, thereby preventing its transformation to α-Fe_2O_3.

1. Introduction

Size-dependent properties are a distinctive characteristic of nanoscopic objects. Among the reasons that account for this behaviour is the fact that at the nanometric scale there is a significant fraction of the atoms that are located at the surfaces or interfaces and consequently, some of the material characteristics such as, for instance, the catalytic activity, melting point or stability of a given polymorph are strongly dependent on the particle size. It is generally accepted that surface energy and surface stress play important roles in the formation and stability of crystalline phases. This was already pointed out more than twenty years ago when it was realised that zirconia microcrystals tended to grow in a metastable tetragonal form rather than in the usual monoclinic crystals.[1] Similar behaviours have been reported thereafter in a variety of systems: metals,[2] semiconductors,[3] organics[4] or the well-studied C allotropes.[5] In the case of the Fe_2O_3 system in the form of nanoparticles it is well known that maghemite (γ-Fe_2O_3) tends to be the stable phase which transform to hematite (α-Fe_2O_3), the stable Fe(III) oxide polymorph, around 400 °C.[6] However, the stability of γ-Fe_2O_3 nanoparticles can be increased up to about 1000 °C when those particles are confined in a silica matrix and above this

---

[a] Institut de Ciència de Materials de Barcelona-CSIC, Campus de la UAB, Bellaterra, 08193, Catalonia, Spain. E-mail: roig@icmab.es; Fax: +34 935805729; Tel: +34 935801853
[b] Centre d'Élaboration de Materiaux et Études Structurales-CNRS, NanoMat group, 29 rue Jeanne Marvig, BP 94347, 31055, Toulouse Cedex 4, France

† Current address: Saint-Gobain Recherche, 39 quai Lucien Lefranc, 93303, Aubervilliers, France.

temperature, the transformation into ε-Fe$_2$O$_3$ starts.[7] In this context, we have recently reported how pure ε-Fe$_2$O$_3$ can be prepared by a controlled sol–gel synthesis inside the pores of a SiO$_2$ xerogel.[8] ε-Fe$_2$O$_3$ is one of the rarest and less studied Fe$_2$O$_3$ polymorphs due to the difficulty in synthesizing it as a single phase. In this work we discuss how the particle confinement in the silica pores, that act as nanovessels, plays an important role in the formation and the high thermal stability of pure ε-Fe$_2$O$_3$ nanoparticles. In particular, a HRTEM study of a series of Fe$_2$O$_3$–SiO$_2$ xerogels annealed at different temperatures reveals that, initially, γ-Fe$_2$O$_3$ nanoparticles are formed at low temperatures which are then transformed into ε-Fe$_2$O$_3$ after subsequent annealing at higher temperatures. This has been complemented by Brunauer–Emmett–Teller (BET) N$_2$ adsorption experiments aiming to evaluate the evolution of the SiO$_2$ matrix porosity upon annealing as well as with calorimetric and structural analysis of the nanoparticles after matrix removal. All these measurements indicate that the amorphous SiO$_2$ matrix acts as a barrier, hindering thermal diffusion and particle growth even at the high temperatures where the formation of ε-Fe$_2$O$_3$ appears to be favoured, thereby preventing its transformation to α-Fe$_2$O$_3$. The interest in exploring phase transformations in spatially restricted fields as a new approach to controlling polymorphism and searching for unknown polymorphs will also be discussed in a more general context.

2. Experimental

Synthesis

SiO$_2$–Fe$_2$O$_3$ nanocomposites are prepared by sol–gel chemistry using iron nitrate (Aldrich, 96%) and tetraethoxysilane (TEOS) (Aldrich, 98%) as precursors with a targeted composition of 30 wt% Fe$_2$O$_3$ in the final SiO$_2$–Fe$_2$O$_3$ composite. First, a gel is obtained through hydrolysis and condensation processes occurring in an acidic hydroethanolic medium at TEOS : H$_2$O : EtOH = 1 : 6 : 6 mole ratio and the appropriate amount of iron nitrate, being the reaction pH ∼0.9. These reactions are self-catalyzed by the nitric acid resulting from the hydrolysis of iron nitrate. The solution is then placed in a beaker covered with pin-holed parafilm to allow the evaporation of the solvent, gelation takes place at room temperature in a time ranging from two weeks to several months depending on the laboratory temperature and humidity. Wet gels are then dried at 60–80 °C for 14 h, crushed in an agate mortar and the resulting xerogels are annealed in air atmosphere for three hours every 100 °C between 300–1100 °C with a 600 °C min^{-1} heating rate between the isotherms. Four different composites, labelled S700, S900, S1000 and S1100 were obtained according to the different temperatures of the final thermal treatment at 700, 900, 1000 and 1100 °C. In addition, part of S1100 was treated in a 12 M NaOH aqueous solution at 80 °C to remove the SiO$_2$ amorphous matrix. After two days the solution was centrifuged and washed several times with distilled water until achieving a neutral pH and finally dried at 60 °C. X-Ray diffraction analysis performed after this treatment revealed that the SiO$_2$ had been efficiently removed while the iron oxide phases present in S1100 remained unaltered. The material containing no silica is labelled S1100B and consists of ε-Fe$_2$O$_3$ nanoparticles.

Transmission electron microscopy (TEM)

Observations were carried out using a Philips CM 30 microscope operating at 300 kV. Before the TEM observations, the samples were crushed and ultrasonically dispersed in ethanol. Drops of the solution were subsequently deposited onto Cu TEM grids that were coated with a conductive polymer, and the ethanol was allowed to evaporate.

Powder X-ray diffraction (XRD)

The formation of the crystalline phases was studied by X-ray diffraction (XRD) in a θ–2θ Bragg–Brentano geometry using a Siemens D5005 powder diffractometer with diffracted beam monochromator and a Brüker D8 diffractometer equipped with a microdiffraction stage. In both cases Cu Kα radiation was used (λ = 1.5406 Å). To quantitatively analyze the structural parameters, *e.g.* phase percentages, crystallite sizes or cell parameters, Rietveld refinements were performed using the *MAUD* program.[9] The orthorhombic (space group *Pna*2_1) structure, described by Tronc *et al.*,[10] and the usual defect inverse spinel structure (space group *Fd-3m*) were adopted for ε-Fe$_2$O$_3$ and γ-Fe$_2$O$_3$, respectively. Modeling of the silica glass is based on the assumption that from the X-ray diffraction pattern it is not possible to distinguish if the structure is completely amorphous or nanocrystalline SiO$_2$.[11] This method approximates the SiO$_2$ amorphous phase as a nanocrystalline solid (cubic structure with space group *P*2$_1$3) in which crystallite size is taken to be of the same order of magnitude as the cell parameters and the disorder is statistically introduced by the microstrain effect.[12]

Surface area measurements

Nitrogen adsorption data were taken at 77 K using an ASAP 2000 surface area analyzer (Micromeritics Instrument Corporation) after heating the samples at 180 °C under vacuum for 24 h to remove the adsorbed species. Surface area determinations were carried out following the BET method.

Thermal analysis

Phase stability in the 300–1800 K range was studied by differential thermal analysis (DTA) on a Perkin-Elmer DTA 7 apparatus at several heating rates under an air atmosphere.

3. Results and discussion

Typically, from the gels prepared by mixing an aqueous solution of a metal salt and an alcoholic solution of a Si–alkoxide, one can obtain a composite of amorphous silica and nanoparticles of the metal oxide by means of appropriate thermal treatments. Thus, it is worth considering both the evolution of the SiO$_2$ matrix and the nanoparticles and trying to correlate them. In Fig. 1, the thermal treatments

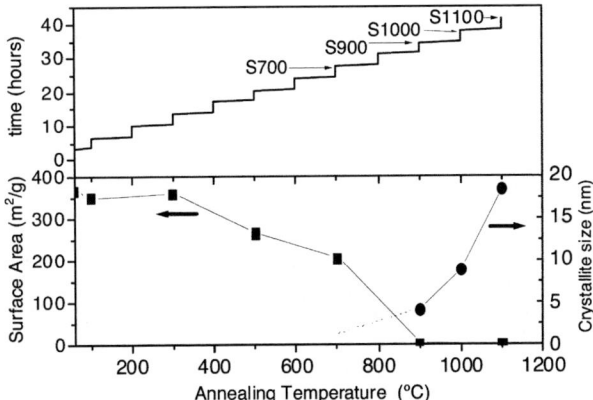

Fig. 1 Thermal treatment of the xerogels yielding the composites studied in this work (upper panel) and its influence on the amorphous SiO$_2$ porosity (■) and the crystallite size of the nanoparticles (●) (lower panel).

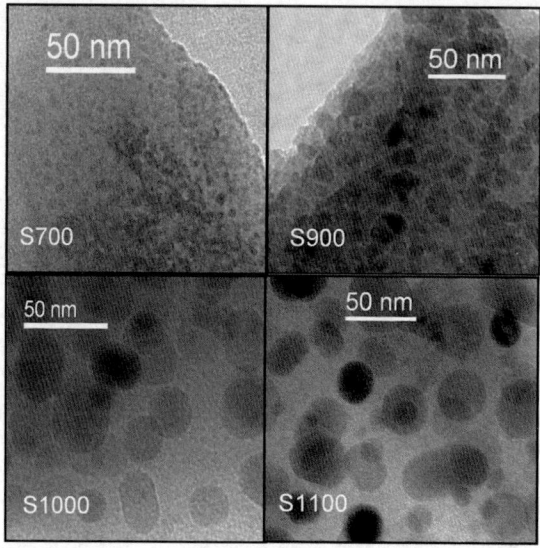

Fig. 2 TEM images of the different SiO_2–Fe_2O_3 composites.

of the different SiO_2–Fe_2O_3 composites that have been studied are represented with the corresponding evolution of the silica matrix porosity, which is considered to be proportional to the surface area measured by nitrogen adsorption. The matrix porosity gradually decreases between 300 and 900 °C, where it vanishes. In the figure we can also observe that for the S900, S1000 and S1100 composites, the mean crystallite size $\langle d \rangle_{XRD}$ of the iron oxide nanoparticles, as obtained from the broadening of the diffraction peaks, increases with the annealing temperature coinciding with the range of annealing temperatures where the silica matrix has no porosity. For S700 it has not been possible to obtain $\langle d \rangle_{XRD}$, since only the characteristic silica glass hump centred around $2\theta = 22°$ was observed in the diffraction pattern of this composite. In Fig. 2, the TEM images of the different composites show a dispersion of nanoparticles in the silica matrix and it can be observed that the nanoparticles coarsen as the annealing temperature is increased. However, in this case the increase in the nanoparticle diameter is especially important between S700 and S1000 and not so evident between S1000 and S1100. Indeed, histograms of the particle size distributions were obtained from several TEM images of S1000 and S1100. The histograms were fitted to log-normal distribution functions which gave average particle sizes of 22 ± 2 nm and 26 ± 4 nm, respectively. Thus, by increasing the annealing temperature between 700 and 1000 °C the nanoparticles grow but this trend is not extended at higher temperatures and a subsequent annealing at 1100 °C has almost no effect on the particle size and its main effect is to increase the crystallinity of the nanoparticles. In fact, $\langle d \rangle_{XRD}$ is to be interpreted as the size of the coherent domains of diffraction within the particles and in general is lower than the particle size observed by TEM. These results combined with the temperature dependence of the silica matrix porosity indicate something which is indeed quite intuitive, namely, that the growth of nanoparticles is favoured by the presence of a porous matrix but hindered as the matrix densifies when decreasing its porosity.

The composites were also studied by high resolution TEM (HRTEM). These experiments revealed interesting features of the crystallisation process and the different iron(III) oxides polymorphs. Fig. 3 presents HRTEM images of nanoparticles corresponding to the S1100 and S1000 composites. The Fourier transform of an HRTEM image allows us to identify the crystallographic phase of the nanoparticle and its orientation. For instance, in Fig. 3 we can see HRTEM images of the

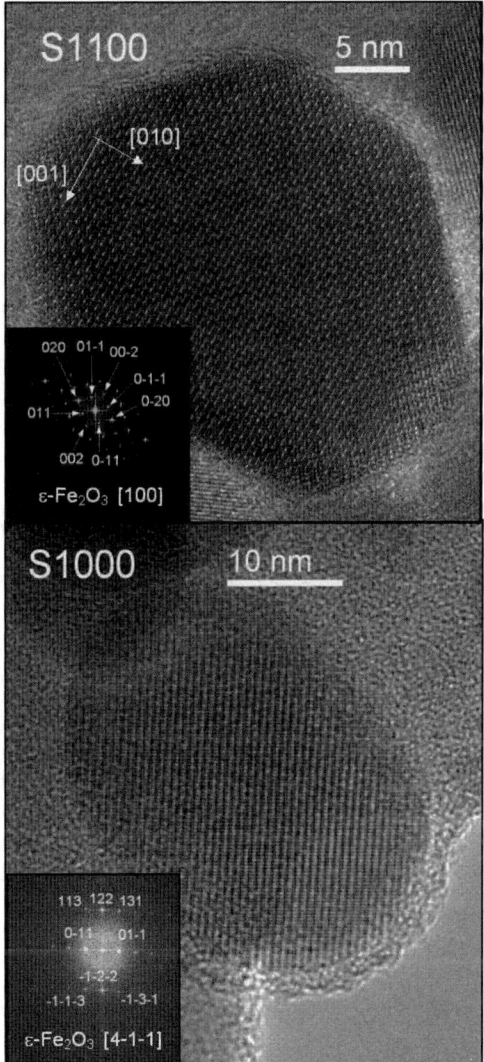

Fig. 3 HRTEM images of nanoparticles from S1100 and S1000 composites. The Fourier transform of the images (insets) corresponds to the [100] and [4-11] zone axes of ε-Fe$_2$O$_3$, respectively.

S1100 and S1000 composites corresponding to ε-Fe$_2$O$_3$ nanoparticles taken along the [100] and [4-1-1] zone axes, respectively. Analysing the HRTEM images of these composites we have only observed the presence of ε-Fe$_2$O$_3$ in accordance with previous XRD results[8] showing that the iron oxide present in composites equivalent to S1100 and S1000 was also ε-Fe$_2$O$_3$ and less than 10 wt% of α-Fe$_2$O$_3$. In that study, relying on the fact that the only XRD patterns with reflections due to the presence of ε-Fe$_2$O$_3$ nanoparticles were those annealed at temperatures above 900 °C, we assumed that the nucleation of the ε-Fe$_2$O$_3$ nanoparticles should occur below this temperature. In particular, the magnetic and Mössbauer spectroscopy measurements indicated that the composite annealed at 700 °C already presented crystalline clusters. However, the analysis of the HRTEM images of composite S700 proves that this first hypothesis was wrong. Indeed, S700 contains γ-Fe$_2$O$_3$ rather than ε-Fe$_2$O$_3$ nanoparticles (see Fig. 4). Thus, it is between 700 and 1000 °C that a

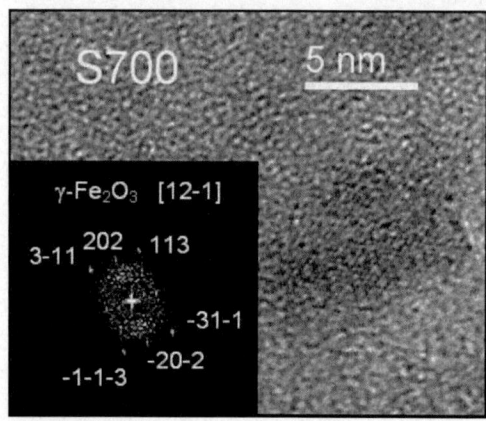

Fig. 4 HRTEM image of a nanoparticle from S700 composite. The Fourier transform of the image (inset) corresponds to the [12-1] zone axis of γ-Fe$_2$O$_3$.

γ → ε phase transformation must occur. In fact, the Fourier transforms of the HRTEM images of S900 could not be ascribed to any of the zone axes of either γ-Fe$_2$O$_3$ or ε-Fe$_2$O$_3$. However, the XRD pattern of S900 is best fitted by assuming that ε-Fe$_2$O$_3$ instead of γ-Fe$_2$O$_3$ is the major iron oxide phase in the composite (see Fig. 5), suggesting that the γ → ε transformation takes place gradually. In fact, many attempts were made to detect the transformation by DTA analysis which turned out to be unfruitful since no distinct exothermal signal could be detected between 700 and 1000 °C. This behaviour would be in accordance with a phase transformation in which the release of energy occurs along a broad temperature range indicating that the transformation consists of a continuous process characterised by the emergence of disorder bringing about a gradual disappearance of the γ polymorph and the subsequent appearance of the ε structure. Interestingly, this

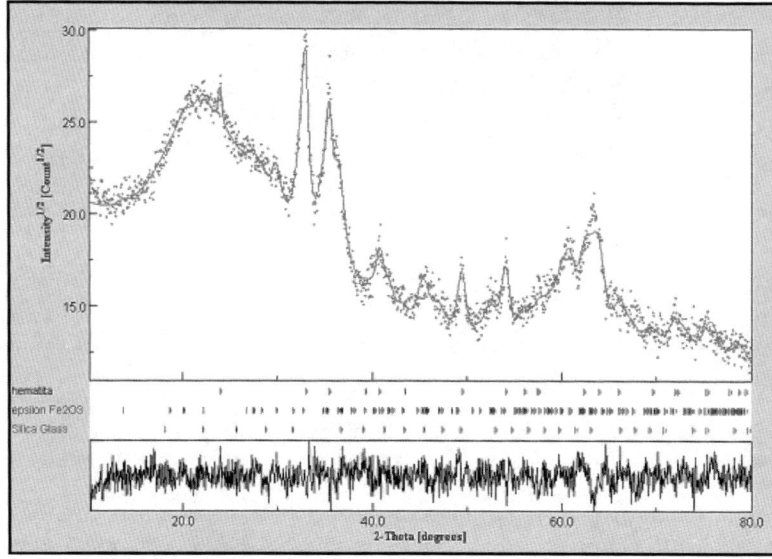

Fig. 5 Experimental (dots), calculated (solid line) and difference plot (lower panel line) for the Rietveld refinement of S900 considering ε-Fe$_2$O$_3$, α-Fe$_2$O$_3$ and amorphous SiO$_2$. Reflection positions of the different phases are indicated by vertical lines.

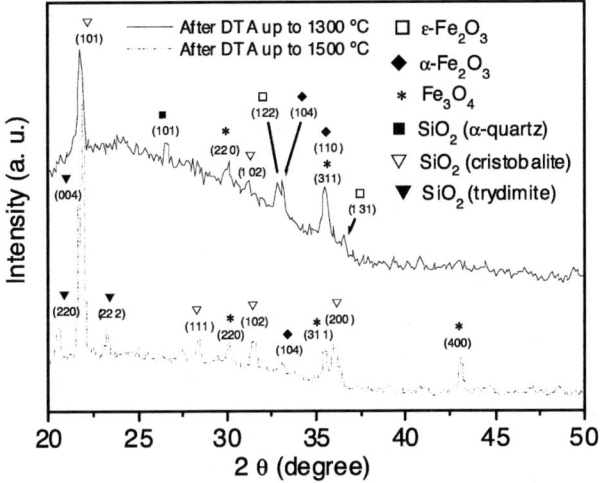

Fig. 6 XRD patterns of composite S1100 collected after heating it up to 1300 °C and 1500 °C in a DTA apparatus under air atmosphere.

kind of processes have been observed in oxide systems such as zirconia[13] or the Fe_2O_3 itself,[14] in which the size-reduction by milling of a stable structure yields a metastable polymorph. Alternatively, one could think of a particular energy landscape presenting local minima between the γ and ε structures in which the phase transition would involve the formation of well defined intermediate polymorphs.

The DTA experiments have also revealed the remarkable thermal stability of the amorphous SiO_2–ε-Fe_2O_3 composite. Actually, the transformation of ε-Fe_2O_3 nanoparticles into stable oxide phases such as hematite or magnetite is preceded by the devitrification of the silica matrix which starts at above 1200 °C. Fig. 6 shows the XRD patterns of the composites recorded after heating the material to 1300 and 1500 °C in DTA runs. It can be observed that after heating the material to 1300 °C, the amorphous silica matrix has been partially transformed into quartz and cristobalite while ε-Fe_2O_3 transforms to α-Fe_2O_3 and Fe_3O_4 (solid curve). Upon further annealing the material to 1500 °C, the crystallisation of amorphous silica occurs, which is stabilised into cristobalite and trydimite, and the disappearance of ε-Fe_2O_3 is completed (dotted curve). These results indicate that the transformation of ε-Fe_2O_3 to the stable Fe oxides is indeed triggered by the devitrification–crystallisation of the silica matrix and points to the prominent role of the latter in both the formation and stability of the ε polymorph. Stronger evidence to support this interpretation is given by the DTA experiment performed with sample S1100B, only containing ε-Fe_2O_3 nanoparticles, in which an exothermic process starting at about 750 °C is clearly detected (see Fig. 7). As expected, the XRD pattern of S1100B collected after the DTA run (see Fig. 8) shows that the ε-Fe_2O_3 nanoparticles have transformed mainly into the α-Fe_2O_3 stable polymorph, a small fraction of Fe_3O_4 has also been detected. Thus, the thermal stability limit of ε-Fe_2O_3 is increased from 750 °C to about 1200 °C by the presence of the silica amorphous matrix, although even in the absence of the latter, the ε-Fe_2O_3 nanoparticles exhibit a better stability at high temperatures compared to γ-Fe_2O_3 nanoparticles which are known to transform to hematite at about 400 °C.[6]

The relative stability of the γ, ε and α polymorphs of Fe_2O_3 can be understood in the framework of a simple model for small-particle phase transformations, *i.e.* one that only considers the surface energies and bulk free energies of the phases involved and neglects other contributions such as stress. In such a model, the size-dependence of the free energy variation per unit volume for the nucleation of a phase b from a phase a, Δg_b^{nuc}, is expressed in the relation (1), where R is the radius of the nucleus,

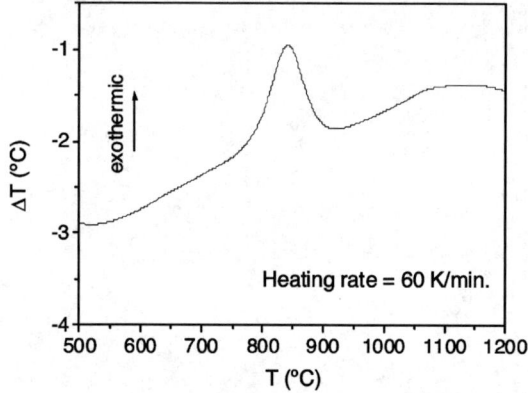

Fig. 7 DTA curve under an air atmosphere of sample S1100B only containing ε-Fe_2O_3 nanoparticles.

Δg_b is negative and represents the bulk free energy variation per unit volume of the transformation and γ_b is the surface energy of an a–b interface. Relation (1) is schematically represented in Fig. 9 for the γ, ε and α Fe_2O_3 polymorphs for a state in which the conditions (i) $\gamma_\gamma < \gamma_\varepsilon < \gamma_\alpha$ and (ii) $\Delta g_\gamma < \Delta g_\varepsilon < \Delta g_\alpha$ are fulfilled. The figure shows that it is the particle size that determines which is the stable polymorph. In this scenario, ε-Fe_2O_3 would form at intermediate sizes from those that stabilise maghemite or hematite. There are several reasons to believe that condition (i) is true. Navrotsky *et al.* have extensively studied the energetics of nanocrystalline aluminas which present polymorphs isostructural with those encountered for Fe_2O_3 and they experimentally established that the surface energy of α-Al_2O_3 is higher than that of γ-Al_2O_3.[15] On the other hand, the ε-Fe_2O_3 structure presents features that are common to both γ-Fe_2O_3 (*e.g.* the presence of tetrahedrally coordinated Fe^{3+} ions) and of the α-Fe_2O_3 (*e.g.* the hexagonal stacking of O planes). In that context one can expect that the surface energy of the ε phase is also intermediate between that of γ and α polymorphs. Regarding condition (ii), although at standard conditions the formation enthalpy of γ-Fe_2O_3 is slightly more negative than that of ε-Fe_2O_3 (−808 and −798 kJ mol^{-1}, respectively),[16] there are no thermodynamic data for ε-Fe_2O_3 available at temperatures above 700 °C. Moreover, at high temperatures the entropic

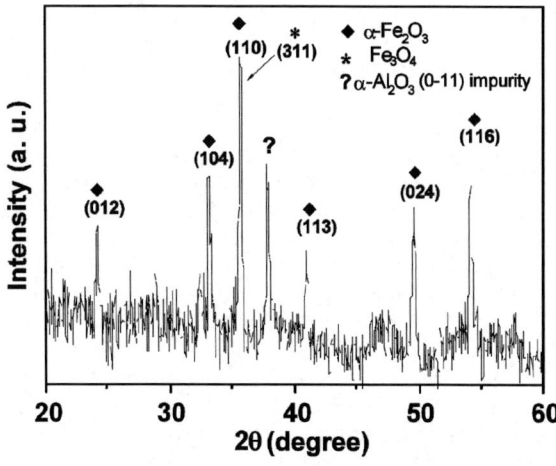

Fig. 8 XRD pattern of material S1100B after a DTA run in an air atmosphere up to 1200 °C.

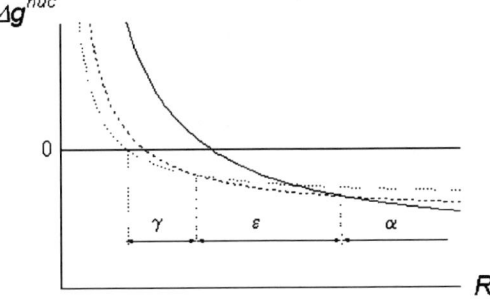

Fig. 9 Schematic representation of the size-dependent free energy variation of a spherical nucleus for γ-Fe_2O_3 (···), ε-Fe_2O_3 (--) and α-Fe_2O_3 (—). The stability ranges for the different polymorphs are also indicated.

term in the free energy becomes more important and a large entropy should be expected for ε-Fe_2O_3 since its asymmetric unit contains four non-equivalent Fe^{3+}.[6] Still, since eqn (1) is more sensitive to surface energy γ_b than to the bulk free energies, (ii) is not a necessary condition for the relative stabilities of Fig. 9 to hold. Somehow, temperatures above 700 °C are necessary for the $\gamma \rightarrow \varepsilon$ transformation to occur and we don't have means of knowing if that is because there is a crossover for the $\Delta g_\gamma < \Delta g_\varepsilon$ condition or because a high activation energy is needed.

$$\Delta g_b^{nuc} = \Delta g_b + \frac{3\gamma_b}{R} \qquad (1)$$

The point that we would like to emphasise here is that the role of the silica matrix is essential to ensure that the somehow antagonistic conditions required for the ε-Fe_2O_3 formation, namely a small particle size and a high temperature, are simultaneously satisfied. Thermal treatments at increasingly higher temperatures provoke an increase in the nanoparticle size which allows the $\gamma \rightarrow \varepsilon$ transformation. However, the gradual reduction in the silica matrix porosity hinders diffusion and ensures that the Fe_2O_3 nanoparticles do not reach the critical size to transform to hematite. The possibility that a significant role is played by the silica matrix due to its interaction with the surface atoms of the Fe oxide nanoparticles has to be ruled out according to the conclusions of an X-ray absorption spectroscopy study on amorphous SiO_2–Fe_2O_3 nanocomposites.[17]

4. Conclusions

The results presented herein for the Fe_2O_3 system show how the confinement of nanoparticles in a matrix yields a polymorphic form otherwise impossible to stabilise. In general, for a phase transition involving the occurrence of a (re)crystallisation, a change in temperature, pressure or pH is needed, usually accompanied by a process of crystalline growth. However, one can imagine for instance a number of nanocrystalline phases or polymorphs that can only be prepared at high temperatures and in a restricted range of particle sizes. In other words, there are probably new nanocrystalline phases and polymorphs that could be prepared if one is able to control the intensive thermodynamic variables by keeping the particle size constant or small enough. As has been shown here for the Fe_2O_3 polymorphs, this can be carried out by performing the nanoparticle synthesis in a matrix and a similar approach has been used to control the synthesis of polymorphs of an organic compound.[4] It is also worth noting that producing the nanoparticles in a matrix offers the advantage of avoiding their agglomeration but at the same time the matrix represents a useful and safe support to keep, manipulate and transport the nanoparticles. In conclusion, the exploration of phase transformations in spatially restricted fields is a new approach to controlling polymorphism and searching for unknown polymorphs is especially appealing at the nanoscale.

References

1. (a) R. C. Gravie, *J. Phys. Chem.*, 1965, **69**, 1238; (b) R. C. Gravie, *J. Phys. Chem.*, 1978, **82**, 218.
2. (a) H. He, G. T. Fei, P. Cui, K. Zheng, L. M. Liang, Y. Li and L. D. Zhang, *Phys. Rev. B*, 2005, **72**, 073310; (b) P. Taneja, R. Banerjee, P. Ayyub and G. K. Dey, *Phys. Rev. B*, 2001, **64**, 033405; (c) M. Gich, E. A. Shafranovsky, A. Roig, A. Ślawska-Waniewska, K. Racka, Ll. Casas, E. Molins, Yu. I. Petrov and M. F. Thomas, *J. Appl. Phys.*, 2005, **98**, 024303.
3. S. H. Tolbert and A. P. Alivisatos, *Science*, 1994, **265**, 373.
4. J.-M. Ha, J. H. Wolf, M. A. Hillmyer and M. D. Ward, *J. Am. Chem. Soc.*, 2004, **126**, 3382.
5. Q. Jiang and Z. P. Chen, *Carbon*, 2006, **44**, 79.
6. P. Ayyub, M. Multani, M. Barma, V. R. Palkar and R. Vijayaraghavan, *J. Phys. C: Solid State Phys.*, 1988, **21**, 2229.
7. C. Chaneac, E. Tronc and J. P. Jolivet, *Nanostruct. Mater.*, 1995, **6**, 715.
8. M. Popovici, M. Gich, D. Nižňanský, A. Roig, C. Savii, Ll. Casas, E. Molins, K. Zaveta, C. Enache, J. Sort, S. de Brion, G. Chouteau and J. Nogués, *Chem. Mater.*, 2004, **16**, 5542.
9. L. Lutterrotti and S. Gialanella, *Acta Mater.*, 1997, **46**, 101.
10. E. Tronc, C. Chanéac and J. P. Jolivet, *J. Solid State Chem.*, 1998, **139**, 93.
11. A. Le Bail, *J. Non-Cryst. Solids*, 1995, **183**, 32.
12. L. Lutterotti, R. Ceccato, R. Dal Maschio and E. Pagani, *Mater. Sci. Forum*, 1998, **278–281**, 87.
13. J. Balley, P. Bills and D. Lewis, *Trans. J. Br. Ceram. Soc.*, 1975, **74**, 247.
14. N. Randrianantoandro, A. M. Mercier, M. Hervieu and J. M. Grenèche, *Mater. Lett.*, 2001, **47**, 150.
15. J. M. McHale, A. Auroux, A. J. Perrotta and A. Navrotsky, *Science*, 1997, **277**, 788.
16. (a) J. Majzlan, K.-D. Grevel, A. Navrotsky and B. F. Woodfield, *Am. Mineral.*, 2003, **88**, 855; (b) J. Majzlan, A. Navrotsky and U. Schwertmann, *Geochim. Cosmochim. Acta*, 2004, **68**, 1049.
17. A. Corrias, G. Ennas, G. Mountjoy and G. Paschina, *Phys. Chem. Chem. Phys.*, 2000, **2**, 1045.

PAPER

Using *in situ* synchrotron radiation wide angle X-ray scattering (WAXS) to study CaCO$_3$ scale formation at ambient and elevated temperature†

Tao Chen,‡[a] Anne Neville,[a] Ken Sorbie[b] and Zhong Zhong[c]

Received 13th November 2006, Accepted 2nd February 2007
First published as an Advance Article on the web 24th April 2007
DOI: 10.1039/b616546j

The formation of calcium carbonate mineral scale is a persistent and expensive problem in oil and gas production, water piping systems, power generator, and batch precipitation. The aim of this paper is to further the understanding of scale formation and inhibition by *in situ* probing of crystal growth by synchrotron radiation wide angle X-ray scattering (WAXS) at ambient and elevated temperature. This novel technique enables *in situ* study of mineral scale formation and inhibition and as such, information on the nucleation and growth processes are accessible. This technique studies bulk precipitation and surface deposition in the same system and will be of great benefit to the understanding of an industrial scaling system. It offers an exciting prospect for the study of scaling. It has been shown that the nucleation and growth of various calcareous polymorphs and their individual crystal planes can be followed in real-time and from this the following conclusions are reached.

• The process of scale deposition on the surface can be divided into an unstable phase and a stable phase. The initial phase of crystallization of calcium carbonate is characterized by instability with individual planes from various vaterite and aragonite polymorphs emerging and subsequently disappearing under the hydrodynamic conditions. After the initial unstable phase, various calcium carbonate crystal planes adhere on the surface and then grow on the surface.

• At 25 °C, the main plane of surface deposit is calcite and a strong (104) peak is detected. The other calcite planes (102), (006), (110) (113) and (202) are hardly detectable under this condition.

• At 80 °C, the main planes in the surface deposit are the (104), (113) and (110) planes of calcite. Stable planes of vaterite and aragonite are also observed.

This paper will discuss how surface scale evolves—exploring the power of the synchrotron *in situ* methodology.

[a] *School of Mechanical Engineering, University of Leeds, Leeds, UK LS2 9JT*
[b] *Institute of Petroleum Engineering, Heriot-Watt University, Edinburgh, UK EH14 4AS*
[c] *Bldg 725D, Brookhaven National Laboratory, Upton NY 11973, USA*

† The HTML version of this article has been enhanced with colour images.
‡ Current address: Clariant Oil Services, Clariant UK Ltd, Howe Moss Place, Dyce, Aberdeen, UK AB21 0GS. E-mail: hawkct@hotmail.com

Introduction

The formation of calcium carbonate mineral scale is a persistent and expensive problem in oil and gas production, water piping systems, power generator, and batch precipitation. The presence of a scale layer can cause a series of problems: impedance of heat transfer, increase of energy consumption and unscheduled equipment shutdown.[1,2]

Calcium carbonate scale formation occurs according to the reaction given in eqn (1).

$$Ca^{2+} + CO_3^{2-} \rightarrow CaCO_3 \quad (1)$$

Calcium and carbonate ions form scale that has very little solubility and are likely to precipitate in water, even if only small amounts of dissolved calcium and carbonate ions are present. One of the driving forces for scale formation is supersaturation and the formula for the supersaturation ratio is given in eqn (2).

$$S_\alpha = (\alpha_1\alpha_2)/K_{(P,T)} \quad (2)$$

where

$$\alpha = \gamma C$$

S_α is the supersaturation ratio and α is the activity of the separate species. α_1 and α_2 are the activity of calcium ion and bicarbonate ion in the solution, respectively. K is normally called the solubility product, which depends on the pressure P and the temperature T. C is concentration of the ions in the solution. γ is the ionic activity coefficient.

Three possibilities exist in terms of scale formation from the solution with thermodynamics. (1) $S_\alpha < 1$: the solution is undersaturated and scale formation is not thermodynamically feasible; (2) $S_\alpha = 1$: the solution is saturated. The scale formation and dissolution rate in the solution is the same and no scale is formed in the solution; (3) $S_\alpha > 1$: the solution is supersaturated and scale formation is thermodynamically possible.

The supersaturation ratio of an aqueous solution of a salt refers to how much more of the salt is currently dissolved in the solution above that which would be present at equilibrium. Some of the salt must precipitate (or crystallise) from a supersaturated solution in order to come to equilibrium. The supersaturation ratio is an important determining factor that controls the precipitation of salts from solution. In a sense, the supersaturation ratio may be thought of as the thermodynamic "driving force" for precipitation; *i.e.* it is more likely that precipitation will occur easily from a higher than from a lower supersaturated solution of a salt. The degree of supersaturation, in turn, is affected by changes in temperature, pressure and pH.

Over the past few decades, great efforts have been made to understand the mechanism of scale formation and inhibition. Traditionally, studies of scale formation have concentrated on bulk scale formation using laboratory beaker tests;[3] turbidity probes, pH measurement or bulk chemical analysis have been used to analyse kinetics of precipitation. The primary focus has been the assessment of the kinetics of homogeneous and heterogeneous precipitation in the bulk solution.[4] It has been demonstrated that there are often wide anomalies between actual deposition and rates estimated by predictive models based on scaling indices and thermodynamics.[4]

Few studies have considered the activity of deposit formation (nucleation and growth) at surfaces. The theoretical aspects of crystal nucleation and growth are only partially understood and information on such processes on component surfaces is especially sparse.

In the last few decades, much effort has been directed at the aspect of scaling on surfaces and has resulted in numerous studies reporting methods to detect and assess scale formation on metal surfaces. Some focus has been turned to this aspect of scaling to attempt to overcome some of the shortfalls of beaker tests.

Hasson et al. studied calcium carbonate scale formation in a falling film system and in a pipe flow system.[5] The thickness of scale deposited on a surface was measured and monitored by the differential pressure between the inlet and outlet of the pipe. Sullivan et al. studied scale formation by monitoring the change in heat transfer.[6] The heat resistance of a scaled pipe is higher than when the same surface is clean. Abdel-All et al.[7] studied the initial stages of calcium carbonate surface adhesion with a quartz crystal microbalance (QCMB). The change in oscillating frequency of the quartz crystal is directly proportional to the change of mass of adhesion on the surface. Teng et al.[8] studied the kinetics of calcite growth using an atomic force microscope (AFM). The microscopic surface processes of calcite were observed. Ramstad et al.[9] studied calcium carbonate deposition using a high pressure and high temperature cell. The scale formed on the surface of sapphire glasses can be observed by microscope. It is possible to simulate the pressure and temperature changes of a fluid being produced from the reservoir up to the surface in the oil and gas industry.[9]

Another means of achieving *surface* scaling (only calcareous) is to apply cathodic protection (CP) where the surface conditions, through production of hydroxyl ions, trigger a sequence of reactions, which result in the deposition of 'calcareous deposits'.[10] Cathodic protection methods provide a way to accelerate scale formation and can give information on the mechanisms of scale formation. Often scale generated by this means is referred to as electrodeposited scale. Because scale generated by electrodeposition forms by a different set of processes that former supersaturation interpreting mechanistic information should be treated with caution.

As mentioned previously, traditional studies on scale formation have been focused on the precipitation formed in the bulk solution or deposition formed on the metal surface. Few studies have been concerned with the difference between bulk precipitation and surface deposition.

Recently, Chen et al.[11,12] studied bulk precipitation and surface deposition simultaneously. The quantity of scale deposited on metal surfaces and the kinetics of precipitate formed in the bulk solution were studied. It was demonstrated that bulk precipitation and surface deposition have different dependencies on the index of supersaturation and so synchrotron radiation wide angle X-ray scattering (WAXS) was used to further the understanding of scale formation and inhibition by *in situ* probing of crystal growth in ambient and elevated temperature in this paper. The crystallization of mineral scale has been followed *in situ* and information on the nucleation and growth processes of both bulk precipitation and surface deposition are accessible. The nucleation and growth of various calcareous polymorphs and their individual crystal planes can be followed in real time. This technique offers an exciting prospect for the study of scaling.

Experiment procedure

Dynamic flow cell tests in conjunction with WAXS measurement

A novel silicon reaction cell was specifically designed for the *in situ* observation of scaling processes under non-ambient conditions. This cell, shown schematically in Fig. 1, allows *in situ* collection of XRD data from a scaling brine mixture. It represents a novel approach to scaling studies. It has been used for preliminary examination of $CaCO_3$ scale formation and inhibition in this study.

The scale formation experiments were carried out at 25 and 80 °C. Calcium carbonate was precipitated spontaneously by mixing brine 1 and brine 2 in the *in situ*

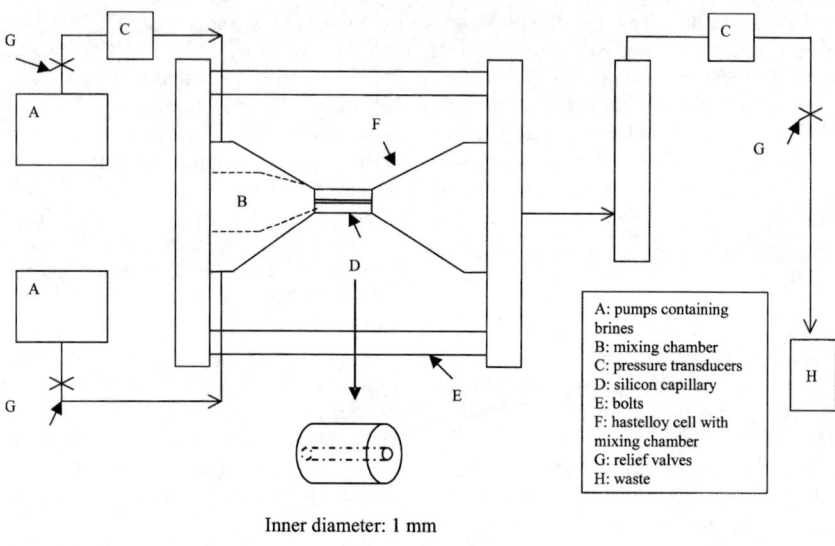

Inner diameter: 1 mm

Length: 9.4 mm

Fig. 1 *In situ* cell for WAXS measurements at high temperature and pressure.

cell for WAXS measurements, as shown in Fig. 1. The composition of brine 1 and brine 2 are given in Table 1. The pH of brine 2 was buffered to 6.8 by acetic acid at the start of experiment and the pH of brine 1 was not adjusted before mixing. A line pressure of 3.4 MPa and a flow rate of 10 ml min^{-1} (5 ml min^{-1} brine 1 and 5 ml min^{-1} brine 2) were used throughout the synchrotron radiation WAXS measurements in this study. An oven was set up between the pumps and the mixing chamber to heat brine 1 and brine 2 and the temperature was controlled at 25 or 80 °C in the mixing chamber. Normally, the experiments were set up to run for 30 frames each taking 120 s to complete with experiment running a total of 60 min (*i.e.* data are collected 30 times every 2 min). Some experiments were stopped before 30 frames due to blockage of the cell by calcium carbonate crystals.

Calcium carbonate crystals deposit on the surface of the silicon capillary cell (as shown in Fig. 1) and precipitate in the bulk solution. The silicon capillary cell is a single crystal silicon cell with 9.4 mm length and 1 mm inner diameter tube. The scaling system is cleaned before experiments by pumping 10% acetic acid, 5% Decon 90 (detergent) and distilled water, respectively to remove the scale deposition and the remaining inhibitors.

The X17B1 beamline with a X-ray energy of 67 keV, at the National Synchrotron Light Source (NSLS) at Brookhaven National Laboratory, USA, was used for the experiments. In contrast to conventional XRD, the synchrotron XRD with high energy takes about 10 s to analyse a calcium carbonate sample and as such real time data can be collected *in situ*. A Siemens CCD detector allows effective two-dimensional data collection.[13]

Table 1 Composition of brine 1 and brine 2 used to study calcium carbonate scale formation detected by *in situ* synchrotron radiation (WAXS)

	Brine 1 conc./mg l^{-1}	Brine 2 conc./mg l^{-1}
NaCl	15 367	15 367
CaCl$_2 \cdot$ 6H$_2$O	15 743	—
NaHCO$_3$	—	6046

Table 2 Main characteristic synchrotron XRD peaks of vaterite, aragonite and calcite[14]

Vaterite		Calcite		Aragonite	
Plane	2-theta/°	Plane	2-theta/°	Plane	2-theta/°
110	5.902	012	5.473	111	6.212
112	6.406	104	6.951	012	6.790
114	7.724	006	7.420	200	7.352
		110	8.426	112	7.723
		113	9.242	021	8.910
		202	10.083	022	10.024

Data analysis

The raw data, from each frame in each WAXS run, are integrated and converted to appropriate file formats by an in-house program at X17B1 at Brookhaven National Laboratory. There is subsequently a range of software available depending on the type of analysis and refinement desired. The combination of the FullProf suite,[14] LMGR suite,[15] and GSAS[16] covers all the analysis required. In broad terms the FullProf suite is used for the initial data analysis, peak fitting, indexing *etc.*, as well as quantitative analysis in the form of peak integrations. The LMGR suite is used for unit cell refinement, and GSAS for any further structural refinements required.

The most commonly chosen 2 theta range for calcium carbonate for $\lambda = 0.368$ Å is approximately 5–10.5°. The main characteristic synchrotron XRD peaks represented for the vaterite, aragonite and calcite (at 2 theta between 5–10.5°) are listed in Table 2.[14]

Results

Growth of calcium carbonate at 25 °C

The growth of calcium carbonate crystals at 25 °C is shown in Fig. 2 as a series of XRD spectra at different times. Nucleation and growth of various calcium carbonate polymorphs and their individual crystal planes can be observed. One of the key features of this system is the initial instability seen in the XRD peaks observed over the initial period (30–40 min). During this phase, individual peaks from $CaCO_3$

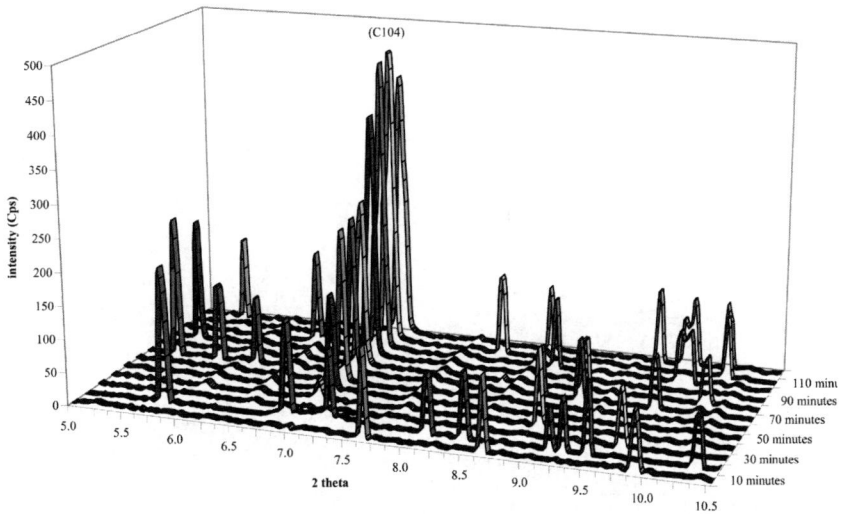

Fig. 2 The growth of calcium carbonate crystals at 25 °C detected by SXRD.

polymorphs are seen to appear and then disappear as they are flushed out of the system. A peak being present then disappearing in the next frame is attributed to crystals forming as bulk precipitate or crystals forming on the cell surface and then being flushed away before the next frame. The majority of these planes can be assigned to vaterite and aragonite polymorphs. For example, the peaks appearing after 10 min at 2 theta values of 8.659 and 9.895° are aragonite (103) and vaterite (008), respectively.

After the initial unstable phase, the first crystal plane to attach to the surface and then grow on the surface is the (104) plane of calcite after 40 min (*i.e.* a peak is present then the intensity and integrated area of this peak increase at the next frame due to growth of the surface deposit).

As the experiment progresses, the growth of the calcite (104) plane is apparent. The other calcite planes (102), (006), (110), (113) and (202) are hardly detectable after 120 min. In addition, various planes from the vaterite and aragonite polymorphs appear but are flushed out of the cell.

Growth of calcium carbonate at 80 °C

The XRD spectra recorded when tracking the growth of calcium carbonate crystals at 80 °C are shown in Fig. 3 for the period up to 18 min. The experiment was stopped after 18 min due to blockage of the cell by calcium carbonate crystals.

The initial phase of crystallization is characterized by instability up to 6 min; evident from the fact that the individual planes from various polymorphs emerge and subsequently disappear under the hydrodynamic conditions. The majority of these planes can be assigned to the aragonite polymorphs. For example, the peak appearing after 4 min at 2 theta 5.697° is the aragonite (110) plane and the peak appearing after 6 min at 2 theta 6.104° is the aragonite (111) plane.

After the initial unstable phase where aragonite is the principal crystal morphology, the crystals then attach to the surface and grow on the surface. Stable (104) and (110) planes of calcite are present after 8 min. The next plane to emerge is the (113) plane of calcite after 10 min. Another three crystal planes are present and grow on the surface after 14 min, which represent (110) and (112) planes of vaterite and the (111) plane of aragonite.

At 80 °C, the growth of calcite (104), (110) and (113) planes is apparent. The calcite (202) plane is observed after 8 min, disappears after 14 min and appears again

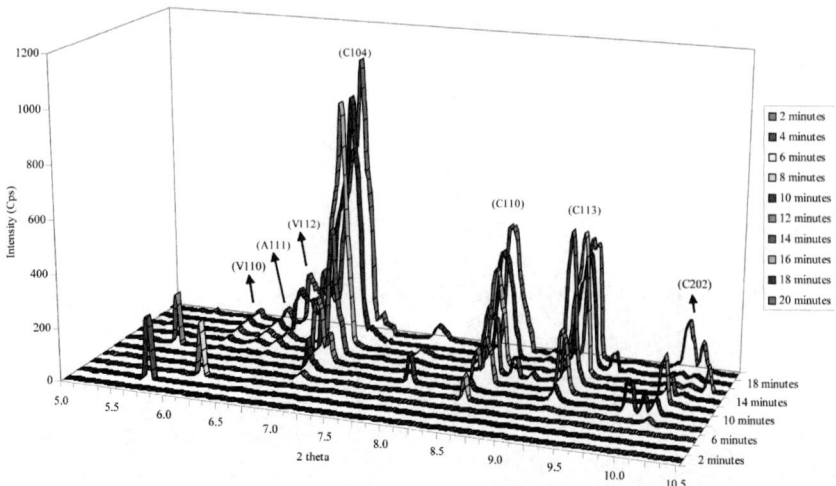

Fig. 3 The growth of calcium carbonate crystals at 80 °C detected by synchrotron radiation (WAXS).

after 20 min, characteristic of an unstable crystal plane. The other calcite planes (102) and (006) are hardly detectable after 20 min. In addition, various planes from the vaterite and aragonite polymorphs start appearing again, as in the initial stage of the experiment, but they are flushed out of the cell. However, after 14 min the first stable vaterite (110) and (112) planes and aragonite (111) plane emerge. This time they are derived from crystals adhering to the surface rather than being flushed out of the cell.

Discussion

Hennessy and co-workers[17] previously demonstrated the potential of *in situ* synchrotron XRD for the study of crystallization mechanisms relevant to oilfield scaling in their work on $BaSO_4$. A key element in this work is a pressure flow cell which is integrated into the beamline. The cell enables high pressure, temperature and various flow regimes to be set, getting towards the kind of conditions seen in the oil industry. Both the evolution of surface deposit and bulk precipitate can be observed. Only preliminary studies were carried out on $BaSO_4$ and in this paper the comprehensive results on $CaCO_3$ are reported.

Growth process of calcium carbonate crystals deposited on a surface

One of the achievements in this study has been to demonstrate that the nucleation and growth of various calcium carbonate polymorphs and their individual crystal planes can be observed. The growth process of calcium carbonate scale can be divided into two parts: (1) the initial phase of crystallization and (2) the growth of crystals on the surface.

The initial phase of crystallization is characterized by instability and individual planes from various polymorphs emerge and subsequently disappear under the hydrodynamic conditions. The majority of these planes can be assigned to the vaterite and aragonite polymorphs.

After the initial unstable phase, the crystals attach to the surface and then grow on the surface and at this point the intensity and integration area of the peaks increase as time elapses. During this process, various planes from the vaterite and aragonite polymorphs start appearing again, but, as in the initial stage of the experiment, they are flushed out of the cell.

Growth process of calcium carbonate crystals at 25 and 80 °C

Temperature is known to be a parameter of particular importance for calcium carbonate crystallization. The effect of temperature on the supersaturation of scaling solutions used in Table 1 is calculated using ScaleSoftPitzer™ Version 4.0 from Rice University[18] and shown in Fig. 4.

The kinetics of calcium carbonate scale formation is a function of temperature, *i.e.* slow kinetics at low temperature. As the temperature increases, the formation of calcium carbonate will accelerates and precipitation may occur at an earlier stage.[19]

Another achievement in this part of study has been to demonstrate that the favoured planes during scale growth are different at ambient temperature (25 °C) and elevated temperature (80 °C).

At 25 °C, the cell was not blocked after 120 min. The growth of calcite (104) plane is apparent. The other pure calcite planes (102), (006), (110) (113) and (202) are hardly detectable after 120 min. It demonstrated that the growth of calcite (104) plane is favoured at 25 °C.

Fig. 5 shows the evolution of X-ray diffraction integrated intensity of the calcite (104) plane at 25 °C. The integrated intensity of X-ray diffraction is in proportion to the crystal volume[17] and the growth rate can be measured by integrated intensity analysis. Before 30 min, the integrated intensities are zero, which means there is no

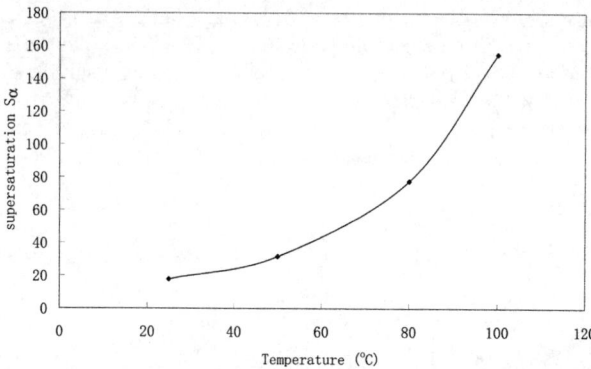

Fig. 4 Effect of temperature on the supersaturation ratio of scaling solutions.[18]

stable calcite (104) plane observed during this period. This period is defined as the induction time of surface adhesion for calcite (104) plane. After 40 min, the integrated intensity is above zero and the integrated intensity increases as time elapses, following an almost linear relationship. It shows the volume of calcite crystals increases and the growth rate of calcite is almost constant under these conditions. The nucleation and growth periods of the calcite (104) plane are difficult to separate at 25 °C. This is likely to be because the formation of new crystalline particles (nucleation) and their subsequent increase in size (crystal growth) processes proceed simultaneously.[20] The only stable plane of calcite formed under 20 °C is the calcite (104) plane. The calcite crystals grow only through the direction of the (104) plane under low supersaturated scaling solution under 25 °C, where the driving force is not high enough for the formation of other calcite planes. The lattice parameter analysis is impossible as only calcite (104) plane was observed. It will be done for the calcium carbonate crystals formed at 80 °C.

At 80 °C, the cell was blocked after 18 min. It showed a faster growth of calcium carbonate scale on the surface at 80 °C than at 25 °C as expected. Calcite (104), (110) and (113) planes are the dominant planes of calcium carbonate scale. In addition, various planes from the vaterite and aragonite polymorphs start appearing again after the initial stage of the experiment, but they are flushed out of the cell, as shown in Fig. 3. However, after 14 min the first stable vaterite (110) and (112) planes and then the aragonite (111) plane emerge. This time they stick to a surface rather than being flushed out of the cell. Considering that these vaterite and aragonite planes were unable to stick to the cell surface at the beginning of the experiment, when no other crystal planes were attached, one might speculate whether the vaterite and

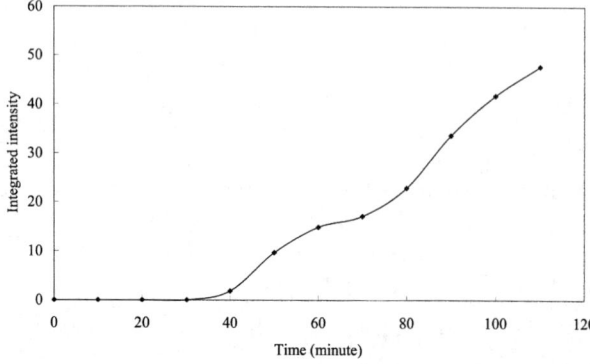

Fig. 5 Evolution of X-ray diffraction integrated intensity of calcite (104) plane at 20 °C.

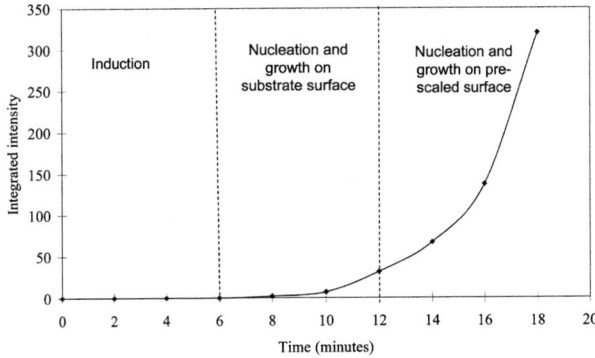

Fig. 6 Evolution of X-ray diffraction integrated intensity peaks of calcite crystals at 80 °C.

Table 3 Summary of SXRD results—formation of stable planes (indicating adherence and growth)

25 °C	80 °C
• Unstable: 30–40 min	• Unstable: 6 min
• Stable: (104) calcite	• Stable: (104) (110) (113) calcite
• No stable vaterite or aragonite	(110) (112) vaterite
	(111) aragonite

aragonite planes are growing on existing calcite planes at the wall of the cell. In addition, it was shown that an increase in temperature favours the formation of metastable aragonite and unstable vaterite in Zhou and Zheng's research.[21]

The growth rate can be measured by integrated intensity analysis and the integrated intensity peaks of calcite crystals are shown in Fig. 6, which shows the growth processes of scale formation and can be divided into three stages: (1) Induction time. Before 6 min, the integrated intensities for calcite crystal planes are zero, which means that there are no stable planes observed during this period. The induction time for surface deposition at 80 °C in the absence of inhibitor is 6 min. (2) Initial stage of nucleation and growth on the surface of substrate. After 8 min, the integrated intensity is above zero and increases slowly as time elapses until 12 min. It represents the first layer of nucleation and growth of scale crystal on the surface of silicon. (3) Nucleation and growth on the pre-scale surface. After 12 min, the integrated intensity increases quickly as time elapses. This represents the nucleation and growth of scale crystals on the pre-scaled surface.

Table 3 summarises the key differences in terms of (i) the stable phase duration and (ii) the stable planes observed. It is clear that increasing temperature increases the surface growth and that calcite growth appears to stimulate adherence of vaterite and aragonite. This agrees with observations in surface deposition tests[2] where aragonite is found to be present at elevated temperature and not at low temperature.

In summary, this study has provided an insight into the mechanism of calcium carbonate scale formation. In particular, it achieved the *in situ* observation of growth of each plane of scale crystals, invaluable for the understanding of mechanisms of growth at the micro level.

Conclusions

• WAXS has been successfully used for the study of calcareous formation and inhibition and enables crystallization mechanisms at engineered surfaces, under realistic conditions, to be investigated.

- The initial phase of crystallization of calcium carbonate is characterized by instability with individual planes from various vaterite and aragonite polymorphs emerging and subsequently disappearing under the hydrodynamic conditions.
- After the initial unstable phase with aragonite, various calcium carbonate crystal planes attach to the surface and then growth on the surface. In addition, various planes from the vaterite and aragonite polymorphs appear again during this stage.
- At 80 °C, the main planes of the surface deposit are the (104), (113) and (110) planes of calcite. The peaks representing the (110) and (112) planes of vaterite and (111) plane of aragonite are detected in the surface deposit after calcite (104), (110) and (103) are observed.
- At 25 °C, the main plane of surface deposit is calcite and a strong (104) peak is detected. The other calcite planes (102), (006), (110) (113) and (202) are hardly detectable under these conditions.
- *In situ* synchrotron XRD can be explored to study the scale formation at micron level, including induction, nucleation and growth process of each face of scale crystals.

Acknowledgements

We thank Brookhaven National Laboratory for the support and the beam time provision for these experiments and the sponsors of the FAST II Joint Industry Project for their financial support for T. Chen.

References

1 Y. Zhang, H. Shaw, R. Farquhar and R. Dawe, The kinetics of carbonate scaling-application for the prediction of downhole carbonate scaling, *J. Pet. Sci. Eng.*, 2001, **29**, 85–95.
2 P. Kjellin, K. Holmberg and M. Nyden, A new method for the study of calcium carbonate growth, *Colloids Surf., A*, 2001, **194**, 49–55.
3 M. Abtahi, K. Baard, J. E. Vindstad and T. Qstvold, Calcium carbonate precipitation and pH variations in oil field waters. A comparison between experimental data and model calculations, *Acta Chem. Scand.*, 1996, **50**, 114–121.
4 W. F. Langelier, The analytical control of the anti-corrosion water treatment, *J. Am. Water Works Assoc.*, 1936, **28**, 1500–1506.
5 D. Hasson, D. Bramsom, B. Limoni-Relis and R. Semiat, Influence of the flow system on the inhibitor action of $CaCO_3$ scale prevention additives, *Desalination*, 1996, **108**, 67–79.
6 P. J. Sullivan, T. Young and J. Carey, Effectiveness of polymer phosphonate blends for inhibition of $CaCO_3$ scale, , *Industrial Water Treatment*, 1996, **26**, 39–44.
7 N. Abdel-All, K. Satoh and K. Sawada, Study of adhesion mechanism of calcareous scaling by using quartz crystal microbalance technique, *Anal. Sci.*, 2001, **17**, 825–828.
8 H. H. Teng, P. M. Dove and J. J. Yoreo, Kinetics of calcite growth: surface processes and relationships to macroscopic rate laws, *Geochim. Cosmochim. Acta*, 2000, **64**(13), 2255–2266.
9 K. Ramstad, T. Tydal, K. M. Askvik and P. Fotland, Predicting carbonate scale in oil producers from high temperature reservoirs, 6th International Symposium on Oilfield Scale, Aberdeen, UK, SPE 87430, May 2004.
10 C. Garcia, G. Courbin, C. Noik, F. Ropital and C. Fiaud, Development of electrochemical quartz cristal microbalance to control carbonate scale deposit, CORROSION/99, Paper N 114, NACE, Houston, 1999.
11 C. Tao, N. Anne and Y. Mingdong, Assessing the effect of Mg^{2+} on $CaCO_3$ scale formation-bulk precipitation and surface deposition, *J. Cryst. Growth*, 2005, **275/1–2**, e1347–e1353.
12 C. Tao, N. Anne and Y. Mingdong, Calcium carbonate scale formation—assessing the initial stages of precipitation and deposition, *J. Pet. Sci. Eng.*, 2005, **46**, 185–194.
13 Z. Zhong, C. C. Kao, D. P. Siddons and J. B. Hastings, Sagittal focusing of high-energy synchrotron X-rays with asymmetric Laue crystals. I. Theoretical considerations, *J. Appl. Crystallogr.*, 2001, **34**, 504–509.
14 T. Roisnel, J. Rodriguez-Carvajal, FullProf, Laboratoire Leon Brillouin (CEA/CNRS).
15 J. Laugier, B. Bochu, LMGP Suite, Laboratoire des Materiaux et du Genie Physique de l'Ecole Superieure de Physique de Grenoble, 2000.

16 A. C. Larson, R. B. von Dreele, GSAS, Los Alamos National Laboratory Report LAUR 86-748, 2000.
17 A. Hennessy, G. Graham, J. Hastings, D. P. Siddons and Z. Zhong, New pressure flow cell to monitor $BaSO_4$ precipitation using synchrotron *in situ* angle-dispersive X-ray diffraction, *J. Synchrontron Radiat.*, 2002, **9**, 323–324.
18 ScaleSoftPitzerTM Vesion 4.0, Rice University, Houston, USA.
19 J. E. Oddo and M. B. Thomson, Simplified calculations of $CaCO_3$ saturation at high temperatures and pressures in brine solutions, *J. Pet. Technol.*, 1982, 1583–1590.
20 J. Garside and R. J. Davey, Secondary contact nucleation: kinetics, growth and scale-up, *Chem. Eng. Commun.*, 1980, **4**, 393–424.
21 G. Zhou and Y. Zheng, Chemical synthesis of $CaCO_3$ minerals at low temperatures and implication for mechanism of polymorphic transition, *Neues Jahrb. Mineral. Abh.*, 2001, **176**(3), 323–343.

PAPER

Nucleation and control of clathrate hydrates: insights from simulation†

C. Moon, R. W. Hawtin and P. Mark Rodger

Received 13th December 2006, Accepted 7th February 2007
First published as an Advance Article on the web 24th May 2007
DOI: 10.1039/b618194p

Clathrate hydrates are important in both industrial and geological settings. They give rise to many technological and environmental applications, including energy production, gas transport, global warming and CO_2 capture and sequestration. In all of these applications there is a need to exert a high degree of control on the crystallisation process, either to promote or inhibit it according to the application. This crystallisation process involves the formation of a tetrahedral hydrogen bonding network (as occurs with ice), but is complicated by mass transport limitations due to the poor mixing of the common guest molecules, such as methane, and the water that forms the host lattice. The net effect is that the mechanisms for hydrate formation and growth are still poorly understood, with the consequence that development of additives to control nucleation and growth is still largely governed by trial-and-error approaches. In this paper we show how classical molecular dynamics simulations can be used to provide a direct simulation of the nucleation process for methane hydrate and consequently to allow direct simulation of the effect of additives on the nucleation and growth process. Data are presented for oligomers of PVP and compared with existing data for PDMAEMA. The results show that the two additives work by very different mechanisms, with PVP increasing the surface energy of the interfacial region and PDMAEMA adsorbing to the surface of hydrate nanocrystals. The surface energy effect is a mechanism that has not previously been considered for hydrate inhibitors.

1. Introduction

Clathrate hydrates are a form of solid solution in which a crystalline, hydrogen-bonding water network (the host) traps a second group of molecules (the guests) within small cages created by the water lattice. Many different molecular species can be incorporated as guests,[1] but probably the most important are components of natural gas—especially methane and carbon dioxide. Since the most common guests are gases in their standard state, clathrate hydrates are often referred to as gas hydrates, or even just hydrates.

The moderately high pressures and low temperatures needed to stabilise clathrate hydrates[2] are common in both natural and industrial settings, with the result that these compounds have a number of important applications.[3] Vast methane hydrate

Chemistry Department, University of Warwick, Coventry, UK CV4 7AL. E-mail: p.m.rodger@warwick.ac.uk

† The HTML version of this article has been enhanced with additional colour images.

deposits are found below the sea floor on the continental margins, providing both a potential fuel reserve[4] and a serious concern for global warming.[5,6] However, the global warming implications are not all adverse. CO_2 forms a hydrate that is potentially more stable than methane, giving rise to the possibility that geological storage involving CO_2 hydrate could provide a *very* long term CO_2 mitigation strategy.[7] Much of the industrial interest arises from the fact that natural gas hydrates can form in oil-field well heads and transmission pipelines, creating blockages that are expensive and sometimes dangerous.[8] Hydrates also possess a number of unusual physical properties, perhaps the most intriguing of which is the so-called self-preservation effect.[9] This is a phenomenon whereby depressurisation of a methane hydrate crystal to pressures below that required for stability can still result in a metastable crystal that persists for months, but only if the depressurisation occurs within a narrow range of temperature (240–270 K) just below the normal freezing point of ice.

A major thrust of hydrate research over the last decade has focused on developing chemical additives to destabilise hydrates or inhibit their formation. Of particular interest has been the discovery of inhibitors that are active at very low concentrations—typically about 1/100th the concentration required for conventional thermodynamic anti-freeze additives such as methanol.[10,11] These low dosage hydrate inhibitors (LDHIs) do not alter the thermodynamic phase boundaries for hydrate stability. Instead they sustain a metastable liquid state, either by suppressing the initial nucleation[12] (kinetic inhibitors, or KIs) or by slowing subsequent growth and preventing crystal aggregation[13] (anti-agglomerants, or AAs). The small concentrations required can give rise to lower operating costs, and with lower environmental impact in oil-field applications; as such, there has been a substantial driving force to try to identify ever more active LDHIs.

An essential prerequisite for designing better LDHIs is a molecular-level understanding of the mechanisms by which they work, and that, in turn, requires a commensurate understanding of the mechanisms for nucleation and growth. It is only in the last few years that the research to provide this molecular-level understanding has begun to be published. Experimental studies with diffraction,[14,15] Raman[16] and NMR[17] methods have been used to study the early stages of growth. These have identified the propensity with which different water cages appear during the early stages of growth, and some of the structural changes that occur as a nano-scale cluster matures into a stable crystal. Molecular simulations are also providing valuable insight into the mechanisms by which hydrates grow[18] and the interaction of typical LDHI monomers with the surface of a hydrate crystal.[19–21]

Of particular relevance in the context of this article has been the publication of a series of direct molecular simulation studies of hydrate nucleation. By simulating the behaviour of a methane/water interface at moderate subcoolings and methane supersaturations, it has been possible to observe the spontaneous nucleation of methane hydrate in multi-nanosecond molecular dynamics simulations.[22,23] These simulations have shown good agreement with the available experimental data, particularly in finding that structure indicative of type II hydrates can be present during the early stages of growth even when the thermodynamically stable phase is the type I structure.[14] Protocols have also been developed to enable an LDHI molecule to be inserted into the simulations without undue perturbation to the nucleating hydrate,[24,25] and thus have enabled direct simulations of the way LDHIs perturb the hydrate growth process. Most recently, an extensive series of simulations were reported for poly(dimethylaminoethylmethacrylate), PDMAEMA (**1**), including multiple replicates of different tacticities and initial conformations for the polymer. These simulations indicated that the PDMAEMA tended to adsorb onto the surface of the hydrate crystals, consistent with the accepted mechanism for LDHI activity.[19,26] Intriguingly, preliminary simulations with poly(vinylpyrrolidone), PVP (**2**), indicated that it induced dissolution of small hydrate clusters without any direct contact between the additive and the hydrate surface.[24,27] The purpose of this article is to present a much more extensive study of PVP to establish the validity of these initial findings and elucidate a mechanism for inhibition that does not involve surface adsorption. Results are presented for different tacticities, molecular weights and initial conformations of the PVP, and for interaction with different sizes of hydrate nanocrystal.

1 **2**
PDMAEMA PVP

2. Methods

2.1 Potentials and protocol

The simulation protocols and potentials were the same as those reported in our earlier studies,[22,24,25] but are outlined here for completeness.

The methane was modelled using a single-site Lennard-Jones potential[28] and water with the SPC potential;[29] this combination has already been shown to give a surprisingly good reproduction of the properties of methane hydrate[30–33] and—most importantly for the current application—to allow nucleation of methane hydrate to be simulated at reasonable temperatures and pressures.[22,23] Additives were modelled using the CHARMM22 force field,[34] with Lorenz–Berthelot combining rules for the water–additive and water–methane Lennard-Jones parameters. Short-range forces were truncated at 12 Å, while EWALD methods were used for electrostatic interactions. All bonds were constrained using SHAKE,[35] as was the bond angle for water.

Nucleation simulations were initiated by taking a film of methane hydrate and heating it under vacuum (see Fig. 1). Trials were made with a number of different heating temperatures and it was found that 300 K gave the most rapid melting without forming methane bubbles within the resulting liquid water. The heating phase was initially carried out for 850 ps, after which time no significant hydrate-like structure was evident within the water, and the concentration of methane had dropped to half that required for the hydrate phase. This system was then cooled to 250 K, pressurised to 300 bar with methane fluid, and then an extended MD simulation (*ca.* 50 ns) performed. This simulation exhibited immediate nucleation and growth of the hydrate phase.[22,23] The melting temperature for this model of methane hydrate is below 280 K at this pressure.[31,36] Additional long timescale simulations were carried out which employed longer initial heating periods, and these led to significant induction periods before nucleation, but otherwise showed similar behaviour.[23]

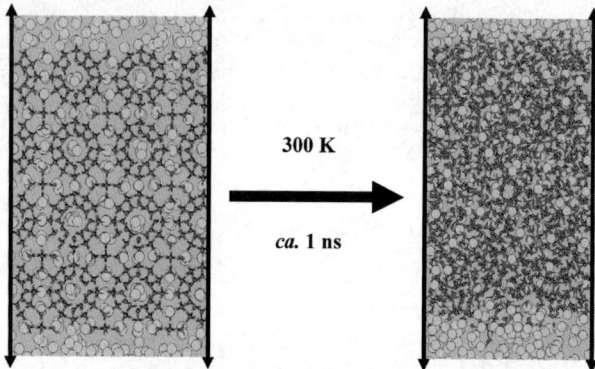

Fig. 1 Schematic representation of the method for preparing the liquid water film. Most of the methane fluid (which was double the thickness of the water film) is not depicted.

The protocol for inserting an inhibitor into a nucleating simulation is shown, schematically, in Fig. 2. An initial configuration was taken from the uninhibited nucleation simulations, and then the polymer inserted into the methane phase near, but not adjacent to, the water film; any methane molecules that overlapped the inserted polymer were removed. The water was then immobilised and the dynamics of the methane and inhibitor simulated for 5–10 ps, to alleviate any strain due to this insertion, before long timescale production simulations were performed as above. Earlier studies examined the influence of insertion height, and found that insertions in the range 5–15 Å above the water made little difference to the results, and that incorporation of the inhibitor into the water film typically occurred in less than 100 ps.[25]

Initial polymer configurations were generated using either helical (all *gauche*) or linear (all *trans*) backbone configurations, with iso- a- and syndio-tactic arrangements of the pyrrolidone rings. Most simulations were performed using a PVP octamer, but some simulations were performed with other molecular weights in the range tetramer–dodecamer; these oligomers are consistent with the molecular weights that have been found, experimentally, to be the most active LDHIs. A summary of the simulated inhibitor systems is given in Table 1.

Overall the simulation box was orthorhombic, with in-plane (x–y) dimensions 33.3 × 29.6 Å and the out-of-plane (z) dimension being typically 150 Å. The resulting water film contained 1656 water molecules (*ca.* 50 Å thick) and in excess of 1000 methane molecules were present in the simulated system. Production simulations were performed using *NPT* MD (Nosé–Hoover) with DL_POLY 2. A 1 fs timestep was used throughout.

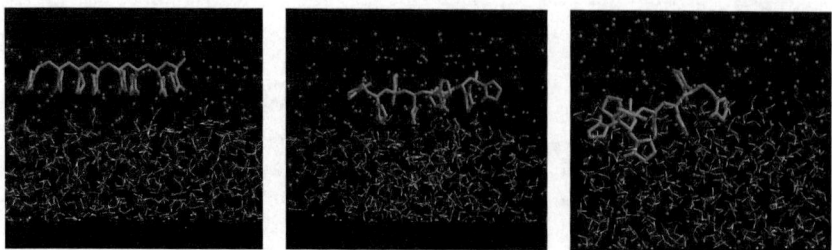

Fig. 2 Insertion process for inhibitor: inhibitor is inserted into the methane phase (left); the water is immobilised and the dynamics of the inhibitor and methane calculated (middle); a full MD simulation is then performed on the relaxed system (right).

Table 1 Summary of the inhibitor systems simulated

Addition of PVP to one surface						Addition of PVP to both surfaces					
t_{sim}/ns[a]	t_{add}/ns[b]	n[c]	tact[d]	conf[e]	label	t_{sim}/ns[a]	t_{add}/ns[b]	n[c]	tact[d]	conf[e]	label
0.6	12	4	I	L	vp4IL	2.4		8	I	L	vp8IL
		6	I	L	vp6IL				A	L	vp8AL
		7	I	L	vp7IL		12			H	vp8AT
		8	I	L	vp8IL			10	I	L	vp10IL
				H	vp8IM			12	I	L	vp12IL
					vp8IP	30		8	I	L	vp8IL
			S	L	vp8SL			12	I	L	vp12IL
					vp8SLy						
				H	vp8SMy						
					vp8SP						
			A	L	vp8AL						
				H	vp8AP						
		9	I	L	vp9IL						
		10	I	L	vp10IL						
		12	I	L	vp12IL						

[a] duration of the inhibited production simulation
[b] time at which the PVP was added to the uninhibited system
[c] number of monomers in the oligomer
[d] tacticity: isotactic (I), syndiotactic (S) or atactic (A)
[e] initial conformation of the polymer backbone: linear (L) or helical (H)

2.2 Analysis

The simulations have been analysed for hydrate structure using the same methods reported in our earlier studies.[33,37] These make use of three order parameters to describe the hydrogen bond network within the water. One of these (F_3) measures deviations from the tetrahedral angle within the hydrogen bond network, while the other two ($F_{4\varphi}$ and F_{4t}) measure the torsion angle and scalar triple product associated with three consecutive bonds within the network. By comparing the set of instantaneous order parameters for any specific water molecule with the distribution of parameters found within various stable phases (see Table 2) it is possible to assign a *local phase* to each water molecule, and thereby to determine both the hydrate content of a system and the extent to which the water molecules with hydrate local phase are clustered. The analysis method correctly identifies 95% of the molecules within stable hydrate and ice phases; for liquid water (SPC) at 270 K

Table 2 Average order parameters for water molecules in various phases. Values for the hydrate phase are the same for CH_4 (type I) and Kr (type II) hydrates

	Environment		
Order Parameter	Liquid	Hydrate	Ice
F_3	0.10	0.01	0.01
$F_{4\varphi}$	0.00	0.70	−0.40
F_{4t}	0.26	0.47	0.29

about 6% of the water molecules are found to experience a hydrate-like local environment, although these show no evidence of clustering together. Hydrate clusters were then identified by searching for sets of water molecules that satisfied the following conditions:

(i) each molecule had the hydrate local phase; and

(ii) each molecule was connected to every other molecule by a contiguous set of hydrate "bonds".

The hydrate "bond" occurred when the distance between the O atoms of two hydrate water molecules was less than 3.5 Å. Additional order parameters monitored included the number of water molecules within 3.5 Å, and number of methane molecules within 5.5 Å, of each water molecule.

The calculation of radial distribution functions (RDFs) for inhomogeneous systems, such as the interfacial systems considered here, raises issues about what is the correct "bulk" density to use in normalising the RDFs. In this work we have chosen to use a self-consistent approach whereby the RDFs are normalised by the average density observed within a sphere of radius R_{max} about each atom, where R_{max} is the maximum radius for which the RDF is calculated. For bulk systems this will give a limiting value of 1, as required, but for interfacial systems it will give an appropriately weighted average of the density across the interface.

3. Results

3.1 Previous nucleation simulations

In order to facilitate discussion of the results in this paper, it is useful to summarise some of the results from our recent simulations of hydrate nucleation[22,23] and how this is affected by LDHIs.[25] The nucleation simulations have been performed at a range of methane supersaturations, corresponding to aqueous concentrations that are 20–50% that of the methane hydrate. The simulations were performed in a thin water film (5 nm) and so concentrations above those for bulk aqueous solution are to be expected. The chemical potential for methane in the aqueous and gas phases was calculated by the Widom method[38] and showed the excess chemical potential for aqueous methane at the beginning of the simulation was 3–7 kcal mol^{-1}, depending on the methane concentration. Nucleation occurred immediately at the highest supersaturation, but induction times of 10–15 ns were observed at the lower supersaturations. Repeat simulations showed that the nucleation and growth processes were stochastic, with the increase in hydrate content during growth sometimes being non-monotonic, but that the general trends were reproducible. Analysis indicated that hydrate clusters of 150–200 water molecules were required for sustained growth under these conditions. To initiate the inhibited simulations in this work we have taken configurations from the uninhibited simulation that showed most rapid and sustained growth, thus giving an extreme test for the LDHIs; this was the system prepared by melting a hydrate film at 300 K for 850 ps (see section 2.1).

Our preliminary study of PVP[24,27] showed that it could be effective in preventing the initial nucleation. This is illustrated in Fig. 3, which depicts how the hydrate content changes with time during several different simulations. In the absence of PVP the hydrate was seen to grow immediately and rapidly. When PVP was added at the very beginning (PVP0) of this simulation, no hydrate growth was observed, and the residual hydrate-like water molecules did not cluster. When the PVP was added after 12 ns (PVP12)—by which stage the largest hydrate cluster contained 200 water molecules—it was found to destabilise the hydrate, with no substantial hydrate clusters remaining after 2 ns. We also report in Fig. 3 a new simulation, in which a configuration was taken from halfway through the PVP12 simulation and the PVP removed. In this case, hydrate growth resumed immediately, confirming that the disruption of the hydrate was due to

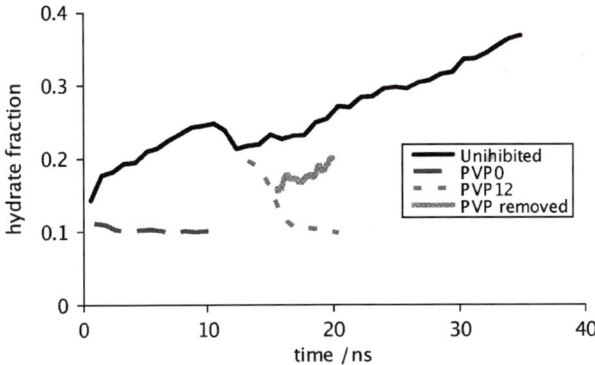

Fig. 3 Hydrate fraction as a function of time in: an uninhibited simulation; simulations in which PVP was added at the beginning of the simulation (PVP0) and after 12 ns (PVP12); and a simulation in which the PVP was removed halfway through the PVP12 simulation. Data have been averaged over consecutive 0.9 ns segments of the trajectories. The hydrate analysis identifies a residual structure of about 6% hydrate in liquid water at 270 K.

the PVP, and not an artefact of any stress created during the insertion process. This behaviour is in stark contrast with the effect of adding PDMAEMA, where the additive was found to act as a seed, enhancing the growth of a crystal to which it then adsorbed.[25]

3.2 PVP oligomers: single molecule insertion

To probe the influence of PVP further, a series of simulations were performed in which different PVP oligomers were added to the growing hydrate simulation. Initially, 15 different oligomers—spanning three different tacticities, two different initial conformations and seven different molecular weights—were used to add to a conformation taken after 22 ns. By this stage, almost 30% of the water was in the hydrate phase, with the dominant cluster containing 260 water molecules. After only 0.6 ns it was apparent that, in contrast with the preliminary study, the PVP was not inhibiting the hydrate growth (see Fig. 4). Only two of the 15 systems showed any net decrease in the amount of hydrate present and in both cases that was less than 0.25% of the total water content. In contrast, half the systems showed a net increase in hydrate content that was an order of magnitude bigger. The preliminary study actually used two oligomers, one inserted at each interface, and so a similar

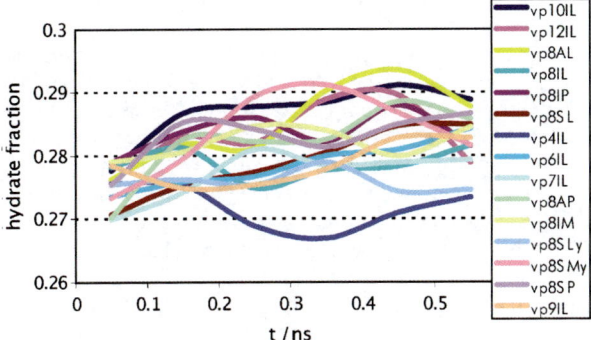

Fig. 4 Hydrate content as a function of time after a single PVP oligomer is added to a configuration taken after 22 ns of the uninhibited simulation. Data have been averaged over consecutive 0.1 ns segments of the trajectories.

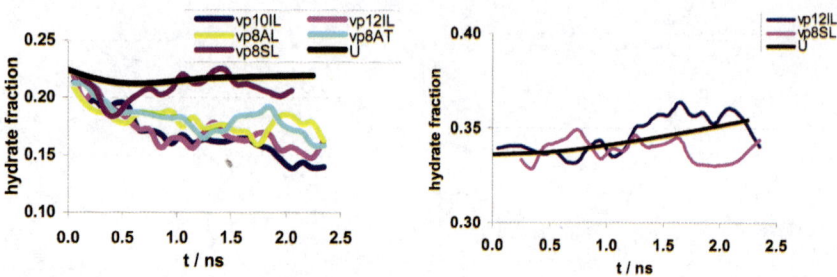

Fig. 5 Hydrate fraction as a function of time when two PVP molecules were inserted after 12 ns (left) or 30 ns (right). Data have been averaged over successive 0.1 ns segments of the trajectories. The corresponding data for the uninhibited system (U) is included for comparison, with data averaged over 0.9 ns segments of the trajectory.

double-insertion protocol was used in subsequent simulations. Further simulations also focused on two insertion times: 12 ns, (hydrate content 20%, largest cluster 200 water molecules) and 30 ns (hydrate content 33%, largest cluster 477 water molecules). The former was the insertion time used in the preliminary study[24] and has a cluster comparable with the critical cluster size, while the latter has clusters that are about twice the critical cluster size. In each case, more than 2 ns of simulation was accumulated following the insertion. In the rest of this section these two sets of simulations will be denoted PVP12 and PVP30, respectively.[39]

3.3 PVP oligomers: two molecule insertion

The oligomers chosen for the dual insertion are listed in Table 1, and the effect on hydrate formation can be seen from the resultant plots of hydrate fraction against time (Fig. 5). For insertion at the earlier time there is a clear inhibitory effect. In four out of five cases the hydrate fraction drops substantially during the simulation, with an average decrease from 21 to 16% over the 2.4 ns. This is in complete agreement with the results of the preliminary study. In the fifth case (a syndiotactic octamer) there is an initial decrease in the hydrate content, but the system subsequently stabilises. It is not clear whether this is due to the chaotic nature of the nucleation and growth events (see, for example, the decrease in the hydrate content of the uninhibited simulation during the period 10–12 ns, Fig. 3) or a real effect of the properties of this oligomer. There were no obvious differences in the trajectories of this oligomer compared with the other systems, and in particular, the vp8SL entered the water phase in a similar time, and at a similar distance from the hydrate cluster, to the other oligomers. We note that similar variations were found in our study of PDMAEMA[25] and so it is likely that the difference is simply due to the stochastic nature of crystal nucleation, but repeat simulations are in progress to confirm this.

In contrast to the 12 ns insertion, no significant influence from the PVP can be identified from the later (30 ns) addition. Given the fluctuations evident in these plots, the hydrate growth in the uninhibited simulation is entirely consistent with that observed in the presence of PVP. Thus, even at double the concentration, the PVP appears to be ineffective in modifying hydrate formation once a critical nucleus has formed. These trends are also consistent with experimental studies with PVP, which show that PVP is effective in delaying the onset of nucleation,[40,41] but that once crystals form its effect on hydrate growth is weak.[42]

With the exception of vp8SL added at 12 ns, the results of these simulations provide no basis for distinguishing between the different oligomers. In our previous study of PDMAEMA it was found that the helical form of the polymer was more active than the linear form. However, in PDMAEMA the monomer group is attached to the backbone at a quaternary carbon, and this hinders twisting of the polymer backbone to the extent that the initial conformation persisted throughout

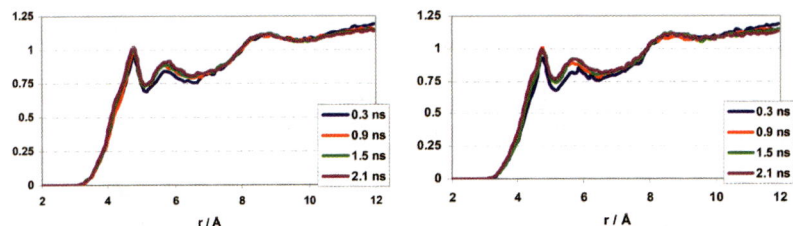

Fig. 6 RDFs for liquid water about the amide N in PVP. Dual insertion after 12 ns (left) and 30 ns (right).

the duration of the simulations. PVP, however, lacks the additional CH_3 side group, so that the pyrrolidone is attached to the backbone at a tertiary carbon atom. This enables relatively free rotation of the backbone bonds, with the result that the initial conformation was disrupted on a timescale of 10^{-1} ns. Consequently, the data in Fig. 4 and Fig. 5 show no evidence that the activity of PVP correlates with its initial conformation. In view of these considerations, further analysis of the simulations has been done by averaging over all oligomers.

Various RDFs were calculated for the inhibited and uninhibited systems. Since the order parameter analysis distinguished three different classes of water molecule according to the geometry of their local environment—hydrate (h), ice (i) and liquid (liq)—we have calculated partial RDFs for these different classes of water. In the analysis presented in this paper we have focused on the distribution involving PVP since our interest is in how the PVP modifies the hydrate formation process. Three types of probe site have been used for the PVP: the amide N and O atoms, which constitute the hydrophilic group, and the CH_2 sites (C) found in the ring and along the backbone. The RDFs shall be denoted $g_{A-B}(r)$, where $A \in \{N, O, C\}$ and $B \in \{h, i, liq, Me\}$ and Me is the centre of the methane molecules.

The time evolution of the PVP–water RDFs have been obtained by averaging the RDFs over oligomers and over successive 0.6 ns segments of the trajectories. Selected examples are given in Fig. 6 and Fig. 7. All the plots show a slight increase in the amount of water found around the polymer during the first segment of the trajectory, but no significant time dependence was observed in any of the PVP–water or PVP–methane distribution functions thereafter. Since the PVP was initially inserted into the methane phase and then allowed to diffuse into the water phase, it is not surprising that the first time segment shows less hydration of the polymer. Conversely, the smallness of this difference and the time independence of the RDFs after 0.6 ns indicate that relaxation of the PVP into the aqueous environment was complete well within the first 0.6 ns. Hence that it is appropriate to average the RDFs over the last 1.8 ns of each simulation.

Averaged RDFs, together with the corresponding integrated coordination number, for the distribution of hydrate water, liquid water and methane about the amide O in PVP are presented in Fig. 8. The discussion is focused on the amide oxygen

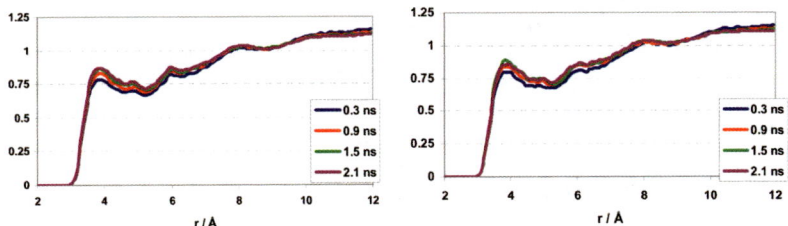

Fig. 7 RDFs for liquid water about the CH_2 groups in PVP. CH_2 groups are found in both the backbone and the pyrrolidone ring. Dual insertion after 12 ns (left) and 30 ns (right).

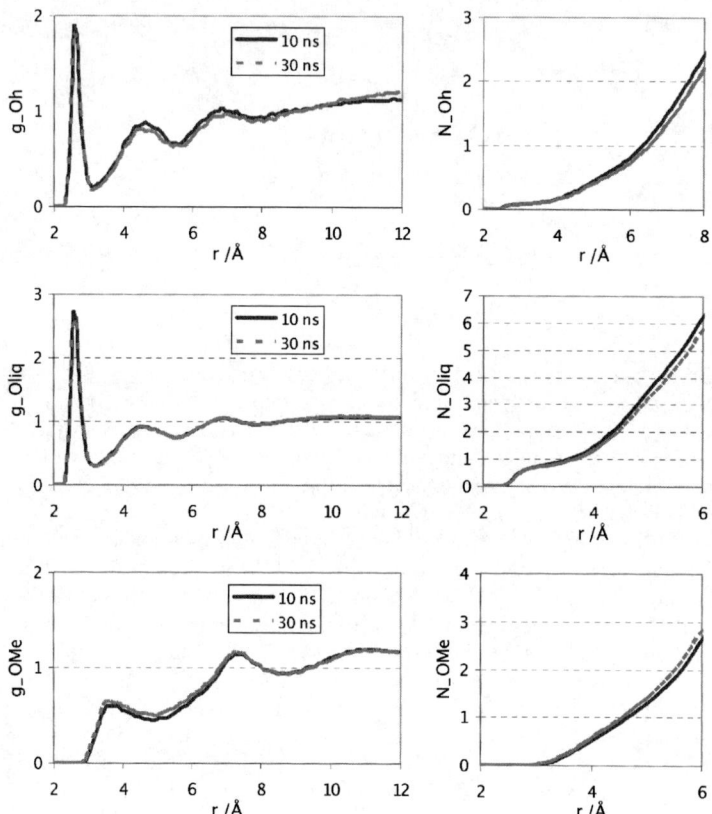

Fig. 8 RDFs (left) and integrated coordination number (right) for hydrate water (h), liquid water (liq) and methane (Me) about the amide O atoms in PVP (O).

because this is the site most likely to be responsible for structuring the water;[43] however the interpretation of the RDFs involving the N and CH$_2$ sites on PVP (not shown) was entirely consistent with the discussion below.

Comparison of the curves for g_{O-h} and g_{O-liq} reveals strong similarities, but also some important differences. The similarities are immediately obvious. Each has three peaks, located at the same radii (2.60, 4.66 and 6.9 Å) and with a similar pattern of peak heights. The differences are more subtle, but also more important. To begin with, the first two peaks are lower for O–h than for O–liq: (2.00, 0.88) and (2.89, 0.92), respectively. Secondly, the limiting behaviour of the two RDFs differs. In g_{O-liq} the RDF attains a limiting value close to 1 by about 8 Å, which indicates that the distribution of liquid water around the polymer is homogeneous. No such limiting value is attained for g_{O-h} at large distances, with the RDF continuing to increase above the bulk value of 1. Such behaviour is typical of systems in which A and B are spatially separated, for example on opposite sides of an interface. Thus the RDFs indicate that the PVP is *not* associated with the hydrate cluster.

This interpretation is confirmed by the integrated coordination numbers (Fig. 8). Whereas there are 0.77 and 4.03 liquid water molecules associated with the first two peaks in g_{O-liq}, respectively, the corresponding values for hydrate waters (g_{O-h}) are just 0.06 and 0.54. Thus only about 8% of the water molecules around the PVP oxygen atoms have a local geometry consistent with hydrate. This proportion is substantially smaller than that found in the system as a whole, where 15% and

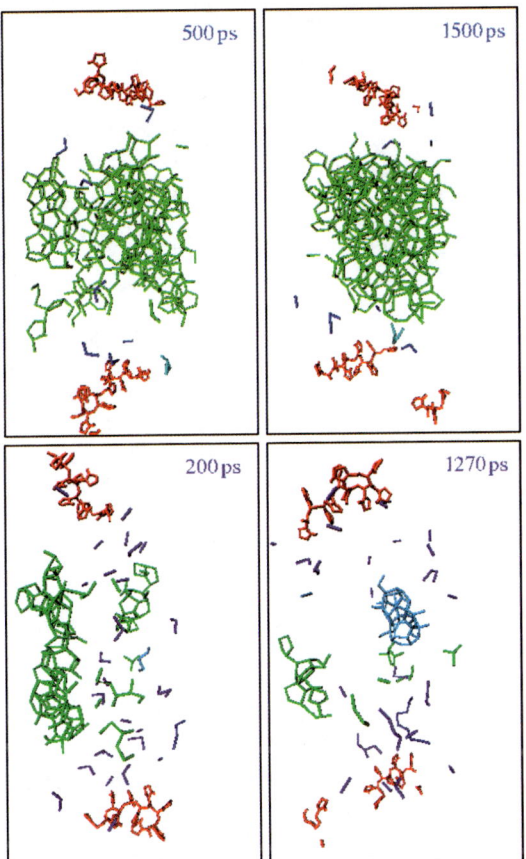

Fig. 9 Representative snapshots of the hydrate network and PVP following insertion of the PVP at 12 ns (bottom) and 30 ns (top). Times given in the pictures indicate the time following insertion. Apparent disjunctions in any cluster/molecule are because they span the x–y periodic boundaries. Colour key: PVP (red); largest hydrate cluster (green); second largest hydrate cluster (cyan); all other hydrate waters (blue).

33% of the water molecules have the hydrate local phase (for the 12 and 30 ns insertions, respectively). Thus the local environment of the PVP is depleted in hydrate water. Intriguingly, the number of waters associated with the first peak is no different between the two insertion times, and increases only marginally for the second peak (0.61 for hydrate and 4.33 for liquid water with the 30 ns insertion), despite the fact that there is much more hydrate present with the later insertion time. Together with the similarity between the RDFs for the two insertion times, this indicates that the local environment of the PVP is independent of the hydrate content of the system as a whole. The local depletion of hydrate water, and its independence from total hydrate content, is not consistent with a surface-adsorption mechanism for the PVP. Rather, the data indicates that the PVP is located typically 2 solvation shells (the distance beyond which the integrated coordination number exceeds one molecule) away from the hydrate cluster. This separation between the PVP and hydrate cluster is also apparent from visual inspection of the trajectories (see Fig. 9). For comparison, the simulations of PDMAEMA[25] reported hydrate : liquid water ratios that were very similar in the vicinity of the polymer and in the system as a whole; surface adsorption was identified in the PDMAEMA simulations.

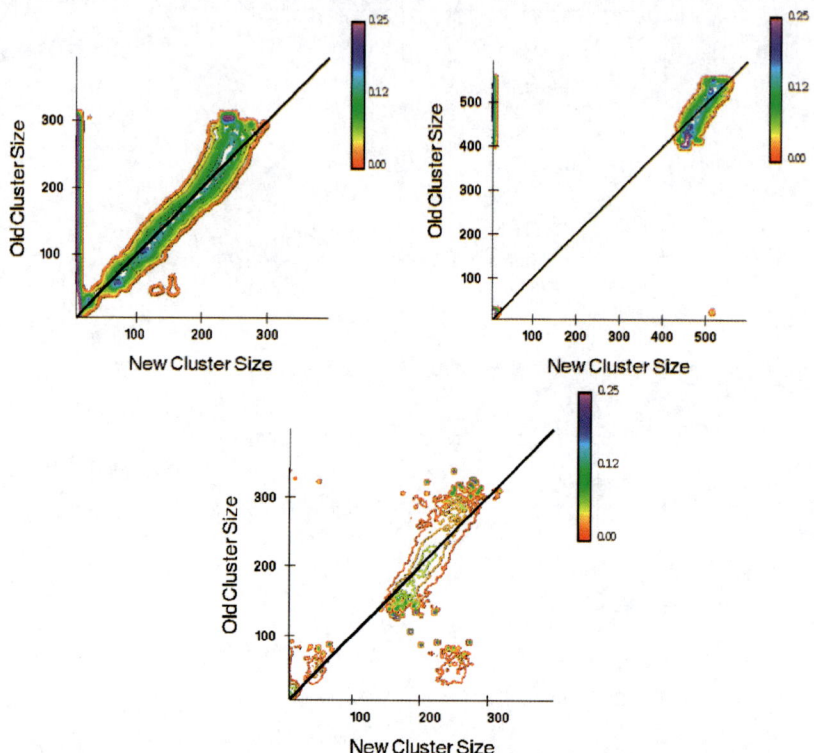

Fig. 10 Transition probabilities, $P(m|n)$, for changes in the size of hydrate clusters, where m is the new cluster size and n is the current (i.e. 'old') cluster size. The line ($m = n$) delineates between growth and decay. (i) PVP insertion at 12 ns, (ii) PVP insertion at 30 ns, and (iii) uninhibited simulation from 12 ns.

The RDFs for methane about the PVP oxygen show a weak, broad first peak at 3.60 Å, with the ensuing minimum occurring at 4.91 Å. Integrating out to the first minimum gives 1.4 methane molecules within the first shell, but only one methane is contained within the first 4.6 Å. One of the suggestions that has been made to explain the activity of LDHIs is that preferential solvation of the LDHI by methane reduces the local concentration of methane. However, we note that the amount of methane found around the PVP added at 30 ns—where the hydrate continued to grow—was slightly greater than was present around the PVP added at 12 ns. Thus our simulations indicate that inhibition by PVP cannot be due to local depletion of methane concentrations.

As a final confirmation that PVP destabilised hydrate early in its growth, but not once clusters in excess of 400 water molecules had formed, we report the transition probability, $P(m|n)$, for changes in the size of hydrate clusters. $P(m|n)$ is defined as the probability that a water molecule located in a cluster of size n at time t will be located in a cluster of size m a time τ later. In this study we found $\tau = 1$ ps to be a suitable time difference, allowing significant changes in cluster size without losing the details of the growth/decay processes.

Contour maps of transition probabilities calculated from the PVP12 and PVP30 simulations are depicted in Fig. 10; the transition probability calculated from 2.7 ns of the inhibited simulation, starting at 12 ns, is also shown for comparison. The top left hand region of each map ($m < n$) represents cluster shrinkage while the bottom right ($m > n$) is cluster growth. It is useful to mention two other points about these contour maps before proceeding to interpret them. Firstly, $P(m|n)$ can only be calculated for those n which are actually observed in the simulation, and so the projection of the contours onto the n (vertical) axis gives an indication of the range of cluster sizes observed. Secondly, since $P(m|n)$ is a conditional probability, it is normalised such that, for every distinct value of n,

$$\sum_m P(m|n) = 1$$

This, in turn, means that the calculated P will be noisy for cluster sizes n that are observed infrequently. This is well illustrated by the extremities of the large cluster contours (ca. 150 and 300 waters) for the uninhibited system (Fig. 10).

The transition probabilities for PVP30 and U are typical of those found during hydrate growth.[23] Both are dominated by a series of contours around the $m = n$ line, and spanning about 150 water molecules. These regions are skewed so that the small cluster end (ca. 150 water molecules for U and 400 for PVP30) favours growth, while shrinkage is more likely at the large cluster end. A tendency for the largest clusters to shrink was seen consistently in uninhibited studies,[23] where it was found to result from an annealing process whereby dendritic networks of water molecules would constantly form and then break up around the surface of the hydrate crystal until, occasionally, a long-lived dendrite would be incorporated into the stable crystal core. In contrast, the contour plots for PVP12 show no region that is biased towards growth. Instead, during the course of the simulation the clusters kept getting smaller, to the extent that the complete region of the transition probability down to $n = 0$ could be calculated without significant noise. Further, whereas the addition of substantial (50–100 molecule) water clusters was common in the uninhibited system, such events were suppressed in the presence of PVP. Intriguingly, such events were found to be enhanced by the presence of PDMAEMA.[25] We conclude that the PVP is effective in destabilising small hydrate clusters.

4. Conclusions

In this paper we have presented a series of MD simulations of the effect of PVP (**1**), a known low dosage hydrate inhibitor (LDHI), on the nucleation and subsequent growth of methane hydrate. A number of analogous simulations were performed, using different oligomers that varied in tacticity, initial conformation, or molecular weight. The results showed that the PVP was effective in preventing nucleation, and in destabilising hydrate clusters that had not grown substantially beyond the critical cluster size (ca. 200 water molecules), but it was ineffective in modifying growth in hydrate clusters that had already grown to double the critical cluster size (ca. 400 water molecules). These results are consistent with experimental data, which suggests that PVP does increase the induction time before hydrate formation is first detected, but that once formed the hydrate will grow rapidly.

Most intriguingly, analysis of the simulations indicated that PVP remained at least 5–10 Å away from the surface of any hydrate clusters/crystals that were present, and that water structure consistent with the hydrate phase was actually suppressed in the vicinity of the PVP. This is in agreement with the results of Anderson et al.,[19] whose calculations indicated that there was no free energy driving force ($\Delta G_{adsorption}$ = +0.4 ± 3.9 kcal mol^{-1}) for the adsorption of vinylpyrrolidone monomers onto the surface of an aqueous methane hydrate crystal.[19] We conclude that the most common mechanism conjectured for LDHI activity—surface adsorption—cannot be responsible for the activity of PVP.

All the results of our simulations can be explained, however, in terms of interfacial energy effects. If the presence of PVP were to increase the interfacial energy, then it would also increase the critical cluster size and decrease the stability of any clusters smaller than the new critical cluster size, thereby increasing the time it takes for a critical nucleus to form. For stable crystal nuclei, however, the energy would be dominated by the favourable bulk crystal energy and so such crystals would not be

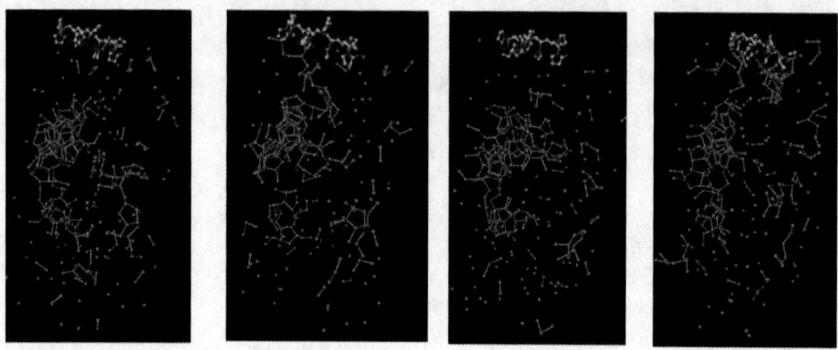

Fig. 11 Four snapshots taken from within a 100 ps portion of one of the PVP12 simulations. The snapshots are presented in chronological order. Hydrate water is depicted in brown, and the PVP carbon, oxygen and nitrogen atoms in green, red and blue, respectively

rendered unstable by the PVP. The hydrate/liquid water interface is known to be diffuse on the molecular length scale, with the structure of the water changing over 10–15 Å.[33] The 5–10 Å separation between the PVP amide groups and the hydrate surface found in the present work indicates that the PVP is present in the interfacial region, though without binding irreversibly to the hydrate crystal. Thus continued growth is feasible once a critical nucleus has formed.

Our simulations also provide some evidence for a molecular mechanism to underpin this explanation. A series of snapshots taken from within a 100 ps interval in one of the PVP12 simulations is shown in Fig. 11. It is apparent that filaments of hydrate water molecules do form from the pyrrolidone groups, but that they are transient. Indeed it is possible for several filaments to form from the same oligomer, but in this case they are likely to be incommensurate with each other and so frustrate, rather than reinforce, hydrate formation. Such transient filaments were also found to form around the hydrate clusters (as discussed in section 3.3) and so create a halo region around the hydrate that would be disrupted by the incommensurate filaments arising from the PVP.

We conclude that this halo effect forms a viable mechanism to explain the kinetic inhibition of methane hydrate formation that is induced by PVP. It leads to increased induction times for the formation of the first hydrate nuclei, but does not slow growth thereafter. The mechanism is not expected to be universal, as surface adsorption has been shown to be viable in PVCap19[19] and PDMAEMA,[25] but may well be responsible for the synergistic effect whereby co-polymers of PVCap and PVP are found to be more effective inhibitors than PVCap alone.[44] It does also provide a new insight into the origins of LDHI activity that should prove beneficial in designing more active inhibitors.

Acknowledgements

The research was supported by ICI through a CASE award, and by INTAS grant 03-51-5537. HPCx time was provided by the Materials Chemistry Consortium under EPSRC grant GR/S13422, and other computational resources were provided by the University of Warwick Centre for Scientific Computing.

References

1 An excellent description of the properties of gas hydrates is given in E. D. Sloan, *Clathrate Hydrates of Natural Gases*, Marcel Dekker, New York, 1998.
2 For example, at 5 °C methane hydrate requires a pressure of at least 4.5 MPa to form, but this drops to 2.5 MPa just above the ice point (0.1 °C); even milder formation conditions

are found with the higher boiling point components of natural gas; extensive details are reviewed in ref. 1, ch. 6.
3 E. D. Sloan, *Nature*, 2003, **426**, 353.
4 Estimates suggest methane hydrate could provide more than twice the energy available from all fossil fuel reserves. See, for example, (*a*) L. Milich, *Glob. Environ. Change Human Policy Dimensions*, 1999, **9**, 179; (*b*) J. Mienert, K. Andreassen, J. Posewang and D. Lukas, *Ann. N. Y. Acad. Sci.*, 2000, **912**, 200.
5 M. J. Benton and R. J. Twitchett, *Trends Ecol. Evolution*, 2003, **18**, 358.
6 K. U. Hinrichs, L. R. Hmelo and S. P. Sylva, *Science*, 2003, **299**, 1214.
7 P. M. Rodger, *Issues underlying the feasibility of storing CO_2 as hydrate deposits*, PH3/25, International Energy Agency, GreenHouse Gas programme, 2000, pp. 1–29, .
8 E. D. Sloan, *Hydrate Engineering*, Soc. Proc. Eng., Richardson, TX, 2000.
9 (*a*) L. A. Stern, S. Circone, S. H. Kirby and W. B. Durham, *J. Phys. Chem. B*, 2001, **105**, 1756; (*b*) E. D. Ershov and V. S. Yakushev, *Cold Reg. Sci. Technol.*, 1992, **20**, 147.
10 C. A. Koh, *Chem. Soc. Rev.*, 2002, **31**, 157.
11 M. A. Kelland, T. M. Svartaas, J. Ovsthus, T. Tomita and J. Chosa, *Chem. Eng. Sci.*, 2006, **61**, 4048.
12 J. P. Lederhos, J. P. Long, A. Sum, R. L. Christiansen and E. D. Sloan, *Chem. Eng. Sci.*, 1996, **51**, 1221.
13 Z. Huo, E. Freer, M. Lamar, B. Sannigrahi, D. M. Knauss and E. D. Sloan, *Chem. Eng. Sci.*, 2001, **56**, 4979.
14 D. K. Staykova, W. F. Kuhs, A. N. Salamatin and T. Hansen, *J. Phys. Chem. B*, 2003, **107**, 10299.
15 H. Thompson, A. K. Soper, P. Buchanan, N. Aldiwan, J. L. Creek and C. A. Koh, *J. Chem. Phys.*, 2006, **124**.
16 T. Uchida, S. Takeya, L. D. Wilson, C. A. Tulk, J. A. Ripmeester, J. Nagao, T. Ebinuma and H. Narita, *Can. J. Phys.*, 2003, **81**, 351.
17 I. L. Moudrakovski, A. A. Sanchez, C. I. Ratcliffe and J. A. Ripmeester, *J. Phys. Chem. B*, 2001, **105**, 12338.
18 J. Vatamanu and P. G. Kusalik, *J. Phys. Chem. B*, 2006, **110**, 15896.
19 B. J. Anderson, J. W. Tester, G. P. Borghi and B. L. Trout, *J. Am. Chem. Soc.*, 2005, **127**, 17852.
20 B. Kvamme, T. Kuznetsova and K. Aasoldsen, *J. Mol. Graphics & Modell.*, 2005, **23**, 524.
21 K. F. Yan, J. G. Mi and C. L. Zhong, *Acta Chim. Sin.*, 2006, **64**, 223.
22 C. Moon, P. C. Taylor and P. M. Rodger, *J. Am. Chem. Soc.*, 2003, **125**, 4706.
23 C. Moon, R. W. Hawtin, P. M. Rodger, manuscript in preparation.
24 C. Moon, P. C. Taylor and P. M. Rodger, *Can. J. Phys.*, 2003, **81**, 451.
25 R. W. Hawtin and P. M. Rodger, *J. Mater. Chem.*, 2006, **16**, 1934.
26 (*a*) U. Karaaslan and M. Parlaktuna, *Proceedings 4th International Conference Natural Gas Hydrates (Yokohama)*, 2002, **2**, 933–937; (*b*) E. Habetinova, A. Lund and R. Larsen, *Proceedings 4th International Conference Natural Gas Hydrates (Yokohama)*, 2002, **2**, 942–946.
27 D. M. Duffy, C. Moon and P. M. Rodger, *Mol. Phys.*, 2004, **102**, 203.
28 J. S. Tse, M. L. Klein and I. R. McDonald, *J. Phys. Chem.*, 1983, **87**, 4198.
29 H. J. C. Berendsen, J. P. M. Postmas, W. F. van Gunsteren and J. Hermans, in *Intermolecular Forces*, ed. B. Pullman, D. Reidel, Dordrecht, Holland, 1981, pp. 331–342.
30 P. M. Rodger, *J. Phys. Chem.*, 1989, **93**, 6850.
31 R. E. Westacott and P. M. Rodger, *Chem. Phys. Lett.*, 1996, **262**, 47.
32 P. M. Rodger, *Ann. N. Y. Acad. Sci.*, 2000, **912**, 474.
33 P. M. Rodger, W. Smith and T. R. Forester, *Fluid Phase Equilib.*, 1996, **116**, 326.
34 A. D. MacKerell, Jr, D. Bashford, M. Bellot, R. L. Dunbrack, Jr, J. D. Evanseck, M. J. Field, S. Fischer, J. Gao, H. Guo, S. Ha, D. Joseph-McCarthy, L. Kuchnir, K. Kuczera, F. T. K. Lau, C. Mattos, S. Michnik, T. Ngo, D. T. Nguyen, B. Prodhom, W. E. Reiher III, B. Roux, M. Schlenkrich, J. C. Smith, R. Stote, J. Straub, M. Watanabe, J. Wiorkiewicz-Kuczera, D. Yin and M. Karplus, *J. Phys. Chem. B*, 1998, **102**, 3586–3616.
35 J. P. Ryckaert, G. Ciccotti and H. J. C. Berendsen, *J. Comput. Phys.*, 1977, **23**, 327.
36 M. T. Storr, PhD thesis, Warwick, 2000.
37 J. Fidler and P. M. Rodger, *J. Phys. Chem. B*, 1999, **103**, 7695.
38 D. Frenkel, B. Smit, *Understanding Molecular Simulation*, Academic Press, San Diego, 1996, p. 157.
39 The notation PVP*n* indicates only the time at which the LDHI was inserted. Additional parameters (such as concentration) will be clear from the context.

40 M. T. Storr, P. C. Taylor, J. P. Monfort and P. M. Rodger, *J. Am. Chem. Soc.*, 2004, **126**, 1569.
41 A. Carstensen, J. L. Creek and C. A. Koh, *Am. Mineral.*, 2004, **89**, 1215.
42 H. Sakaguchi, R. Ohmura and Y. H. Mori, *J. Cryst. Growth*, 2003, **247**, 631.
43 T. J. Carver, M. G. B. Drew and P. R. Rodger, *Phys. Chem. Chem. Phys.*, 1999, **1**, 1807.
44 R. Larsen, C. A. Knight, K. T. Rider and E. D. Sloan, *J. Cryst. Growth*, 1999, **204**, 376.

Crystallization of carbon tetrachloride in confined geometries

Adil Meziane,[a] Jean-Pierre E. Grolier,[b] Mohamed Baba[b] and Jean-Marie Nedelec*[c]

Received 7th November 2006, Accepted 7th February 2007
First published as an Advance Article on the web 11th April 2007
DOI: 10.1039/b616128f

The thermal behaviour of carbon tetrachloride confined in silica gels of different porosities was studied by differential scanning calorimetry. Both the melting point and the low temperature phase transition were measured and found to be inextricably dependant on the degree of confinement. The amount of solvent was varied through two sets of experiments, sequential addition and original progressive evaporation allowing the measurement of DSC signals for the various transitions as a function of the amount of CCl_4. These experiments allowed the determination of the transition enthalpies in the confined state, which in turn allowed the determination of the exact quantities of solvent undergoing these transitions. A clear correlation was found between the amounts of solvent (both free and confined) undergoing the two transitions, demonstrating that the formation of the adsorbed layer t does not interfere with the second transition. The thickness of this layer and the porous volumes of the two silica samples were measured and found to be in very close agreement with the values determined by gas sorption.

1. Introduction

The peculiar behaviour of liquids in confined geometries has attracted a lot of interest over the past ten years, and a comprehensive review was published in 2001.[1] The case of water[2,3] is particularly relevant with obvious practical applications and numerous works deal with this substance. The revival of interest in transitions in confined geometries undoubtedly comes from the considerable progress in the preparation of nanoporous materials with controlled pore sizes and with spatially controlled pore distribution and connectivity. In this context the discovery of MCM (Mobil Corporation Materials) type materials[4] has played an important role. The use of organized molecular systems (surfactants) for the spatial limitation of the condensation of alkoxide precursors is now common and has been extended to various systems and diverse pore organizations. The availability of such porous materials with controlled porosity, and to some extent with tuneable porosity, has led to an increased interest in the study of crystallisation in confined geometries. The

---

[a] Laboratoire de Photochimie Moléculaire et Macromoléculaire, UMR CNRS 6505, Université Blaise Pascal, 24 Avenue des Landais, 63177 Aubière Cedex, France
[b] Laboratoire de Thermodynamique des Solutions et des Polymères, UMR CNRS 6003, Université Blaise Pascal, 24 Avenue des Landais, 63177 Aubière Cedex, France
[c] TransChiMiC, Laboratoire des Matériaux Inorganiques, UMR CNRS 6002, Université Blaise Pascal, 24 Avenue des Landais, 63177, Aubière, France. E-mail: j-marie.nedelec@univ-bpclermont.fr; Fax: 00 33 (0)4 73 40 71 95; Tel: 00 33 (0)4 73 40 71 08

practical interest of liquids in porous materials is also very widespread with the case of oil recovery as a major example. The chemistry of water in clouds is also greatly affected by confinement effects.

More importantly, the research devoted to the preparation of nanocrystals with good control of both crystal size and size distribution has been expanding incredibly in the last twenty years.[5] In particular, semi-conducting nanocrystals or quantum dots have been the subject of many research papers[6,7] due to the possible observation in these materials of a direct quantum effect correlated to the size of the crystals.

Porous materials appeared to be ideal candidates for the preparation of such nanocrystals, utilizing the pores as nanoreactors where the crystallization of the desired material could be confined. In particular, many examples concerning MCM-41 and SBA-15 mesoporous silica templates can be found in the literature, for instance see ref. 8 and 9. Another very interesting example of crystallization in confined geometries is biomineralization.[10,11] Biomineralization is a complex process in which the solution conditions, organic template, and crystal confinement coordinate to yield nanostructured composite materials with controlled morphology and mechanical and structural properties. Over the past few decades, various aspects of this mineralization process have been examined both by characterizing those found in nature and creating synthetic composites.

Another field in which crystallization in confined geometries play a major role is polymer science. Numerous examples demonstrate how the confinement can modify the kinetics of crystallization of polymers and also the morphology of the crystals.[12]

All these selected examples demonstrate how crucial it is to get information concerning crystallization in confined geometries. In particular the energetics of crystallization in confining media is not well documented.

The well known modification of the freezing point of liquids in confined geometries has led to the development of techniques for the characterization of the porosity of solids. Such techniques are based upon the Gibbs–Thomson equation[13,14] which relates the shift ΔT of the temperature of crystallization to the pore size of the confining material according to:[15]

$$\Delta T = T_p - T_0 = \frac{2\sigma_{SL}\cos\theta T_0}{\Delta H_m \rho_s R_p} \approx \frac{k}{\Delta H_m R_p} \quad (1)$$

where T_p is the melting point of a liquid confined in a pore of radius R_p, T_0 is the normal melting point of the liquid, σ_{SL} is the surface energy of the solid/liquid interface, θ the contact angle, ΔH_m the melting enthalpy, ρ_s the density of the solid and k a constant.

The measurement of ΔT using calorimetry or NMR spectroscopy leads to thermoporosimetry[16] and NMR cryoporometry,[17] respectively. The advantages of both techniques have been discussed extensively.[18]

First proposed by Kuhn et al.[19] in the 1950s, thermoporosimetry can also be of great value for the characterization of soft networks like polymeric gels.[20] In this case, the confinement is created by the meshes defining the 3-dimensional polymer network. The study of the modification of the polymer architecture using thermoporosimetry requires knowledge of the behaviour of liquids capable of causing these organic materials to swell. We recently developed reference porous materials for the calibration of thermoporosimetry using various solvents.[21,22] In this systematic work, we observed that some solvents presenting a low temperature phase transition in the solid state offered additional interest.[23] Indeed, this transition is also affected by confinement and is an interesting alternative to the use of the liquid to solid transition since it is usually much more energetic. From a practical point of view the use of these transitions does not change the procedure requiring the calibration of the technique with samples of known porosity. But from a fundamental point of view, this observation raises some questions about the underlying thermodynamics. The objective of this paper is to discuss the transitions of carbon tetrachloride in confined geometries, firstly because CCl_4 is an effective solvent for polymer swelling and secondly because it has an observed solid state phase transition.[24,25]

2. Theoretical considerations

According to eqn (1), the shift of the transition temperature of a confined liquid ΔT is inversely proportional to the radius of the pore in which it is confined. In fact it is well known that not all of the solvent takes part in the transition and that a significant part of it remains adsorbed on the surface of the pore. The state of this adsorbed layer has been discussed extensively in the case of water. Consequently, the radius measured by the application of the Gibbs–Thomson equation should be written $R = R_p - t$ where t is the thickness of the adsorbed layer leading to a reformulation[7] of eqn (1) as

$$R_p = \frac{k}{\Delta H_m \Delta T} + t \quad (2)$$

The value of t can be determined by the calibration procedure using materials containing various pore sizes. This is the traditionally adopted procedure. The problem in doing so is that the underlying hypothesis is that the thickness of the adsorbed layer t does not vary with pore size. For small pores, the error in t can lead to large errors in the measurement of R_p.

We proposed an alternative method of measuring t by adding sequentially various amounts of liquid into the porous material.[26] As stated before, this layer t represents the part of the solvent which does not crystallize. For solvents like CCl_4 which exhibit a further transition at low temperature, the behaviour of this adsorbed layer is an open question. Does this solvent participate in the second transition? Is a new adsorbed layer created on the top of the first one? To obtain further insight into these questions we studied the behaviour of CCl_4 in mesoporous silica gels as described in the following section.

3. Experimental

3.1 Mesoporous silica gels

Mesoporous monolithic silica gels (2.5 mm × 5.6 mm diameter cylinders) were prepared by the acid catalysed hydrolysis and condensation of a silicon alkoxide, following procedures reviewed elsewhere.[27] Careful control of the aging time performed at 900 °C allowed the production of samples with controlled textural properties. In this study two samples (A and B) with different textural properties (specific surface area (SSA), total pore volume (V_p) and pore size distribution (PSD)) were used. The textural characteristics of the samples were determined by N_2 sorption.

3.2 Gas sorption measurements

Textural data for the silica gels were determined using a Quantachrome Autosorb 1 apparatus. The instrument permits a volumetric determination of the isotherms using a discontinuous static method at 77.4 K. The adsorptive gas was nitrogen with a purity of 99.999%. The cross sectional area of the adsorbate was taken to be 0.162 nm^2 for the SSA calculations. Prior to N_2 sorption, all samples were degassed at 100 °C for 12 h under reduced pressure. The masses of the degassed samples were used in order to estimate the SSA. The BET[28] SSA was determined by taking at least 4 points in the $0.05 < P/P_0 < 0.3$ relative pressure range. The pore volume was obtained from the amount of nitrogen adsorbed on the samples up to a partial pressure taken in the range $0.994 < P/P_0 < 0.999$. Pore size distributions were calculated from the

Table 1 Porous characteristics of the silica gel samples

Sample	SSA/m^2 g^{-1}	V_p/cm^3 g^{-1}	R_{av}/nm	R_p/nm
A	183	1.327	14.5	14.25
B	166	0.991	11.9	8.7

desorption isotherm by the BJH method.[29] The mean pore radius R_{av} was calculated according to

$$R_{av} = \frac{2V_p}{S_{BET}} \quad (3)$$

corresponding to a cylindrical shape for the pores. This is also the underlying hypothesis in eqn (1).

Textural data for the two samples are displayed in Table 1. In this table, the modal pore diameter R_p is also shown. This value fairly matches the R_{av} derived from S_{BET} measurements with the assumption of a cylindrical shape, thus confirming the validity of the hypothesis on the pore shape.

3.3 DSC measurements

A Mettler-Toledo DSC821 instrument calibrated (both for temperature and enthalpy) with metallic standards (In, Pb, Zn) and n-heptane was used to record the thermal curves. It was equipped with an intracooler set allowing a scanning range of temperatures between −70 and 600 °C. About 10 or 20 mg of the studied material was introduced into an aluminium DSC pan to undergo an appropriate temperature program. To allow the system to be in an equilibrium state, a slow freezing rate is required.[30] A rate of −0.7 °C min^{-1} was chosen. Other slower cooling rates were tested, but no significant discrepancy was detected. CCl$_4$ (Aldrich) of HPLC quality was used without any supplementary purification.

4. Results and discussion

4.1 Thermal behavior of free CCl$_4$

Bulk CCl$_4$ has been studied before and its thermal phase transitions are well characterized.[24,25] Its complex thermal transition system is shown in Fig. 1

As it is cooled down, liquid CCl$_4$ crystallizes first into a face centered cubic phase (FCC) which upon further cooling follows a phase transition into a rhombohedral phase (R) which, in turn, transforms into a monoclinic crystalline structure (M) around −48 °C. Heating the M phase leads reversibly to R; however, upon heating R melts directly without transforming into the FCC phase. Observation of the transition heats (Fig. 1), shows that the R-to-liquid transition releases an enthalpy (13.6 J g^{-1}) equivalent to the combined heats liberated by the liquid-to-FCC (9.6 J g^{-1}) and FCC-to-R (3.8 J g^{-1}) transitions. Takei et al.[24] showed that both solid-to-solid and liquid-to-solid transitions of CCl$_4$ were strongly dependent on the average pore size of the material in which the liquid is confined. In particular, they demonstrated that the FCC-to-R transition is no longer observed when the pore radius is smaller than 16.5 nm, as is the case for our silica samples (see Table 1).

Because of the complex behaviour of CCl$_4$ upon cooling, we chose to use the heating of the solvent and so limit the study to the M-to-R and R-to-liquid transitions.

These two transitions were studied for CCl$_4$ confined in the two porous samples A and B.

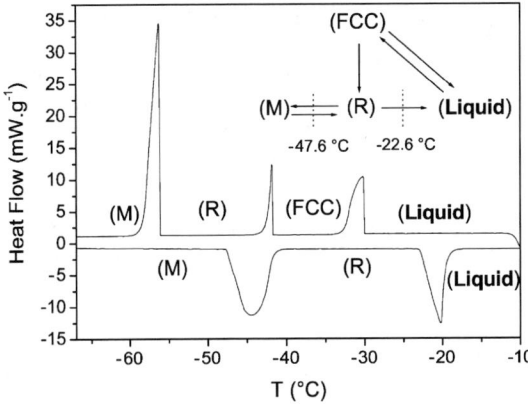

Fig. 1 DSC thermogram of pure CCl$_4$ showing the different transitions.

4.2 Thermal behaviour of CCl$_4$ confined in sample A

The objective is to obtain quantitative information on the solvent undergoing both transitions (both confined and free solvent). In order to do so, we performed sequential additions of precise quantities of CCl$_4$ in the sample as described in ref. 26. Briefly, a known mass of silica gel (about 20 mg) is set in the DSC pan which is sealed. A small hole is drilled in the cover allowing further injection of known masses of CCl$_4$. This procedure allows a precise control of the mass of solvent added. After each thermal cycle, a new injection is performed.

For the first time to our knowledge, we also performed some experiments in the reverse direction, by progressively evaporating the solvent starting with a large excess. This was performed by inert gas flushing in the DSC pan at 25 °C. The subsequent evaporation of the solvent is controlled by the flushing time. Obviously in this case we do not know the mass of CCl$_4$ remaining, but we can calculate it from the measured enthalpies.

The thermograms recorded for various quantities of CCl$_4$ added to sample A are shown in Fig. 2. In this figure, 4 peaks can be observed which are labelled from 1 to 4 going from low temperature to room temperature. The assignments of all peaks are presented in Table 2.

Fig. 3 presents the DSC curves recorded upon desorbing the CCl$_4$ by gas flushing. As can be seen in the figure, the control of the flushing time makes it

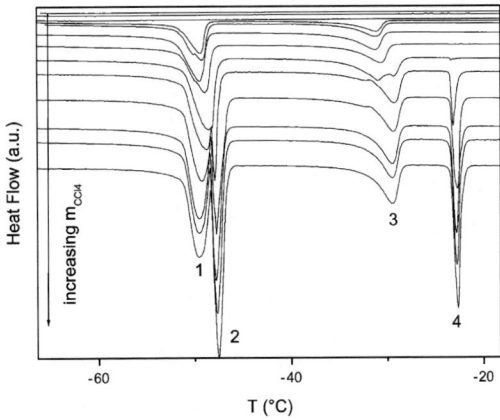

Fig. 2 Thermograms recorded for various amounts of CCl$_4$ added to sample A.

Table 2 Labelling of the different peaks observed in the DSC curves

Peak 1	Peak 2	Peak 3	Peak 4
M → R Confined solvent	M → R Free solvent	Melting of R Confined solvent	Melting of R Free solvent

possible to evaporate the liquid slowly and to discriminate all the steps. What is interesting in this experiment is that the whole set of experiments can be programmed automatically, thus giving a considerable amount of experimental data.

In Fig. 2, it can be seen that for small quantities of added CCl_4, no transition is observed. This first step corresponds to the creation of the adsorbed layer t onto the surface of the porous silica gel. For higher amounts of CCl_4, peak 3 appears at a temperature shifted with respect to the normal melting temperature of solid CCl_4. At about the same time, peak 1 also appears corresponding to the M-to-R transition for the confined solvent. The intensities of these two peaks increase upon further addition of solvent until they remain constant coinciding with the appearance of peaks 2 and 4 which correspond to excess free solvent. A plot of the heats corresponding to peaks 3 and 4 ($H3$ and $H4$) as a function of the mass of CCl_4 added (m_{CCl_4}) is presented in Fig. 4. The different steps are clearly observable. The point at which $H3$ is different from zero corresponds to the end of the creation of the adsorbed layer. This point allows the determination of the quantity of solvent (m_t) participating in the formation of this layer. At the point when $H3$ remains constant the pores are totally filled ($H3_{max}$) thus allowing the determination of the porous volume of the sample (see section 3.4).

To measure the amounts of CCl_4 involved in each transition precisely, we need to know the transition enthalpy at a given temperature. These values are known for free solvent transiting at regular temperatures but not for confined solvent which undergoes transitions at lower temperatures. From Fig. 4, we can measure $H3_{max}$ at the point at which all pores are filled, in the constant part of the curve. In this case the enthalpy corresponds to a mass of solvent equal to the total mass added (m_{vp}) minus the mass required for the creation of the adsorbed layer (m_t) namely $m = m_{vp} - m_t$. We can then deduce the enthalpy of melting per gram for the confined solvent $\Delta H3 = 13.67$ J g^{-1}.

For the M-to-R transition, the situation is different. Because of the overlapping of peaks 1 and 2 we can only use the sum $H1 + H2$. If we plot the evolution of $H3$ and $H4$ as a function of ($H1 + H2$) we obtain the curves presented in Fig. 5.

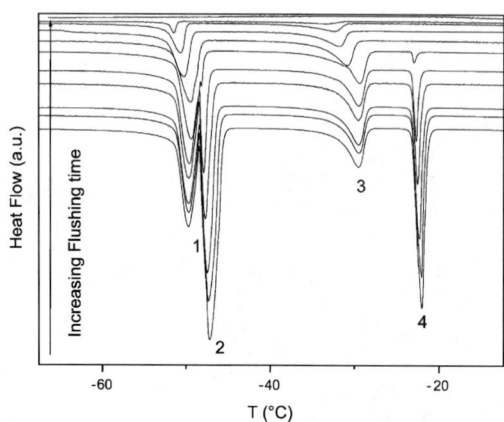

Fig. 3 Thermograms recorded for various flushing times for sample A filled with CCl_4.

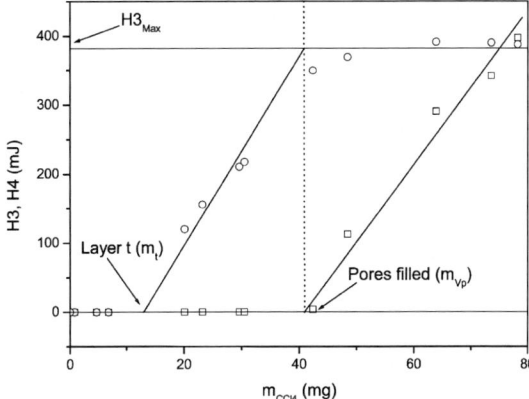

Fig. 4 Evolution of $H3$ (circles) and $H4$ (squares) as a function of the mass of CCl_4 added to sample A.

It is worth noting that the points corresponding to desorption experiments complete nicely the points corresponding to sequential addition of solvent (empty and full symbols, respectively).

The point where $H4$ differs from zero corresponds to the $H1_{max}$ value corresponding to the totality of solvent undergoing the transition (in this case $H2 = 0$). The enthalpy of transition $\Delta H1$ can then be deduced: $\Delta H1 = 27.22$ J g^{-1}. Knowing $\Delta H1$ and $\Delta H3$, we can now calculate the masses of solvent which undergo the various transitions for all points. Fig. 6 presents the correlation between these masses $M3$ and $M1$ (the indexes correspond to the different peaks).

A clear correlation is observed between the confined solvent which undergoes the R-to-liquid and the M-to-R transitions. This correlation is observed both for addition and evaporation experiments. This clearly confirms that all solvent undergoing the first transition also undergoes the second one.

This observation is further confirmed by the plot in Fig. 7 showing the correlation between $M2$ and $M4$, the masses of free solvent which undergo the transitions 2 and 4. $M2$ is determined through the following equation:

$$M2 = \frac{(H1 + H2) - H1_{max}}{\Delta H2} \quad (4)$$

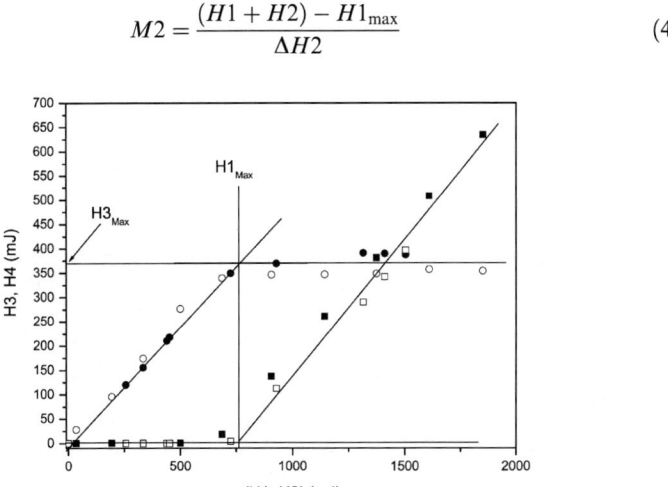

Fig. 5 Evolution of $H3$ (circles) and $H4$ (squares) for progressive filling (full symbols) and desorption (empty symbols) as a function of $(H1 + H2)$.

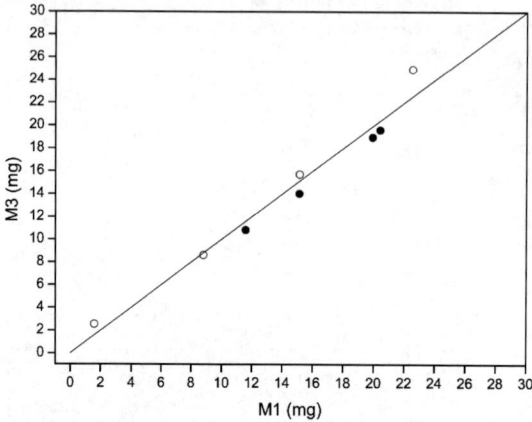

Fig. 6 Evolution of the mass of confined solvent undergoing transition M-to-R ($M1$) as a function of the mass of confined solvent undergoing the R-to-liquid transition ($M3$). Progressive filling (●) and desorption (○) experiments. The line $y = x$ is also plotted.

where $H1_{max}$ is the enthalpy required for the M-to-R transition of the liquid totally filling the pores (see Fig. 5) and $\Delta H2 = 46.6$ J g^{-1}, the specific enthalpy for the M-to-R transition of free CCl$_4$.

Once again a clear correlation is observed between the two quantities confirming that all the solvent which has crystallised outside the pores undergoes the second transition at a regular temperature (no confinement). These conclusions also demonstrate that the layer t remains adsorbed and does not participate in the low-temperature transition.

4.3 Thermal behaviour of CCl$_4$ confined in sample B

The same experiments and calculations were applied to sample B which presents smaller pores *i.e.* higher confinement.

The thermograms recorded for sample B filled with CCl$_4$ upon progressive evaporation are displayed in Fig. 8. Because of the higher degree of confinement, the two peaks 1 and 2 are well resolved and can be discriminated.

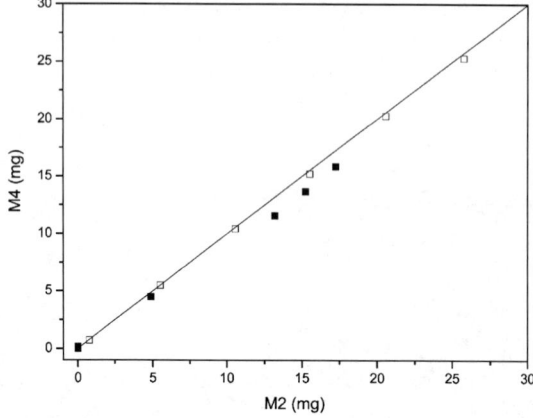

Fig. 7 Correlation between the masses of free solvent undergoing the R-to-liquid ($M2$) and the M-to-R transitions ($M4$). Progressive filling (■) and desorption (□) experiments. The $y = x$ line is also plotted.

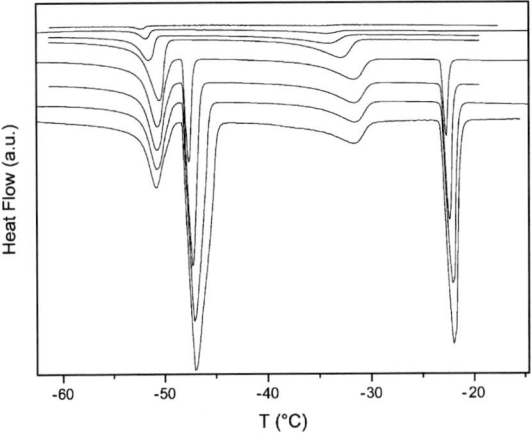

Fig. 8 Thermograms recorded for various flushing times for sample B filled with CCl$_4$.

Following the same procedure, we can plot the evolution of $H3$ and $H4$ as a function of m_{CCl_4} as performed in Fig. 9.

The plot of $H1$ and $H2$ as a function of m_{CCl_4} (not shown here) can also be performed.

From these curves, $\Delta H1$ and $\Delta H3$ can be derived for sample B ($\Delta H1 = 22.19$ J g^{-1} and $\Delta H3 = 10.13$ J g^{-1}). Together with the known values of $\Delta H2$ (46.6 J g^{-1}) and $\Delta H4$ (25.07 J g^{-1}) they allow the calculation of the quantities of solvent which undergo the different transitions for the various experiments.

The quantities $M1$, $M2$, $M3$ and $M4$ are plotted in Fig. 10 both for progressive filling and for evaporation experiments. A clear correlation between $M3$ and $M1$ on the one hand and between $M4$ and $M2$ on the other hand is observed, confirming the conclusions drawn from the study of sample A. Scheme 1 depicts the global behaviour of CCl$_4$ inside the porous silica recapitulating the different steps and the main conclusions of this work.

4.4 Calculation of porous volumes and thicknesses of adsorbed layers

Considering the curves in Fig. 4 and 9, we can measure the mass of solvent corresponding to total filling of the pores m_{vp}. The porous volume of the gel can

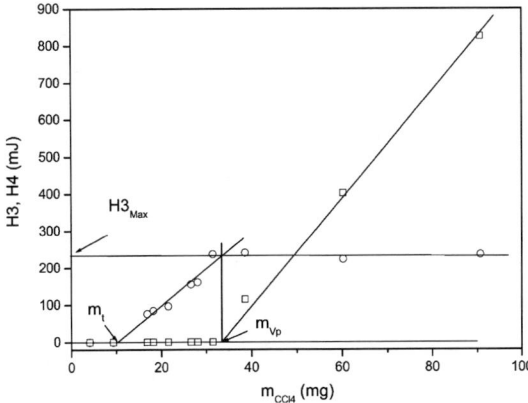

Fig. 9 Evolution of $H3$ (circles) and $H4$ (squares) as a function of the mass of CCl$_4$ added in sample B.

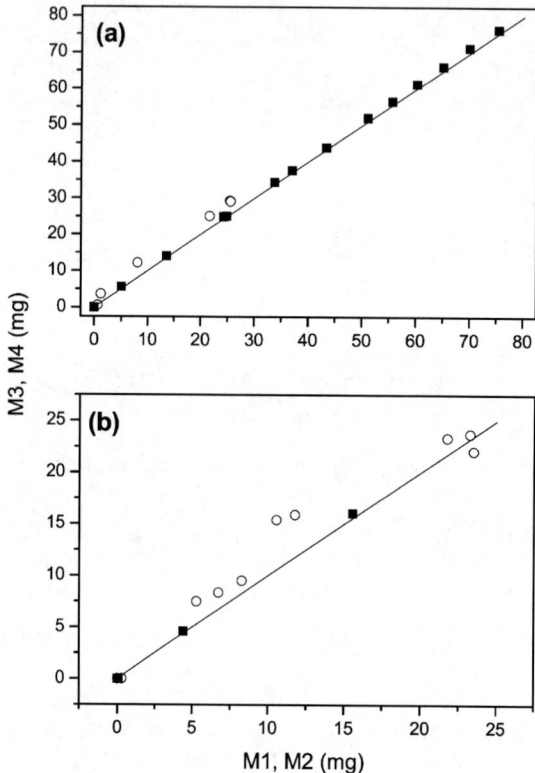

Fig. 10 Correlation between $M3$ and $M1$ (circles) and $M2$ and $M4$ (squares) for progressive filling of the pores (b) and for evaporation (a).

then be calculated according to:

$$V_p = \frac{m_{vp}}{\rho m_{SiO_2}} \quad (5)$$

where ρ is the density of CCl_4. We took the value at $-20\ °C$ ($\rho = 10.85$ kmol m^{-3}).[31]

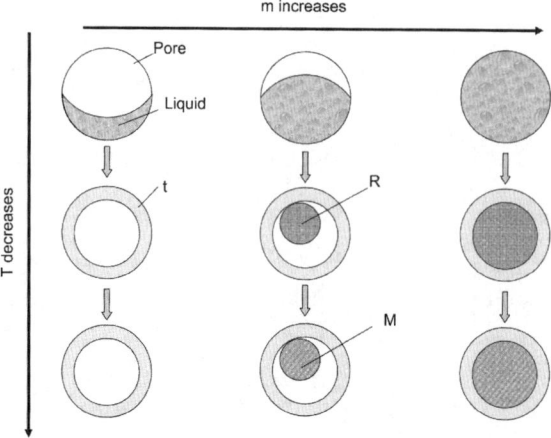

Scheme 1 Global behaviour of confined CCl_4 in porous materials, R is the rhombohedral phase, M the monoclinic phase, and t is the adsorbed layer.

Table 3 Porous volumes measured by DSC and gas sorption and thickness t of the adsorbed layer for the two silica gels A and B

Sample	V_p/cm^3 g^{-1}	V_{N2}/cm^3 g^{-1}	t/nm
A	1.35	1.327	2.3
B	0.99	0.991	1.9

From the same figures, we can also measure the mass of the adsorbed layer m_t, and the thickness of this layer can be calculated according to:

$$t = \frac{m_t}{\rho(SSA)(m_{SiO_2})} \qquad (6)$$

SSA being the specific surface area of the silica sample given in Table 1.

The results are summarized in Table 3.

The calculated values of V_p are in very good agreement with the values measured by nitrogen sorption, with an error less than 2%. The values of t determined for samples A and B are also in good agreement with average values given in ref. 25 after the calibration procedure with samples of various pore size.

All calculations were performed with a constant value of ρ_{CCl_4} measured at −20 °C. Obviously no information can be found in the literature for densities of CCl_4 at lower temperatures since it is usually solid at these temperatures. Nevertheless, using the value at −20 °C, the error must be small. Furthermore, with the validity of such an approach demonstrated, we can now consider the exact porous volume to calculate the exact density of the confined solvent at various low temperatures.

5. Conclusions

The thermal behavior of CCl_4 confined in two mesoporous silica gels of different porosities was studied. The two transitions (solid to liquid and monoclinic to rhombohedral) were measured and are found to be affected by the confinement. The enthalpies of these two transitions were determined for the first time at the temperatures corresponding to confined solvent. Using these enthalpies, a clear correlation between the solvent undergoing the first and the second transitions has been shown. Consequently, the adsorbed layer which is created during the intrusion of CCl_4 inside the pores of the silica gels is kept constant and does not participate in the two transitions. The thickness of this layer was measured for both samples and is found to depend slightly on the pore radius. Finally the porous volumes of the silica gels have been measured and the values agree very closely with those derived from nitrogen sorption isotherms. It has been demonstrated that using samples of known porosity (measured by mercury intrusion porosimetry or gas sorption analysis) could allow the measurement of thermodynamical data of confined liquids (enthalpy of transition, density, contact angle, *etc.*).

Acknowledgements

Financial support from the French ANR under project Nanothermomécanique (ACI Nanosciences No. 108) is gratefully acknowledged. The authors would like to thank A. Gordon and Prof. S. Turrell for careful reading of the paper.

References

1. H. K. Christenson, *J. Phys.: Condens. Matter*, 2001, **13**, 95.
2. J. Dore, *Chem. Phys.*, 2000, **258**, 327.
3. B. Webber and J. Dore, *J. Phys.: Condens. Matter*, 2004, **16**, 5449.

4 C. T. Kresge, M. E. Leonowicz, W. J. Roth, J. C. Vartuli and J. S. Beck, *Nature*, 1992, **359**, 710.
5 M. P. Pileni, *J. Phys. Chem. B*, 2001, **105**(17), 3358.
6 C. B. Murray, C. R. Kagan and M. G. Bawendi, *Annu. Rev. Mater. Sci.*, 2000, **30**, 545.
7 B. Capoen, T. Gacoin, J. M. Nedelec, S. Turrell and M. Bouazaoui, *J. Mater. Sci.*, 2001, **36**(10), 2565.
8 F. Gao, Q. Y. Lu, X. Y. Liu, Y. S. Yan and D. Y. Zhao, *Nano Lett.*, 2001, **1**(12), 743.
9 J. Kim, J. E. Lee, J. Lee, J. H. Yu, B. C. Kim, K. An, Y. Hwang, C. H. Shin, J. G. Park, J. Kim and T. Hyeon, *J. Am. Chem. Soc.*, 2006, **128**(3), 688.
10 S. Mann, *Nature*, 1993, **365**(6446), 499.
11 A. P. Alivisatos, *Science*, 2000, **289**(5480), 736.
12 X. N. Yang, A. Alexeev, M. A. J. Michels and J. Loos, *Macromolecules*, 2005, **38**(10), 4289.
13 J. Gibbs, *Collected Works*, Yale University Press, New Haven, CT, 1928.
14 S. W. Thomson, *Philos. Mag.*, 1871, **42**, 448.
15 C. L. Jackson and G. B. Mc Kenna, *J. Chem. Phys.*, 1990, **93**(12), 9002.
16 M. Brun, A. Lallemand, J.-F. Quinson and C. Eyraud, *Thermochim. Acta*, 1977, **21**, 59.
17 J. Strange, M. Rahman and E. Smith, *Phys. Rev. Lett.*, 1993, **71**(21), 3589.
18 J. M. Nedelec, J. P. E. Grolier and M. Baba, *J. Sol-Gel Sci. Technol.*, 2006, **40**, 191.
19 W. Kunh, E. Peterli and H. Majer, *J. Polym. Sci.*, 1955, **16**, 539.
20 M. Baba, J. M. Nedelec, J. Lacoste, J. L. Gardette and M. Morel, *Polym. Degrad. Stab.*, 2003, **80**(2), 305.
21 N. Billamboz, M. Baba, M. Grivet and J. M. Nedelec, *J. Phys. Chem. B*, 2004, **108**, 12032.
22 N. Bahloul, M. Baba and J. M. Nedelec, *J. Phys. Chem. B*, 2005, **109**, 16227.
23 M. Baba, J. M. Nedelec, J. Lacoste and J. L. Gardette, *J. Non-Cryst. Solids*, 2003, **315**, 228.
24 T. Takei, Y. Ooda, M. Fuji, T. Watanabe and M. Chikazawa, *Thermochim. Acta*, 2000, **352–353**, 199.
25 B. Husár, S. Commereuc, L. Lukáč, S. Chmela, J. M. Nedelec and M. Baba, *J. Phys. Chem. B*, 2006, **110**, 5315.
26 M. Baba, J. M. Nedelec and J. Lacoste, *J. Phys. Chem. B*, 2003, **107**, 12884.
27 L. L. Hench, *Sol-Gel Silica:Processing, Properties and Technology Transfer*, Noyes Publications, New York, 1998.
28 S. Brunauer, P. H. Emmet and E. Teller, *J. Am. Chem. Soc.*, 1938, **60**, 309.
29 E. P. Barret, L. G. Joiner and P. P. Halenda, *J. Am. Chem. Soc.*, 1951, **73**, 373.
30 R. Landry, *Thermochim. Acta*, 2005, **433**(1–2), 27.
31 Dipper, NIST Standard References Database v.9.0 (1985).

General Discussion

Dr Schön opened the discussion of Professor Parkin's paper: In order to try to simulate the process, it would be important to know where the reactions between WCl6 and water/ethanol take place. Do they occur in the gas phase, on the surface, or in both regions? What is the effect of the choice of glass on the process and the morphology, and how does the surface influence the formation of tungsten oxide in the first place?

Professor Parkin replied: Water–WCl6 has a significant gas phase component, with the film "snowing" onto the surface. The ethanol–WCl6 is largely surface reaction based—conventional CVD deposition. The choice of glass is important, however other work we have done on other glass surfaces and coated glass surfaces, alumina membranes and gold electrodes indicate that the same range of crystal morphologies seen for tungsten oxide can be formed on virtually any surface and that it is the reaction chemistry that is the most important determinant of crystal growth and structure.

Professor Roberts asked: It would be useful to probe the inter-relationship between the crystal chemistry and the bulk morphology. What surface chemistry is exposed on the (010) needle end facets and what role does the non-stoichiometry play ($W^{6+} \rightarrow W^{5+}$ reduction) in changing the surface chemistry? Is the reduced W^{5+} form exposed more on one surface than another and what role does the formation of oxygen vacancies play?

Professor Parkin answered: These are very interesting and pertinent comments. it is entirely possible that having W(v) on one face could promote the chemistry and morphologies seen. We have done a quite detailed XPs study, however not on single oriented crystals and hence cannot address that question. However, it would be an experiment we would be keen to try.

Professor Roberts said: X-Ray absorption spectroscopy at the W L-edge using synchrotron radiation might provide a way forward for such studies. This technique is particularly useful for probing oxidation state changes and the local changes in metal coordination associated with this.

Professor Catlow asked: How precisely are you able to control the stoichiometry of your films? This will clearly have a substantial influence on their properties.

Professor Parkin replied: The stoichiometry could be controlled quite well and fully stoichiometric WO_3 formed from reactions with water as the primary oxygen source and reduced tungsten oxide with ethanol. The use of mixed oxidants, ethanol and water, allowed some control of the value of x in WO_{3-x}.

Professor Sankar asked: What is the time scale of your reaction?
Is the glass plate is perfectly flat and does non homogeneity at micro scale influence the formation of the material?

Professor Parkin responded: The glass plate is virtually atomically flat. Standard deviations of roughness are typically less than a nanometer for float glass. The reaction timescale is very short, on the order of a fraction of a second. Precursor residence times are very short in the reactor, typically 0.1–0.2 s.

The introduction of a different surface or surface roughness would change the nature of the film produced. CVD normally gives conformal deposition. We use our reactor in a laminar flow regime, high surface roughness can introduce a turbulent gas flow and give rise to patterned coatings. We have not investigated these affects in this paper.

Dr Burley communicated: For the difference in crystal sizes, is the effect purely due to differences in partial pressures of oxygen, or is it a molecular-based effect?

Professor Parkin communicated in reply: The differences in crystal size are due to the relative rates of reactions of the precursors, partial oxygen pressure is important but it is not possible to quantify. Hence the molecular chemistry is fundamental. It is thought that the tungsten oxide formed by reaction with water is at least partially from reactions in the gas phase, whereas the tungsten oxide formed using ethanol is a more conventional CVD surface growth phenomena.

Professor Anwar opened the discussion of Dr Gich's paper by asking a general question: How can we determine surface free energies (or surface tensions) experimentally for nanoparticles? Could we for instance determine surface free energies for well-formed surfaces and then extrapolate to nanoparticles? If so, how?

Professor Jones replied: We are currently trying to establish experimental procedures to use AFM measurements to map differences in surface energy as one moves across the surface of organic crystals.[1]

1 C. Gardner and W. Jones, unpublished work.

Professor Roberts replied: I think it would be problematic to extrapolate *e.g.* simulations at large sizes into the nano size ranges as this would neglect the effects of surface relaxation which is more prominent for the smaller size ranges. A potentially better approach would be to use grid-based molecular modelling methods (see ref. 1) to model solution binding to surface relaxed nano-clusters. This way, perhaps, the inherent disorder of the clusters would be better taken into account.

1 R. B. Hammond, K. Pencheva, V. Ramachandran and K. J. Roberts, Application of grid-based molecular methods for modelling solvent-dependent crystal growth morphology: Aspirin crystallised from aqueous ethanolic solution, *Cryst. Growth Des.*, 2007, in press.

Professor Addadi addressed Dr Gich: You stressed in your interpretation mainly space confinement. As the particles are so small the interface should have overwhelming importance on the fate of the bulk phase. Are there interactions between the silica matrix and the Fe oxide phase? Interestingly, limpet teeth are composed of ferrihydrite crystals in a silica matrix. The mechanism of formation is completely different, but the interactions between the matrix and the crystals would greatly contribute to the mechanical properties of the material.

Dr Gich replied: The interaction between matrix and nanoparticles could be indeed the mechanism of the polymorph stabilisation. However, Corrias *et al.* studied the Fe_2O_3–SiO_2 interactions on iron oxide nanoparticles in silica xerogel matrices by X-ray absorption spectroscopy in order to explain the formation of either hematite or maghemite nanoparticles and they did not reach a conclusive answer.[1] Also, once the silica matrix is removed the ε-phase nanoparticles are still stable, pointing out that the role of interactions between matrix and nanoparticle surface do not seem to be important for the polymorph stability. Moreover, Mössbauer studies did not identify any fingerprint that could be assigned to Si–O–Fe species.

1   A. Corrias, G. Ennas, G. Mountjoy and G. Paschina, *Phys. Chem. Chem. Phys.*, 2002, **2**, 1045.

Dr Schön asked: Concerning the issue of how a matrix can affect the nucleation and growth of different phases, I would like to mention work using the low-temperature atom beam deposition method,[1] where at liquid nitrogen temperature a compound is deposited from the gas phase as an amorphous matrix with relatively low density. Upon heating the system slowly, nucleation and growth of crystalline phases takes place. Due to the fact that the density of the nuclei is higher than that of the matrix, and the surface of the nuclei is connected to the matrix, there exists an effective negative pressure, and the phase with the lowest density tends to grow first. Could such an effect be involved in your system, *i.e.* is there a sufficiently high difference in density between the α-, γ- and ε-modifications of Fe_2O_3?

1   D. Fischer and M. Jansen, *Angew. Chem., Int. Ed.*, 2002, **41**, 3746; D. Fischer, Z. Cancarevic, J. C. Schön and M. Jansen, *Z. Anorg. Allg. Chem.*, 2004, **630**, 156.

Dr Gich replied: This is a very interesting mechanism which could be of use to understand the stabilisation of polymorphs since the densities of maghemite, ε-Fe_2O_3 and hematite are 4.90, 5.0 and 5.23 g cm^{-3}, respectively. In fact, both density and surface tension or interfacial energy depend on the bond strength. At least in the case of α-Fe_2O_3 and γ-Fe_2O_3 this positive correlation seems to hold since the former has a larger surface tension: the surface modifications of α and γ nanoparticles have been investigated by XANES,[1] revealing that the local symmetry of the Fe^{3+} ions is lower in the surface than in the bulk for hematite whereas no substantial differences were found for maghemite.

1   X. Chen, T. Liu, M. C. Thurnauer, R. Csencsits and T. Rajh, *J. Phys. Chem. B*, 2002, **106**, 8539–8546.

Professor Catlow said: I think that it is entirely plausible that the surface energy of the γ phase is lower than that of the α or ε phases, as the γ structure is defective and defects tend to stabilise surfaces.

It should be possible to use modelling techniques to calculate and rationalise the differences between the surface energies of the three phases, which will be related to their crystal structures.

Dr Gich replied: It makes sense, simulation studies may be of great help to fully understand the problem.

Professor Breu asked: Did you control the oxygen fugacity, which certainly is a crucial factor when annealing transition metal oxides? Are you sure that your product is valence pure Fe(III)? Did you record Mössbauer spectra and have you checked for traces of Fe(II)?

Dr Gich answered: We did not control the oxygen fugacity but a significant presence of Fe^{2+} can be ruled out thanks to the Mössbauer spectroscopy and X-ray and neutron diffraction experiments performed in a study of a low temperature phase transition in ε-Fe_2O_3.[1] Indeed, the fitting of the Mössbauer spectra and the Rietveld refinement of the X-ray diffractograms are consistent with a ferric oxide. However, as the crystallisation of SiO_2 occurs at about 1200 °C, ε-Fe_2O_3 is in turn transformed to the stable high temperature iron oxides: hematite and the Fe^{2+}-containing magnetite (see Fig. 6 of our paper).

1   M. Gich, C. Frontera, A. Roig, E. Taboada, E. Molins, H. R. Rechenberg, J. D. Ardisson, W. A. A. Macedo, C. Ritter, V. Hardy, J. Sort, V. Skumryev and J. Nogués, *Chem. Mater.*, 2006, **18**, 3889–3897.

Dr Hughes asked: The paper states that conversion from $\varepsilon\text{-}Fe_2O_3$ to $\alpha\text{-}Fe_2O_3$ occurs as the SiO_2 matrix crystallizes at 1200 °C. Is this because the particles increase in size during quartz formation and, if so, is there direct evidence for this size increase?

Dr Gich answered: We believe that the increased density of the silica matrix after transforming from an amorphous to a crystalline structure gives free volume around the iron oxide nanoparticles which together with the high temperature enabling diffusion would allow particle growth. However, this has not been experimentally confirmed and we should definitely try to do this.

Professor Harding commented: One needs to remember that the relevant energies are interfacial energies and so, particularly in the vitrified core, there may be strong constraints on what nuclei can form due to the necessity of epitaxial accommodation.

Professor Catlow commented: Obviously the interfacial energy is the key quantity, although it is likely that the differences in the interfacial energies are dominated by the differences in surface energies.

Professor Anderson asked: Growing particles in confined matrices is something that has been done for decades. You alluded to zeolites but also, now, there is a problem of extremely well-defined mesoporous materials whereby both cage size and surface functionality may be controlled. The sizes are very similar to the sizes of the iron oxide particles you are making. Would using such matrices give you a greater degree of control over your particle formation?

Dr Gich replied: Yes indeed, and in fact the point I wanted to stress with this work on Fe_2O_3 polymorphs is the potential of crystal growth in spatially restricted fields as a route to discover new polymorphs or to stabilize metastable ones.

Professor Roberts said: This is very important work as *via* confinement we do not allow domain size to get large hence 'freezing' in metastable nanocrystalline phases. Arguably these such phases are generic to many solids but never seen as *e.g.* X-ray techniques do not permit structural studies in the nano-size range. Electron diffraction studies reveal that nano particles of platinum-black prepared at 10 nm size from the pyrolytic decomposition of chloroplatinic acid has a perfect fcc structure. Thus, by the time the crystals have grown into the micro-sized range, surface stabilisation effects become less important and the material transforms into a more 'stable' phase (see, *e.g.*, ref. 1). There is, I believe, huge potential to develop this work *via* designing both the different nano confinement size scales/shapes and the nature of the templating structure of the nucleating templating interface.

The links to the paper by Hammond *et al.*[2] with respect to modelling nano-cluster crystallography and the paper by Watanabe *et al.*[3] on magnetic levitation should be noted. In the latter technique the templating environment is removed and the purely homogeneous case can be explored. Potentially both techniques can be used to help differentiate mechanistic components and this facilitates a better understanding.

1   R. B. Hammond, K. Pencheva and K. J. Roberts, *J. Phys. Chem. B*, 2005, **109**, 19550–19552.
2   R. B. Hammond, K. Pencheva and K. J. Roberts, *Faraday Discuss.*, 2007, **136**, DOI: 10.1039/b616757b.
3   M. Watanabe, M. Adachi, T. Morishita, K. Higuchi, H. Kobatake and H. Fukuyama, *Faraday Discuss.*, 2007, **136**, DOI: 10.1039/b616394g.

Dr Murray asked: Heat will be released during crystallisation. This heat will be dissipated to the environment more efficiently in smaller particles, hence will heat up less during crystallisation. The more the droplet heats up (bigger droplets will heat up more) the more likely metastable phases will be to transform to stable phases. Do you think that this is potentially an important process in your system?

Dr Gich answered: This mechanism should be taken into account for systems in which slight differences in temperature are enough to stabilise a given polymorph. However, in the case of these Fe oxides the differences in temperature to transform from one phase to another are of the order of hundreds of Kelvin.

Professor Rodger commented: There has been considerable discussion of the importance of surface *vs.* interfacial energy in determining either the kinetic or the thermodynamic control of the phase of the nanoparticles. In fact, what should matter is the interfacial free energy, and free energies are ensemble properties. For liquid droplets, even at the nanodrop level, one would imagine that each droplet samples the ensemble of interfacial configurations over time, and thus one would have a well defined interfacial free energy for each droplet. For solid particles, however, each particle is likely to sample a different state from the ensemble. Thus, if interfacial free energies are the constraining force in allowing the ε phase to form, then for sufficiently small pore sizes one might expect different particles to form different phases. In any case, the issue of whether it is merely size or size and interface that enables the ε phase to form can be resolved easily by forming the nanoparticles in xerogels other than pure SiO_2.

Dr Gich responded: Interfacial energy is important but there might be systems in which you can neglect it and just think of the surface energy vacuum if the matrix–particle surface interaction is much smaller than the interaction between the particle surface and particle bulk atoms.

Professor Roberts said: The critical issue here is that surface energy can be used, rather than an interfacial tension when there is no interfacial wetting effect (*i.e.* for a contact angle of 180°). However, when there is a degree of surface wetting it is necessary to use an interfacial tension (which can be estimated from modelling, see *e.g.* ref. 1) to model the process fundamentals. The key issue to be addressed is to understand the structural role played by the crystal surface, *i.e.* is it simply a templating matrix or does it wet and hence interact chemically with the crystallising medium?

1   R. B. Hammond, K. Pencheva and K. J. Roberts, *Cryst. Growth Des.*, 2006, **6**, 1324–1334.

Professor Rodger asked: If, as suggested in the paper, the stabilisation of the ε-phase is purely due to the size constraints on the nanoparticles, then the nanoparticles should be independent of the material used to create the matrix. Have you tried forming this phase in any other matrix?

Dr Gich answered: No, this has not been tried in another matrix since this is not a straightforward process, the precursors for both the matrix and the ε-Fe_2O_3 nanoparticles being mixed together in a sol–gel synthesis. Perhaps an similar synthesis could be undertaken substituting the Si-alkoxide by another metal alkoxide (*i.e.* Ti).

Dr Schön addressed Dr Gich and Professor Rodger: Following up on some earlier comments, clearly there is a straightforward definition of the interface energy between two phases (where the surface energy constitutes the special case of one of the two components being the vacuum; if we want to define free energies of the interface at non-zero temperatures, one would have to replace the vacuum by a well-defined atmosphere, of course). In particular, it should be possible, in principle, to measure the value of such an interface energy for large quasi-bulk systems by using calorimetric measurements. Could this not give us some insight into the issue of how the differences in interface energies control the formation of the various phases in the experiment under "cluster in vacuum", "inclusion in amorphous SiO_2 matrix" and "inclusion inside crystallizing SiO_2 matrix" conditions? Do you think you could apply such measurements to the finite-size surfaces/interfaces such as the ones seen in your experiment, *e.g. via* "averages" over many inclusions?

Dr Gich responded: I think that the main problem in what you propose is that the calorimetric measurements you mention may not be straightforward due to the presence of the amorphous silica matrix hindering the heat conduction once it is released from the nanoparticle.

Professor Harding opened the discussion of Dr Chen's paper, presented by Ms Martinod: Is it possible that an amorphous phase of $CaCO_3$ is involved (which might be harder to detect) which could help explain the appearance and disappearance of the peaks?

Professor Neville replied on behalf of Dr Chen: It is possible that an amorphous phase of $CaCO_3$ forms and, in fact, there has been recent discussions in the scaling community about the relevance of amorphous phases of $CaCO_3$. However, it is felt that in this case the peaks, since they correspond very well to the 2θ values listed for the calcite, vaterite and aragonite polymorphs, that these are crystalline phases which are present at one particular time frame.

Dr Hare communicated: To follow up Professor Harding's query, could the authors find a way of capturing the "flushings-out" (downstream, once they are "out"), and examining these? Might we assume they would have remained unchanged structurally by the flushing?

Professor Neville communicated in reply: Yes, this is relatively easy and would lead to us capturing a lot of useful information. It is likely that not only will these have grown in size, they could have changed to a different polymorph. We do have ways of capturing the waste at a given time and quenching in a solution of high concentration of inhibitor which then maintains the structure and so we could do this to ensure that no further changes occur by remaining in the waste for a prolonged period.

Professor Vlieg asked: Why would you not be able to do a quantitative WAXS analysis? In principle, the technique is very quantitative. If you have problems with preferred orientation, you could use a 2D detector.

Professor Neville replied: We are looking into this.

Dr Schön asked: How big are the nuclei at various times along the growth process?

Professor Neville replied: Visible nuclei on the surface can be sub 50 nm as determined by *in situ* atomic force microscopy.

Professor Hodnett asked: Can you indicate the linear flow rate in your capillary reactor, its Reynolds number and the supersaturation? Do you have any information on the nature and strength of the adhesive forces between the calcite scale and capillary surface?

Professor Neville replied: The flow rate is 20 ml min^{-1} in the capillary which has a diameter of 1 mm. SI ($= \log S_a$) $= 1.25$, Re ≈ 450.

Dr Liu asked: When repeating the experiments, is it always the same CaCO$_3$ diffraction peaks that appear and disappear in the unstable phase? The peak disappearance is attributed to crystals being flushed away, as stated in the paper; these crystals should still exist in the waste, can they be detected in the waste?

Professor Neville responded: The answer is that the stable phases are always the same but the unstable peaks are taken at a snapshot in time. To be honest because of the constraints in access to the synchrotron beamline it has been impossible to repeat all experiments but we have confirmed for a number that the stable phase is repeatable and the length of the unstable phase is repeatable. The crystals can clearly been seen in the waste but until now we have not analysed these—this is something we are thinking of in future work. To analyse the waste would also give us access to the residual inhibitor amount.

Professor Heyes asked: Have your work and results had any consequences for the design of inhibitor strategies, or new inhibitor formulations? My question is about the possible applications of your work in the area of scale inhibition.

Professor Neville replied: To date, scale inhibitors have been found mainly on an *ad hoc* basis. It is only in recent years that there has been mechanistic information available in the literature to explain how inhibitors work. The scaling community are mainly dissociated from the industrial crystallization community and this means that they are losing out in having some of the real experts in crystal growth to communicate with. There is a shift in the thinking of some of the scaling community to focus on understanding how inhibitors work and this is primarily driven by the need to shift away from current phosphonate chemicals to more green chemicals. To "design" the scaling inhibitors of the future it is important to understand how the current good and poor inhibitors work.

Dr ter Horst opened the discussion Professor Rodger's paper: Did you simulate homogenous nucleation? Isn't the decaying cluster in simulation bad luck? Due to large concentrations of additive in the simulation box, the observed effect could be of thermodynamic origin. The additive changes the phase diagram and supersaturation. An inhibition of nucleation seems difficult to explain with classical nucleation theory: unhindered nucleation can take place at any position where the additive molecule is not present.

Professor Rodger replied: The nucleation simulated in this system is not truly homogeneous, since it occurs near to a methane/water interface, but certainly no seed particles were introduced into the simulation. As detailed in the paper, considerable care was taken to ensure that destabilisation of the crystal nucleus was not a direct result of inserting the inhibitor. Numerous repeat simulations were also performed, differing in the specific initial position of all the methane and inhibitor atoms and in the velocities of all atoms, to ensure that the results were not simply an accident of the chaotic nature of crystal nucleation. It is particularly noteworthy that crystal growth resumed when the inhibitor was removed, thus we conclude that the destabilisation of the hydrate is causally linked to the effect of the inhibitor. Simulations with PDMAEMA using the same protocol[1] showed that the PDMAEMA actually promoted growth of the hydrate nanoparticles at a statistical confidence level in excess of 95%, so disruption of the crystal is not an automatic consequence of the insertion protocol.

The effect on the phase diagram is not possible to determine from these calculations, but we note that we are suggesting that the PVP actually works by adjusting the thermodynamics of the interfacial regions. The formation of methane hydrate is, predominantly, an interfacial reaction and so is governed by the free energetics of the hydrocarbon/water interface. The concentrations used in this work were similar to those used in experimental studies, and so there will not be vast regions of the interface where the additive is not present.

1 R. W. Hawtin and P. M. Rodger, *J. Mater. Chem.*, 2006, **16**, 1934–1942.

Professor Anderson asked: Can you say as a general rule that molecules that will bind to a crystal facet are likely to be nucleation enhancers and at the same time growth inhibitors?

Professor Rodger answered: It is certainly our experience, working with both hydrate and wax inhibitors, that molecules which bind specifically to a crystal facet also tend to act as nucleation sites. The effectiveness with which they do this does depend on factors such as molecular weight for polymeric inhibitors. Specifically, we have seen this for PDMAEMA[1] which docks to the hydrate nanoparticles and enhances the *initial* growth phase, while PVP (the subject of this paper) does not dock to the surface of the hydrate particles and was observed to destabilise the hydrate.

1 R. W. Hawtin and P. M. Rodger, *J. Mater. Chem.*, 2006, **16**, 1934–1942.

Professor Davey said: Firstly, a comment on growth inhibition versus nucleation catalysis. It has been well demonstrated, particularly by the work of Leiserowitz, Weissbuch and Lahav *et al.*[1] that an inhibitor becomes a catalyst when located in the 2D array of a Langmuir film.

PVP seems to have a non-specific mechanism, somehow disturbing the aqueous layers at the crystal surface. Is this why it is active in so many other systems? What about the activity of polyvinyl alcohol?

1 I. Weissbuch, R. Popvitz-Biro, L. Leiserowitz and M. Lahav, in *The Lock and Key Principle*, ed. J. P. Behr, John Wiley and Sons, Chichester, 1994, ch. 6, pp. 171–246.

Professor Rodger replied: The vast majority of growth and nucleation inhibitors that have been investigated for controlling hydrate formation are synthetic polymers, and as such most show, in themselves, little evidence of long-range spatial ordering. As such, differences are to be expected between the hydrate LDHIs and a 2D array on a Langmuir film. Nevertheless, there are reports of polymeric inhibitors acting as crystallisation seeds in similar systems (see, for example, ref. 1 for wax additives). The activity we find for PVP is non-specific, and so as you suggest, should therefore be general. Indeed, PVP has long been used as a polymer that would alter water activity.[2] Although this generality is not necessarily advantageous: PVP is considered to be one of the weaker LDHIs.

PVA has certainly been reported to be a growth inhibitor for ice.[3] I understand there have been some studies that have examined PVA as part of an inhibitor blend, but am not aware of any published studies.

1 R. Kern and R. Dassonville, *J. Cryst. Growth*, 1992, **116**, 1.
2 See, for example, P. Molyneux, in *Water: A Comprehensive Treatise*, ed. F. Franks, Plenum Press, New York, 1975, vol. 4, chap. 7, pp. 617–801 or H. H. G. Jellinek and S. Y. Fok, *Kolloid Z. Z. Polym.*, 1967, **220**, 122.
3 T. Inada and P. R. Modak, *Chem. Eng. Sci.*, 2006, **61**, 3149–3158.

Professor Mazzotti asked: With reference to the Introduction:

(i) What is the concern for global warming due to methane clathrate (CH_4 release, etc.)?

(ii) What is the storage potential (and issues) of CO_2 hydrates (in relation to the corresponding phase diagram)?

Professor Rodger responded: Most of the concerns for global warming arise from the combination of two facts: (i) the amount of methane locked in natural methane hydrate deposits is huge, with reports suggesting that as much as half the organic carbon in the world exists as methane in methane hydrate; and (ii) methane is more than 20 times more potent than CO_2 as a greenhouse gas.

Increases in global temperatures would destabilise the existing deposits, and so there is a risk of a substantial positive feedback mechanism. There are numerous reports in the geological literature linking massive methane hydrate releases with periods of greenhouse-effect warming.[1] Other environmental hazards have also been proposed from the geological records. For example, one mechanism for the mass extinction at the Permian/Triassic boundary[2] was that global warming triggered the release of methane from sub-sea methane hydrate deposits; this methane was subsequently oxidised to CO_2 and resulted in "an oceanic acid bath" as well as reduced oxygen levels in the ocean.

At sea-floor temperatures, CO_2 hydrate will form at considerably lower pressures than CH_4 hydrate, and so the existence of vast methane hydrate deposits that have been stable on geological timescales raises the possibility that geological CO_2 hydrate storage could be a feasible, long-term sequestration strategy. One of the major issues relates to the phase diagram of carbon dioxide. In particular, the critical temperature is about 284 K, and above this temperature very much larger pressures are needed to stabilise the hydrate. Since geological storage is likely to be significantly below the sea floor (and hence warmer than 275 K), thermal stability of the CO_2 hydrates may be a difficult issue. Depressurisation associated with tectonic activity is also an issue.[3]

1 See, for example, A. K. Tripati and H. Elderfield, *Geochem. Geophys. Geosyst.*, 2004, **5**, Q02006.
2 E. Heydari and J. Hassanzadeh, *Sediment. Geol.*, 2003, **163**, 147–163.
3 M. Maslin, M. Owen, S. Day and D. Long, *Geology*, 2004, **32**, 53–56.

Professor Hyne said: Important as crystalline clathrate hydrates of methane may be in future supplies of natural gas from deep sea locations, it has already been shown (see *e.g.* ref. 1) that the presence of hydrogen sulfide in admixture with the methane (sour gas) can significantly affect the stability of the hydrate structure. Do we know how the hydrogen sulfide might be involved at the hydrate crystal structure level? Do we have a "thiohydrate" type substitution in the lattice structure?

1 V. A. Khoroshilov and E. B. Buchielmer, *J. Phys. Chem. USSR*, 1973, **47**(9), 2393 (in Russian).

Professor Rodger replied: The conventional explanation of how help-gases, such as H_2S, work does not require subsitution of the help-gas into the water lattice. This is usually explained in terms of the van der Waals–Platteeuw (vdWP) theory, which assumes that the hydrate becomes the thermodynamically stable state for water only under conditions in which configurational entropy from arranging guests (nonstoichiometrically) amongst the cavities outweighs the free energy penalty of forming the hydrate water lattice in the first place. This effect is very subtle, with estimates of the free energy difference between ice and an empty hydrate lattice being only about 1 kJ mol^{-1}. Within this theory, the guests do not directly perturb the water lattice, and guest–host interactions are important only in determining the extent to which the cavities are filled (and hence in determining the additional configurational entropy).

Within this framework, the enhanced attraction of the help-gas for water (compared with guests such as methane) leads it to fill many of the cavities even at very low partial pressures of the help-gas; the consequence is that less uptake of the methane (or other guest) is required and so formation occurs at less extreme thermodynamic states. Thermodynamic prediction programs that incorporate the vdWP theory are able to predict help-gas effects quantitatively, and so suggest that H_2S is primarily incorporated into the hydrate cavities.

I am unaware of any evidence that H_2S substitutes for water in the host lattice. Indeed, crystal structure determinations involving H_2S do not report any such incorporation[1] and recent calculations on gas phase water clusters consider only geometries in which the H_2S sits in the centre of an (H_2O), sub > 20 cage.[2]

1 K. A. Udachin, C. I. Ratcliffe, G. D. Enright and J. A. Ripmeester, *Supramol. Chem.*, 1997, **8**, 173–176.
2 V. N. McCarthy and K. D. Jordan, *Chem. Phys. Lett.*, 2006, **429**, 166–168.

Professor Vlieg asked: The nucleation inhibition mechanism you demonstrate is very interesting and very different from the standard picture that the additive needs to bond to the crystal. Did you simulate whether PVP bonds to your crystal surfaces?

Professor Rodger answered: We have studied the interaction between various oligomers of PVP and idealised, rigid, hydrate surfaces in the past[1] and found that there is some attraction between the pyrrolidone ring and partial hydrate cavities at the surface of the hydrate, but that this attraction is considerably weakened by competition between neighbouring monomer units in the oligomers. A more recent study[2] calculated the free energy of adsorption of vinyl pyrrolidone to a hydrate surface under water and found the free energy of this process to be maginally unfavourable ($+0.5$ kcal mol^{-1}).

1 T. J. Carver, M. G. B. Drew and P. M. Rodger, *J. Chem. Soc., Faraday Trans.*, 1995, **91**, 3449–3460; T. J. Carver, M. G. B. Drew and P. M. Rodger, *J. Chem. Soc., Faraday Trans.*, 1996, **92**, 5029–5033; T. J. Carver, M. G. B. Drew and P. M. Rodger, *Anal. N. Y. Acad. Sci.*, 2000, **912**, 658–668.
2 B. J. Anderson, J. W. Tester, G. P. Borghi and B. L. Trout, *J. Am. Chem. Soc.*, 2005, **127**, 17852–17862.

Professor Bensch asked: Do you expect changes/alteration of the results if additional ions which are present in sea water are added/considered in the calculations?

Professor Rodger replied: Most definitely, but this is largely a thermodynamic effect. The activity of water in a brine is lower than that of pure water, and so stronger thermodynamic driving forces are required to move the water from the brine to the hydrate. This effect is included in the common thermodynamic modelling packages for hydrates that are used in engineering applications (see, for example, the work of E. D. Sloan at the Colorado School of Mines, or B. Tohidi at Heriot-Watt University).

Professor Catlow asked: If PVP inhibits predominantly by water structuring, do all water structuring solutes have an inhibiting effect?

Professor Rodger responded: I think it would be more appropriate to suggest that all water structuring solutes will *alter* the kinetics of hydrate nucleation and growth, but whether it inhibits or promotes depends on *how* the solute structures the water. In the extreme, there are some solutes—such as the small cyclic ethers (THF, TMO and DMO)—which are soluble in water and which form single hydrates in their own right. One must conclude that such solutes (under appropriate conditions) actually promote hydrate-like structure in the water. Even if one focuses on hydrate growth inhibitors, it is still possible for the inhibitor to promote hydrate structure in the water, but not allow it to propagate. Our own simulations studies of PDMAEMA[1] suggest that it falls into this category. In the present paper we propose that PVP acts by frustrating the growth of hydrate water networks within the interfacial region. This is a new mechanism for kinetic inhibition of hydrates and I believe that it will pertain to additives other than PVP, but I do not believe it will prove to a universal mechanism.

1 R. W. Hawtin and P. M. Rodger, *J. Mater. Chem.*, 2006, **16**, 1934–1942.

Dr Ristic asked:
(1) Is this simulation experimentally trackable?
(2) Can one get a feeling for the timescale of the self-assembly process?

Professor Rodger replied:
(1) The data presented in this paper relate to nano-scale hydrate particles, and experimental studies on these lengthscales are difficult. Most lab-based experiments only register hydrates once they grow to *ca.* µm size. There is possibly some evidence on this lengthscale implicit in recent diffraction studies,[1] and our simulations are consistent with those studies. Experimental studies have also tracked the relative proportion of occupied large and small cages during nucleation, but data is again limited to much larger particles than studied in this work.

(2) In most experimental systems the induction time to nucleation takes hours, or even days, with clean systems and methane/water mixtures. Much faster times are seen in this study partly because of the supersaturation of methane (chemical potential difference between methane in aqueous and fluid phases was about 4–7 kcal mol^{-1} in our simulations) and because of the focus on a thin aqueous film. In many laboratory experiments, once nucleation occurs the subsequent growth can be extremely rapid. We reported nucleation rates for ethane hydrate (277 K, *ca.* 12 bar) of $\sim 10^3$ ml min^{-1} and growth rates of ~ 1 µm min^{-1},[2] but these were from granulometer studies that could not detect particles smaller than 2 µm diameter.

1 See, *e.g.*, P. Buchanan, A. K. Soper, H. Thompson, R. E. Westacott, J. L. Creek, G. Hobson and C. A. Koh, *J. Chem. Phys.*, 2005, 123 or W. F. Kuhs, D. K. Staykova and A. N. Salamatin, *J. Phys. Chem. B*, 2006, **110**, 13283–13295.
2 M. T. Storr, P. C. Taylor, J. P. Monfort and P. M. Rodger, *J. Am. Chem. Soc.*, 2004, **126**, 1569–1576.

Dr Wakisaka said: Some of the organic molecules works as hydrate promoters. The most typical one is THF. What kind of difference do you expect between the hydrate promoter and the kinetic inhibitor from the viewpoint of interaction with water?

Also, THF is miscible with water, but the H-bond interaction between THF and H$_2$O is very weak. Therefore, the H-bonding network of water will not be influenced by the addition of THF. This should be related to the promotion of hydrate.[1]

1 M. Ohtake, Y, Yamamoto, T. Kawamura, A. Wakisaka, W. F. de Souza and A. M. V. de Freitas, *J. Phys. Chem. B*, 2005, **109**, 16879.

Professor Rodger replied: One needs to be careful with the terminology. It would be more appropriate to call THF a hydrate former rather than a hydrate promoter. If one is contrasting inhibitors with promoters, the implication should be that the promoter acts more as a catalyst, whereas the hydrate former will be incorporated into the resulting hydrate. THF will form THF hydrate in its own right, and as such it will affect the thermodynamics of hydrate stability, not just the kinetics. Hydrate formation mechanisms are, potentially, very different for methane hydrate and THF hydrate precisely because THF is miscible with water while the CH_4 is not. This means that methane hydrate formation is predominantly an interfacial phenomenon whereas THF formation occurs in a bulk aqueous phase. With respect to the interaction between THF and water, recent *ab initio* calculations by Belosludov have indicated that there is considerable polarisation of the both the etheric O and water in the host lattice associated when THF is enclosed in a clathrate cage, and that this does lead to significant perturbations of the water network.

Professor Roberts asked: The potential for gas hydrate mining and transportation to market was alluded to. In the latter case the higher temperature for transport for the gas hydrate *versus* liquid petroleum gas (LPG) make this approach potentially attractive. Are there advances in fundamental gas-hydrate science still needed to exploit this energy source, *i.e.* thermal properties for product isolation, or is this simply a case of economic competitiveness of the process?

Professor Rodger answered: There are two aspects here, and both have technological bottlenecks associated with them. With respect to mining, most of known methane hydrate deposits are biogenic subsea deposits. This has the advantage that they are almost pure methane, but also that they tend to be finely dispersed, thus requiring substantial technological developments in mining methods before exploitation is commercially viable. There are, however, some deposits that are more concentrated, either because the gas source was thermogenic (and hence more localised) or due to annealing of the biogenic deposits over geological time periods, and these deposits are beginning to be developed. Methane from hydrate has been extracted from the Messoyakha field (Siberia) for many years, while commercial trials have also begun at the Mallik well (Mackenzie Delta, Canada).

The second aspect relates to transport. The technology to produce methane hydrate and transport it as a slurry does exist, but considerable efforts are being put into developing technologies that will allow methane hydrate to be transported under ambient conditions by exploiting the "anomalous preservation" effect.[1]

1 L. A. Stern, S. Circone, S. H. Kirby and W. B. Durham, *J. Phys. Chem. B*, 2001, **105**, 1756–1762.

Professor Mazzotti asked: With regard to the existence of natural methane clathrates and non-existence of natural carbon dioxide clathrates, is this due to the difference in solubility of the two species?

Professor Rodger answered: Both hydrates can be formed in laboratories, and so it is unlikely that the absence of natural CO_2 deposits is simply due to solubility. Indeed, experiments have been performed that form CO_2 hydrate on the sea floor. It is more likely that because CO_2 can react with water, diffusion pathways are available to CO_2 that are not present in CH4, and that this leads to instability of CO_2 hydrate on geological timescales.

Dr Slater asked: Following on from the discussion in an earlier session about the identity and distribution of solution species in the mother liquor from which microporous materials are formed, have you examined the distribution of rings within solution in the presence and absence of the hydrate inhibitors?

Professor Rodger responded: This analysis is currently in progress.

Professor Anderson opened the discussion of Dr Nedelec's paper: Could you please clarify whether the mesoporous silicas used in your work are ordered as in MCM type materials or disordered?

Dr Nedelec replied: The materials that have been used in our work are indeed completely disordered materials. Nevertheless, they present narrow pore size distribution which is obviously very important for the calibration procedure. Furthermore, it is possible to tune to some extent the mean pore size of the silica gels (2–20 nm) by simply varying the densification conditions. The surface chemistry is thus kept constant for all samples. Also, these gels are monolithic which make them very easy to handle compared to powders.

Dr De Leeuw asked: With regard to the "free" CCl_4 mentioned in the paper, is this bulk material within the pores or a phase surrounding the confined lattice?

Dr Nedelec replied: The "free" solvent is outside the pores. It is in fact excess solvent corresponding to overfilling of the porosity. This solvent is not confined and exhibits regular transition temperatures.

Professor Catlow asked: Could you give more details of the other systems that have been studied with your approach?

Dr Nedelec replied: Historically, thermoporosimetry was first proposed in the 1950s by Kuhn[1] and further popularized by Brun[2] in the late 1970s. At the beginning studies dealt essentially with water and benzene. More recently other solvents have been studied. Thanks to the use of our sol–gel derived reference materials we have been able to publish data for cyclohexane which also display a low temperature solid phase transition, various substituted benzenes, linear alkanes with different chain length, acetonitrile, dioxane and acetone.[3–11] The interest of these solvents lies in their ability to swell organic and hybrid gels.

1   W. Kuhn, E. Peterli and H. Majer, *J. Polym. Sci*, 1955, **16**, 539.
2   M. Brun, A. Lallemand, J.-F. Quinson and C. Eyraud, *Thermochim. Acta*, 1977, **21**, 59.
3   M. Baba, J. M. Nedelec, J. Lacoste and J. L. Gardette, *J. Non-Cryst. Solids*, 2003, **315**(3), 228–238.
4   M. Baba, J. M. Nedelec, J. Lacoste, J. L. Gardette and M. Morel, *Polym. Degrad. Stab.*, 2003, **80**(2), 305–313.
5   M. Baba, J. M. Nedelec and J. Lacoste, *J. Phys. Chem. B*, 2003, **107**, 12884–12890.
6   J. M. Nedelec and M. Baba, *J. Sol–Gel Sci. Technol.*, 2004, **31**, 169–173.
7   N. Billamboz, M. Baba, M. Grivet and J. M. Nedelec, *J. Phys. Chem. B*, 2004, **108**(32), 12032–12037.
8   N. Billamboz, J. M. Nedelec, M. Grivet and M. Baba, *ChemPhysChem*, 2005, **6**(6), 1126–1132.
9   N. Bahloul, M. Baba and J. M. Nedelec, *J. Phys. Chem. B*, 2005, **109**(34), 16227–16229.
10 B. Husar, S. Commereuc, I. Lukac, S. Shmela, J. M. Nedelec and M. Baba, *J. Phys. Chem. B*, 2006, **110**(11), 5315–5320.
11 J. M. Nedelec, J. P. E. Grolier and M. Baba, *J. Sol–Gel Sci. Technol.*, 2006, **40**, 191–200.

PAPER

Concluding Remarks
Crystal growth and nucleation: tracking precursors to polymorphs†

Patrick R. Unwin

Received 21st May 2007, Accepted 21st May 2007
First published as an Advance Article on the web 15th June 2007
DOI: 10.1039/b707653n

Standing on the shoulders of giants

One of the earliest and legendary Faraday Discussions took place in Bristol in April 1949 and considered the topic of crystal growth. In reviewing the present Discussion (136), it is illuminating to look back at Faraday Discussion 5, revisit the burning issues of the time, and see how the field has developed in the intervening half-century. Faraday Discussion 5 was chaired by the then President of the Faraday Society, Prof. Sir John Lennard-Jones, KBE, and it must have been a fascinating meeting with 300 participants, a sizeable number from industry as well as academia. The Discussion is famous for seminal contributions by: Frank on the importance of dislocations in crystal growth;[1] Burton and Cabrera on the effect of surface structure on crystal growth;[2,3] and treatments of the kinetics of crystallization from Dunning and co-workers.[4-6] A primary focus of the meeting was inorganic systems and sessions reflected the important topics of the day: "theory of crystal growth", "nucleation and normal growth", "abnormal and modified crystal growth" and, illustrating the commercial importance of crystallization, "mineral synthesis and technical aspects".

Turning to the concluding remarks from Faraday Discussion 5, which was given separately by Bunn[7] and Rowland,[8] the following comments are particularly revealing:

• " . . . we need a dynamic theory, which treats crystal growth as a dynamic event." (Bunn)
• " . . . the gap between theory and the experimental approaches has been too wide." (Rowland)
• "It is essential that we learn as much as possible about the surface of the substrate." (Rowland)
• "The subject is still at an alchemical stage." (Rowland)

In light of the above comments, it is opportune to ask how much has the field changed in the intervening period?

The plethora of high resolution spectroscopy and microscopy techniques that subsequently developed have provided a wealth of information on surface structure and composition, but, surprisingly, the paper of Vleig et al.[9] in the present Discussion suggested that, half a century on, there is still a "large gap between experiment and theory"! The present Discussion has served to illustrate that the gap is closing, although much remains to be done. While the field of crystal growth has largely moved beyond the alchemical stage, Discussion 136 has highlighted that there are emerging and fascinating processes, such as the formation of co-crystals from grinding, where the basic elementary processes are, as yet, elusive.[10]

Department of Chemistry, University of Warwick, Coventry, UK CV4 7AL

† The HTML version of this article has been enhanced with colour images.

Between Discussion 5 and Discussion 136, Faraday Discussion 95—held in Strathclyde in April 1993—considered crystal growth from the point of view of surface structure, kinetics and the role of defects. Chaired by Prof. John Sherwood, it attracted 99 participants and was a showcase for emerging in situ surface and structural techniques, such as atomic force microscopy (AFM),[11] interferometry[12] and surface X-ray diffraction.[13] The meeting was also a forum for the emergence of powerful simulation methods,[14,15] which have clearly developed strongly in the past decade or so. As in 1949, the meeting considered inorganic crystals, but also embraced semi-conductors, metals and polymers.

From surfaces to solutions

The implementation of instrumental techniques, apparent in Faraday Discussion 95, has been a very strong feature of Discussion 136, with an unprecedented variety in evidence and a trend towards multi-technique studies.[16–18] Strikingly, the techniques have addressed the solution as much as, if not more than, the surface, and the importance of solution entities and precursors in determining the outcome of crystallisation has been an important theme. This aspect has been seen in studies of molecular materials,[16,19] where spectroscopic techniques have afforded detailed information on molecular interactions in solution. However, it is also clear from fast-quench electron microscopy[18] and electrospray mass spectrometry[20] studies that there are significant clustering processes that occur initially in solution in inorganic systems that need to be elucidated.

As solution entities aggregate and grow, the size-dependence of phase stability is also an important consideration for the ultimate outcome of a crystallisation experiment. This issue was considered in Faraday Discussion 95,[21,22] and is pertinent to aspects of this Discussion which have considered influences on the stability of nanostructures.

The classical view of nucleation and growth in solution is embodied in the simple schematic in Fig. 1(a). This summarises the traditional view that crystals nucleate and grow from elementary species (ions, molecules) in a supersaturated solution, although phase transformations may also occur in the later stages. The evidence

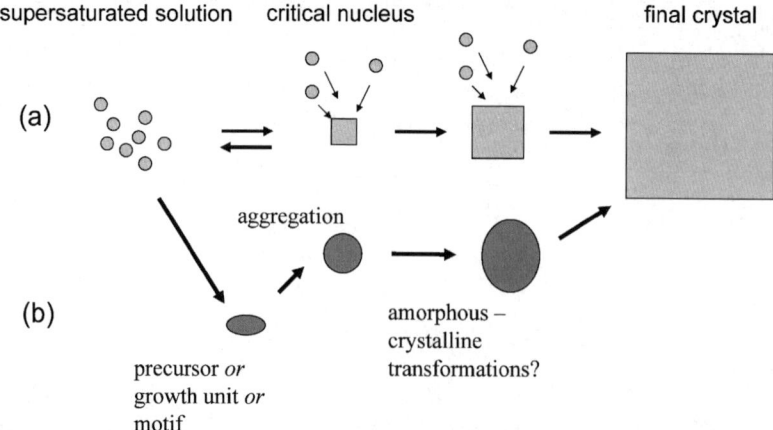

Fig. 1 (a) Simplified schematic of the nucleation and growth of a crystal, involving the formation of nuclei from elementary species in solution (ions, molecules). For nuclei which attain the critical size, growth continues (fed by the elementary species) until a final crystal is achieved. Phase transformations are not shown for simplicity. Part (b) summarises the emerging view of other key steps and entities that may be involved, such initial association in solution to form precursors, growth units or motifs, followed by aggregation, transformation, etc. Additionally (but not shown), some of these entities may be key feedstocks when the final crystal is growing.

from several papers presented in this Discussion, for both inorganic and organic systems, is that the situation in solution is considerably more complex. We have seen that the association of solution species to form "precursors", "growth units" or "motifs" is an important initial step. These entities may then aggregate, with the local loss of solvent. As the aggregates grow in size, there are various amorphous–crystalline transformations or phase transformations that can occur *en route* to a macro-crystal (Fig. 1(b)). Additionally, it is important to note that the solution assemblies and aggregates, which form in this way, may themselves be a direct feedstock for later macro-crystals, as evident in the biomineralisation of calcite and aragonite, where amorphous calcium carbonate appears to feed the growth of mineral phases.[23,24] The extent to which these different entities are important, and the particular pathways that operate, will clearly depend, among other factors, on the timescale of the growth process, its nature (*e.g.* batch *versus* flow), mass transport rates, the supersaturation level and its time-dependence (fixed *versus* free drift).

From precursors to polymorphs

This Discussion has provided evidence for a significant link between nanoscopic entities formed in solution and the resulting crystal phase. The important role of solution precursors in determining polymorph formation was particularly evident in two papers. First, Davey *et al.* clearly showed that the formation of α-inosine was linked to the presence of α-like dimers in solution.[19] Second, the simulations of Roberts *et al.*[25] revealed that clusters of β-benzophenone have more disorder than those of the α-form, providing an understanding as to why the formation of β-benzophenone requires significant undercooling.

The paper by Hughes *et al.*[16] demonstrated how precursors in molecular systems can be tracked all the way through to polymorphs *via* an elegant combination of complementary experimental techniques and simulation methods. In particular, NMR measurements of chemical shifts difference and diffusion coefficients, together with molecular dynamics simulations, revealed early aggregation events. Small angle neutron scattering (SANS) was proposed as a method for monitoring the growth of aggregates, while SANS and solid state NMR could be used to provide key information on the structural form of evolving microcrystals. Finally, the end product could be characterised comprehensively with X-ray diffraction (and other structural techniques). In this way, it is possible to rationalise competing pathways and events *en route* to crystallisation.

A different set of techniques was used powerfully by Rieger *et al.*[18] to obtain major new insights on $CaCO_3$ crystallisation at high supersaturation. This work highlighted the formation of an emulsion-like material at short (millisecond) times, followed by an evolution to nanoparticles, then vaterite and finally calcite. The work raises the question as to the suitability of existing models and theories for crystal nucleation and growth from solution. From an experimental viewpoint, it would be desirable to be able to follow the evolution of solution speciation (including nanostructures) from very short times onwards. There is evidently a challenge to physical-analytical scientists to develop methodologies which will improve our understanding of solution entities and their dynamics on a wide range of length- and time-scales.

Characteristics of clusters

The nature of nanoscale clusters, important in crystal growth, was the focus of several interesting papers. The observation of magic numbers in clusters of alkali metal halides,[20] identified using electrospray mass spectrometry, is particularly exciting from the point of view of identifying key nanostructures. Likewise, we have seen that *in situ* scattering techniques can offer information on the precursors involved in the formation of microporous materials.[26] The power of simulation techniques in identifying and characterising key clusters was also evident. Notably, the stabilisation of small water clusters by Cu was an aspect of the paper from Michalides,[27] while *ab initio* molecular dynamics simulations provided new insights into the interaction of water with calcite.[28] The detailed information available from simulation on the action of additives in controlling nuclei formation in clathrate hydrates attracted a great deal of debate and excitement.[29]

The Discussion has identified many fundamental questions that need to be addressed to develop further this area, such as: what is the relationship between clusters and "growth units" and how do they interconvert (dynamically)? What are the key properties of clusters, *e.g.* interfacial energies and embryonic structure and how do we characterise these? What is the role of clusters in the growth of larger crystals? As highlighted above, are they a feedstock? What is their distribution in the concentration boundary layer at larger crystals? How do clusters formed in solution interact with larger crystals? An underlying question to be addressed is what is meant by the labels "active" and "spectator" species, used in some areas, when the various species and entities involved in crystal growth will interconvert?

Turning to the surface

The complexity of growth interfaces was revealed by X-ray surface diffraction. The use of this technique by Vleig *et al.*[9] provided new insights on the nature of the interface between potassium dihydrogen phosphate and aqueous solution as a function of pH, although measurements were necessarily averaged over large areas. X-ray microdiffraction allows measurements to be made at higher spatial resolution. Cherezov and Caffrey used this technique to investigate whether the lamellar phase in membrane protein crystallisation in mesophases acts as a conduit for the transport of material.[30] Promising data were obtained, but the experiments were complicated by radiation damage. Clearly these techniques have much to offer in the future, although technical issues remain to be resolved.

In the field of nanoporous materials, we learned how modelling and high resolution structural measurements inform each other to advance understanding of surface processes. Notably, high resolution transmission electron microscopy and theoretical studies have given insights into the growth mechanisms of zeolites,[31] while Anderson *et al.* showed that AFM provided key information on growth processes which could be used in phenomenological modelling.[17]

As mentioned above, recognition of the importance of dislocations in crystal growth was an important breakthrough of Discussion 5 and a key aspect of Discussion 95. Although dislocations featured less prominently in this Discussion than one might have anticipated, the importance of screw dislocations in the growth of zeolites was elegantly demonstrated.[17] Moreover, Kahr *et al.* carried out a huge number of studies on potassium hydrogen phthalate to test the 'crystals-as-genes' hypothesis of Cairns-Smith.[32] This paper focused on growth at screw dislocations as a mechanism for the transfer of imperfections from one crystal to another. An outcome of this work was a plea by the author for further studies to aid understanding of the molecular mechanisms of dislocation nucleation.

Progress in AFM

The work of Anderson *et al.*,[17] in particular, highlighted the importance of AFM as a technique to study dynamics on crystal surfaces, but this technique was arguably under represented. A key challenge when using AFM is to control mass transport and the supersaturation level adjacent to the growing crystal. *In situ* AFM measurements have typically been carried out in a commercially-available fluid cell, but it is only comparatively recently that a full fluid dynamics model has become available to treat convection in this type of cell.[33] This should lead to the more quantitative application of AFM for kinetic measurements. A potential complication when using AFM on large crystals, however, is that only a small area is investigated and this will often be some way downstream in the flow stream across the crystal surface. Clearly, if there are changes in supersaturation along the crystal surface due to the growth process, it may be difficult to quantitatively interpret images. One solution to this problem is to work with microcrystals, where essentially all the surface of interest can be visualized.[34]

It is also important to highlight recent developments in AFM technology which allow true video rate imaging (15 frame/s or so). The use of a resonant scanner for one dimension and a passive mechanical feedback loop yields an increase in imaging rate of three orders of magnitude compared to conventional AFM.[35] The technique has already provided major insights into the kinetics and mechanism of polyethylene oxide crystallization from the melt, where spiral growth was visualized in real time.[36]

Growth at chemically-functionalised interfaces

In the Introductory Lecture, Addadi revealed how nature's templates promote crystal growth.[24] Progress in this area in the last few years has been spectacular and, in parallel, more sophisticated chemically-functionalised interfaces have been applied for crystal growth in the laboratory. This was particularly evident in one of the posters presented at this meeting.[37] The symposium on "Polymer Directed Mineralisation" at the recent 233rd ACS Annual Meeting also highlighted the fast-moving nature of this area, with significant contributions on *in vitro* biomineralisation,[38] the clear identification of amorphous calcium carbonate as an intermediate during the formation of calcite on self-assembled monolayers,[39] and the clever design of 3-D templates to promote the formation of amorphous calcium carbonate.[40] The use of immiscible liquid/liquid (oil/water) interfaces for crystallization is a further promising area, well-illustrated in Faraday Discussion 129.[41] In the present Discussion we learned of developments in the characterisation of glasses used for biomedical and related applications.[42,43]

A role for electrochemistry?

There was little on electrochemistry at this Discussion, yet there is a long history of studies of nucleation and growth on electrode surfaces. The advent of *in situ* video-rate STM, for example, has revealed incredible insights into step growth and dissolution on copper single crystal surfaces.[44,45] The beauty of studying electrocrystallisation is that the driving force can be changed simply *via* the potential applied to the electrode. Moreover, the current is a direct measure of the interfacial flux and so current–time measurements, for example, can provide tremendous insight on the kinetics and mechanism of nucleation.[46] By employing microscale interfaces it is even possible to measure induction times and the nucleation and growth of individual nuclei.[47,48] Electrochemical methods thus have much to offer in the investigation of nucleation and growth dynamics, particularly at short length and timescales.

Bridging the gap between theory and experiment: addressing the mesoscale

Advances in simulation methods mean that more complex processes can be treated, which can increasingly be linked with experiment. Molecular dynamics simulations are particularly powerful, as they allow calculations of an ensemble of configurations, and permit a system to be followed as it evolves. The power of such approaches was evident at this Discussion in the nucleation studies of Rodger *et al.*[29] Moreover, it is apparent that these methods provide key information that can be used to inform and interpret experiments. However, since such methods are

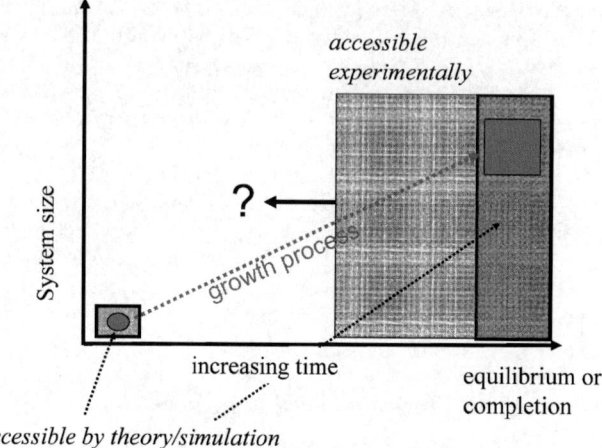

Fig. 2 Simplified summary of the time and length scales in crystal growth that can be addressed experimentally and by theory/simulation, and their relationship to the overall size/time scale of the crystal growth process (represented as an initial nucleus and final crystal, coloured dark grey). There is good coverage of experiment and theory at long times (approaching equilibrium) for a wide range of length scales. Sub-millisecond timescale events are difficult to address experimentally in condensed phase systems, whereas molecular dynamics simulation methods are typically restricted to nanoscopic and nanosecond (or shorter) timescales. The key challenge is to address the critical mesoscopic regime.

typically limited to small scale systems (a few thousand atoms) and short times (up to a few nanoseconds), it is imperative to appropriately translate the key information that molecular dynamics provides to larger length scales and longer times relevant to real crystal growth systems. Recent studies from Gale *et al.* have shown that it is still possible to use molecular dynamics simulations to obtain key information on crystal growth at the molecular level and apply this to kinetic Monte Carlo simulations, which operate on much longer length and time scales.[49] This appears to be one route which will enable the gap between small/short time systems and large/long time systems to be bridged.

Theory has long been able to inform about surface features on crystals and equilibrium crystal shapes.[50,51] One of the exciting aspects of the advent of AFM was the ability to observe directly many of the surface features and growth mechanisms predicted much earlier theoretically.[52–54] As a consequence, there is generally good overlap between experiment and theory for long times (approaching equilibrium) and over a range of length scales, as illustrated schematically in Fig. 2.

This Discussion has highlighted several experiments which have pushed the timescale on which precipitation and crystallization phenomena can be studied into the millisecond time domain, but there is still a considerable way to go to shorten further the timescales of experiments in order to address critical dynamic processes as a crystal nucleates and grows. As mentioned above, progress in simulation methods suggest a method to link nanoscopic and macroscopic systems, but other approaches are also likely to be developed which expand the length and timescales of traditional molecular dynamics simulation methods.

At present there is essentially zero overlap on theory and experiment in the critical mesoscopic regime (see Fig. 2) and the exciting challenge will be to develop new experiments and complementary simulation methods which are able to address the length and timescales characteristic of this regime. The evidence emerging from this Discussion is that those studies which involve both simulation and experiment have the potential to provide the deepest insight into crystal growth and nucleation.

References

1. F. C. Frank, *Discuss. Faraday Soc.*, 1949, **5**, 48.
2. W. K. Burton and N. Cabrera, *Discuss. Faraday Soc.*, 1949, **5**, 33.
3. N. Cabrera and W. K. Burton, *Discuss. Faraday Soc.*, 1949, **5**, 40.
4. W. J. Dunning, *Discuss. Faraday Soc.*, 1949, **5**, 79.
5. S. H. Bransom, W. J. Dunning and B. Millard, *Discuss. Faraday Soc.*, 1949, **5**, 83.
6. S. H. Bransom and W. J. Dunning, *Discuss. Faraday Soc.*, 1949, **5**, 96.
7. C. W. Bunn, *Discuss. Faraday Soc.*, 1949, **5**, 364.
8. P. R. Rowland, *Discuss. Faraday Soc.*, 1949, **5**, 364.
9. E. Vlieg, M. Deij, D. Kaminski, H. Meekes and W. van Enckevort, *Faraday Discuss.*, 2007, **136**, DOI: 10.1039/b618566p.
10. T. Fričič and W. Jones, *Faraday Discuss.*, 2007, **136**, DOI: 10.1039/b616399h.
11. P. E. Hillner, S. Manne, P. K. Hansma and A. J. Gratz, *Faraday Discuss.*, 1993, **95**, 191.
12. T. Tsukamoto, *Faraday Discuss.*, 1993, **95**, 183.
13. K. M. Robinson and W. E. O'Grady, *Faraday Discuss.*, 1993, **95**, 55.
14. S. C. Parker, E. T. Kelsey, P. M. Uliver and J. O. Titiloye, *Faraday Discuss.*, 1993, **95**, 75.
15. R.-F. Xiao, J. I. D. Alexander and F. Rosenberger, *Faraday Discuss.*, 1993, **95**, 85.
16. C. E. Hughes, K. D. M. Harris, P. C. Griffiths, S. Hamad and C. R. A. Catlow, *Faraday Discuss.*, 2007, **136**, DOI: 10.1039/b616757h.
17. M. W. Anderson, J. R. Agger, L. I. Meza, C. B. Chong and C. S. Cundy, *Faraday Discuss.*, 2007, **136**, DOI: 10.1039/b617782b.
18. J. Rieger, T. Frechen, G. Cox, W. Heckmann, C. Schmidt and J. Thieme, *Faraday Discuss.*, 2007, **136**, DOI: 10.1039/b701450c.
19. R. A. Chiarella, A. L. Gillon, R. C. Burton, R. J. Davey, G. Sadiq, A. Auffret, M. Cioffi and C. A. Hunter, *Faraday Discuss.*, 2007, **136**, DOI: 10.1039/b616164m.
20. A. Wakisaka, *Faraday Discuss.*, 2007, **136**, DOI: 10.1039/b615977j.
21. A. Keller, G. Goldbeck-Wood and M. Hikosaka, *Faraday Discuss.*, 1993, **95**, 109.
22. R. J. Davey, *Faraday Discuss.*, 1993, **95**, 160.
23. N. Nassif, N. Pinna, N. Gehrke, M. Antonietti, C. Jäger and H. Cölfen, *Proc. Natl. Acad. Sci. U. S. A.*, 2005, **102**, 12653.
24. L. Addadi, *Faraday Discuss.*, 2007, **136**, DOI: 10.1039/b704418f.
25. R. B. Hammond, K. Pencheva and K. J. Roberts, *Faraday Discuss.*, 2007, **136**, DOI: 10.1039/b616757h.
26. G. Sankar, T. Okuba, W. Fan and F. Meneau, *Faraday Discuss.*, 2007, **136**, DOI: 10.1039/b700090c.
27. A. Michaelides, *Faraday Discuss.*, 2007, **136**, DOI: 10.1039/b616689j.
28. J. Lardge, D. Duffy and M. Gillan, *Faraday Discuss.*, 2007, **136**, Poster 24.
29. C. Moon, R. W. Hawtin and P. M. Rodger, *Faraday Discuss.*, 2007, **136**, DOI: 10.1039/b618194p.
30. V. Cherezov and M. Caffrey, *Faraday Discuss.*, 2007, **136**, DOI: 10.1039/b618173b.
31. B. Slater, T. Ohsuma, Z. Lui and O. Terasaki, *Faraday Discuss.*, 2007, **136**, DOI: 10.1039/b618677g.
32. T. Bullard, J. Freundethal, S. Avagyan and B. Kahr, *Faraday Discuss.*, 2007, **136**, DOI: 10.1039/b616612c.
33. D. Gasperino, A. Yeckel, B. K. Olmsted, M. D. Ward and J. J. Derby, *Langmuir*, 2006, **22**, 6578.
34. P. S. Dobson, L. A. Bindley, J. V. Macpherson and P. R. Unwin, *Langmuir*, 2005, **21**, 1255.
35. A. D. Humphris, M. Miles and J. K. Hobbs, *Appl. Phys. Lett.*, 2005, **86**, 034106.
36. J. K. Hobbs, C. Vasilev and A. D. Humphris, *Polymer*, 2005, **46**, 10226.
37. N. B. J. Hetherington, M. F. Butler and F. C. Meldrum, *Faraday Discuss.*, 2007, **136**, Poster 16.
38. S. I. Stupp, Abstract PMSE 101. 233rd ACS National Meeting, Chicago, IL, March 25–29, 2007.
39. J. J. de Yoreo, J. R. I. Lee, S. Elhadj, D. Wang, Y.-J. Han, T. M. Willey, R. W. Meulenberg, L. J. Terminello, T. van Buuren and P. M. Dove, Abstract PMSE 102, 233rd ACS National Meeting, Chicago, IL, March 25–29, 2007.
40. J. Aizenberg, Abstract PMSE 139, 233rd ACS National Meeting, Chicago, IL, March 25–29, 2007.
41. D. Rautaray, R. Kavathekar and M. Sastry, *Faraday Discuss.*, 2005, **129**, 205.
42. A. R. Boccaccini, Q. Chen, L. Lefebvre, L. Gremillard and J. Chevalier, *Faraday Discuss.*, 2007, **136**, DOI: 10.1039/b616539g.
43. A. Tilocca, N. H. de Leeuw and A. N. Cormack, *Faraday Discuss.*, 2007, **136**, DOI: 10.1039/b617540f.

44 O. M. Magnussen, L. Zitzler, B. Gleich and R. J. Behm, *Electrochim. Acta*, 2001, **46**, 3725.
45 O. M. Magnussen, W. Polewska, L. Zitzler and R. J. Behm, *Faraday Discuss.*, 2001, **121**, 43.
46 R. M. Penner, *J. Phys. Chem. B*, 2002, **106**, 3339.
47 M. Fleischmann, L. J. Li and L. M. Peter, *Electrochim. Acta*, 1989, **34**, 475.
48 J. D. Guo, T. Tokimoto, R. Othman and P. R. Unwin, *Electrochem. Commun.*, 2003, **5**, 1005.
49 S. Piana, M. Reyhani and J. D. Gale, *Nature*, 2005, **438**, 70.
50 W. K. Burton, N. Cabrera and F. C. Frank, *Philos. Trans. R. Soc. London, Ser. A*, 1951, **243**, 299.
51 S. Amelinckx, *Nature*, 1952, **170**, 760.
52 H. H. Teng, P. M. Dove, C. A. Orme and J. J. de Yoreo, *Science*, 1998, **282**, 724.
53 J. V. Macpherson, P. R. Unwin, A. C. Hillier and A. J. Bard, *J. Am. Chem. Soc.*, 1996, **118**, 6445.
54 P. R. Unwin and J. V. Macpherson, *Chem. Soc. Rev.*, 1995, **24**, 109, and references therein.

Poster titles

Influence of lysozyme on the formation of spherical amorphous calcium carbonate nanoparticles **Alina E. Voinescu, Didier Touraud, Alois Lecker, Arno Pfitzner, Barry Ninham** and **W. Kunz**, *University of Regensburg, Germany*

Ice nucleation in emulsions: towards a direct measurement of critical nucleus size **Jian Liu, Catherine E. Nicholson** and **Sharon J. Cooper**, *University of Durham, UK*

Growth and microstructure of InAs on GaAs: the evolution of the micro structure beyond the pseudomorphic limit **S. E. Babcock, G. Suryanarayanan, A. A. Khandekar, X. Song, M. Rathi, J. Webb** and **T. F. Kuech**, *University of Wisconsin – Madison, USA*

Microscopic examination of crystal growth on heat transfer surfaces **Timo Geddert, Stephan Kipp, Wolfgang Augustin** and **Stephan Scholl**, *Technical University of Braunschweig, Germany*

Crystallisation of pharmaceutical salts in organic solvents: the effects of solvent and temperature on solution speciation **Helen Jones, Brian Cox** and **Roger Davey**, *AstraZeneca, UK*

Synthesis and structural studies of thaumasite and ettringite related materials **S. Cairns, C. Wilson** and **C. A. Kirk**, *University of Glasgow, UK*

The effect of reactants onto the formation of $Ni(C_6N_4H_{18})SnS_3$. *In situ* EXAFS investigations under solvothermal conditions **N. Pienack, M.-E. Ordolff** and **W. Bensch**, *University of Kiel, Germany*

Drop-based microfluidic process for the production of crystals of controlled size and shape **Richard D. Dombrowski, Yinghe He, Norman J. Wagner** and **James D. Litster**, *University of Queensland, Australia*

High-throughput computing of additives on crystal surfaces **V. Thorne, P. Murray-Rust, M. Heppenstall-Butler** and **M. Butler**, *Unilever Centre for Molecular Science Informatics, UK*

Effects of impurites on nucleation and growth of taranabant drug substance **Alex M. Chen, Yadan Chen, Yong Liu, George Zhou, Brian Phenix, Yaling Wang, Robert Wenslow** and **Narayan Variankaval**, *Merck and Co. Inc, USA*

Nucleation of zeolites: the big nano-jigsaw **Miguel J. Mora-Fonz, C. Richard A. Catlow** and **Dewi W. Lewis**, *University College London, UK*

An artist's impression of a truly primitive gene **A. G. Cairns-Smith**, *Glasgow University, UK*

Influence of organic templates on the crystal growth of nanoporous materials **Kim Jelfs, Ben Slater, David J. Willock** and **Dewi W. Lewis**, *The Royal Institution of Great Britain and University College London, UK*

The effects of deuteration on the polymorphism of glycine **Colan E. Hughes, Kenneth D. M. Harris** and **Peter C. Griffiths**, *Cardiff University, UK*

The crystallisation of aqueous droplets **Benjamin J. Murray** and **Allan K. Bertram**, *Leeds University, UK*

Hard and soft templating of calcium carbonate **N. B. J. Hetherington**, **A. N. Kulak**, **E. H. Noel** and **F. C. Meldrum**, *University of Bristol, UK*

Ammonium dinitramide: optimization of the unit cell and morphology calculation **Indra Fuhr**, **Paul Bernd Kempa**, **Michael Herrmann** and **Joachim Ulrich**, *Fraunhofer Institute for Chemical Technology, Germany*

Simulating scale formation in desalination processes **M. Gloede**, **M.-W. Lumey** and **T. Melin**, *RWTH Aachen University, Germany*

Protein crystallization studies using dual polarisation interferometry **A. Boudjemline**, **M. J. Swann**, **N. J. Freeman**, **D. T. Clarke** and **G. R. Jones**, *Farfield Scientific Ltd, UK*

Nano particle production by using an emulsification/crystallization process **Ran Huo**, **Malcolm Povey** and **Yulong Ding**, *University of Leeds, UK*

Real time monitoring of conversions between crystal polymorphs in aqueous solution **Karl Box**, **Mariotte Corson**, **John Comer** and **Yin-Chao Tseng**, *Sirius Analytical Ltd, UK*

A novel experiment for assessing precipitation and dissolution rates for ionizable drugs in aqueous solutions **Karl Box**, **John Comer** and **Martin Stuart**, *Sirius Analytical Ltd, UK*

The nucleation of polymorphs in simulations **J. H. ter Horst**, *Delft University of Technology, The Netherlands*

Ab initio molecular dynamics study of the calcite–water interaction **Jennifer Lardge**, **Dorothy Duffy** and **Mike Gillan**, *University College London, UK*

Predicting properties of mixed metal fluorides and oxides for optical and electronic applications **Robert A. Jackson**, **Elizabeth M. Maddock** and **Mario E. G. Valerio**, *Keele University, UK*

Nucleation of soft particle liquids **D. M. Heyes**, *University of Surrey, UK*

Molecular dynamics to probe molecular clustering and polymorph selectivity as a function of solvent composition: case of 2,6-dihydroxybenzoic acid **Robert B. Hammond**, **Klimentina Pencheva** and **Kevin J. Roberts**, *University of Leeds, UK*

Nano-domain formation in Sc-doped zirconia and Gd-doped ceria **A. E. MacHale** and **A. N. Cormack**, *Alfred University, USA*

Computer simulations of silicate nano-tubes: hydration, dissolution and self-assembly **Zhimei Du** and **Nora H. De Leeuw**, *University College London, UK*

Models for the nucleation and growth of calcium carbonate **Colin Freeman**, **Mingjun Yang**, **John Harding**, **Dmytro Antypov**, **David Cooke**, **James Elliott**, **Dorothy Duffy**, **Michael Gillan**, **Jennifer Lardge**, **Michael Allen**, **David Quigley**, **Mark Rodger** and **Tiffany Walsh**, *University of Sheffield, UK*

The role of water in the synthesis of cocrystals and cocrystal hydrates **Shyam Karki, Tomislav Friščić** and **William Jones**, *University of Cambridge, UK*

Evolution of supramolecular halogen-bonded structures in cocrystallisation *via* grinding: from a finite assembly to an infinite chain, **Tomislav Friščić, Dominik Cinčić** and **William Jones**, *University of Cambridge, UK*

Nucleation and growth of ZnS nanoclusters **Said Hamad Gomez, Eleonora Spano, Sylvain Cristol** and **C. Richard A. Catlow**, *The Royal Institution of Great Britain, UK*

Asymmetric crystal growth of resorcinol along the polar axis: are surface reconstruction and conformational change the culprits? **Jamshed Anwar, Jittima Chatchawalsaisin** and **John Kendrick**, *University of Bradford, UK*

Simulating synthesis **Dean C. Sayle, Shawn Feng, Zhong Lin Wang, Stephen C. Parker, Jacek Klinowski** and **Thi X. T. Sayle**, *Cranfield University, UK*

Investigation of calcite crystal growth mechanism in the presence of poly-L-glutamic acid and poly-L-aspartic acid, **Damir Kralj, Branka Njegić Džakula, Ljerka Brečević** and **Giuseppe Falini**, *Ruđer Bošković Institute, Croatia*

In situ monitoring of crystal growth and dissolution of oriented layered double hydroxide crystals immobilized on silicon **H. C. Greenwell, R. L. Anderson, L. A. Bindley, P. R. Unwin, P. J. Holliman, W. Jones, P. V. Coveney** and **S. L. Barnes**, *University of Wales, Bangor, UK*

Towards mixed metal compounds for nano oxide layers. Or: how to put metal ions in prison **Fabienne Gschwind** and **Katharina M. Fromm**, *University of Fribourg, Switzerland*

Atomistic simulation of the role of water on adsorption and growth processes at mineral surfaces **Stephen C. Parker, Jeremy Allen, Corinne Arrouvel, Dino Spagnoli** and **Wojlek Grén**, *University of Bath, UK*

The Skinner Prize for the best poster was awarded to Nicola Hetherington from University of Bristol, UK, for her poster on hard and soft templating of calcium carbonate.

List of Participants

Professor L. Addadi, *Weizmann Institute of Science, Israel*
Professor M. Anderson, *University of Manchester, United Kingdom*
Dr R. Anderson, *University of Wales, Bangor, United Kingdom*
Professor J. Anwar, *University of Bradford, United Kingdom*
Dr W. Augustin, *Technical University of Braunschweig, Germany*
Dr S. Batten, *Royal Society of Chemistry, United Kingdom*
Mr A. Baumgartner, *Universität Bayreuth, Germany*
Miss S. Beckett, *Kenyon International, United Kingdom*
Professor W. Bensch, *University of Kiel, Germany*
Dr A. Boccaccini, *Imperial College London, United Kingdom*
Mr Boudjemline, *Farfield Scientific, United Kingdom*
Professor J. Breu, *Universität Bayreuth, Germany*
Dr J. Burley, *De Montfort University, United Kingdom*
Professor M. Caffrey, *University of Limerick, Ireland*
Mr S. Cairns, *University of Glasgow, United Kingdom*
Dr A. Cairns-Smith, *University of Glasgow, United Kingdom*
Professor R. Catlow, *University College London, United Kingdom*
Mr A. Chen, *Merck and Co., United Kingdom*
Mr J. Comer, *Sirius Analytical Instruments Ltd, United Kingdom*
Dr T. Cooper, *University of Cambridge, United Kingdom*
Dr M. Copsey, *Royal Society of Chemistry, United Kingdom*
Professor A. Cormack, *Alfred University, USA*
Miss D. Croker, *University of Limerick, Ireland*
Mr T. Daff, *Birkbeck College, United Kingdom*
Dr J. Dalton, *Johnson Matthey, United Kingdom*
Professor R. Davey, *University of Manchester, United Kingdom*
Dr De Leeuw, *University College London, United Kingdom*
Dr Z. Du, *University College London, United Kingdom*
Dr D. Duffy, *University College London, United Kingdom*
Mr V. Dusastre, *Nature Publishing Group, United Kingdom*
Dr D. Fischer, *Max Planck Institute for Solid State Research, Germany*
Dr T. Friščić, *University of Cambridge, United Kingdom*
Dr I. Fuhr, *Fraunhofer Institut Chemische Technologie, Germany*
Mr T. Geddert, *Technical University of Braunschweig, Germany*
Dr M. Gich, *Saint-Gobain Recherche, France*
Miss F. Gschwind, *University Fribourg, Switzerland*
Dr S. Hamad, *Davy Faraday Research Laboratory, United Kingdom*
Dr R. Hammond, *University of Leeds, United Kingdom*
Professor J. Harding, *University of Sheffield, United Kingdom*
Dr A. Hare, *United Kingdom*
Professor K. Harris, *Cardiff University, United Kingdom*
Miss N. Hetherington, *University of Bristol, United Kingdom*
Professor D. Heyes, *University of Surrey, United Kingdom*
Professor K. Hodnett, *University of Limerick, Ireland*
Dr J. ter Horst, *Delft University of Technology, The Netherlands*
Dr C. Hughes, *Cardiff University, United Kingdom*
Miss R. Huo, *University of Leeds, United Kingdom*
Professor J. Hyne, *University of Calgary, Canada*
Dr R. Jackson, *University of Keele, United Kingdom*
Miss K. Jelfs, *University College London, United Kingdom*
Professor B. Jones, *University of Cambridge, United Kingdom*
Miss H. Jones, *AstraZeneca, United Kingdom*
Professor B. Kahr, *University of Washington, USA*

Mr S. Karki, *University of Cambridge, United Kingdom*
Dr D. Kralj, *Ruđer Bošković Institute, Croatia*
Mr T. Kuech, *University of Wisconsin, USA*
Professor R. Kuroda, *University of Tokyo, Japan*
Miss J. Lardge, *University College London, United Kingdom*
Dr D. Lewis, *University College London, United Kingdom*
Dr K. Lewtas, *Infineum UK Ltd, United Kingdom*
Professor J. Litster, *University of Queensland, Australia*
Dr J. Liu, *University of Durham, United Kingdom*
Mr N. Loges, *University of Mainz, Germany*
Dr M. Lumey, *RWTH Aachen, Germany*
Mr D. Lynham, *BPB Plc, United Kingdom*
Mr J. Majors, *Swenson Technology*
Ms A. Martinod, *University of Leeds, United Kingdom*
Professor M. Mazzotti, *ETH Zurich, Switzerland*
Dr A. Michaelides, *University College London, United Kingdom*
Dr C. Moore, *Royal Society of Chemistry, United Kingdom*
Mr M. Mora Fonz, *University College London, United Kingdom*
Ms A. Munroe, *University of Limerick, Ireland*
Dr B. Murray, *University of Leeds, United Kingdom*
Mr J. Murray, *University of Limerick, Ireland*
Mr D. Musumeci, *University of Sheffield, United Kingdom*
Dr J. Nedelec, *Universite Blaise Pascal, France*
Dr G. Nehrke, *Alfred Wegener Institute, Germany*
Ms L. O'Shea, *Unversity of Limerick, Ireland*
Mr T. O'Sullivan, *University of Limerick, Ireland*
Professor S. Parker, *University of Bath, United Kingdom*
Professor I. Parkin, *University College London, United Kingdom*
Professor G. Parkinson, *Alcoa World Alumina, Australia*
Miss J. Pienack, *University of Kiel, Germany*
Dr J. Popplewell, *Farfield Scientific, United Kingdom*
Professor S. Price, *University College London, United Kingdom*
Dr J. Rieger, *BASF AG, Germany*
Dr R. Ristic, *University of Sheffield, United Kingdom*
Professor K. Roberts, *University of Leeds, United Kingdom*
Professor P. Roberts, *ICI R&I, United Kingdom*
Professor M. Rodger, *University of Warwick, United Kingdom*
Dr L. Ruston, *AstraZeneca, United Kingdom*
Professor G. Sankar, *The Royal Institution of Great Britain, United Kingdom*
Dr D. Sayle, *Cranfield University, United Kingdom*
Dr C. Schön, *Max Planck Institute for Solid State Research, Germany*
Dr A. Sheikh, *Abbott Labs, USA*
Dr B. Slater, *Royal Institution of Great Britain, United Kingdom*
Dr M. Swann, *Farfield Sensors Ltd, United Kingdom*
Professor O. Terasaki, *Stockholm University, Sweden*
Dr V. Thome, *Unilever Centre for Molecular Science Informatics, United Kingdom*
Dr A. Tilocca, *University College London, United Kingdom*
Professor P. Unwin, *University of Warwick, United Kingdom*
Professor E. Vlieg, *Radboud University, The Netherlands*
Dr A. Voinescu, *University of Regensburg, Germany*
Dr P. Vonk, *DSM Research, The Netherlands*
Dr A. Wakisaka, *National Institute of Advanced Industrial Science and Technology, Japan*
Professor M. Watanabe, *Gakushuin University, Japan*
Dr I. Wilkes, *Saint Gobain Gypsum, United Kingdom*
Mr D. Wray, *Centre for Process Innovation, United Kingdom*
Dr S. Wren, *AstraZeneca, United Kingdom*

Index of contributors*

Adachi, M., **279**
Addadi, L., **9**, 107, 213, 309, 395
Agger, J. R., **143**
Anderson, M. A., 107, **143**, 213, 395
Anwar, J., 107, 395
Auffret, A., **179**
Avagyan, S., **231**
Baba, M., **383**
Bensch, W., 107, 309, 395
Blackman, C. S., **329**
Boccaccini, A. R., **27**, 107
Bonafos, C., **345**
Breu, J., 309, 395
Brozio, J., **247**
Bullard, T., **231**
Burley, J., 395
Burton, R. C., **179**
Caffrey, M., 107, 213, **195**
Cairns-Smith, A., 309
Catlow, C. R. A., **71**, 107, 213, 309, 395
Chen, H. H., **9**
Chen, Q., **27**
Chen, T., **355**
Cherezov, V., **195**
Chevalier, J., **27**
Chiarella, R. A., **179**
Chong, C. B., **143**
Cioffi, M., **179**
Comer, J., 107
Cormack, A. N., **45**
Cox, G., **265**
Cundy, C. S., **143**
Davey, R. J., 107, **179**, 213, 309, 395
Deij, M., **57**
de Leeuw, N. H., **45**, 395
Fan, W., **157**
Frechen, T., **265**
Freudenthal, J., **231**
Friščić, T., **167**
Fukuyama, H., **279**
Gich, M., 107, **345**, 395
Gillon, A. L., **179**
Goldberg, H. A., **9**
Gremillard, L., **27**
Griffiths, P. C., **71**
Grolier, J.-P. E., **383**
Hamad, S., **71**, 107
Hammond, R. B., **91**, 309
Harding, J., 107, 309, 395

Hare, A., 107, 213, 309, 395
Harris, K. D. M., **71**
Hawtin, R. W., **367**
Heckmann, W., **265**
Heyes, D., 107, 309, 395
Higuchi, K., **279**
Hodnett, K., 213, 309, 395
Hughes, C. E., **71**, 107, 213, 309, 395
Hunter, C. A., **179**
Hyett, G., **329**
Hyne, J., 107, 309, 395
Jones, W., **167**, 213, 309, 395
Kahr, B., 107, 213, **231**, 309
Kaminski, D., **57**
Kobatake, H., **279**
Kuroda, R., 107, 213
Lefebvre, L., **27**
Lewis, D., 213
Lewtas, K., 309
Lindenberg, C., **247**
Litster, J., 213
Liu, J., 395
Liu, Z., **125**
Mazzotti, M., 107, 213, **247**, 309, 395
Meekes, H., **57**
Meneau, F., **157**
Meza, L. I., **143**
Meziane, A., **383**
Michaelides, A., **287**, 309
Molins, E., **345**
Moon, C., **367**
Morishita, T., **279**
Murray, B., 107, 309, 395
Nedelec, J.-M., **383**, 395
Neville, A., **355**, 395
Nudelman, F., **9**
Ohsuna, T., **125**
Okubo, T., **157**
Parkin, I. P., **329**, 395
Pencheva, K., **91**
Rieger, J., 107, **265**, 309
Ristic, R., 107, 309, 395
Roberts, K. J., **91**, 107, 213, 309, 395
Rodger, P. M., 107, 213, **367**, 395
Roig, A., **345**
Sadiq, G., **179**
Sankar, G., **157**, 213, 395
Schmidt, C., **265**
Schöll, J., **247**
Schön, C., 107, 213, 309, 395

Snoeck, E., **345**
Sorbie, K., **355**
Slater, B., **125**, 213, 395
Taboada, E., **345**
Terasaki, O., **125**, 213
ter Horst, J., 107, 213, 395
Thieme, J., **265**
Tilocca, A., **45**, 107
Unwin, P., 213, 309, **409**

van Enckevort, W., **57**
Vicum, L., **247**
Vonk, P., 107, 213, 309
Vlieg, E., **57**, 107, 213, 309, 395
Wakisaka, A., **299**, 309, 395
Watanabe, M., **279**, 309
Weiner, S., **9**
Zhong, Z., **355**

* The page numbers in **bold** type indicate papers submitted for discussions.